Parkinson's Disease: From Clinical Aspects to Molecular Basis

Parkinson's Disease: From Clinical Aspects to Molecular Basis

Editor: John Stewart

AMERICAN
MEDICAL PUBLISHERS
www.americanmedicalpublishers.com

AMERICAN
MEDICAL PUBLISHERS
www.americanmedicalpublishers.com

Cataloging-in-Publication Data

Parkinson's disease : from clinical aspects to molecular basis / edited by John Stewart.
p. cm.
Includes bibliographical references and index.
ISBN 978-1-63927-778-0
1. Parkinson's disease. 2. Parkinson's disease--Diagnosis. 3. Parkinson's disease--Treatment.
4. Parkinson's disease--Molecular aspects. 5. Brain--Diseases. I. Stewart, John.
RC382 .P37 2023
616.833--dc23

American Medical Publishers,
41 Flatbush Avenue,
1st Floor, New York,
NY 11217, USA

ISBN 978-1-63927-778-0 (Hardback)

Contents

Preface

Parkinson's disease (PD) refers to a type of long-term degenerative disorder of the central nervous system which primarily impacts the motor system. It is clinically categorized through movement disorders, including tremor, muscular stiffness and akinesia. The study of its molecular and biochemical mechanisms aids in better understanding of the normal brain ageing. The cause of PD is still unknown, but both environmental and inherited factors are assumed to have a role in the disease. A separate set of genes whose dysfunctional regulation is significantly linked to the disease. These genes comprise DJ-1, ATP13A2, alpha-synuclein, PTEN induced putative kinase 1 (PINK1), leucine rich repeat kinase 2 (LRRK2) and parkin. In clinical practice, a range of medications are utilized for treating PD, the majority of which work by raising dopamine levels in the midbrain. This book provides significant information to help develop a good understanding of Parkinson's disease. It aims to shed light on clinical aspects and molecular basis of this disease. This book will serve as a reference to a broad spectrum of readers.

The researches compiled throughout the book are authentic and of high quality, combining several disciplines and from very diverse regions from around the world. Drawing on the contributions of many researchers from diverse countries, the book's objective is to provide the readers with the latest achievements in the area of research. This book will surely be a source of knowledge to all interested and researching the field.

In the end, I would like to express my deep sense of gratitude to all the authors for meeting the set deadlines in completing and submitting their research chapters. I would also like to thank the publisher for the support offered to us throughout the course of the book. Finally, I extend my sincere thanks to my family for being a constant source of inspiration and encouragement.

Editor

1

Mitophagy in Parkinson's Disease: From Pathogenesis to Treatment

Jia Liu [1], Weijin Liu [1], Ruolin Li [1] and Hui Yang [1,2,*]

[1] Department of Neurobiology School of Basic Medical Sciences, Beijing Institute for Brain Disorders, Capital Medical University, Beijing 100069, China

[2] Center of Parkinson's Disease Beijing Key Laboratory of Neural Regeneration and Repair, Beijing Key Laboratory on Parkinson's Disease, Key Laboratory for Neurodegenerative Disease of the Ministry of Education, Beijing 100069, China

* Correspondence: huiyang@ccmu.edu.cn

Abstract: Parkinson's disease (PD) is the second most common neurodegenerative disease. The pathogenesis of PD is complicated and remains obscure, but growing evidence suggests the involvement of mitochondrial and lysosomal dysfunction. Mitophagy, the process of removing damaged mitochondria, is compromised in PD patients and models, and was found to be associated with accelerated neurodegeneration. Several PD-related proteins are known to participate in the regulation of mitophagy, including PINK1 and Parkin. In addition, mutations in several PD-related genes are known to cause mitochondrial defects and neurotoxicity by disturbing mitophagy, indicating that mitophagy is a critical component of PD pathogenesis. Therefore, it is crucial to understand how these genes are involved in mitochondrial quality control or mitophagy regulation in the study of PD pathogenesis and the development of novel treatment strategies. In this review, we will discuss the critical roles of mitophagy in PD pathogenesis, highlighting the potential therapeutic implications of mitophagy regulation.

Keywords: mitophagy; Parkin; Parkinson's disease; PINK1; treatment

---

1. Introduction

Parkinson's disease (PD) is one of the most common neurodegenerative disorders in the world, and is characterized by typical motor deficits, including bradykinesia, tremors, rigidity, and postural instability, and a series of non-motor symptoms such as dysosmia, constipation, and depression. The pathological features of PD include the progressive loss of dopaminergic neurons in the substantia nigrapars compacta (SNpc) and the formation of Lewy bodies [1].

PD can be subdivided into familial and sporadic PD. Epidemiological studies have shown that about 10% of PD cases are inherited, whereas the remaining cases are sporadic. Although the etiology of PD remains unknown, it is believed to involve both genetic and environmental factors, and is associated with aging [2,3]. Thus far, several proteins have been identified as contributing to PD pathogenesis, including α-synuclein (α-syn), Parkin, PTEN-induced putative kinase (PINK)1, DJ-1, Leucine-rich repeat kinase (LRRK) 2, and others (Table 1). Many of these are known to participate in mitochondrial quality control or lysosomal functions [4]. Furthermore, familial PD patients with gene mutations usually show mitochondrial defects and impairment of the autophagic pathway [5–8], indicating that these two elements are critical components of PD pathogenesis.

Table 1. Overview of Parkinson's disease (PD)-related genes. Genes related to PD pathogenesis are listed in Table 1. The locus of genes, hereditary properties, and the onset of disease are described. AD, autosomal dominant; AR, autosomal recessive.

Loci	Gene	Protein	Position	Inheritance	Onset
PARK1	*SNCA*	Alpha-synuclein	4q21	AD, sporadic	Early or late
PARK2	*PRKN*	Parkin	6q25–q27	AR, sporadic	Early
PARK3	*Unknown*	Unknown	2p13	AD	Late
PARK5	*UCHL1*	Ubiquitin C-Terminal Hydrolase L1	4p14	AD	Late
PARK6	*PINK1*	PTEN-induced putative kinase 1	1p35–p36	AR	Early
PARK7	*DJ-1*	Protein DJ-1	1p36	AR	Early
PARK8	*LRRK2*	Leucine-rich repeat kinase 2	12q12	AD, sporadic	Early or late
PARK9	*ATP13A2*	ATPase 13A2	1p36	AR	Early
PARK10	*Unknown*	Unknown	1p32	Unknown	Unknown
PARK11	*GIGYF2*	GRB10 interacting GYF protein 2	2q36–q37	AD	Late
PARK12	*Unknown*	Unknown	Xq21–q25	Unknown	Unknown
PARK13	*HTRA2*	Serine peptidase 2	2p13	AD	Late
PARK14	*PLA2G6*	Phospholipase A2 Group VI	22q12–q13	AR	Early
PARK15	*FBX07*	F-Box protein 7	22q12–q13	AR	Early
PARK17	*VPS35*	Vacuolar protein sorting 35	16q11.2	AD	Late
PARK18	*EIF4G1*	Eukaryotic translation initiation factor 4 gamma, 1	3q27.1	AD	Late
PARK19	*DNAJC6*	DNAJ subfamily C member 6	1p31.3	AR	Early
PARK20	*SYNJ1*	Synaptojanin-1	21q22.11	AR	Early
PARK21	*DNAJC13*	DNAJ subfamily C member 13	3q22.1	AD	Early
PARK22	*CHCHD2*	Coiled coil-helix-coiled coil-helix domain 2	7p11.2	AD	Late
PARK23	*VPS13C*	Vacuolar protein sorting 13 homolog C	15q22.2	AR	Early
-	*GBA*	Glucocerebrosidase	1q21	AD	Unknown

2. Mitophagy Pathways

Mitophagy, the selective degradation of mitochondria via autophagy, is a key process for maintaining mitochondrial homeostasis. Mitochondrial turnover via this mechanism is regarded as a significant mechanism for maintaining neuronal health [9]. However, abnormal mitophagy accompanies neurodegeneration. [10] A growing number of studies have shown that autophagy dysfunction impairs mitochondrial homeostasis, and in turn, mitochondrial defects also impact lysosomal functions, suggesting a complex relationship between these processes. [11] Mitophagy impairment results in the progressive accumulation of defective mitochondria, leading to neuronal death and eventual neurodegeneration. In this review, we discuss the critical roles of mitophagy in PD pathogenesis, highlighting the potential therapeutic implications of mitophagy regulation.

Cells possess several mitophagy mechanisms, and different stresses promote mitophagy in distinct cellular contexts. Indeed, mitophagy can be divided into Parkin-dependent or independent pathways, with some crosstalk between them (Figure 1) [12,13].

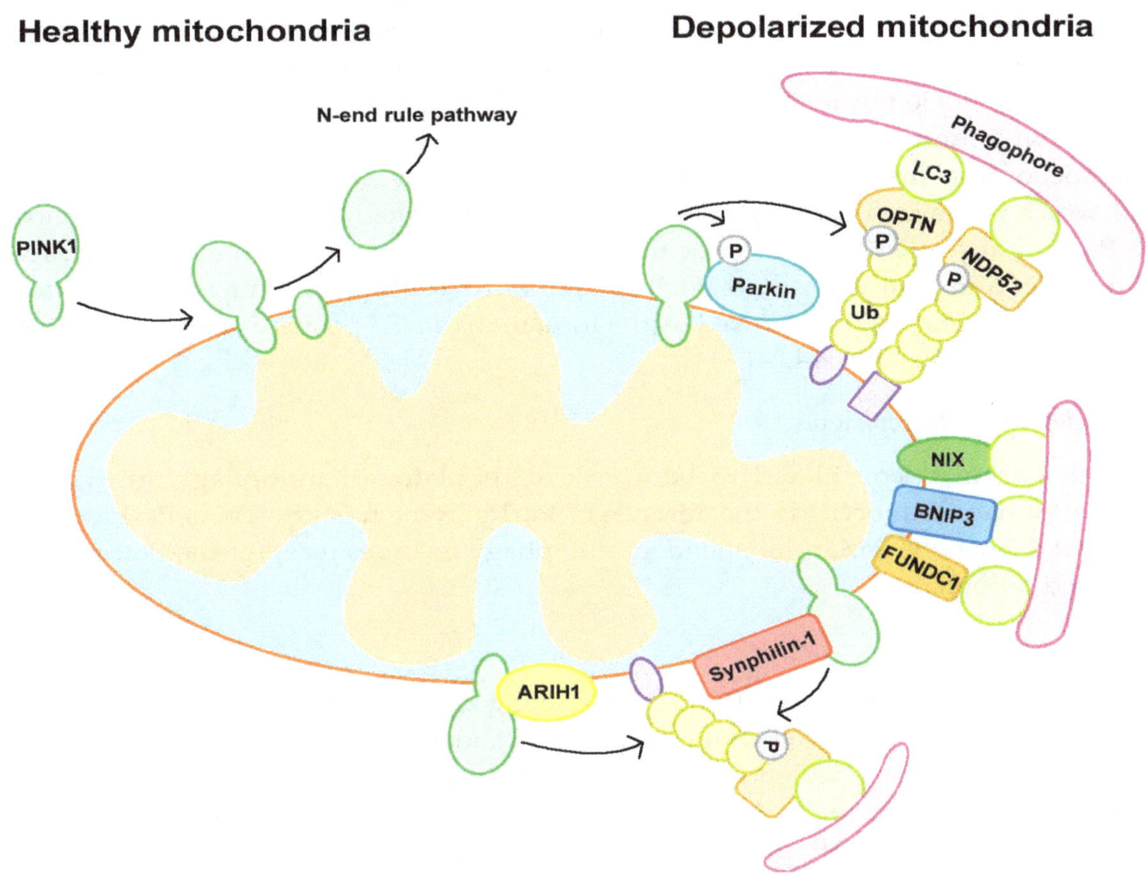

Figure 1. Mitophagy pathways. Mitophagy can be divided into Parkin-dependent or independent pathways. Under normal conditions, PINK1 localizes to mitochondria and is translocated to the mitochondrial inner membrane (MIM), where it is cleaved and subsequently degraded by an N-end rule pathway. However, when mitochondria become depolarized, PINK1 accumulates at the outer mitochondrial membrane (OMM) and recruits Parkin. Activated Parkin leads to the ubiquitination of substrates and the recruitment of autophagy receptors to initiate mitophagy. In addition, Parkin-independent mitophagy includes receptor-mediated and ubiquitin ligase-mediated mitophagy. BNIP3, BCL2/adenovirus E1B 19 kDa interacting protein 3; FUNDC1, FUN14 domain-containing protein 1; NDP52, nuclear dot protein 52 kDa; NIX, BCL2/adenovirus E1B 19 kDa interacting protein 3 like; OPTN, optineurin; Ub, ubiquitin.

2.1. Parkin-Dependent Mitophagy

In 2008, the Youle lab first identified the relationship of the PD-related gene Parkin and mitophagy, which was regarded as a landmark study in mitophagy [14]. Subsequently, several studies showed that another PD-related gene, PINK1, also participated in this process [15–17]. PINK1 is a serine/threonine kinase encoded by the *PARK6* gene. Parkin is an E3 ubiquitin ligase encoded by the *PARK2* gene. Mutations in PINK1 and Parkin result in autosomal recessive PD [18,19]. Loss of function of PINK1 or Parkin causes prominent mitochondrial pathology and loss of dopaminergic neurons [20–22], and subsequent studies discovered a crucial role for PINK1 and Parkin in mitophagy. In the current decade, much research related to PINK1/Parkin-mediated mitophagy has emerged, and this pathway is regarded as the most common and important pathway in mitophagy [23].

The PINK1/Parkin pathway regulates Ub-dependent mitophagy. The assembly of ubiquitin chains on mitochondria is important for the removal of damaged mitochondria via this pathway. This assembly requires three significant elements: PINK1 as a mitochondrial damage sensor, Parkin as a signal amplifier, and ubiquitin chains as the signal effector [24]. PINK1 has a mitochondrial targeting sequence that can guide it to mitochondria, and it triggers mitophagy by sensing mitochondrial depolarization or the accumulation of reactive oxygen species (ROS). Under normal conditions, PINK1 localizes to mitochondria and is translocated to the mitochondrial inner membrane (MIM), where it is cleaved and inactivated by the MIM protease presenilin-associated rhomboid-like protein (PARL) and subsequently degraded by an N-end rule pathway [25,26]. However, when mitochondria become depolarized, PINK1 cannot be translocated to the MIM and cleaved, resulting in its accumulation at the outer mitochondrial membrane (OMM) and the subsequent recruitment and phosphorylation of Parkin [16], which is localized in the cytoplasm and in an inactivated form in healthy mitochondria [27]. Activated Parkin subsequently leads to the ubiquitination of mitochondrial membrane proteins and the recruitment of autophagy receptors such as optineurin (OPTN) and nuclear dot protein 52 kDa (NDP52) to mitochondria [28,29], followed by the formation of LC3-positive phagophores that degrade mitochondria via the lysosome [14,24].

2.2. Parkin-Independent Mitophagy

Although Parkin is considered to be a crucial regulator of mitophagy, growing evidence shows that mitophagy can occur in the absence of Parkin, which is known as Parkin-independent mitophagy [13]. Generally, Parkin-independent mitophagy includes receptor-mediated and ubiquitin ligase-mediated mitophagy.

2.2.1. Receptor-Mediated Mitophagy

Several protein receptors have a LIR (LC3-interacting region) motif, which allows them to bind to LC3 to induce mitophagy [30]. B-cell lymphoma 2 (BCL2)/adenovirus E1B 19 kDa interacting protein 3 (BNIP3) and its homolog BCL2/adenovirus E1B 19 kDa interacting protein 3 like (NIX/BNIP3L) are BH3-only proteins belonging to the B-cell lymphoma 2 (BCL-2) family, and both of them can induce mitophagy in HeLa cells lacking Parkin expression. They can insert into the OMM via their C-terminus, and their N-terminus has an LIR domain to facilitate binding to LC3 or GABARAP [31,32]. FUN14 domain-containing protein 1 (FUNDC1) is a mitophagy receptor that responds to hypoxia-induced mitophagy; it can localize to the OMM and has an LIR domain allowing its association with LC3 [33]. Autophagy and Beclin 1 regulator 1 (AMBRA1), an upstream autophagy regulator, can interact with LC3 through an LIR motif; interestingly, AMBRA1 can trigger both Parkin-dependent and Parkin-independent mitophagy [34,35]. In addition to these receptors, some lipids can also bind to LC3 to induce mitophagy, such as cardiolipin. Under normal conditions, cardiolipin is localized in the IMM, and it is exported to the OMM upon mitochondrial stress and binds to LC3 to trigger mitophagy directly [36].

2.2.2. Ubiquitin Ligase-Mediated Mitophagy

An earlier study suggested that Parkin is indispensable for mitophagy induction, but in 2015, researchers found that PINK1 can recruit NDP52 and optineurin to mitochondria to trigger mitophagy directly, independent of Parkin; these data indicate that Parkin is not indispensable for mitophagy, but rather acts to amplify this signal [37]. In addition to Parkin, there are several other ubiquitin E3 ligases that function in mitophagy. A novel E3 ligase called ARIH1 was found to participate in mitophagy in a PINK1-dependent manner [38]. Mitochondrial ubiquitin ligase activator of NF-kB1 (MUL1) is another ubiquitin E3 ligase on the OMM, and it can compensate for the loss of Parkin/PINK1 loss in a PD model to rescue their mutation-induced phenotypes [39]. Synphilin-1 can interact with PINK1 and be recruited to the mitochondria to promote PINK1-dependent mitophagy. This mitophagy pathway was independent of PINK1-mediated phosphorylation and Parkin [40].

3. Mitophagy in PD Pathogenesis

The mitochondrion is the critical organelle for generating energy for cellular processes, and mitochondrial functionality determines whether a cell survives or dies. Disturbances to mitochondrial homeostasis are known to contribute to several neurodegenerative diseases, including PD [41,42]. Mitochondrial respiratory chain deficits, especially reductions in the activity of complex I, were found in post-mortem brains from sporadic PD patients [43], indicating a significant role of mitochondria in PD pathogenesis. In addition, one of the earliest studies observed mitochondria within autophagosomes in the neurons of PD patients, indicating a potential link between autophagy, damaged mitochondria, and PD pathogenesis [44]. Subsequently, abnormal mitophagy was observed in several PD models, including environmental or genetic forms [36,45–50].

Most PD-associated gene mutations participate in mitochondrial dysfunction and mitophagy disorder, including PINK1 and Parkin [51] (Table 2). Therefore, a thorough understanding of how these genes participate in mitochondria quality control or mitophagy modulation is critical in the study of PD pathogenesis and for developing new treatment strategies (Figure 2).

Table 2. Overview of mitophagy or mitochondrial dynamics deficits in PD transgenic models. Overexpression or deficiency of PD-related genes, including *SNCA, PRKN, PINK1, DJ-1, LRRK2, and GBA*, contributes to mitophagy or defects in mitochondrial dynamics. α-syn, α-synuclein; DA, dopaminergic; GBA, glucocerebrosidase; LRRK2, Leucine-rich repeat kinase 2; PINK1, PTEN-induced putative kinase 1.

Gene	Model	Motor Deficits	Loss of DA Neurons	α-syn Pathology	Mitophagy Defect	Mitochondrial Dynamics Defect
SNCA	Wide-type α-syn overexpression	+	+	+	+	+
	A53T α-syn overexpression	+	+	+	+	+
	A30P α-syn overexpression	+	+	+	-	-
PRKN	Parkin deficiency	-	-	-	+	+
PINK1	PINK1 deficiency	-	-	-	+	+
DJ-1	DJ-1 deficiency	+	-	-	+	+
LRRK2	G2019S LRRK2 overexpression	+	+	-	+	+
GBA	L1444P GBA overexpression	+	unknown	+	+	+
	GBA deficiency	+	unknown	+	+	+

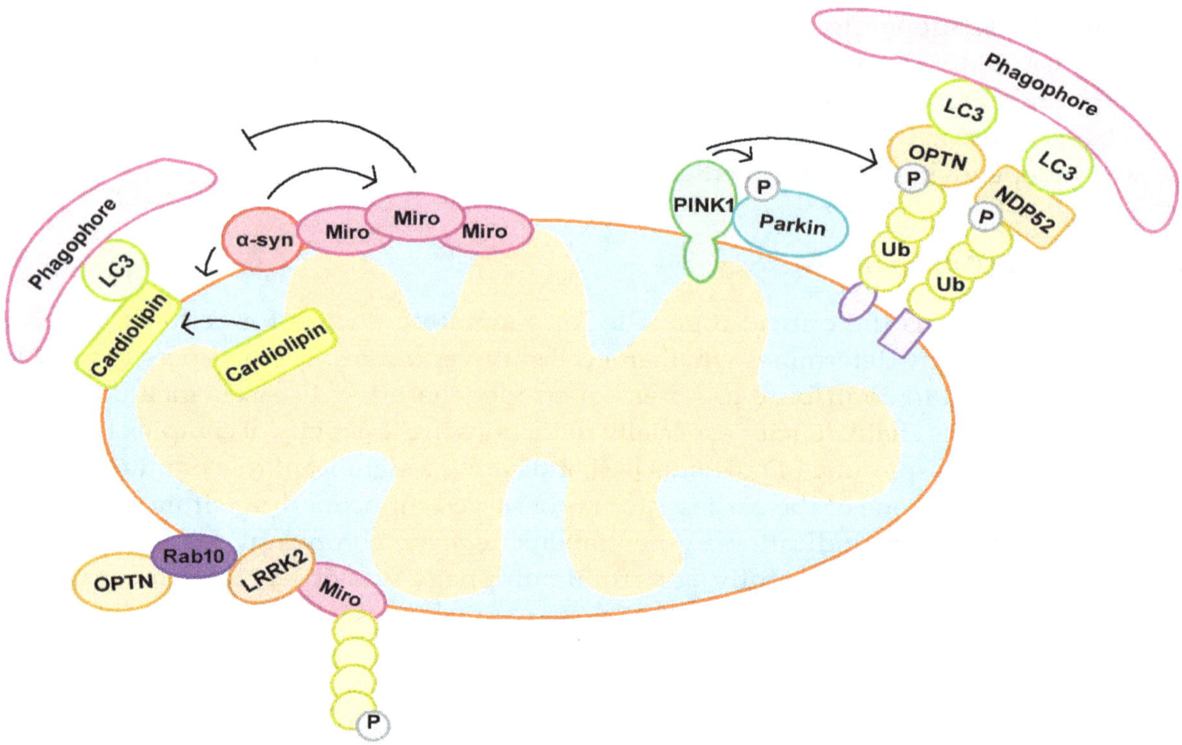

Figure 2. PD-related proteins participate in mitophagy. PINK1 accumulates at the outer mitochondrial membrane (OMM) and recruits Parkin to initiate mitophagy. α-syn interacts with Miro and upregulates Miro protein levels, leading to excessive, abnormal Miro accumulation on the mitochondrial surface and delayed mitophagy. Mitochondrial localized α-syn also promotes cardiolipin exposure on OMM; the latter further recruited LC3 to mitochondria and induced mitophagy. LRRK2 interacts with RAB10, which binds OPTN to induce mitophagy. LRRK2 interacted with Miro and contributed to its removal via mitophagy. NDP52, nuclear dot protein 52 kDa; OPTN, optineurin; Ub, ubiquitin.

3.1. PINK1 and Parkin

As mentioned above, PINK1 and Parkin play critical roles in the process of mitophagy. In the following, we will discuss the canonical mitophagy regulation and non-canonical mitophagy regulation mediated by PINK1 and/or Parkin.

3.1.1. Canonical Mitophagy Regulation

PINK1-dependent activation of Parkin is the major pathway leading to mitophagy, as described above. PINK1 plays a dual role of phosphorylating Ub and Parkin on damaged mitochondria. PINK1 phosphorylates Ub or poly-Ub chains at serine 65 (Ser65), and Parkin mediates a feed-forward mechanism to produce poly-Ub chains, thereby amplifying mitophagy signals [52]. Therefore, PINK1 and Parkin cooperatively recognize damaged mitochondria with phosphorylated poly-Ub (p-Ub) chains under stress. Studies on human post-mortem brain specimens have shown that distinct pools of p-Ub-positive structures co-localized with markers of mitochondria, autophagy, and lysosomes. Furthermore, p-Ub structures accumulated in the brains of Lewy body disease patients in an age-dependent and Braak stage-dependent manner, suggesting that p-Ub may be a biomarker for mitochondrial impairment in aging and disease [53]. A recent study showed that a novel mutant of PINK1, I368N, cannot be stabilized on the OMM upon mitochondrial stress due to conformational changes in its active site that fail to allow for polyubiquitination [54]. The RING1-IBR (in-between-RING) domain of Parkin preferentially binds to ubiquitin in a phosphorylation-dependent manner [55]. Another report showed that p-Ub binds to the RING1 of Parkin at His302 and Arg305, and promotes the disengagement of the UBL from RING1 and Parkin phosphorylation [56].

Ubiquitination is a reversible process, and several deubiquitinating enzymes (DUBs) are known to act in mitophagy. In addition to the roles of PINK1 and Parkin, the formation of p-Ub or poly-Ub chains is balanced by the activities of deubiquitinases. DUBs such as USP30 and USP35 oppose mitophagy by eliminating Ub chains generated by Parkin on OMM [57,58]. Parkin interacts with USP8 and removes K6-linked ubiquitin chains to regulate its own activity [59,60]. USP15 is another antagonist of Parkin; it does not affect the ubiquitination or translocation of Parkin, but rather inhibits Parkin-mediated mitochondrial ubiquitination [61]. A recent study showed that PTEN-long (PTEN-L), a novel PTEN isoform, regulated negatively mitophagy by dephosphorylating p-Ub via its protein phosphatase activity. PTEN-L prevented Parkin mitochondrial translocation, allevieated Parkin phosphorylation, inactivated Parkin activity, and further disrupted the feedforward of mitophagy [62,63]. Therefore, the balance between ubiquitination and deubiquitination regulates mitophagy and mitochondrial homeostasis.

Parkin has an equivalent Ser65 residue that is similar to ubiquitin, and is located within its N-terminal ubiquitin-like (UBL) domain; this residue is phosphorylated by PINK1, resulting in an open and active conformation [27,64]. Three substitutions in the UBL domain of Parkin (G12R, R33Q, and R42P) were found to significantly decrease PINK1 ability to phosphorylate Parkin. Two other UBL domain substitutions (G12R and T55I) increased the autoubiquitination of Parkin, suggesting that these substitutions increase Parkin degradation [65]. Parkin S65N, a PD-associated mutation, cannot be activated by PINK1. In addition, mice overexpressing Parkin S65A, which is a mutant that cannot be phosphorylated by PINK1, exhibit selective motor deficits, highlighting the critical role of Parkin Ser65 phosphorylation in PD pathogenesis [66]. Furthermore, growing evidence has shown that PINK1–Parkin signaling is prominent in dopaminergic neurons compared with other neurons, suggesting the vulnerability of dopaminergic neurons to mitochondrial stress [67].

The activity of PINK1 can be regulated by its post-translational modification. PINK1 can phosphorylate itself to regulate its kinase activity; Ser228 and Ser402 sites can be autophosphorylated on truncated PINK1, but not on full-length PINK1, and these phosphorylated PINK1 residues further regulate the phosphorylation of Parkin and Ub, and are involved in the induction of mitophagy [68]. Mitochondrial dysfunction arises from increased levels of S-nitrosylated PINK1 (SNO-PINK1), which is a specific post-translational modification on PINK1 that inhibits its kinase activity. SNO-PINK1 formation was found to disrupt mitophagy by decreasing Parkin translocation to mitochondria, further contributing to neuron death [69].

Despite post-translational modification, PINK1 and Parkin are also modulated by other factors. Several studies demonstrated the interaction between p53 and parkin [70–72]; another study showed that PINK1 can be down-regulated by p53 directly via its transcriptional activity, and further modulated mitophagy [73]. Interestingly, one recent study for the first demonstrated that Parkin could also act upstream to PINK1 via its transcription factor function by the activation of presenilins promoters. It is regarded as a novel feedback loop between Parkin and PINK1 in the control of mitophagy [74].

3.1.2. Non-Canonical Mitophagy Regulation

In addition to their known roles in mitophagy, Parkin and PINK1 may modulate mitophagy in other ways. The autophagy protein Beclin1 can interact with Parkin in the cytosol, and is involved in Parkin translocation to mitochondria [75]. Similarly, PINK1 also interacts with Beclin1 [76] at the mitochondria-associated membrane (MAM), which is a specific region between the ER and mitochondria involved in mitochondrial quality control, suggesting a novel role for PINK1 in mitophagy regulation [77]. In addition, a recent study showed that dopaminergic neurons and microglia exhibit a high degree of mitophagy, while basal mammalian mitophagy occurs independently of PINK1, suggesting the presence of other, yet-to-be-discovered mitophagy pathways [78].

3.1.3. PINK1/Parkin-Mediated Mitophagy and Mitochondrial Dynamics

Mitophagy mediated by PINK1 and Parkin also involves changes in mitochondrial dynamics, which play a significant role in maintaining mitochondrial homeostasis, especially in neurons. Miro is an adaptor located on the outer mitochondrial membrane; it mediates mitochondrial motility under normal conditions, and is removed from damaged mitochondria upon stress to facilitate mitochondrial clearance via mitophagy. Miro turnover on damaged mitochondria is altered in PD patient-derived fibroblasts with Parkin mutations. Mitochondrial dysfunction triggers the Lys27-type ubiquitination of Miro on the OMM in a PINK1-dependent and Parkin-dependent manner. Additionally, Miro can stabilize phosphomutant versions of Parkin on the OMM, suggesting its role as a member of the Parkin receptor complex [79]. PINK1 was found to phosphorylate Miro, and phosphorylated Miro activated the proteasomal degradation of Miro in a Parkin-dependent manner [80]. Interestingly, the PINK1/Parkin pathway can quarantine damaged mitochondria prior to their clearance by preventing mitochondrial movement [81].

Fission and fusion control mitochondrial morphology, and their balance is critical for the mitochondrial network. Growing evidence suggests that PINK1 and Parkin are critical for modulating mitochondrial fission and fusion. Indeed, the mitochondrial fusion proteins mitofusins (MFN) are substrates of PINK1 and Parkin [82]. Parkin can ubiquitinate MFN1/2, and these polyubiquitin chains are phosphorylated by PINK1 in return [83]. The loss of PINK1 and Parkin causes increases in MFN abundance and damaged mitophagy processes in Drosophila [84]. PINK1 and Parkin mutations in PD patient-derived fibroblasts also impaired MFN1/2 ubiquitination [85]. Subsequently, these signals recruited autophagy receptors to mitochondria, such as P62 and optineurin (OPTN), leading to eventual degradation by lysosomes [86]. Parkin also interacts with and ubiquitinates dynamin-related protein 1 (Drp1), which is a critical fission-related protein, to promote its proteasome-dependent degradation independent of mitophagy [87].

MFN2 also functions as an endoplasmic reticulum (ER)–OMM tether, and its phosphorylation and ubiquitination can trigger the disassembly of MFN2 complexes from the OMM to dissociate mitochondria from the ER; these findings raise the possibility of a regulatory mechanism of mitochondria–ER contact related to PINK1/Parkin that is independent of mitophagy [88]. ER and mitochondria are more closely associated in primary fibroblasts from PD patients with Parkin mutations and Parkin knockout (KO) mice compared with controls. Moreover, the abundance of MFN2 in the MAM was found to be elevated in PARK2 KO tissue and was accompanied by increased Ca^{2+} transfer from the ER to mitochondria, suggesting that Parkin is directly involved in regulating ER–mitochondria contacts [89].

3.2. α-Synuclein

α-Synuclein (α-syn) is the main component of Lewy bodies, and its mutation, duplication, or triplication results in autosomal-dominant PD [90]. α-Syn accumulation contributes to dopaminergic neuron death; although the underlying mechanism remains obscure, growing evidence suggests that mitochondrial dysfunction plays a significant role [41]. Many studies have found that α-syn could translocate to mitochondria via its N-terminus, and impaired mitochondrial function [91,92]. Moreover, α-syn was also found to impair autophagy, particularly mitophagy, and the latter process was further exacerbated by this detrimental mechanism through the impaired removal of dysfunctional mitochondria [93,94].

α-Syn can impair mitophagy in numerous ways. In the neurons of PD patients, α-syn interacts with Miro via its N-terminus and upregulates Miro protein levels, leading to excessive, abnormal Miro accumulation on the mitochondrial surface and delayed mitophagy, suggesting that Miro is a target of α-syn-associated mitochondrial injury [95]. Overexpression of the A53T α-syn mutant results in p38 MAPK activation, and this mutant directly phosphorylated Parkin at serine 131 to disturb its function and mitophagy [96]. In A53T α-syn-overexpressing mice, α-syn accumulates on mitochondria to cause increased mitophagy and neuronal death, while these mitochondrial deficits can be rescued by silencing Parkin and overexpressing Mfn2 or a dominant-negative variant of Drp1 [46,97]. In A53T and E46K α-syn transgenic mice, α-syn accumulates on the mitochondrial membrane and promotes cardiolipin exposure on the mitochondrial surface. Cardiolipin exposure recruited LC3 to mitochondria and induced mitophagy [98]. Yeast overexpressing both the human wild-type SNCA gene and A53T mutant showed enhanced mitophagy activities [99]. These studies indicate the role of abnormal mitophagy in α-syn-mediated toxicity.

3.3. LRRK2

Mutations in LRRK2, which is a member of the leucine-rich kinase family, are a cause of autosomal dominant PD [100]. Growing evidence suggests that mutations in LRRK2 result in abnormal mitophagy, although the mechanism remains controversial. One study showed that the levels of autophagy markers p62 and LC3 were increased in induced pluripotent stem cell-derived dopaminergic neurons from PD patients with the G2019S mutation in LRRK2, which is the most common LRRK2 mutation related to PD, suggesting the involvement of abnormal autophagy in G2019S-induced neurotoxicity [101]. Another study showed that the number of fragmented mitochondria increased and mitophagic clearance was reduced in human neuroepithelial stem cells from PD patients carrying the G2019S mutation [102]. However, another study suggested that G2019S LRRK2 mutations increase mitophagy due to histone deacetylase activation [103]. In addition, a study of human iPSC-derived neurons showed that LRRK2 interacted with Miro and contributed to its removal. The G2019S mutation delayed mitophagy initiation, while the knockdown of Miro rescued the injury caused by G2019S, indicating that Miro is also a target in the mitophagy injury induced by the LRRK2 mutation [104,105].

Several studies have suggested a role for LRRK2 in PINK1 and Parkin-mediated mitophagy. A recent study showed that RAB10, a substrate of LRRK2, accumulated on damaged mitochondria in a PINK1-dependent and Parkin-dependent manner. Subsequently, RAB10 was found to bind OPTN to promote its accumulation and subsequently induced mitophagy, indicating that LRRK2 is involved in PINK1-mediated and Parkin-mediated mitophagy via RAB10 [106]. However, a separate study suggested that LRRK2 attenuated PINK1-dependent and Parkin-dependent mitophagic clearance via its kinase activity [107].

3.4. DJ-1

Mutations in DJ-1, which is encoded by the *PARK6* gene, cause a rare form of autosomal recessive PD [108]. Upon stress, DJ-1 localizes to mitochondria and acts as a redox sensor/reductase, and its depletion led to mitochondrial deficits and an increase in ROS levels. DJ-1 is regarded as a neuroprotective factor, and mitochondria-localized DJ-1 regulates the clearance of endogenous ROS [109]. DJ-1 deficiency led to an increased level of oxidative stress [110], which is usually associated with mitophagy impairment [111]. The loss of DJ-1 caused increased Parkin recruitment to damaged mitochondria and increased mitophagy, and DJ-1 levels accumulated on mitochondria under oxidative stress conditions dependent on Parkin and PINK1, suggesting a link between DJ-1 and the PINK1/Parkin-mediated pathway [112]. Parkin regulated DJ-1 levels via a signaling cascade implying p53, indicating that p53 and DJ-1 acted acting downstream of parkin [113]. Additionally, a separate study suggested that DJ-1 functioned in parallel with the PINK1/Parkin pathway to maintain mitochondrial function under an oxidative environment [114]. Another study showed mitochondrial defects in DJ-1 knockout flies, which is similar to PINK1 and Parkin mutants. Interestingly, DJ-1 overexpression rescues the phenotype of flies that are deficient for PINK1, but not Parkin. These data also suggest that DJ-1 is critical for mitochondrial function and acts in parallel to or downstream of PINK1 [115].

3.5. GBA1

GBA1 encodes the lysosomal enzyme glucosylceramidase beta/β-glucocerebrosidase, and its heterozygous mutations are among the most common genetic risk factors of PD [116]. GBA1 homozygous mutations cause Gaucher's disease (GD), which is the most frequent lysosomal storage disorder, and some GD patients and their relatives show parkinsonian manifestations [117]. It has been estimated that GBA1 mutations lead to a 20-fold to 30-fold increased risk of PD, and at least 7–10% of PD patients have a GBA1 mutation [118]. Autophagy defects have been confirmed in iPSC-derived neurons from GBA1-associated PD patients [119], and mitochondrial function was also impaired in GD patients [120]. Additionally, GBA1 deficiencies resulted in α-syn aggregation in PD or GD models and patients [121–123], indicating that GBA1 is involved in α-syn pathology and PD pathogenesis, likely by impairing autophagy and mitochondrial function.

In primary neurons from GBA1-knockout mice, autophagy was impaired and mitochondrial function was profoundly compromised with a reduced membrane potential [123]. In $GBA^{L444P/WT}$ knock-in mice, the L444P mutation results in mitochondrial dysfunction by inhibiting mitochondrial priming and autophagy, which are two critical steps for mitophagy. In type II neuronopathic GD, the downregulation of mitophagy resulted in the accumulation of insoluble α-syn deposits [48]. Furthermore, impaired mitophagy and excessive oxidative stress were found in post-mortem brain tissue from PD patients carrying heterozygous GBA mutations, suggesting a link between mitophagy dysfunction and GBA heterozygous mutations [124].

3.6. Other PD-Related Proteins

Vacuolar protein sorting-associated protein 35 (Vps35), which is encoded by *PARK17*, causes autosomal-dominant, late-onset PD. Vps35 deficiency or mutation resulted in mitochondrial dysfunction and the loss of dopaminergic neurons [8,125]. Vps35 interacts with Parkin, but not with Pink1; furthermore, its overexpression rescues several Parkin-mutant phenotypes [126].

F-box only protein 7 (Fbxo7), which is encoded by *PARK15*, is a PD-related gene. Mutations of Fbxo7 cause autosomal recessive juvenile atypical PD. Studies showed that Fbxo7 mutation impairs mitophagy, suggesting that Fbxo7 participated in the modulation of mitochondrial homeostasis [127]. Wild-type Fbxo7 can promote mitophagy in response to stress, while mutant Fbxo7 was found to inhibit mitophagy [128]. Another study found that Fbxo7 participates in mitophagy by interacting with PINK1 and Parkin directly, while its PD-related mutations interfered with this process [129]. A PD-related Fbxo7 mutation also recruited Parkin to damaged mitochondria and promoted its aggregation [130].

4. Modulation of Mitophagy in PD Treatment

Mitochondrial deficits and autophagy impairment are critical aspects of PD pathogenesis, with impaired mitophagy found in the brains of PD patients and models. Therefore, correcting mitophagy is a promising avenue for the development of efficient treatments for PD. Pharmacological agents that selectively modulate mitophagy are currently lacking, and thus the clinical applicability of this approach remains limited [131]. Although some compounds such as trifluorocarbonylcyanide phenylhydrazone and the combination of antimycin/oligomycin were found to trigger mitophagy, their effects were toxic and non-specific. Therefore, these agents are not suitable for PD treatment [132]. Campanella et al. developed a compound, P62-mediated mitophagy inducer (PMI), that activates endogenous mitophagy without Parkin recruitment or the dissipation of mitochondrial membrane potential. Therefore, PMI has been regarded as a promising chemical candidate [133]. However, whether PMI exerts a therapeutic effect in PD models remains unclear. One recent study suggested that several pathogenic Parkin variants impaired mitophagy, and thus targeting these variants in the design of genotype-specific drugs represents a promising direction [134].

Thus far, several synthetic and natural chemical compounds have been used to trigger mitophagy for treating PD models [10,135]. An interesting study showed that a carrier with a homozygous Parkin mutation resulting in the loss of functional Parkin had not developed PD by her eighth decade of life, indicating the potential existence of a putative mechanism of protection against PD. Further study showed that this carrier had preserved mitochondrial function, and mitophagy could be mediated by mitochondrial receptor Nip3-like protein X (Nix), which is a pathway independent of

PINK1/Parkin [136]. This finding suggests that Nix may be a promising target for PINK1/Parkin-related PD treatment, because it could serve as an alternative mediator of mitophagy [137,138].

The phosphorylation of Parkin and Ub mediated by PINK1 and ubiquitination-mediated Parkin form a feedforward mechanism of mitophagy [52]. Therefore, regulation of the PINK1–Parkin–Ub feedforward loop is a strategy for mitophagy modulation and PD treatment. PTEN-L can dephosphorylate p-Ub and suppress mitophagy via blockage of the feedforward mechanism [63], so the inhibitor of PTEN-L may contribute to mitophagy activation. As mentioned above, USP30, USP35, and USP15 can counteract Parkin activity and regulate mitophagy negatively. The depletion of USP30 enhanced mitochondria degradation in neurons [57]. Similarly, the knockdown of USP35 or USP15 also promoted the mitophagy pathoway [58,61]. However, another DUB, USP8, may regulate Parkin and mitophagy positively, because it removed K6-linked ubiquitin chains from Parkin preferentially to promote the efficient recruitment of Parkin and induction of mitophagy [60]. Although research targeting the PINK1–Parkin–Ub feedforward loop is still limited, it is a valuable direction in PD treatment (Figure 3).

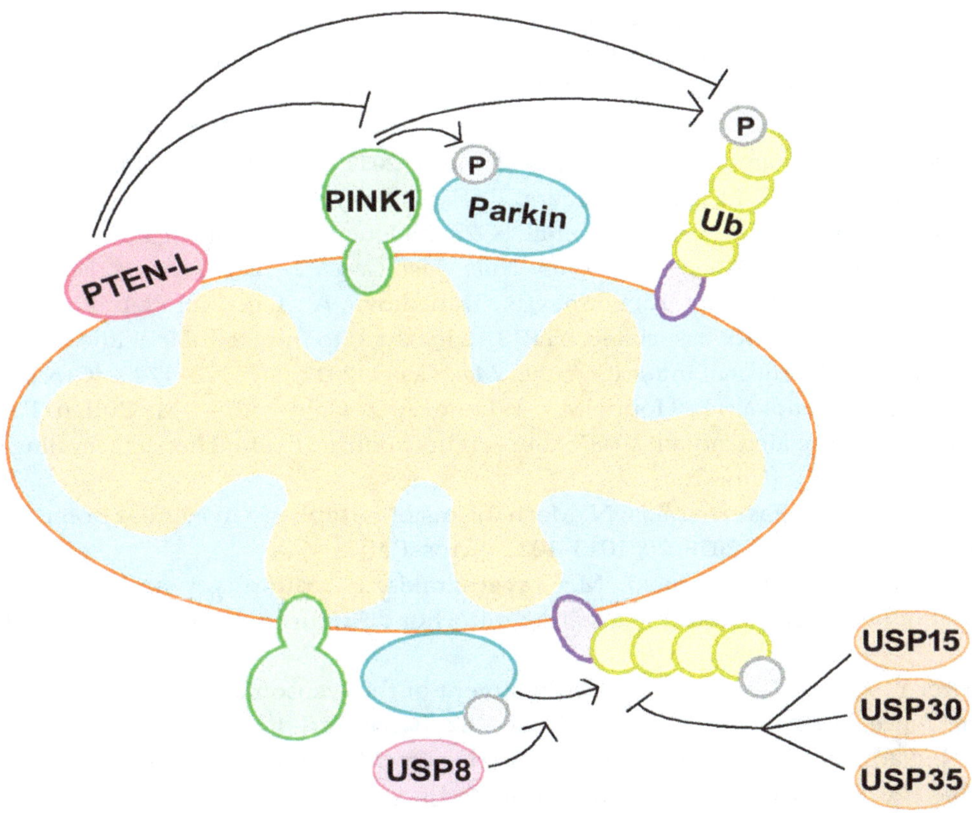

Figure 3. The PINK1–Parkin–Ub feedforward loop. PINK1, Parkin, and Ub form a feedforward mechanism of mitophagy. Regulation of this feedforward loop is a strategy for mitophagy modulation and PD treatment. PTEN-L dephosphorylate p-Ub and suppress mitophagy via blockage of the feedforward mechanism. USP30, USP35, and USP15 can counteract Parkin activity and regulate mitophagy negatively, while USP8 regulate Parkin and mitophagy positively. PTEN-L, phosphate and tension homology deleted on chromsome ten-long; Ub, ubiquitin; USP, ubiquitin specific protease.

5. Conclusions

Many PD-related genes or risk factors are associated with mitophagy defects. There is a strong reciprocal relationship between mitochondria and autophagy, with the impairment of one process usually resulting in damage to the other, and this vicious cycle eventually contributes to the pathogenesis of PD. However, there are some remaining questions. Although the modulation of mitophagy is regarded as a potential approach in PD treatment, whether mitopagy can be specifically targeted is always a question. Although the PINK1–Parkin–Ub feedforward loop is critical for the regulation of mitophagy, its mechanism is still obscure. Despite these, there is a a critical need to understand mitochondrial and lysosomal dysfunction and how their interplay participate in PD pathogenesis, as targeting these crosstalks and restoring mitophagy represent a promising therapeutic approach for PD.

Author Contributions: J.L., W.L., R.L. and H.Y. wrote the manuscript.

References

1. Homayoun, H. Parkinson Disease. *Ann. Intern. Med.* **2018**, *169*, ITC33–ITC48. [CrossRef] [PubMed]
2. Kalia, L.V.; Lang, A.E. Parkinson disease in 2015: Evolving basic, pathological and clinical concepts in PD. *Nat. Rev. Neurol.* **2016**, *12*, 65–66. [CrossRef] [PubMed]
3. Yoon, H.H.; Min, J.; Jeon, S.R. Optogenetics to restore neural circuit function in Parkinson's disease. *J. Neurorestoratol.* **2018**, *6*, 88–92. [CrossRef]
4. Verstraeten, A.; Theuns, J.; Van Broeckhoven, C. Progress in unraveling the genetic etiology of Parkinson disease in a genomic era. *Trends Genet.* **2015**, *31*, 140–149. [CrossRef] [PubMed]
5. Nakamura, K.; Nemani, V.M.; Azarbal, F.; Skibinski, G.; Levy, J.M.; Egami, K.; Munishkina, L.; Zhang, J.; Gardner, B.; Wakabayashi, J.; et al. Direct membrane association drives mitochondrial fission by the Parkinson disease-associated protein alpha-synuclein. *J. Biol. Chem.* **2011**, *286*, 20710–20726. [CrossRef] [PubMed]
6. Papkovskaia, T.D.; Chau, K.Y.; Inesta-Vaquera, F.; Papkovsky, D.B.; Healy, D.G.; Nishio, K.; Staddon, J.; Duchen, M.R.; Hardy, J.; Schapira, A.H.; et al. G2019S leucine-rich repeat kinase 2 causes uncoupling protein-mediated mitochondrial depolarization. *Hum. Mol. Genet.* **2012**, *21*, 4201–4213. [CrossRef] [PubMed]
7. Ramonet, D.; Podhajska, A.; Stafa, K.; Sonnay, S.; Trancikova, A.; Tsika, E.; Pletnikova, O.; Troncoso, J.C.; Glauser, L.; Moore, D.J. PARK9-associated ATP13A2 localizes to intracellular acidic vesicles and regulates cation homeostasis and neuronal integrity. *Hum. Mol. Genet.* **2012**, *21*, 1725–1743. [CrossRef]
8. Wang, W.; Wang, X.; Fujioka, H.; Hoppel, C.; Whone, A.L.; Caldwell, M.A.; Cullen, P.J.; Liu, J.; Zhu, X. Parkinson's disease-associated mutant VPS35 causes mitochondrial dysfunction by recycling DLP1 complexes. *Nat. Med.* **2016**, *22*, 54–63. [CrossRef]
9. Palikaras, K.; Lionaki, E.; Tavernarakis, N. Mechanisms of mitophagy in cellular homeostasis, physiology and pathology. *Nat. Cell Biol.* **2018**, *20*, 1013–1022. [CrossRef]
10. Palikaras, K.; Daskalaki, I.; Markaki, M.; Tavernarakis, N. Mitophagy and age-related pathologies: Development of new therapeutics by targeting mitochondrial turnover. *Pharm. Ther.* **2017**, *178*, 157–174. [CrossRef]
11. Park, J.T.; Lee, Y.S.; Cho, K.A.; Park, S.C. Adjustment of the lysosomal-mitochondrial axis for control of cellular senescence. *Ageing Res. Rev.* **2018**, *47*, 176–182. [CrossRef] [PubMed]
12. Nguyen, T.N.; Padman, B.S.; Lazarou, M. Deciphering the Molecular Signals of PINK1/Parkin Mitophagy. *Trends Cell Biol.* **2016**, *26*, 733–744. [CrossRef] [PubMed]
13. Villa, E.; Marchetti, S.; Ricci, J.E. No Parkin Zone: Mitophagy without Parkin. *Trends Cell Biol.* **2018**, *28*, 882–895. [CrossRef] [PubMed]
14. Narendra, D.; Tanaka, A.; Suen, D.F.; Youle, R.J. Parkin is recruited selectively to impaired mitochondria and promotes their autophagy. *J. Cell Biol.* **2008**, *183*, 795–803. [CrossRef] [PubMed]
15. Matsuda, N.; Sato, S.; Shiba, K.; Okatsu, K.; Saisho, K.; Gautier, C.A.; Sou, Y.S.; Saiki, S.; Kawajiri, S.; Sato, F.; et al. PINK1 stabilized by mitochondrial depolarization recruits Parkin to damaged mitochondria and activates latent Parkin for mitophagy. *J. Cell Biol.* **2010**, *189*, 211–221. [CrossRef] [PubMed]
16. Narendra, D.P.; Jin, S.M.; Tanaka, A.; Suen, D.F.; Gautier, C.A.; Shen, J.; Cookson, M.R.; Youle, R.J. PINK1 is selectively stabilized on impaired mitochondria to activate Parkin. *Plos Biol.* **2010**, *8*, e1000298. [CrossRef]
17. Vives-Bauza, C.; Zhou, C.; Huang, Y.; Cui, M.; de Vries, R.L.; Kim, J.; May, J.; Tocilescu, M.A.; Liu, W.; Ko, H.S.; et al. PINK1-dependent recruitment of Parkin to mitochondria in mitophagy. *Proc. Natl. Acad. Sci. USA* **2010**, *107*, 378–383. [CrossRef]

18. Valente, E.M.; Abou-Sleiman, P.M.; Caputo, V.; Muqit, M.M.; Harvey, K.; Gispert, S.; Ali, Z.; Del Turco, D.; Bentivoglio, A.R.; Healy, D.G.; et al. Hereditary early-onset Parkinson's disease caused by mutations in PINK1. *Science* **2004**, *304*, 1158–1160. [CrossRef]
19. Kitada, T.; Asakawa, S.; Hattori, N.; Matsumine, H.; Yamamura, Y.; Minoshima, S.; Yokochi, M.; Mizuno, Y.; Shimizu, N. Mutations in the parkin gene cause autosomal recessive juvenile parkinsonism. *Nature* **1998**, *392*, 605–608. [CrossRef]
20. Morais, V.A.; Haddad, D.; Craessaerts, K.; De Bock, P.J.; Swerts, J.; Vilain, S.; Aerts, L.; Overbergh, L.; Grunewald, A.; Seibler, P.; et al. PINK1 loss-of-function mutations affect mitochondrial complex I activity via NdufA10 ubiquinone uncoupling. *Science* **2014**, *344*, 203–207. [CrossRef]
21. Clark, I.E.; Dodson, M.W.; Jiang, C.; Cao, J.H.; Huh, J.R.; Seol, J.H.; Yoo, S.J.; Hay, B.A.; Guo, M. Drosophila pink1 is required for mitochondrial function and interacts genetically with parkin. *Nature* **2006**, *441*, 1162–1166. [CrossRef] [PubMed]
22. Khymenets, O.; Ortuno, J.; Fito, M.; Covas, M.A.; Farre, M.; de la Torre, R. Evaluation of RNA isolation procedures from human blood and its application for gene expression studies (Sod-1, Sod-2). *Anal. Biochem.* **2005**, *347*, 156–158. [CrossRef] [PubMed]
23. McWilliams, T.G.; Muqit, M.M. PINK1 and Parkin: Emerging themes in mitochondrial homeostasis. *Curr. Opin. Cell. Biol.* **2017**, *45*, 83–91. [CrossRef] [PubMed]
24. Harper, J.W.; Ordureau, A.; Heo, J.M. Building and decoding ubiquitin chains for mitophagy. *Nat. Rev. Mol. Cell Biol.* **2018**, *19*, 93–108. [CrossRef] [PubMed]
25. Yamano, K.; Youle, R.J. PINK1 is degraded through the N-end rule pathway. *Autophagy* **2013**, *9*, 1758–1769. [CrossRef] [PubMed]
26. Meissner, C.; Lorenz, H.; Hehn, B.; Lemberg, M.K. Intramembrane protease PARL defines a negative regulator of PINK1- and PARK2/Parkin-dependent mitophagy. *Autophagy* **2015**, *11*, 1484–1498. [CrossRef]
27. Trempe, J.F.; Sauve, V.; Grenier, K.; Seirafi, M.; Tang, M.Y.; Menade, M.; Al-Abdul-Wahid, S.; Krett, J.; Wong, K.; Kozlov, G.; et al. Structure of parkin reveals mechanisms for ubiquitin ligase activation. *Science* **2013**, *340*, 1451–1455. [CrossRef]
28. Wong, Y.C.; Holzbaur, E.L. Temporal dynamics of PARK2/parkin and OPTN/optineurin recruitment during the mitophagy of damaged mitochondria. *Autophagy* **2015**, *11*, 422–424. [CrossRef]
29. Wong, Y.C.; Holzbaur, E.L. Optineurin is an autophagy receptor for damaged mitochondria in parkin-mediated mitophagy that is disrupted by an ALS-linked mutation. *Proc. Natl. Acad. Sci. USA* **2014**, *111*, E4439–E4448. [CrossRef]
30. Liu, L.; Sakakibara, K.; Chen, Q.; Okamoto, K. Receptor-mediated mitophagy in yeast and mammalian systems. *Cell Res.* **2014**, *24*, 787–795. [CrossRef]
31. Hanna, R.A.; Quinsay, M.N.; Orogo, A.M.; Giang, K.; Rikka, S.; Gustafsson, A.B. Microtubule-associated protein 1 light chain 3 (LC3) interacts with Bnip3 protein to selectively remove endoplasmic reticulum and mitochondria via autophagy. *J. Biol. Chem.* **2012**, *287*, 19094–19104. [CrossRef] [PubMed]
32. Rogov, V.V.; Suzuki, H.; Marinkovic, M.; Lang, V.; Kato, R.; Kawasaki, M.; Buljubasic, M.; Sprung, M.; Rogova, N.; Wakatsuki, S.; et al. Phosphorylation of the mitochondrial autophagy receptor Nix enhances its interaction with LC3 proteins. *Sci. Rep.* **2017**, *7*, 1131. [CrossRef] [PubMed]
33. Liu, L.; Feng, D.; Chen, G.; Chen, M.; Zheng, Q.; Song, P.; Ma, Q.; Zhu, C.; Wang, R.; Qi, W.; et al. Mitochondrial outer-membrane protein FUNDC1 mediates hypoxia-induced mitophagy in mammalian cells. *Nat. Cell Biol.* **2012**, *14*, 177–185. [CrossRef] [PubMed]
34. Strappazzon, F.; Nazio, F.; Corrado, M.; Cianfanelli, V.; Romagnoli, A.; Fimia, G.M.; Campello, S.; Nardacci, R.; Piacentini, M.; Campanella, M.; et al. AMBRA1 is able to induce mitophagy via LC3 binding, regardless of PARKIN and p62/SQSTM1. *Cell Death Differ.* **2015**, *22*, 419–432. [CrossRef] [PubMed]
35. Van Humbeeck, C.; Cornelissen, T.; Hofkens, H.; Mandemakers, W.; Gevaert, K.; De Strooper, B.; Vandenberghe, W. Parkin interacts with Ambra1 to induce mitophagy. *J. Neurosci.* **2011**, *31*, 10249–10261. [CrossRef] [PubMed]
36. Chu, C.T.; Ji, J.; Dagda, R.K.; Jiang, J.F.; Tyurina, Y.Y.; Kapralov, A.A.; Tyurin, V.A.; Yanamala, N.; Shrivastava, I.H.; Mohammadyani, D.; et al. Cardiolipin externalization to the outer mitochondrial membrane acts as an elimination signal for mitophagy in neuronal cells. *Nat. Cell Biol.* **2013**, *15*, 1197–1205. [CrossRef]

37. Lazarou, M.; Sliter, D.A.; Kane, L.A.; Sarraf, S.A.; Wang, C.; Burman, J.L.; Sideris, D.P.; Fogel, A.I.; Youle, R.J. The ubiquitin kinase PINK1 recruits autophagy receptors to induce mitophagy. *Nature* **2015**, *524*, 309–314. [CrossRef]
38. Villa, E.; Proics, E.; Rubio-Patino, C.; Obba, S.; Zunino, B.; Bossowski, J.P.; Rozier, R.M.; Chiche, J.; Mondragon, L.; Riley, J.S.; et al. Parkin-Independent Mitophagy Controls Chemotherapeutic Response in Cancer Cells. *Cell Rep.* **2017**, *20*, 2846–2859. [CrossRef]
39. Yun, J.; Puri, R.; Yang, H.; Lizzio, M.A.; Wu, C.; Sheng, Z.H.; Guo, M. MUL1 acts in parallel to the PINK1/parkin pathway in regulating mitofusin and compensates for loss of PINK1/parkin. *Elife* **2014**, *3*, e01958. [CrossRef]
40. Szargel, R.; Shani, V.; Abd Elghani, F.; Mekies, L.N.; Liani, E.; Rott, R.; Engelender, S. The PINK1, synphilin-1 and SIAH-1 complex constitutes a novel mitophagy pathway. *Hum. Mol. Genet.* **2016**, *25*, 3476–3490. [CrossRef]
41. Grunewald, A.; Kumar, K.R.; Sue, C.M. New insights into the complex role of mitochondria in Parkinson's disease. *Prog. Neurobiol.* **2019**, *177*, 73–93. [CrossRef] [PubMed]
42. Huang, H.; Sharma, H.S.; Chen, L.; Saberi, H.; Mao, G. 2018 Yearbook of Neurorestoratology. *J. Neurorestoratol.* **2019**, *7*, 8–17.
43. Parker, W.D., Jr.; Boyson, S.J.; Parks, J.K. Abnormalities of the electron transport chain in idiopathic Parkinson's disease. *Ann. Neurol.* **1989**, *26*, 719–723. [CrossRef] [PubMed]
44. Zhu, J.H.; Guo, F.; Shelburne, J.; Watkins, S.; Chu, C.T. Localization of phosphorylated ERK/MAP kinases to mitochondria and autophagosomes in Lewy body diseases. *Brain Pathol.* **2003**, *13*, 473–481. [CrossRef] [PubMed]
45. Cherra, S.J., 3rd; Steer, E.; Gusdon, A.M.; Kiselyov, K.; Chu, C.T. Mutant LRRK2 elicits calcium imbalance and depletion of dendritic mitochondria in neurons. *Am. J. Pathol.* **2013**, *182*, 474–484. [CrossRef] [PubMed]
46. Chinta, S.J.; Mallajosyula, J.K.; Rane, A.; Andersen, J.K. Mitochondrial alpha-synuclein accumulation impairs complex I function in dopaminergic neurons and results in increased mitophagy in vivo. *Neurosci. Lett.* **2010**, *486*, 235–239. [CrossRef] [PubMed]
47. Dagda, R.K.; Cherra, S.J., 3rd; Kulich, S.M.; Tandon, A.; Park, D.; Chu, C.T. Loss of PINK1 function promotes mitophagy through effects on oxidative stress and mitochondrial fission. *J. Biol. Chem.* **2009**, *284*, 13843–13855. [CrossRef] [PubMed]
48. Osellame, L.D.; Duchen, M.R. Defective quality control mechanisms and accumulation of damaged mitochondria link Gaucher and Parkinson diseases. *Autophagy* **2013**, *9*, 1633–1635. [CrossRef]
49. Dagda, R.K.; Zhu, J.; Kulich, S.M.; Chu, C.T. Mitochondrially localized ERK2 regulates mitophagy and autophagic cell stress: Implications for Parkinson's disease. *Autophagy* **2008**, *4*, 770–782. [CrossRef]
50. Zhu, J.H.; Horbinski, C.; Guo, F.; Watkins, S.; Uchiyama, Y.; Chu, C.T. Regulation of autophagy by extracellular signal-regulated protein kinases during 1-methyl-4-phenylpyridinium-induced cell death. *Am. J. Pathol.* **2007**, *170*, 75–86. [CrossRef]
51. Pickrell, A.M.; Youle, R.J. The roles of PINK1, parkin, and mitochondrial fidelity in Parkinson's disease. *Neuron* **2015**, *85*, 257–273. [CrossRef] [PubMed]
52. Ordureau, A.; Sarraf, S.A.; Duda, D.M.; Heo, J.M.; Jedrychowski, M.P.; Sviderskiy, V.O.; Olszewski, J.L.; Koerber, J.T.; Xie, T.; Beausoleil, S.A.; et al. Quantitative proteomics reveal a feedforward mechanism for mitochondrial PARKIN translocation and ubiquitin chain synthesis. *Mol. Cell* **2014**, *56*, 360–375. [CrossRef] [PubMed]
53. Hou, X.; Fiesel, F.C.; Truban, D.; Castanedes Casey, M.; Lin, W.L.; Soto, A.I.; Tacik, P.; Rousseau, L.G.; Diehl, N.N.; Heckman, M.G.; et al. Age- and disease-dependent increase of the mitophagy marker phospho-ubiquitin in normal aging and Lewy body disease. *Autophagy* **2018**, *14*, 1404–1418. [CrossRef] [PubMed]
54. Ando, M.; Fiesel, F.C.; Hudec, R.; Caulfield, T.R.; Ogaki, K.; Gorka-Skoczylas, P.; Koziorowski, D.; Friedman, A.; Chen, L.; Dawson, V.L.; et al. The PINK1 p.I368N mutation affects protein stability and ubiquitin kinase activity. *Mol. Neurodegener.* **2017**, *12*, 32. [CrossRef] [PubMed]
55. Yamano, K.; Queliconi, B.B.; Koyano, F.; Saeki, Y.; Hirokawa, T.; Tanaka, K.; Matsuda, N. Site-specific Interaction Mapping of Phosphorylated Ubiquitin to Uncover Parkin Activation. *J. Biol. Chem.* **2015**, *290*, 25199–25211. [CrossRef] [PubMed]

56. Sauve, V.; Lilov, A.; Seirafi, M.; Vranas, M.; Rasool, S.; Kozlov, G.; Sprules, T.; Wang, J.; Trempe, J.F.; Gehring, K. A Ubl/ubiquitin switch in the activation of Parkin. *EMBO J.* **2015**, *34*, 2492–2505. [CrossRef] [PubMed]
57. Bingol, B.; Tea, J.S.; Phu, L.; Reichelt, M.; Bakalarski, C.E.; Song, Q.; Foreman, O.; Kirkpatrick, D.S.; Sheng, M. The mitochondrial deubiquitinase USP30 opposes parkin-mediated mitophagy. *Nature* **2014**, *510*, 370–375. [CrossRef] [PubMed]
58. Wang, Y.; Serricchio, M.; Jauregui, M.; Shanbhag, R.; Stoltz, T.; Di Paolo, C.T.; Kim, P.K.; McQuibban, G.A. Deubiquitinating enzymes regulate PARK2-mediated mitophagy. *Autophagy* **2015**, *11*, 595–606. [CrossRef]
59. Durcan, T.M.; Fon, E.A. USP8 and PARK2/parkin-mediated mitophagy. *Autophagy* **2015**, *11*, 428–429. [CrossRef]
60. Durcan, T.M.; Tang, M.Y.; Perusse, J.R.; Dashti, E.A.; Aguileta, M.A.; McLelland, G.L.; Gros, P.; Shaler, T.A.; Faubert, D.; Coulombe, B.; et al. USP8 regulates mitophagy by removing K6-linked ubiquitin conjugates from parkin. *EMBO J.* **2014**, *33*, 2473–2491. [CrossRef]
61. Cornelissen, T.; Haddad, D.; Wauters, F.; Van Humbeeck, C.; Mandemakers, W.; Koentjoro, B.; Sue, C.; Gevaert, K.; De Strooper, B.; Verstreken, P.; et al. The deubiquitinase USP15 antagonizes Parkin-mediated mitochondrial ubiquitination and mitophagy. *Hum. Mol. Genet.* **2014**, *23*, 5227–5242. [CrossRef] [PubMed]
62. Wang, L.; Wang, J.; Tang, Y.; Shen, H.M. PTEN-L puts a brake on mitophagy. *Autophagy* **2018**, *14*, 2023–2025. [CrossRef] [PubMed]
63. Wang, L.; Cho, Y.L.; Tang, Y.; Wang, J.; Park, J.E.; Wu, Y.; Wang, C.; Tong, Y.; Chawla, R.; Zhang, J.; et al. PTEN-L is a novel protein phosphatase for ubiquitin dephosphorylation to inhibit PINK1-Parkin-mediated mitophagy. *Cell Res.* **2018**, *28*, 787–802. [CrossRef] [PubMed]
64. Aguirre, J.D.; Dunkerley, K.M.; Mercier, P.; Shaw, G.S. Structure of phosphorylated UBL domain and insights into PINK1-orchestrated parkin activation. *Proc. Natl. Acad. Sci. USA* **2017**, *114*, 298–303. [CrossRef] [PubMed]
65. Aguirre, J.D.; Dunkerley, K.M.; Lam, R.; Rusal, M.; Shaw, G.S. Impact of altered phosphorylation on loss of function of juvenile Parkinsonism-associated genetic variants of the E3 ligase parkin. *J. Biol. Chem.* **2018**, *293*, 6337–6348. [CrossRef] [PubMed]
66. McWilliams, T.G.; Barini, E.; Pohjolan-Pirhonen, R.; Brooks, S.P.; Singh, F.; Burel, S.; Balk, K.; Kumar, A.; Montava-Garriga, L.; Prescott, A.R.; et al. Phosphorylation of Parkin at serine 65 is essential for its activation in vivo. *Open Biol.* **2018**, *8*, 180108. [CrossRef] [PubMed]
67. Shiba-Fukushima, K.; Ishikawa, K.I.; Inoshita, T.; Izawa, N.; Takanashi, M.; Sato, S.; Onodera, O.; Akamatsu, W.; Okano, H.; Imai, Y.; et al. Evidence that phosphorylated ubiquitin signaling is involved in the etiology of Parkinson's disease. *Hum. Mol. Genet.* **2017**, *26*, 3172–3185. [CrossRef]
68. Aerts, L.; Craessaerts, K.; De Strooper, B.; Morais, V.A. PINK1 kinase catalytic activity is regulated by phosphorylation on serines 228 and 402. *J. Biol. Chem.* **2015**, *290*, 2798–2811. [CrossRef]
69. Oh, C.K.; Sultan, A.; Platzer, J.; Dolatabadi, N.; Soldner, F.; McClatchy, D.B.; Diedrich, J.K.; Yates, J.R., 3rd; Ambasudhan, R.; Nakamura, T.; et al. S-Nitrosylation of PINK1 Attenuates PINK1/Parkin-Dependent Mitophagy in hiPSC-Based Parkinson's Disease Models. *Cell Rep.* **2017**, *21*, 2171–2182. [CrossRef]
70. Da Costa, C.A.; Sunyach, C.; Giaime, E.; West, A.; Corti, O.; Brice, A.; Safe, S.; Abou-Sleiman, P.M.; Wood, N.W.; Takahashi, H.; et al. Transcriptional repression of p53 by parkin and impairment by mutations associated with autosomal recessive juvenile Parkinson's disease. *Nat. Cell Biol.* **2009**, *11*, 1370–1375. [CrossRef]
71. Zhang, C.; Lin, M.; Wu, R.; Wang, X.; Yang, B.; Levine, A.J.; Hu, W.; Feng, Z. Parkin, a p53 target gene, mediates the role of p53 in glucose metabolism and the Warburg effect. *Proc. Natl. Acad. Sci. USA* **2011**, *108*, 16259–16264. [CrossRef] [PubMed]
72. Viotti, J.; Duplan, E.; Caillava, C.; Condat, J.; Goiran, T.; Giordano, C.; Marie, Y.; Idbaih, A.; Delattre, J.Y.; Honnorat, J.; et al. Glioma tumor grade correlates with parkin depletion in mutant p53-linked tumors and results from loss of function of p53 transcriptional activity. *Oncogene* **2014**, *33*, 1764–1775. [CrossRef] [PubMed]
73. Goiran, T.; Duplan, E.; Rouland, L.; El Manaa, W.; Lauritzen, I.; Dunys, J.; You, H.; Checler, F.; Alves da Costa, C. Nuclear p53-mediated repression of autophagy involves PINK1 transcriptional down-regulation. *Cell Death Differ.* **2018**, *25*, 873–884. [CrossRef] [PubMed]
74. Goiran, T.; Duplan, E.; Chami, M.; Bourgeois, A.; El Manaa, W.; Rouland, L.; Dunys, J.; Lauritzen, I.; You, H.; Stambolic, V.; et al. beta-Amyloid Precursor Protein Intracellular Domain Controls Mitochondrial Function

by Modulating Phosphatase and Tensin Homolog-Induced Kinase 1 Transcription in Cells and in Alzheimer Mice Models. *Biol. Psychiatry* **2018**, *83*, 416–427. [CrossRef] [PubMed]
75. Choubey, V.; Cagalinec, M.; Liiv, J.; Safiulina, D.; Hickey, M.A.; Kuum, M.; Liiv, M.; Anwar, T.; Eskelinen, E.L.; Kaasik, A. BECN1 is involved in the initiation of mitophagy: It facilitates PARK2 translocation to mitochondria. *Autophagy* **2014**, *10*, 1105–1119. [CrossRef] [PubMed]
76. Michiorri, S.; Gelmetti, V.; Giarda, E.; Lombardi, F.; Romano, F.; Marongiu, R.; Nerini-Molteni, S.; Sale, P.; Vago, R.; Arena, G.; et al. The Parkinson-associated protein PINK1 interacts with Beclin1 and promotes autophagy. *Cell Death Differ.* **2010**, *17*, 962–974. [CrossRef] [PubMed]
77. Gelmetti, V.; De Rosa, P.; Torosantucci, L.; Marini, E.S.; Romagnoli, A.; Di Rienzo, M.; Arena, G.; Vignone, D.; Fimia, G.M.; Valente, E.M. PINK1 and BECN1 relocalize at mitochondria-associated membranes during mitophagy and promote ER-mitochondria tethering and autophagosome formation. *Autophagy* **2017**, *13*, 654–669. [CrossRef] [PubMed]
78. McWilliams, T.G.; Prescott, A.R.; Montava-Garriga, L.; Ball, G.; Singh, F.; Barini, E.; Muqit, M.M.K.; Brooks, S.P.; Ganley, I.G. Basal Mitophagy Occurs Independently of PINK1 in Mouse Tissues of High Metabolic Demand. *Cell Metab.* **2018**, *27*, 439–449.e5. [CrossRef] [PubMed]
79. Birsa, N.; Norkett, R.; Wauer, T.; Mevissen, T.E.; Wu, H.C.; Foltynie, T.; Bhatia, K.; Hirst, W.D.; Komander, D.; Plun-Favreau, H.; et al. Lysine 27 ubiquitination of the mitochondrial transport protein Miro is dependent on serine 65 of the Parkin ubiquitin ligase. *J. Biol. Chem.* **2014**, *289*, 14569–14582. [CrossRef]
80. Shlevkov, E.; Kramer, T.; Schapansky, J.; LaVoie, M.J.; Schwarz, T.L. Miro phosphorylation sites regulate Parkin recruitment and mitochondrial motility. *Proc. Natl. Acad. Sci. USA* **2016**, *113*, E6097–E6106. [CrossRef]
81. Wang, X.; Winter, D.; Ashrafi, G.; Schlehe, J.; Wong, Y.L.; Selkoe, D.; Rice, S.; Steen, J.; LaVoie, M.J.; Schwarz, T.L. PINK1 and Parkin target Miro for phosphorylation and degradation to arrest mitochondrial motility. *Cell* **2011**, *147*, 893–906. [CrossRef] [PubMed]
82. McLelland, G.L.; Fon, E.A. MFN2 retrotranslocation boosts mitophagy by uncoupling mitochondria from the ER. *Autophagy* **2018**, *14*, 1658–1660. [CrossRef] [PubMed]
83. Chen, Y.; Dorn, G.W., 2nd. PINK1-phosphorylated mitofusin 2 is a Parkin receptor for culling damaged mitochondria. *Science* **2013**, *340*, 471–475. [CrossRef] [PubMed]
84. Ziviani, E.; Tao, R.N.; Whitworth, A.J. Drosophila parkin requires PINK1 for mitochondrial translocation and ubiquitinates mitofusin. *Proc. Natl. Acad. Sci. USA* **2010**, *107*, 5018–5023. [CrossRef] [PubMed]
85. Rakovic, A.; Grunewald, A.; Kottwitz, J.; Bruggemann, N.; Pramstaller, P.P.; Lohmann, K.; Klein, C. Mutations in PINK1 and Parkin impair ubiquitination of Mitofusins in human fibroblasts. *PLoS ONE* **2011**, *6*, e16746. [CrossRef] [PubMed]
86. Karbowski, M.; Youle, R.J. Regulating mitochondrial outer membrane proteins by ubiquitination and proteasomal degradation. *Curr. Opin. Cell Biol.* **2011**, *23*, 476–482. [CrossRef] [PubMed]
87. Nguyen, S.T.; Bain, A.M.; Bedimo, R.G.; Hall, R.G.; Busti, A.J. Acute elevation of triglycerides after initiation of fosamprinavir/ritonavir in an HIV-negative patient with baseline hypertriglyceridemia. *J. Clin. Lipidol.* **2008**, *2*, 398–400. [CrossRef]
88. McLelland, G.L.; Goiran, T.; Yi, W.; Dorval, G.; Chen, C.X.; Lauinger, N.D.; Krahn, A.I.; Valimehr, S.; Rakovic, A.; Rouiller, I.; et al. Mfn2 ubiquitination by PINK1/parkin gates the p97-dependent release of ER from mitochondria to drive mitophagy. *Elife* **2018**, *7*, e32866. [CrossRef]
89. Gautier, C.A.; Erpapazoglou, Z.; Mouton-Liger, F.; Muriel, M.P.; Cormier, F.; Bigou, S.; Duffaure, S.; Girard, M.; Foret, B.; Iannielli, A.; et al. The endoplasmic reticulum-mitochondria interface is perturbed in PARK2 knockout mice and patients with PARK2 mutations. *Hum. Mol. Genet.* **2016**, *25*, 2972–2984. [CrossRef]
90. Polymeropoulos, M.H.; Lavedan, C.; Leroy, E.; Ide, S.E.; Dehejia, A.; Dutra, A.; Pike, B.; Root, H.; Rubenstein, J.; Boyer, R.; et al. Mutation in the alpha-synuclein gene identified in families with Parkinson's disease. *Science* **1997**, *276*, 2045–2047. [CrossRef]
91. Zhang, H.; Liu, J.; Wang, X.; Duan, C.; Wang, X.; Yang, H. V63 and N65 of overexpressed alpha-synuclein are involved in mitochondrial dysfunction. *Brain Res.* **2016**, *1642*, 308–318. [CrossRef] [PubMed]
92. Shen, J.; Du, T.; Wang, X.; Duan, C.; Gao, G.; Zhang, J.; Lu, L.; Yang, H. alpha-Synuclein amino terminus regulates mitochondrial membrane permeability. *Brain Res.* **2014**, *1591*, 14–26. [CrossRef] [PubMed]
93. Martinez, J.H.; Alaimo, A.; Gorojod, R.M.; Porte Alcon, S.; Fuentes, F.; Coluccio Leskow, F.; Kotler, M.L. Drp-1 dependent mitochondrial fragmentation and protective autophagy in dopaminergic SH-SY5Y cells overexpressing alpha-synuclein. *Mol. Cell Neurosci.* **2018**, *88*, 107–117. [CrossRef] [PubMed]

94. Liu, J.; Wang, X.; Lu, Y.; Duan, C.; Gao, G.; Lu, L.; Yang, H. Pink1 interacts with alpha-synuclein and abrogates alpha-synuclein-induced neurotoxicity by activating autophagy. *Cell Death Dis.* **2017**, *8*, e3056. [CrossRef] [PubMed]
95. Shaltouki, A.; Hsieh, C.H.; Kim, M.J.; Wang, X. Alpha-synuclein delays mitophagy and targeting Miro rescues neuron loss in Parkinson's models. *Acta Neuropathol.* **2018**, *136*, 607–620. [CrossRef]
96. Chen, J.; Ren, Y.; Gui, C.; Zhao, M.; Wu, X.; Mao, K.; Li, W.; Zou, F. Phosphorylation of Parkin at serine 131 by p38 MAPK promotes mitochondrial dysfunction and neuronal death in mutant A53T alpha-synuclein model of Parkinson's disease. *Cell Death Dis.* **2018**, *9*, 700. [CrossRef] [PubMed]
97. Choubey, V.; Safiulina, D.; Vaarmann, A.; Cagalinec, M.; Wareski, P.; Kuum, M.; Zharkovsky, A.; Kaasik, A. Mutant A53T alpha-synuclein induces neuronal death by increasing mitochondrial autophagy. *J. Biol. Chem.* **2011**, *286*, 10814–10824. [CrossRef]
98. Ryan, T.; Bamm, V.V.; Stykel, M.G.; Coackley, C.L.; Humphries, K.M.; Jamieson-Williams, R.; Ambasudhan, R.; Mosser, D.D.; Lipton, S.A.; Harauz, G.; et al. Cardiolipin exposure on the outer mitochondrial membrane modulates alpha-synuclein. *Nat. Commun* **2018**, *9*, 817. [CrossRef]
99. Sampaio-Marques, B.; Felgueiras, C.; Silva, A.; Rodrigues, M.; Tenreiro, S.; Franssens, V.; Reichert, A.S.; Outeiro, T.F.; Winderickx, J.; Ludovico, P. SNCA (alpha-synuclein)-induced toxicity in yeast cells is dependent on sirtuin 2 (Sir2)-mediated mitophagy. *Autophagy* **2012**, *8*, 1494–1509. [CrossRef]
100. Zimprich, A.; Biskup, S.; Leitner, P.; Lichtner, P.; Farrer, M.; Lincoln, S.; Kachergus, J.; Hulihan, M.; Uitti, R.J.; Calne, D.B.; et al. Mutations in LRRK2 cause autosomal-dominant parkinsonism with pleomorphic pathology. *Neuron* **2004**, *44*, 601–607. [CrossRef]
101. Schwab, A.J.; Sison, S.L.; Meade, M.R.; Broniowska, K.A.; Corbett, J.A.; Ebert, A.D. Decreased Sirtuin Deacetylase Activity in LRRK2 G2019S iPSC-Derived Dopaminergic Neurons. *Stem Cell Rep.* **2017**, *9*, 1839–1852. [CrossRef] [PubMed]
102. Walter, J.; Bolognin, S.; Antony, P.M.A.; Nickels, S.L.; Poovathingal, S.K.; Salamanca, L.; Magni, S.; Perfeito, R.; Hoel, F.; Qing, X.; et al. Neural Stem Cells of Parkinson's Disease Patients Exhibit Aberrant Mitochondrial Morphology and Functionality. *Stem Cell Rep.* **2019**, *12*, 878–889. [CrossRef] [PubMed]
103. Yakhine-Diop, S.M.S.; Niso-Santano, M.; Rodriguez-Arribas, M.; Gomez-Sanchez, R.; Martinez-Chacon, G.; Uribe-Carretero, E.; Navarro-Garcia, J.A.; Ruiz-Hurtado, G.; Aiastui, A.; Cooper, J.M.; et al. Impaired Mitophagy and Protein Acetylation Levels in Fibroblasts from Parkinson's Disease Patients. *Mol. Neurobiol.* **2019**, *56*, 2466–2481. [CrossRef] [PubMed]
104. Hsieh, C.H.; Shaltouki, A.; Gonzalez, A.E.; Bettencourt da Cruz, A.; Burbulla, L.F.; St Lawrence, E.; Schule, B.; Krainc, D.; Palmer, T.D.; Wang, X. Functional Impairment in Miro Degradation and Mitophagy Is a Shared Feature in Familial and Sporadic Parkinson's Disease. *Cell Stem Cell* **2016**, *19*, 709–724. [CrossRef] [PubMed]
105. Wang, X. Destructive cellular paths underlying familial and sporadic Parkinson disease converge on mitophagy. *Autophagy* **2017**, *13*, 1998–1999. [CrossRef]
106. Wauters, F.; Cornelissen, T.; Imberechts, D.; Martin, S.; Koentjoro, B.; Sue, C.; Vangheluwe, P.; Vandenberghe, W. LRRK2 mutations impair depolarization-induced mitophagy through inhibition of mitochondrial accumulation of RAB10. *Autophagy* **2019**, 1–20. [CrossRef]
107. Bonello, F.; Hassoun, S.M.; Mouton-Liger, F.; Shin, Y.S.; Muscat, A.; Tesson, C.; Lesage, S.; Beart, P.M.; Brice, A.; Krupp, J.; et al. LRRK2 impairs PINK1/Parkin-dependent mitophagy via its kinase activity: Pathologic insights into Parkinson's disease. *Hum. Mol. Genet.* **2019**. [CrossRef]
108. Bonifati, V.; Rizzu, P.; van Baren, M.J.; Schaap, O.; Breedveld, G.J.; Krieger, E.; Dekker, M.C.; Squitieri, F.; Ibanez, P.; Joosse, M.; et al. Mutations in the DJ-1 gene associated with autosomal recessive early-onset parkinsonism. *Science* **2003**, *299*, 256–259. [CrossRef]
109. Andres-Mateos, E.; Perier, C.; Zhang, L.; Blanchard-Fillion, B.; Greco, T.M.; Thomas, B.; Ko, H.S.; Sasaki, M.; Ischiropoulos, H.; Przedborski, S.; et al. DJ-1 gene deletion reveals that DJ-1 is an atypical peroxiredoxin-like peroxidase. *Proc. Natl. Acad. Sci. USA* **2007**, *104*, 14807–14812. [CrossRef]
110. Meulener, M.C.; Xu, K.; Thomson, L.; Ischiropoulos, H.; Bonini, N.M. Mutational analysis of DJ-1 in Drosophila implicates functional inactivation by oxidative damage and aging. *Proc. Natl. Acad. Sci. USA* **2006**, *103*, 12517–12522. [CrossRef]
111. Krebiehl, G.; Ruckerbauer, S.; Burbulla, L.F.; Kieper, N.; Maurer, B.; Waak, J.; Wolburg, H.; Gizatullina, Z.; Gellerich, F.N.; Woitalla, D.; et al. Reduced basal autophagy and impaired mitochondrial dynamics due to loss of Parkinson's disease-associated protein DJ-1. *PLoS ONE* **2010**, *5*, e9367. [CrossRef] [PubMed]

112. Joselin, A.P.; Hewitt, S.J.; Callaghan, S.M.; Kim, R.H.; Chung, Y.H.; Mak, T.W.; Shen, J.; Slack, R.S.; Park, D.S. ROS-dependent regulation of Parkin and DJ-1 localization during oxidative stress in neurons. *Hum. Mol. Genet.* **2012**, *21*, 4888–4903. [CrossRef] [PubMed]
113. Duplan, E.; Giaime, E.; Viotti, J.; Sevalle, J.; Corti, O.; Brice, A.; Ariga, H.; Qi, L.; Checler, F.; Alves da Costa, C. ER-stress-associated functional link between Parkin and DJ-1 via a transcriptional cascade involving the tumor suppressor p53 and the spliced X-box binding protein XBP-1. *J. Cell Sci.* **2013**, *126*, 2124–2133. [CrossRef] [PubMed]
114. Thomas, K.J.; McCoy, M.K.; Blackinton, J.; Beilina, A.; van der Brug, M.; Sandebring, A.; Miller, D.; Maric, D.; Cedazo-Minguez, A.; Cookson, M.R. DJ-1 acts in parallel to the PINK1/parkin pathway to control mitochondrial function and autophagy. *Hum. Mol. Genet.* **2011**, *20*, 40–50. [CrossRef] [PubMed]
115. Hao, L.Y.; Giasson, B.I.; Bonini, N.M. DJ-1 is critical for mitochondrial function and rescues PINK1 loss of function. *Proc. Natl. Acad. Sci. USA* **2010**, *107*, 9747–9752. [CrossRef] [PubMed]
116. Sidransky, E.; Nalls, M.A.; Aasly, J.O.; Aharon-Peretz, J.; Annesi, G.; Barbosa, E.R.; Bar-Shira, A.; Berg, D.; Bras, J.; Brice, A.; et al. Multicenter analysis of glucocerebrosidase mutations in Parkinson's disease. *N. Engl. J. Med.* **2009**, *361*, 1651–1661. [CrossRef] [PubMed]
117. Tayebi, N.; Callahan, M.; Madike, V.; Stubblefield, B.K.; Orvisky, E.; Krasnewich, D.; Fillano, J.J.; Sidransky, E. Gaucher disease and parkinsonism: A phenotypic and genotypic characterization. *Mol. Genet. Metab.* **2001**, *73*, 313–321. [CrossRef] [PubMed]
118. Migdalska-Richards, A.; Schapira, A.H. The relationship between glucocerebrosidase mutations and Parkinson disease. *J. Neurochem.* **2016**, *139* (Suppl. 1), 77–90. [CrossRef]
119. Valley, N.A.; Blower, P.G.; Wood, S.R.; Plath, K.L.; McWilliams, L.E.; Richmond, G.L. Doubling down: Delving into the details of diacid adsorption at aqueous surfaces. *J. Phys. Chem. A* **2014**, *118*, 4778–4789. [CrossRef]
120. De la Mata, M.; Cotan, D.; Oropesa-Avila, M.; Garrido-Maraver, J.; Cordero, M.D.; Villanueva Paz, M.; Delgado Pavon, A.; Alcocer-Gomez, E.; de Lavera, I.; Ybot-Gonzalez, P.; et al. Pharmacological Chaperones and Coenzyme Q10 Treatment Improves Mutant beta-Glucocerebrosidase Activity and Mitochondrial Function in Neuronopathic Forms of Gaucher Disease. *Sci. Rep.* **2015**, *5*, 10903. [CrossRef]
121. Choi, J.H.; Stubblefield, B.; Cookson, M.R.; Goldin, E.; Velayati, A.; Tayebi, N.; Sidransky, E. Aggregation of alpha-synuclein in brain samples from subjects with glucocerebrosidase mutations. *Mol. Genet. Metab.* **2011**, *104*, 185–188. [CrossRef] [PubMed]
122. Cullen, V.; Sardi, S.P.; Ng, J.; Xu, Y.H.; Sun, Y.; Tomlinson, J.J.; Kolodziej, P.; Kahn, I.; Saftig, P.; Woulfe, J.; et al. Acid beta-glucosidase mutants linked to Gaucher disease, Parkinson disease, and Lewy body dementia alter alpha-synuclein processing. *Ann. Neurol.* **2011**, *69*, 940–953. [CrossRef] [PubMed]
123. Osellame, L.D.; Rahim, A.A.; Hargreaves, I.P.; Gegg, M.E.; Richard-Londt, A.; Brandner, S.; Waddington, S.N.; Schapira, A.H.; Duchen, M.R. Mitochondria and quality control defects in a mouse model of Gaucher disease–links to Parkinson's disease. *Cell Metab.* **2013**, *17*, 941–953. [CrossRef] [PubMed]
124. Li, H.; Ham, A.; Ma, T.C.; Kuo, S.H.; Kanter, E.; Kim, D.; Ko, H.S.; Quan, Y.; Sardi, S.P.; Li, A.; et al. Mitochondrial dysfunction and mitophagy defect triggered by heterozygous GBA mutations. *Autophagy* **2019**, *15*, 113–130. [CrossRef] [PubMed]
125. Tang, F.L.; Liu, W.; Hu, J.X.; Erion, J.R.; Ye, J.; Mei, L.; Xiong, W.C. VPS35 Deficiency or Mutation Causes Dopaminergic Neuronal Loss by Impairing Mitochondrial Fusion and Function. *Cell Rep.* **2015**, *12*, 1631–1643. [CrossRef] [PubMed]
126. Malik, B.R.; Godena, V.K.; Whitworth, A.J. VPS35 pathogenic mutations confer no dominant toxicity but partial loss of function in Drosophila and genetically interact with parkin. *Hum. Mol. Genet.* **2015**, *24*, 6106–6117. [CrossRef]
127. Delgado-Camprubi, M.; Esteras, N.; Soutar, M.P.; Plun-Favreau, H.; Abramov, A.Y. Deficiency of Parkinson's disease-related gene Fbxo7 is associated with impaired mitochondrial metabolism by PARP activation. *Cell Death Differ.* **2017**, *24*, 120–131. [CrossRef]
128. Zhou, Z.D.; Xie, S.P.; Sathiyamoorthy, S.; Saw, W.T.; Sing, T.Y.; Ng, S.H.; Chua, H.P.; Tang, A.M.; Shaffra, F.; Li, Z.; et al. F-box protein 7 mutations promote protein aggregation in mitochondria and inhibit mitophagy. *Hum. Mol. Genet.* **2015**, *24*, 6314–6330. [CrossRef]
129. Burchell, V.S.; Nelson, D.E.; Sanchez-Martinez, A.; Delgado-Camprubi, M.; Ivatt, R.M.; Pogson, J.H.; Randle, S.J.; Wray, S.; Lewis, P.A.; Houlden, H.; et al. The Parkinson's disease-linked proteins Fbxo7 and Parkin interact to mediate mitophagy. *Nat. Neurosci.* **2013**, *16*, 1257–1265. [CrossRef]

130. Zhou, Z.D.; Sathiyamoorthy, S.; Angeles, D.C.; Tan, E.K. Linking F-box protein 7 and parkin to neuronal degeneration in Parkinson's disease (PD). *Mol. Brain* **2016**, *9*, 41. [CrossRef]
131. Moors, T.E.; Hoozemans, J.J.; Ingrassia, A.; Beccari, T.; Parnetti, L.; Chartier-Harlin, M.C.; van de Berg, W.D. Therapeutic potential of autophagy-enhancing agents in Parkinson's disease. *Mol. Neurodegener.* **2017**, *12*, 11. [CrossRef] [PubMed]
132. East, D.A.; Campanella, M. Mitophagy and the therapeutic clearance of damaged mitochondria for neuroprotection. *Int. J. Biochem. Cell. Biol.* **2016**, *79*, 382–387. [CrossRef]
133. East, D.A.; Fagiani, F.; Crosby, J.; Georgakopoulos, N.D.; Bertrand, H.; Schaap, M.; Fowkes, A.; Wells, G.; Campanella, M. PMI: A DeltaPsim independent pharmacological regulator of mitophagy. *Chem. Biol.* **2014**, *21*, 1585–1596. [CrossRef] [PubMed]
134. Yi, W.; MacDougall, E.J.; Tang, M.Y.; Krahn, A.I.; Gan-Or, Z.; Trempe, J.F.; Fon, E.A. The Landscape of Parkin Variants Reveals Pathogenic Mechanisms and Therapeutic Targets in Parkinson's Disease. *Hum. Mol. Genet.* **2019**. [CrossRef] [PubMed]
135. Georgakopoulos, N.D.; Wells, G.; Campanella, M. The pharmacological regulation of cellular mitophagy. *Nat. Chem. Bio.l* **2017**, *13*, 136–146. [CrossRef] [PubMed]
136. Koentjoro, B.; Park, J.S.; Sue, C.M. Nix restores mitophagy and mitochondrial function to protect against PINK1/Parkin-related Parkinson's disease. *Sci. Rep.* **2017**, *7*, 44373. [CrossRef] [PubMed]
137. Park, J.S.; Koentjoro, B.; Sue, C.M. Commentary: Nix restores mitophagy and mitochondrial function to protect against PINK1/Parkin-related Parkinson's disease. *Front. Mol. Neurosci.* **2017**, *10*, 297. [CrossRef]
138. Olszewska, D.A.; Lynch, T. Nature's Parkin experiment: Nix-a novel protective mechanism in Parkinson's disease. *Mov. Disord.* **2017**, *32*, 992. [CrossRef]

2

Peripheral Delivery of Neural Precursor Cells Ameliorates Parkinson's Disease-Associated Pathology

George Edwards III [1,†], Nazaret Gamez [1,2,†], Enrique Armijo [1,3], Carlos Kramm [1,3], Rodrigo Morales [1,4], Kathleen Taylor-Presse [1], Paul E. Schulz [1], Claudio Soto [1,3] and Ines Moreno-Gonzalez [1,2,4,5,*]

1 The Mitchell Center for Alzheimer's Disease and Related Brain Disorders, Department of Neurology, The University of Texas Houston Health Science Center at Houston, 77030 Houston, TX, USA; George.A.Edwards@uth.tmc.edu (G.E.); Nazaret.GamezRuiz@uth.tmc.edu (N.G.); Enrique.Antonio.ArmijoFuentes@uth.tmc.edu (E.A.); Carlos.F.KrammBarria@uth.tmc.edu (C.K.); Rodrigo.MoralesLoyola@uth.tmc.edu (R.M.); Kathleen.B.TaylorPresse@uth.tmc.edu (K.T.-P.); Paul.E.Schulz@uth.tmc.edu (P.E.S.); Claudio.Soto@uth.tmc.edu (C.S.)
2 Dpto. Biologia Celular, Genetica y Fisiologia, Instituto de Investigacion Biomedica de Malaga-IBIMA, Facultad de Ciencias, Universidad de Malaga, 29010 Malaga, Spain
3 Universidad de los Andes, Facultad de Medicina, Av. San Carlos de Apoquindo, 2200, Las Condes, 7620001 Santiago, Chile
4 Centro Integrativo de Biología y Química Aplicada (CIBQA), Universidad Bernardo O'Higgins, 8370993 Santiago, Chile
5 Networking Research Center on Neurodegenerative Diseases (CIBERNED), 29010 Malaga, Spain
* Correspondence: inesmoreno@uma.es
† Denotes equal contribution.

Abstract: Parkinson's disease (PD) is a progressive neurodegenerative disorder characterized by loss of motor control due to a wide loss of dopaminergic neurons along the nigro-striatal pathway. Some of the mechanisms that contribute to this cell death are inflammation, oxidative stress, and misfolded alpha-synuclein-induced toxicity. Current treatments are effective at managing the early motor symptoms of the disease, but they become ineffective over time and lead to adverse effects. Previous research using intracerebral stem cell therapy for treatment of PD has provided promising results; however, this method is very invasive and is often associated with unacceptable side effects. In this study, we used an MPTP-injected mouse model of PD and intravenously administered neural precursors (NPs) obtained from mouse embryonic and mesenchymal stem cells. Clinical signs and neuropathology were assessed. Female mice treated with NPs had improved motor function and reduction in the neuroinflammatory response. In terms of safety, there were no tumorigenic formations or any detectable adverse effect after treatment. Our results suggest that peripheral administration of stem cell-derived NPs may be a promising and safe therapy for the recovery of impaired motor function and amelioration of brain pathology in PD.

Keywords: neuronal precursors; stem cells; Parkinson's disease; therapy; intravenous; inflammation; clinical symptoms

1. Introduction

Parkinson's disease (PD) is the second most common neurodegenerative disease. It is characterized by the loss of dopaminergic neurons of the substantia nigra pars compacta (SNpc) leading to a progressive movement impairment that can include tremors, bradykinesia, rigidity, and postural instability [1,2]. PD patients also display non-motor symptoms such as anxiety and depression, psychosis, gastrointestinal problems, and sleep dysregulation [3–5]. The pathophysiology underlying PD includes the aggregation of misfolded alpha-synuclein protein that generates intracellular Lewy bodies [2,6,7]. In addition, there is an associated neuroinflammatory process governed by the action of microglia and astrocytes, which has been postulated to play a crucial role in PD pathogenesis [8,9]. The presence of reactive microglia and an elevation in the number of astrocytes have been detected in the SNpc of PD patients and in mouse and primate models of PD systemically injected with 1-methyl-4-phenil-1,2,3,6-tetrahydropyridine (MPTP) [10–13].

Some treatments, such as deep-brain stimulation and pharmacotherapy, have shown benefit in reducing the symptoms, but there is no cure for PD [14–16]. Stem cell (SC) therapy using neuronal transplantation has shown some promise in combating PD [17–19]. Embryonic stem cells (ESCs) are pluripotent cells derived from the inner mass of the blastocyst. Although promising, their use in research and for therapeutic purposes has generated ethical concerns. Mesenchymal stem cells (MSCs) are multipotent stromal cells that can be obtained from the individual's bone marrow and have the ability to differentiate into a variety of cells, such as connective tissue cells (muscle, fat, bone, and cartilage) [20,21]. MSCs offer advantages over ESCs, such as being derived from the patient to whom they are given, a low risk of tumor formation, no risk of an immune response, and no ethical dilemmas. Both ESCs and MSCs, are capable of generating neural precursors (NPs), which are a promising therapeutic strategy for neurological disorders due to their capacity to differentiate into functional astrocytes, oligodendrocytes, and mature neurons. Their genetic stability, limited propagation, lack of tumorigenicity, and their circumvention of ethical concerns make them ideal tools for therapeutic interventions [22].

Previous studies have shown that MSCs and dopaminergic neurons derived from human pluripotent stem cells can ameliorate PD symptoms when administered directly into the brain of PD models [23,24]. Although the mechanism is not well understood, it is known that MSCs can produce neurotrophic factors, which could facilitate cell survival, mitigate oxidative stress, increase the production of anti-inflammatory cytokines, inhibit glial activation, and suppress cellular apoptosis [25–27].

Clinical trials in humans have tested the safety and efficacy of intracerebral transplantation of human ESCs-derived NPs into the striatum in PD patients. Nevertheless, attempts to get stem cells into the brain to replace the cells lost in neurodegenerative diseases, such as PD, have been unsuccessful. In addition, post-mortem studies in brains of treated patients show presence of very few stem cells with no evidence that they differentiated into cells that replaced the lost ones. In addition, surviving cells contained Lewy bodies [28,29].

Studies performed using an intravenous (i.v.) infusion of stem cells in stroke patients, however, showed that it is not necessary to deliver stem cells into the brain to have a positive effect on the clinical outcome [30]. Furthermore, intravascular infusion techniques are safe, less invasive than brain grafting, and could constitute a more convenient administration approach for stem cell therapy in the clinic. Here, we investigate the effect of the peripheral inoculation of NPs, derived from ESCs and MSCs, as a non-invasive therapy to ameliorate the motor symptoms and modulate the disease-associated neuroinflammation in a drug-induced (MPTP) mouse model of PD. Safety, in terms of tumor formation, was also evaluated.

2. Results

2.1. Generation of Neural Precursors for Cell Therapy

To develop a systemic, stem cell-based treatment for PD, we first produced NPs from both mouse ESCs and MSCs (Figure 1A,G). ESCs were characterized using the pluripotency markers E-cadherin and Sox2 (Figure 1B,C), whereas MSCs were positive for the markers CD44 and βI integrin (Figure 1H,I).

NPs can be derived in vitro under the influence of mitogens, such as EGF and FGF-2, and they retain the potential to differentiate into neurons, astrocytes, and oligodendrocytes [31–33]. ESCs and MSCs were subjected to a multistep protocol for converting them into a monolayer of self-renewing NPs. ESCs and MSCs were adapted for at least three passages on gelatin in order to expose them to neural induction stimuli. These cells, designated as ESC-NPs and MSC-NPs, were continuously and homogenously propagated, with no significant differences in morphology or growth rate for over 30 passages. Generated NPs showed elongated bipolar morphology, end feet, and oval nuclei (Figure 1D,J). NPs maintained in self-renewing conditions were positive for markers typically observed in NPs by immunocytochemistry, including musashi (Figure 1E,K) and vimentin (Figure 1F,L).

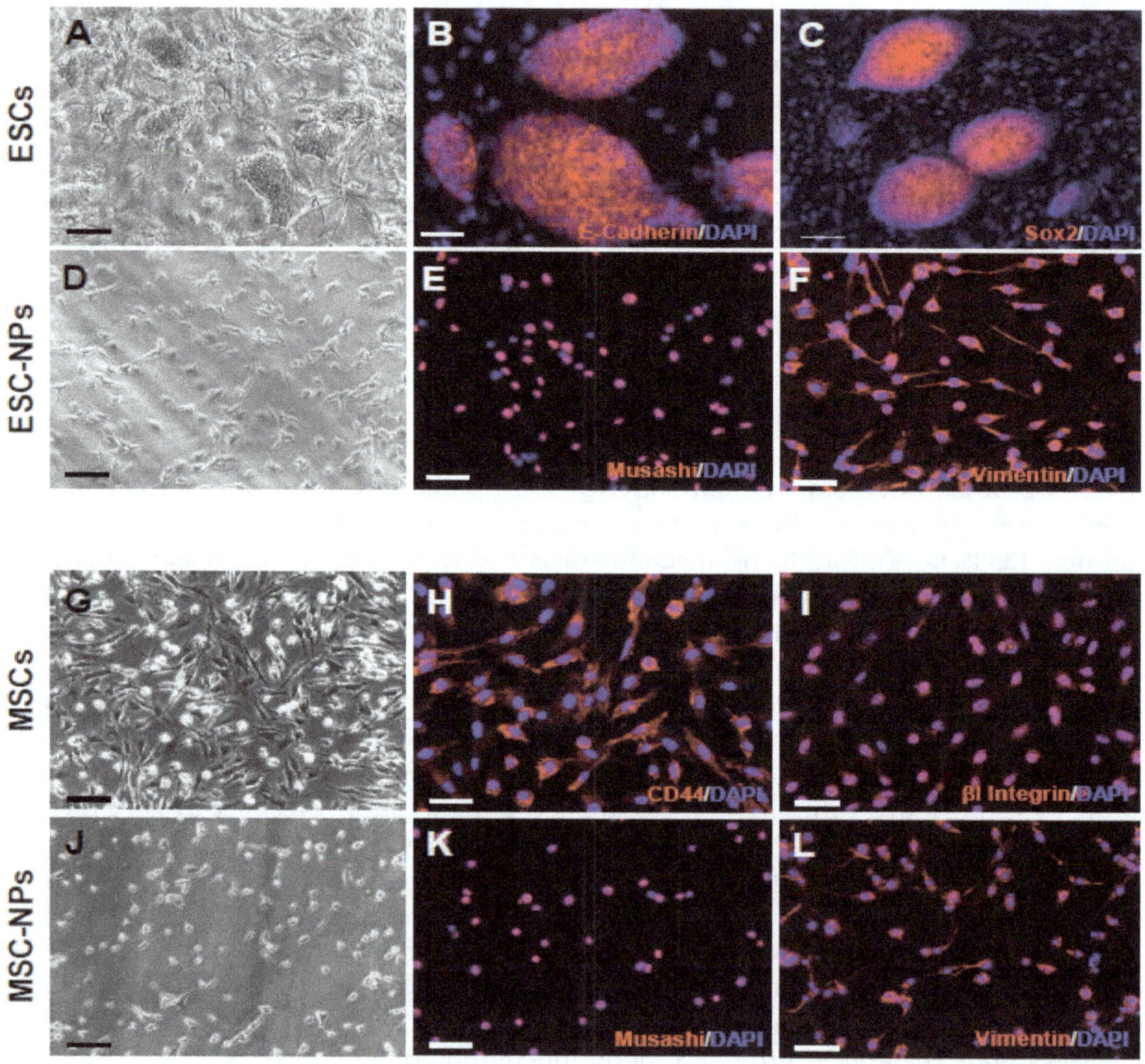

Figure 1. Neural precursors obtained from embryonic and mesenchymal stem cells. Morphology of ESCs at passage number 10 (**A**). Immunocytochemistry for E-cadherin (**B**) and SOX2 (**C**). NPs have elongated bipolar morphology, end feet, and oval nuclei (**D**). Immunocytochemistry for the NP markers musashi (**E**) and vimentin (**F**). Morphology of MSCs at passage number 7 (**G**). Immunocytochemistry for CD44 (**H**) and βI integrin (**I**). NPs showed similar morphology as we observed on ESC-NPs (**J**). MSC-NPs were positive for musashi (**K**) and vimentin (**L**). Nuclei were stained with DAPI (blue). Scale bar A, D, G, J: 25 μm; B, C, E, F, H, I, K, L: 50 μm.

2.2. Short-Term Attenuation of PD-Induced Motor Deficits After Stem Cell Therapy

To test the efficacy of an i.v. injection of NPs to treat PD pathology, we used MPTP-treated mice. This is perhaps the most widely used model of PD, which is characterized by loss of dopaminergic neurons in the nigrostriatal pathway with accompanying Parkinsonism. C57BL/6J mice were treated intraperitoneally with MPTP hydrochloride (30 mg/kg) administered daily for 5 consecutive days to generate the PD-like phenotype. These MPTP mice demonstrated impairment in the motor specific tasks of wire hanging and rotarod as compared to non-MPTP injected mice by student t-test ($p < 0.001$ and $p < 0.05$, respectively) (Figure S1). Ten days after the last MPTP injection, mice were inoculated twice over a week with 1×10^6 cells of either ESC-NPs or MSC-NPs by tail vein injection. Animals injected with PBS were used as controls. Ten days after the last cell injection, mouse groups were behaviorally assessed by the wire hang and rotarod tests to determine the effect of the inoculation in clinical signs (Figure 2).

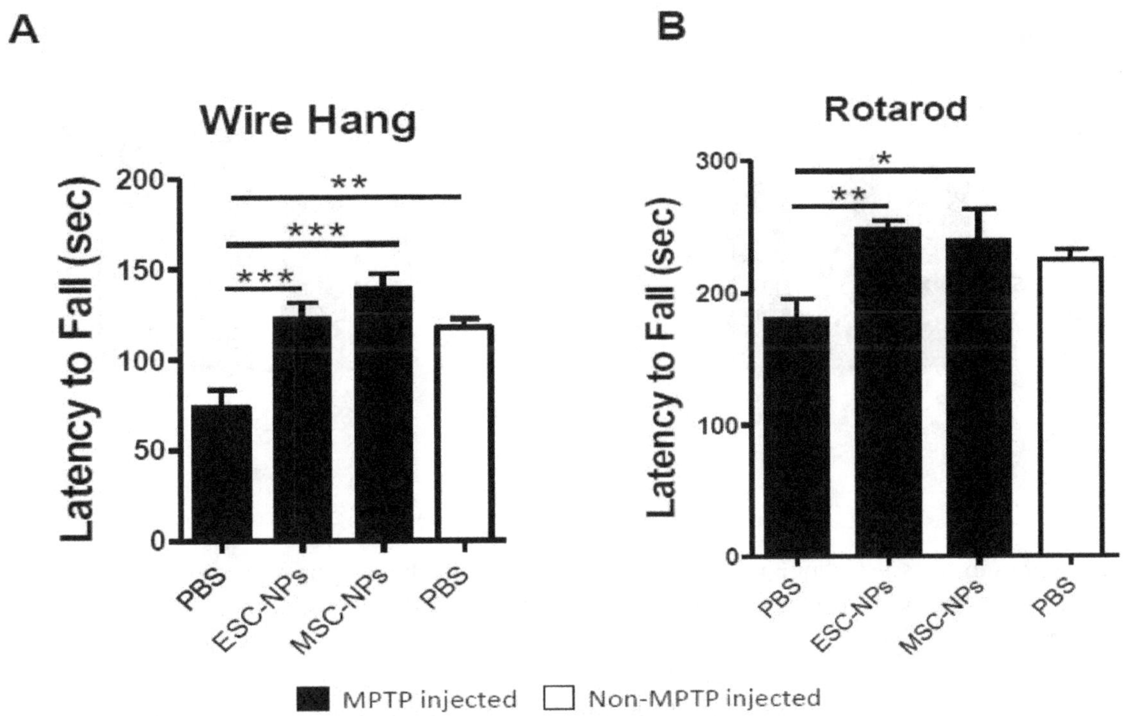

Figure 2. Behavioral assessment of MPTP-induced mice 10 days after NP treatment. Bar graphs reveal the latency to fall in seconds (sec) in the wire hang (**A**) and rotarod (**B**) tests. In both tests, we analyzed MPTP-induced animals (black) inoculated with ESC-NPs, MSC-NPs, and PBS 10 days after the last injection, and age matched non-MPTP-injected mice (white) treated with PBS. Data are expressed as mean ± s.e.m. ANOVA, Dunnett's multiple comparison test, * $p < 0.05$, ** $p < 0.01$, *** $p < 0.001$. N = 9–10 mice/group.

ESC-NP- and MSC-NP-inoculated mice revealed a better wire hanging performance compared to age-matched PBS-injected mice and displayed better grip strength and motor coordination by Dunnett's multiple comparison test ($p < 0.0001$ and $p < 0.0001$, respectively), as shown in Figure 2A. One-way ANOVA demonstrated an overall effect among the groups by wire hang and rotarod tests ($p < 0.0001$ and $p < 0.05$, respectively). Moreover, the rotarod test (Figure 2B) showed a significant difference in the latency to fall between ESC-NPs and MSC-NPs compared to PBS-injected mice ($p < 0.01$ and $p < 0.05$, respectively). The level of performance was similar to animals not treated with MPTP as observed by no main differences in behavior between the control non-MPTP mice treated with PBS and the MPTP-injected mice treated with NPs. Therefore, NP peripheral inoculation induced amelioration of clinical symptoms observed in MPTP-exposed mice 10 days after treatment.

In an effort to determine the long-lasting effects of the NP peripheral treatment, another set of animals was subjected to the same experimental conditions but sacrificed 3 months after the NP inoculation. Figure 3A illustrates their behavioral performance in the wire hanging test. We observed that there was no significant difference in latency to fall from the wire between ESC-NPs or MSC-NPs and PBS-inoculated mice (one-way ANOVA, $p > 0.05$). However, there was also no difference between animals injected with MPTP and untreated wild type mice, indicating that at this time, the deleterious effect of MPTP was not detectable by the wire hanging test. Conversely, there was a significant difference among the groups (one-way ANOVA, $p < 0.05$), where motor performance on the rotarod (Figure 3B) was significantly better in the MSC-NP mice and non-MPTP PBS mice, than the PBS-injected mice ($p < 0.05$).

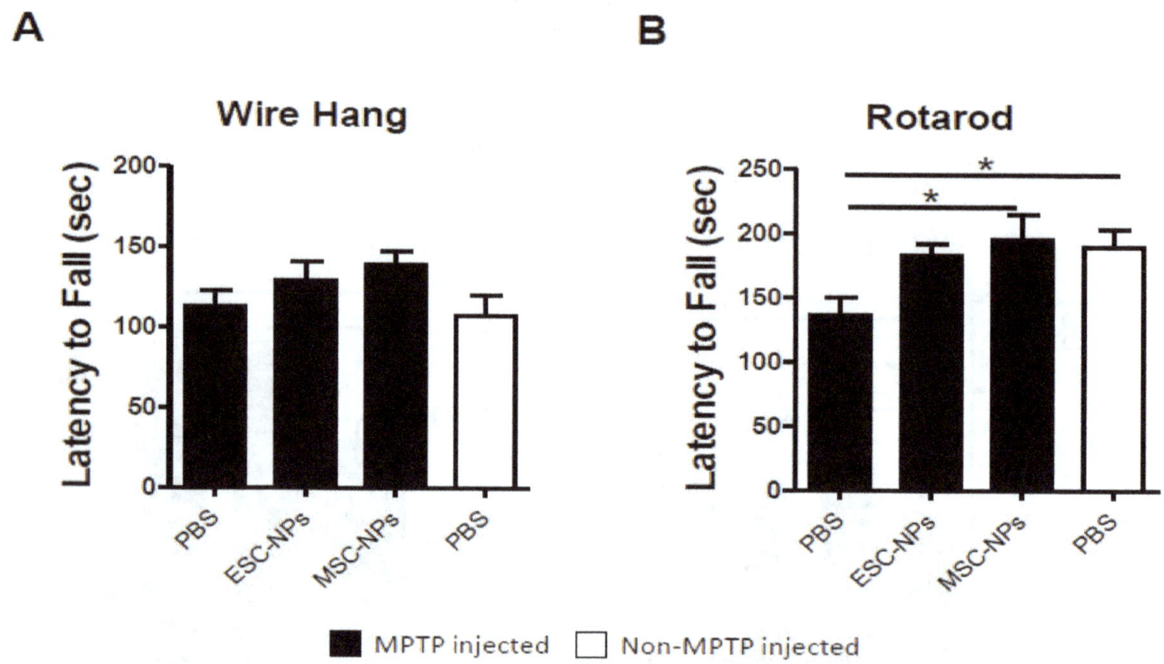

Figure 3. Behavioral assessment 3 months after NP treatment. Bar graphs represent wire hang (**A**) and rotarod (**B**) performance represented as the mean latency to fall (seconds, sec). MPTP-injected mice (black) were treated with ESC-NPs, MSC-NPs, and PBS 3 months after the last injection with NPs. Age matched non-MPTP-injected mice (white) treated with PBS were used as a control. Data are expressed as mean ± s.e.m. ANOVA, Dunnett's multiple comparison test, * $p < 0.05$. N = 9–10 mice/group.

These results indicate that NPs from ESC and MSC origin have an early beneficial effect on motor symptoms in motor-dependent behavior paradigms. Because the MPTP-induced model does not present a progressive pathology, behavioral data demonstrate that the treatment is less robust over time but still significant.

2.3. NP Therapy Does Not Increase Dopamine Levels but Modulates Neuroinflammation

MPTP gets into CNS efficiently and induces the loss of dopaminergic neurons. These mice display reduced levels of tyrosine hydroxylase (TH) expression, a measurement of dopamine levels, in the SNpc compared to non-injected mice. To determine whether infusion of NPs modified levels of dopaminergic neurons, we analyzed the immunoreactivity (IR) of TH in NP and PBS-treated animals. Histological analysis of TH-IR demonstrated no differences in the levels of TH in the experimental and control groups at short (10 days after treatment) and long (3 months after treatment) terms, suggesting that the changes observed in the motor tests were not due to modifications in the dopamine levels (Figure S2). Therefore, we focused our study on whether other hallmarks of PD pathology were influenced by the peripheral inoculation of NPs.

A chronic inflammatory response is thought to contribute to the pathogenic processes underlying PD. For that reason, we next tested the effects of stem cells on neuroinflammation. To investigate the effects of NPs on astrogliosis and microgliosis in the MPTP-induced PD model, immunohistochemistry analysis and quantification was performed for inflammatory markers throughout the brain. The non-MPTP PBS injected group was also subjected to neuroinflammation analysis (Figure 4A). We found a significant difference among the groups in GFAP and BLBP-expressing astrocyte levels in the brain (one-way ANOVA, $p < 0.0001$). MPTP-injected mice inoculated with PBS displayed significant astrogliosis compared to the non-MPTP PBS control group ($p < 0.01$). Treatment with both ESC- and MSC-NPs significantly decreased GFAP expression in MPTP-injected mice when compared to the PBS control group and to the non-MPTP injected group ($p < 0.001$) (Figure 4C). Then, immunostaining of reactive astroglia was performed using the BLBP (brain lipid-binding protein) antibody. BLBP is a marker of radial glial and reactive astrocytes as well as adult neural stem cells [34]. We quantified cells that were positive for both GFAP (in green) and BLBP (in red) to determine the amount of reactive astrocytes. Histological analysis demonstrated that BLBP-expressing astroglial cells were reduced in MPTP-injected mice receiving the ESC-NP ($p < 0.001$) and MSC-NP ($p < 0.001$) inoculation relative to PBS-injected animals (Figure 4B). Furthermore, BLBP-positive astrocyte levels in NP-treated mice were similar to non-MPTP PBS control mice. There was a significant difference among the evaluated groups by one-way ANOVA ($p < 0.0001$) (Figure 4D). For the expression of Iba-1, a microglia marker. MPTP induced a significant increase in Iba-1 burden with respect to non-MPTP injected mice (Figure 5A1–4). The i.v. treatment using both ESC- and MSC-derived NPs significantly reduced the burden of Iba1-immunopositive microglia compared to the PBS injected mice (Figure 5C). Interestingly, the reduction in reactive microglia was more prominent in animals treated with NPs derived from MSCs when compared to ESCs ($p < 0.001$), while non-MPTP PBS animals had less Iba-1 IR compared to ESC-NPs mice (one-way ANOVA, $p < 0.05$). Monocytes can infiltrate the brain and differentiate into macrophages. To determine whether the reduction of Iba1 was due to a decrease in microglial cells (Iba1/CD45^{lo}) or monocyte/macrophage (Iba1/CD45^{hi}) infiltration, we immunostained brain sections using the CD45 antibody [35]. The CD45 signal was shown mostly in animals treated with MPTP without treatment of NPs (Figure 5B1–4). The representative pictures and immunoreactivity (IR) quantification data demonstrate that the levels of CD45 were significantly reduced in animals treated with MSC-NPs (Figure 5D), showing that there is a decrease in monocyte/macrophage infiltration. CD45-positive cells were found mainly around the third ventricle and a significant difference between the groups was found (one-way ANOVA, $p = 0.014$).

2.4. *Intravenous Treatment with NPs is Not Tumorigenic*

In order to test the safety of the i.v. inoculation with NPs, we carefully checked for the development of tumors from injected NPs by palpation and visual screening once every 2 weeks after the last inoculation. No tumor formation was observed during the execution of these experiments. In addition, hematoxylin-eosin general staining was performed in the brain, lung, spleen, kidney, and liver to determine the presence of tumor formation in animals sacrificed 10 days, 3 months, and 9 months after the treatment. No evidence for tumor cell masses was detected in any of the animals analyzed, as observed in the representative pictures in Figure 6A–D. We also analyzed the tissue samples using the proliferation antibody Ki67, and no tumor formation was found (Figure 6E–H).

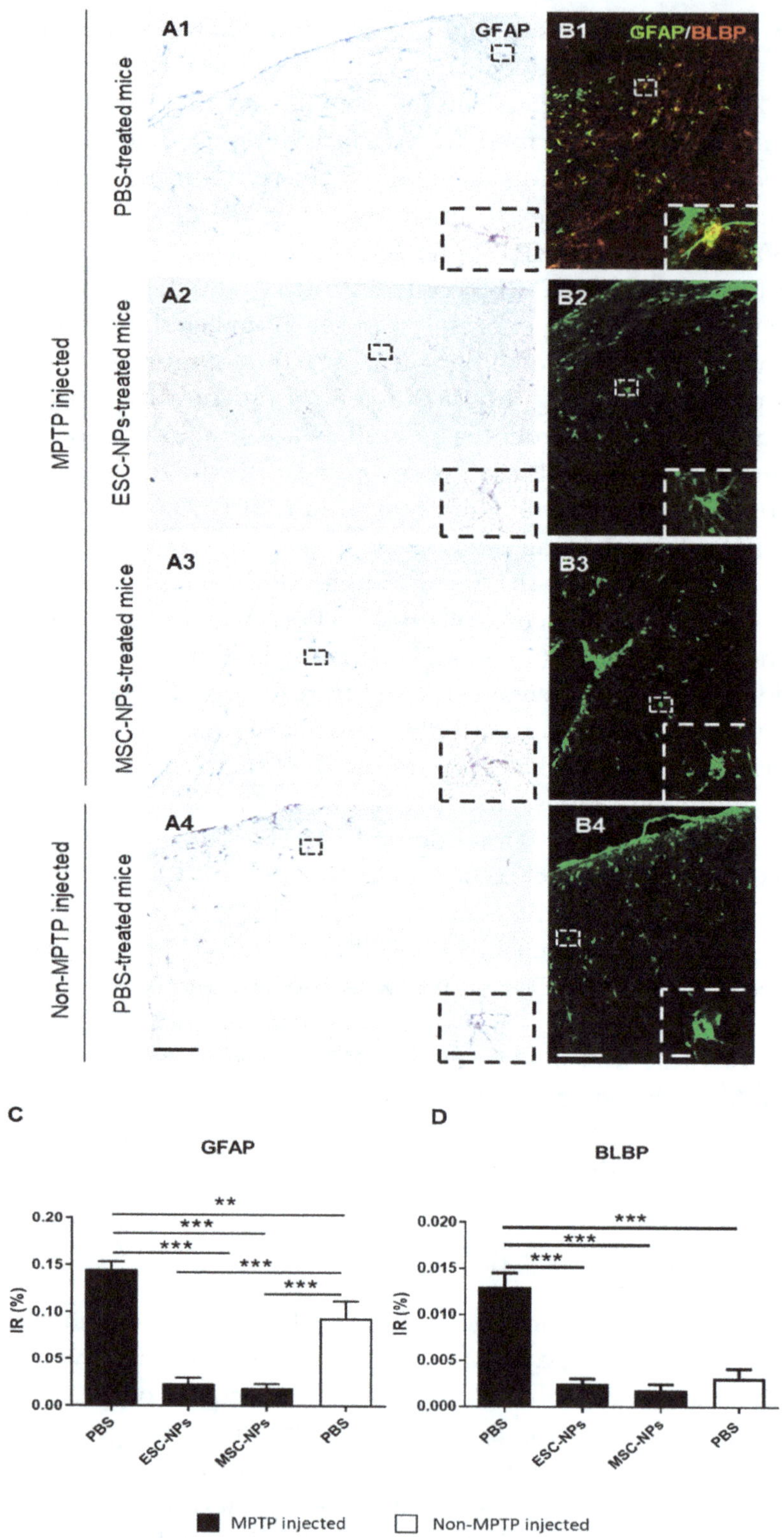

Figure 4. MPTP-injected mice show reduced astrogliosis after NP peripheral treatment. MPTP inoculated ESC-NPs and MSC-NPs and PBS-treated mice, along with non-MPTP PBS mice, were sacrificed 10 days after the infusion. Brains were collected and astrocytosis was evaluated by histology. Representative images for bright field GFAP immunostaining (**A1**–**A4**) and double immunofluorescence for GFAP and BLBP (**B1**–**B4**) in the striatum of MPTP-injected and non-MPTP-injected age-matched mice are presented. Insets show details of immunopositive astrocytes depicted in dashed squares. Image analysis quantification of the GFAP (**C**) and GFAP/BLBP-positive cells (**D**) in the whole brain (ANOVA, ** $p < 0.01$; ***$p < 0.001$). Scale bar: 100µm (A1–A4, B1–B4); 10µm (insets). N = 6 mice/group, 8 sections/mouse.

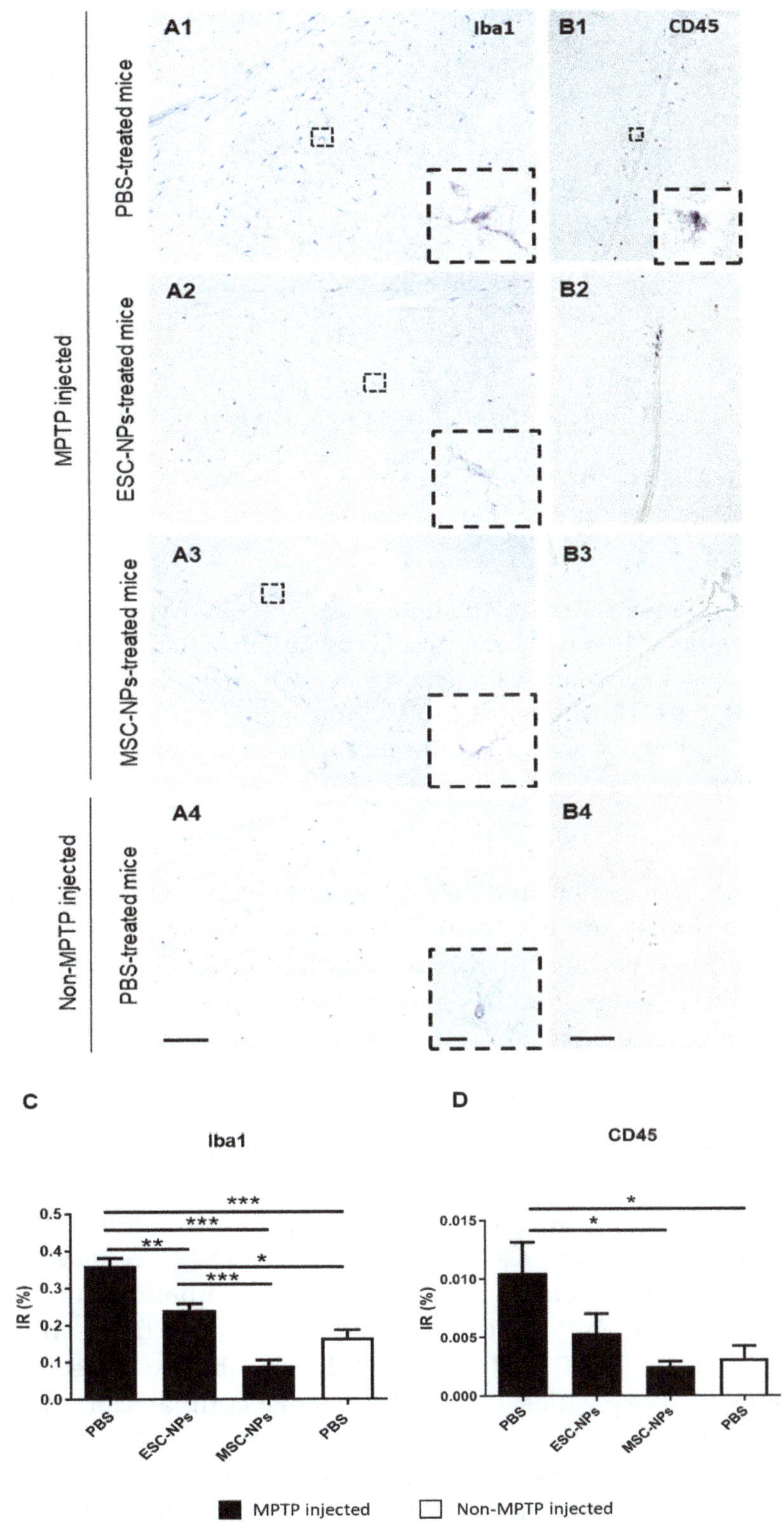

Figure 5. NPs treatment reduces early microglial and macrophage load in PD mice. MPTP-injected mice were evaluated 10 days after treatment with PBS, ESC-NPs and MSC-NPs. Representative bright field microphotography in the striatum (**A**) and third ventricle (**B**) are shown. Immunostaining against Iba1 for microglial cells (A1–A4) and CD45 for macrophages (B1–B4) was performed in 8 brain sections throughout the brain. Insets show details of Iba1 and CD45 positive cells depicted in dashed squares. Image analysis quantification of Iba1 (**C**) in the brain of MPTP-injected animals treated with either ESC-NPs, MSC-NPs, PBS, or non-MPTP PBS injected controls (ANOVA, ** $p < 0.01$, *** $p < 0.001$). Immunoreactive (IR) quantification of monocyte-derived macrophages (**D**) of all groups (ANOVA, * $p < 0.05$). Scale bar: 100 μm (A1–A4, B1–B4); 10 μm (insets). N = 6 mice/group, 8 sections/mouse.

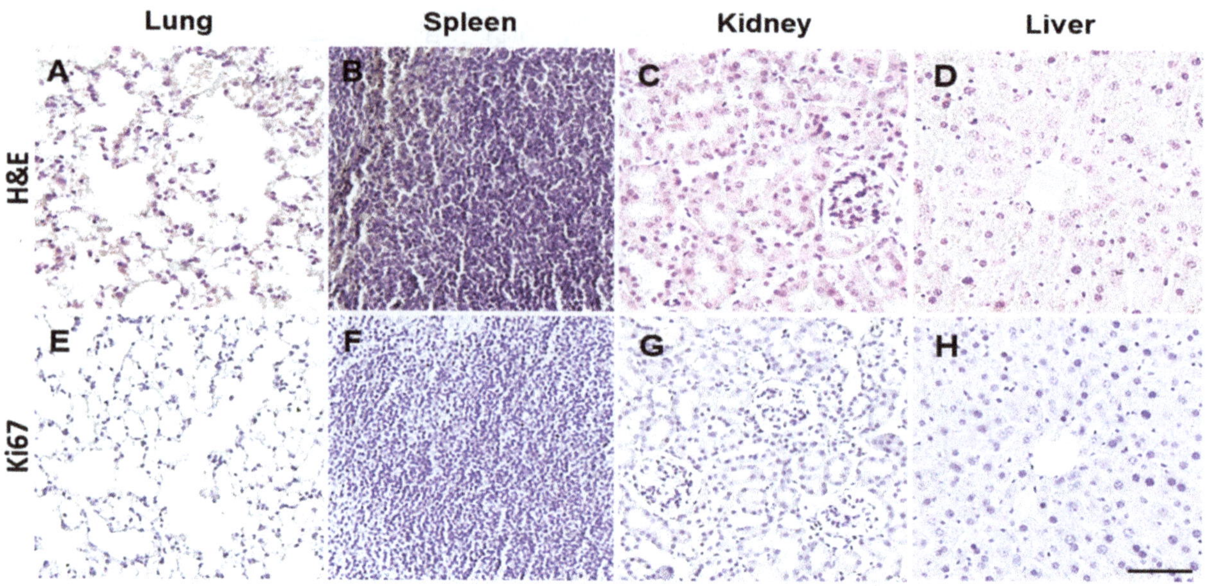

Figure 6. Tumorigenicity assessment of NP treatment. ESC-NPs and MSC-NPs infused animals were assessed for tumor formation 10 days, 3 months, and 9 months after the injection. Representative microphotographs of histological analysis using general hematoxylin-eosin staining (**A–D**) and immunohistochemistry for Ki67 counterstained with hematoxylin (**E–H**). Representative pictures of lung (**A**), spleen (**B**), kidney (**C**), and liver (**D**) were absent of any tumorigenic formation. Scale bar: 50 μm. N = 8 mice/group.

3. Discussion

PD is a devastating neurodegenerative disease associated with loss of dopaminergic neurons in the striatum, brain inflammation and accumulation of α-synuclein aggregates in the form of Lewy bodies [1]. Currently, there is no cure for this disease and available treatments only mitigate the symptoms [36]. One potential therapy that has gained support involves the use of stem cells. The goal of this study was to test whether the peripheral injection of NPs obtained from ESCs and MSCs could safely ameliorate the motor abnormalities and associated neuropathology in a mouse model of PD.

To this end, NPs obtained from embryonic and mesenchymal stem cells were inoculated into MPTP-induced mice. We found that animals treated with both cell types showed an amelioration in their behavioral impairment compared to PBS-treated control animals in the rotarod and wire hanging tasks as early as 10 days after stem cell therapy. Importantly, motor-related benefits somewhat persisted 3 months following treatment. Chiefly, MSC-NP treated mice had a restored ability to stay on the rotarod compared to PBS-inoculated mice, indicating an amelioration of the motor symptoms of MPTP-induced PD, while ESC-NP-treated rodents presented a trend in improvement, as well. Furthermore, 3 months after ESC-NP and MSC-NP therapy by i.v. injection, experimental mice demonstrated improved motor capacities, though less robust compared to the 10 day post-treatment time period, compared to PBS-treated mice.

This study was performed in 2-month-old female C57BL/6J mice. Gender could play an important role in the clinical treatment and ultimate outcome of PD. In fact, clinical data has reported a slight prevalence and incidence in males and quicker onset of PD. On the other hand, females tend to develop tremor-dominant PD [37]. Thus, further evaluation of the gender-specific effect of this therapy needs to be determined.

Previous reports using SC therapy by intracerebral injection noted that the introduced cells were able to survive and were present in brain-yielding novel neurons, as evaluated by the increased expression of TH in PD brains compared to controls [23,38–40]. Survival was very low; however, it remains unclear whether the injected cells were synaptically integrated and whether the changes observed were due to replacement of lost neurons by these remaining cells. Another alternative is that the cells release SC-derived factors that enhance endogenous cells for support. The amelioration of clinical symptoms observed in our cell-treated MPTP-induced mice was not due to an increase in TH levels after the treatment, suggesting an alternative mechanism of action for the injected cells. Thus, other factors associated with the MPTP treatment may be improved after NP treatment leading to an amelioration of PD clinical manifestations.

As a better performance in behavioral tasks was observed 10 days after the treatment with NPs, the brain of these mice were analyzed for potential mechanisms of action. Currently, there are two proposed mechanisms responsible for neuronal cell death in PD and other neurodegenerative disorders. The "cell-autonomous" mechanism refers to an accumulation of intrinsic damage in the degenerating neurons, whereas the "non-cell-autonomous" mechanism results from an indirect degeneration of the affected neurons caused by pathological interactions with neighboring cells, mostly resident glial cells (microglia and astroglia). These glial cells govern the inflammatory process that has been repeatedly linked to neurodegeneration in PD, but whether the neuroinflammation is a cause or a consequence of neuronal degeneration remains unanswered [8].

The MPTP model for PD is known to generate excessive cellular reactive oxygen species (ROS), which induce oxidative stress and neuroinflammation linked to degeneration of nigrostriatal DA neurons [41]. In fact, neuroinflammation is suggested as the cause and the effector of the neurodegenerative process in PD animal models. A growing body of evidence has suggested that higher glial cell activation is connected to DA neuronal loss in the SNpc [42]. The present study demonstrated that intravascular infusion of both ESCs-NPs and MSCs-NPs in MPTP-induced PD models significantly reduced the burden of reactive astroglia and microglia cells throughout the brain compared to the PBS-injected mice 10 d.p.i. Even more, the activation of astroglia and macrophages was drastically decreased in the MPTP mouse model after the i.v. treatment with NPs, indicating a potential mechanism of amelioration of motor impairment in treated animals. More studies need to be done to determine the direct effect of infused NPs on glial activation and the potential effect on motor behaviors and neurodegenerative diseases.

Another viable hypothesis is that SC-induced release of trophic factors could contribute to their efficacy. It has been suggested that excretion of certain trophic and growth factors, such as brain-derived neurotrophic factor (BDNF), glial-derived neurotrophic factor (GDNF), stromal cell-derived factor (SDF), and transforming growth factor-β (TGF-β), provide benefit following SC treatment in injury and disease models. GDNF delivery in PD animal models and patients has had some encouraging results [43,44].

Therefore, peripheral delivery of NPs could be considered as a therapeutic strategy to ameliorate PD symptoms using a non-invasive approach with no detectable side effects. For instance, NPs can be derived from MSC present in the adipose tissue from the same individual to avoid any immune rejection. In addition, the use of ESC-NPs would involve ethical issues, and their effect seems to be lower than that obtained from MSC-derived NPs in our model. An alternative therapy could be the use of NPs from induced pluripotent SCs (iPSCs) that can be obtained from the patient´s isolated fibroblasts. The similarity between mouse and human NPs also remains to be evaluated and if those obtained from humans have the ability to produce the same derived factors able to reduce the inflammatory process. Transcriptomic and proteomic analyses of both mouse and human NPs are currently ongoing in our lab to determine the potential translation of this therapy into humans. Another therapeutic approach could involve the use of conditioned media obtained from NPs in culture in an effort to enrich the inoculum with the released factors in a cell-free treatment, but it could well be that factors produced in vivo and responsible for the observed effects are not similar to the ones released in vitro.

The present study suggests that the intravascular infusion of NPs obtained from both ESC and MSC is an innocuous, non-invasive treatment—as compared to intracerebral injection of SCs—that may be used as a treatment for PD. Effects seen here could be mostly due to the modulation of the neuroinflammatory response and amelioration of motor impairment without tumorigenicity. As brain inflammation is commonly involved in many other neurodegenerative diseases, the present strategy could be a therapeutic candidate for Alzheimer's disease, multiple sclerosis, amyotrophic lateral sclerosis, and Huntington's disease, among others.

4. Materials and Methods

4.1. Preparation of Mouse NPs from ESC and MSCs

Mouse embryonic stem cells (ESC; derived from a 129sv mouse) were kindly provided by Dr. Eva Zsigmond (Director of Transgenic and Stem Cell Service Unit at the Brown Foundation Institute of Molecular Medicine for the Prevention of Human Diseases). ESCs were plated on irradiated mouse embryonic fibroblasts cells CF-1 (iMEFs; MTI-GlobalStem, Rockville, MD, USA) and cultured in mouse ESC medium (DMEM (Invitrogen, Carlsbad, CA, USA), 15% FBS (Atlanta Biologicals, Flowery Branch, GA, USA), 1× nonessential amino acids (Invitrogen), 1× Glutamax (Invitrogen), 1mM sodium pyruvate (Invitrogen), 10 mM HEPES (Invitrogen), 1000 U/mL leukemia inhibitory factor (LIF) (R&D Systems, Minneapolis, MN, USA), 0.1% 1-thioglycerol (Sigma, St. Louis, MO, USA) and 1× antibiotic and antimitotic stock solution (Invitrogen). The culture medium was replaced daily. Mouse mesenchymal stem cells (MSC; derived from a C57BL/6 mouse, Cyagen, Santa Clara, CA, USA) were cultured using the mouse mesenchymal stem cell growth medium (Cyagen). The culture medium was replaced every three days. Derivation of neuropotent self-renewing NPs was performed as previously described [45]. ESC and MSCs were harvested, resuspended in neural induction medium (DMEM/F12 and Neurobasal medium (1:1, Invitrogen), 0.5× N2 supplement (Invitrogen), 1× B27 supplement (Invitrogen) and 1× antibiotic and antimitotic stock solution (Invitrogen) and transferred to 0.1% gelatin-coated plates. Culture medium was replaced every day. After 12 days, cells were harvested by Accutase (Millipore, Burlington, MA, USA), resuspended in neural progenitor expansion media (DMEM/F12, 1% N2 supplement, 20 ng/mL EGF and 20 ng/mL bFGF) (R&D Systems) and 1× antibiotic and antimitotic stock solution and transferred to Geltrex LDEV-free reduced growth factor basement membrane matrix (1:100, Invitrogen)-treated dishes. Medium was changed as needed.

4.2. Immunocytochemistry

ESCs and MSCs were fixed with Bouin's solution (Sigma-Aldrich, St. Louis, MO, USA) for 15 min at 4 °C and stored. Fixed cells were blocked, permeabilized and incubated with the primary antibodies overnight in TCT buffer (0.25% Carrageenan and 0.1% Tween-20 in Tris-HCl buffer pH 7.8). The primary antibodies used were: Sox2 (1:100, Abcam, Cambridge, UK), E-Cadherin (1:15, BD Biosciences, San Jose, CA, USA), CD44 (1:100, Abcam), βI Integrin (1:100, Abcam), Musashi (1:200, Abcam) and Vimentin (1:100, Abcam). After rinsing, appropriate secondary antibodies conjugated to Alexa fluorophores 488 or 598 (Molecular Probes, Invitrogen, Carlsbad, CA, USA) were diluted at 1:500 in TCT solution and incubated for 1 hr at room temperature. After rinsing, cells were mounted in Vectashield hardset mounting medium with 4',6-diamidino-2-phenylindole (DAPI) (Vector laboratories) to counterstain nuclei. Images were acquired with a Leica DMI 6000B microscope.

4.3. Induction of Parkinson's Disease Phenotype

As a model of PD, we used the well-established and widely used mouse PD model based on the administration of MPTP (1-methyl-4-phenyl-1,2,3,6-tetrahydropyridine). Two-month-old female C57BL/6J mice (Jackson Laboratories, Bar Harbor, ME, USA) were treated intraperitoneally with MPTP hydrochloride (30 mg/kg) administered daily for 5 consecutive days, as previously reported, to induce selective destruction of dopaminergic neurons of the nigrostriatal tract [46]. The cumulative dose was 150 mg/kg and control mice received the same volume of PBS vehicle. Immediate Parkinsonism traits were perceived following MPTP induction. C57BL/6J animals were housed in groups of up to five in individually ventilated cages under standard conditions (22 °C, 12 h light–dark cycle) receiving food and water ad libitum. All animal experiments were carried out in accordance with UTHealth IACUC regulations (AWC-14-0094 and AWC-17-0140).

4.4. NP Treatment

Following a recovery period of 10 days after MPTP injection, experimental mice were inoculated with 1×10^6 cell count (300 μL) of either ESC or MSC differentiated into mouse NPs or PBS for controls by tail vein injection. Inoculation was done twice in a one-week span. ESC-NPs, MSC-NPs, and PBS mouse groups were behaviorally assessed before sacrifice at 10 days, 3 months, and 9 months after treatment. Animals were sacrificed by CO_2 inhalation, perfused with PBS-EDTA, and their brains, lungs, liver, kidneys, spleen, and pancreas removed, post-fixed into formalin solution, and embedded in paraffin for histological analysis. Animals were visually inspected for tumor formation. Tissues were visually analyzed for any anomaly, including changes in size, color or shape of the organ and existence of tumorigenic masses.

4.5. Wire Hang

Motor coordination and strength was measured by the wire hang task as previously described [47]. Briefly, a 40-cm long 2-mm thick metallic wire was placed 38 cm above a clean home-cage. The animal was placed in the middle of the wire with all four limbs and gently turned upside-down along the axis of the wire. The time until the mouse completely releases its grasp and falls was recorded in 3 minutes using 3 trials. The animal had a 30 second recovery interval until the next trial. The animal was prevented from balancing and limited the use of the tail by the observer.

4.6. Rotarod

Motor coordination and activity, grip-strength, and fatigue are all measurements that are assessed by the commonly used behavioral task of rotarod (Med Associates Inc., Fairfax, VT, USA). As previously described [48], the animal was placed on a horizontal oriented, rotating cylinder that accelerated from 4 to 40 rpms over a 5 minute trial. The latency to fall and best trial were quantified over 3 trials with 2 minutes interval rest. Six minutes was used as the final cut-off latency to fall measurement. In addition, one difficulty of the task is the rodent's desire to grip the rotating rod to prevent falling; thus, 2 consecutive rotations of the animal clinging was considered as a latency to fall measurement. The animals were aided for 30 seconds in the first trial to prevent unnecessary behavior (i.e., turning and false-positive falls/jumps).

4.7. Immunohistochemistry

Paraffin coronal sections of 10 μm thickness were first treated with 3% H_2O_2–10% v/v methanol in PBS at a pH of 7.4 for 20 min to inhibit endogenous peroxidase. For general antigen retrieval, sections were previously heated at 80 °C for 30 min in 10 mM citrate buffer at pH 6.0 and were incubated with the primary antibody overnight at room temperature. The following primary antibodies were used for this study: rabbit polyclonal anti-tyrosine hydroxylase (1:1000; Millipore Sigma), rabbit polyclonal anti-Iba1 (1:1000; Wako Chemicals, Richmond, VA, USA), mouse monoclonal anti-GFAP (1:1000; Millipore Sigma), rabbit polyclonal anti-BLBP (1:1000; Abcam), rabbit polyclonal anti-CD45 (1:1000; Abcam), and goat polyclonal anti-Ki67 (1:200, Santa Cruz Biotechnology, Dallas, TX, USA). The tissue-bound primary antibody was detected by the incubation with the corresponding HRP-linked secondary antibody (1:500; Vector Laboratories) for 1.5 hr and visualized using diaminobenzidine (DAB) substrate. Samples were dehydrated in graded alcohol, cleared in xylene, and covered with DPX for bright field analysis. For double immunofluorescence labeling, sections were first sequentially incubated with the indicated primary antibodies (GFAP and BLBP) followed by the corresponding Alexa 488/594 secondary antibodies (1:500; Invitrogen). Sections were then rinsed and covered with Vectashield hardset mounting medium with DAPI (Vector laboratories) for immunofluorescent evaluation.

4.8. *Quantitative Image Analysis*

Glial burden was defined as the percentage of area stained with anti-GFAP (astrocytes), anti-BLBP (reactive astrocytes), anti-Iba1 (microglia) and anti-CD45 (macrophages) by the total area analyzed, as previously described [49,50]. Six to eight sections per mouse were immunostained and examined under the microscope (DMI6000B, Leica). Digital images were acquired with the program Surveyor and were analyzed using ImageJ version 1.43 software (NIH). The GFAP, Iba1, CD45 and double GFAP/BLBP immunopositive signal in the brain was converted into 8-bit gray scale and their intensity was thresholded to quantify the immunoreactivity (IR) area per total area analyzed. TH optical density (OD) was determined as the intensity of the immunoreactive signal (0–255) divided by the area analyzed (arbitrary units). The TH-IR area was defined as the area stained with anti-TH antibody in pixel units (px^2).

4.9. *Statistical Analysis*

Graphs are expressed as means ± standard error of the mean (s.e.m.). After confirming normal distribution with Skewness and Kurtosis statistic tests, student t-test or one-way analysis of variance (ANOVA), followed by a post-hoc Dunnett's multiple comparison or Tukey's tests, were used to analyze differences among groups. Statistical differences for all tests were considered significant at the confidence level of $p < 0.05$. Statistical analysis was performed using GraphPad Prism 5.0 software (GraphPad Software Inc).

Author Contributions: G.E. performed behavioral tests and collected the tissue, analyzed the data, and wrote the initial draft. N.G. performed all the histological analysis, analyzed the data, and helped in the manuscript writing. E.A. produced the neuronal precursors used for the treatment. C.K. and R.M. participated in the generation of P.D. model and the cell treatment. K.T.-P. processed the tissue for toxicity assays. P.E.S. and C.S. provided financial support for the project, conceptualized the research aims, and contributed to the manuscript. I.M.-G. performed some experiments, designed the experimental plan, oversighted the project, and wrote the final version of the manuscript.

References

1. Kalia, L.V.; Lang, A.E. Parkinson's disease—The lancet. *Lancet* **2015**, *86*, 896–912. [CrossRef]
2. Dawson, T.M.; Dawson, V.L. Molecular pathways of neurodegeneration in Parkinson's disease. *Science* **2003**, *302*, 819–822. [CrossRef] [PubMed]
3. Chaudhuri, K.R.; Healy, D.G.; Schapira, A.H.V. Non-motor symptoms of Parkinson's disease: Diagnosis and management. *Lancet Neurol.* **2006**, *5*, 235–245. [CrossRef]
4. Cummings, J.L. Depression and parkinson's disease: A review. *Am. J. Psychiatry* **1992**, *149*, 443–454. [PubMed]
5. Fasano, A.; Visanji, N.P.; Liu, L.W.C.; Lang, A.E.; Pfeiffer, R.F. Gastrointestinal dysfunction in Parkinson's disease. *Lancet Neurol.* **2015**, *14*, 625–639. [CrossRef]
6. Soto, C. Unfolding the role of protein misfolding in neurodegenerative diseases. *Nat. Rev. Neurosci.* **2003**, *4*, 49–60. [CrossRef] [PubMed]
7. Polymeropoulos, M.H.; Lavedan, C.; Leroy, E.; Ide, S.E.; Dehejia, A.; Dutra, A.; Pike, B.; Root, H.; Rubenstein, J.; Boyer, R.; et al. Mutation in the alpha-synuclein gene identified in families with Parkinson's disease. *Science* **1997**, *276*, 2045–2047. [CrossRef] [PubMed]
8. Gelders, G.; Baekelandt, V.; Van der Perren, A. Linking neuroinflammation and neurodegeneration in Parkinson's disease. *J. Immunol. Res.* **2018**, *2018*, 4784268. [CrossRef] [PubMed]

9. Dexter, D.T.; Jenner, P. Parkinson disease: From pathology to molecular disease mechanisms. *Free Radic. Biol. Med.* **2013**, *62*, 132–144. [CrossRef]
10. McGeer, P.L.; Itagaki, S.; Boyes, B.E.; McGeer, E.G. Reactive microglia are positive for HLA-DR in the substantia nigra of Parkinson's and Alzheimer's disease brains. *Neurology* **1988**, *38*, 1285–1291. [CrossRef]
11. Braak, H.; Sastre, M.; Del Tredici, K. Development of α-synuclein immunoreactive astrocytes in the forebrain parallels stages of intraneuronal pathology in sporadic Parkinson's disease. *Acta Neuropathol.* **2007**, *114*, 231–241. [CrossRef] [PubMed]
12. Liberatore, G.T.; Jackson-Lewis, V.; Vukosavic, S.; Mandir, A.S.; Vila, M.; Mcauliffe, W.G.; Dawson, V.L.; Dawson, T.M.; Przedborski, S. Inducible nitric oxide synthase stimulates dopaminergic neurodegeneration in the MPTP model of Parkinson disease. *Nat. Med.* **1999**, *5*, 1403–1409. [CrossRef] [PubMed]
13. McGeer, P.L.; Schwab, C.; Parent, A.; Doudet, D. Presence of reactive microglia in monkey substantia nigra years after 1-methyl-4-phenyl-1,2,3,6-tetrahydropyridine administration. *Ann. Neurol.* **2003**, *54*, 599–604. [CrossRef] [PubMed]
14. Outeiro, T.F.; Ferreira, J. Current and future therapeutic strategies for Parkinson's disease. *Curr. Pharm. Des.* **2009**, *15*, 3968–3976. [CrossRef]
15. Bloem, B.R.; de Vries, N.M.; Ebersbach, G. Nonpharmacological treatments for patients with Parkinson's disease. *Mov. Disord.* **2015**, *30*, 1504–1520. [CrossRef]
16. Okun, M.S. Deep-brain stimulation for Parkinson's disease the clinical problem. *N. Engl. J. Med.* **2012**, *367*, 1529–1538. [CrossRef]
17. Olanow, C.W.; Kordower, J.H.; Lang, A.E.; Obeso, J.A. Dopaminergic transplantation for Parkinson's disease: Current status and future prospects. *Ann. Neurol.* **2009**, *66*, 591–596. [CrossRef]
18. Anisimov, S. V Cell-based therapeutic approaches for Parkinson's disease: Progress and perspectives. *Rev. Neurosci.* **2009**, *20*, 347–382. [CrossRef]
19. Correia, A.S.; Anisimov, S.V.; Li, J.Y.; Brundin, P. Stem cell-based therapy for Parkinson's disease. *Ann. Med.* **2005**, *37*, 487–498. [CrossRef]
20. Caplan, A.I.; Bruder, S.P. Mesenchymal stem cells: Building blocks for molecular medicine in the 21st century. *Trends Mol. Med.* **2001**, *7*, 259–264. [CrossRef]
21. Yamanaka, K.T. Shinya induction of pluripotent stem cells from mouse embryonic and adult fibroblast cultures by defined factors. *Cell* **2006**, *126*, 663–676.
22. Kim, S.U.; Lee, H.J.; Kim, Y.B. Neural stem cell-based treatment for neurodegenerative diseases. *Neuropathology* **2013**, *33*, 491–504. [CrossRef] [PubMed]
23. Li, Y.; Chen, J.; Wang, L.; Zhang, L.; Lu, M.; Chopp, M. Intracerebral transplantation of bone marrow stromal cells in a 1-methyl-4-phenyl-1,2,3,6-tetrahydropyridine mouse model of Parkinson's disease. *Neurosci. Lett.* **2001**, *316*, 67–70. [CrossRef]
24. Kriks, S.; Shim, J.W.; Piao, J.; Ganat, Y.M.; Wakeman, D.R.; Xie, Z.; Carrillo-Reid, L.; Auyeung, G.; Antonacci, C.; Buch, A.; et al. Dopamine neurons derived from human ES cells efficiently engraft in animal models of Parkinson's disease. *Nature* **2011**, *480*, 547–551. [CrossRef]
25. Galindo, L.T.; Filippo, T.R.M.; Patricia, S.; Ariza, C.B.; Moreira, C.M.; Camara, N.O.S.; Porcionatto, M.A. Mesenchymal stem cell therapy modulates the inflammatory response in experimental traumatic brain injury. *Neurol. Res. Int.* **2011**, *2011*, 564089. [CrossRef]
26. Drago, D.; Cossetti, C.; Iraci, N.; Gaude, E.; Musco, G.; Bachi, A.; Pluchino, S. The stem cell secretome and its role in brain repair. *Biochimie* **2013**, *95*, 2271–2285. [CrossRef]
27. Joyce, N.; Annett, G.; Wirthlin, L.; Olson, S.; Bauer, G.; Nolta, A. Mesenchymal stem cells for the treatment of neurodegenerative disease. *Regen. Med.* **2010**, *5*, 933–946. [CrossRef]
28. Kordower, J.H.; Chu, Y.; Hauser, R.A.; Freeman, T.B.; Olanow, C.W. Lewy body-like pathology in long-term embryonic nigral transplants in Parkinson's disease. *Nat. Med.* **2008**, *14*, 504–506. [CrossRef]
29. Li, J.-Y.; Englund, E.; Holton, J.L.; Soulet, D.; Hagell, P.; Lees, A.J.; Lashley, T.; Quinn, N.P.; Rehncrona, S.; Björklund, A.; et al. Lewy bodies in grafted neurons in subjects with Parkinson's disease suggest host-to-graft disease propagation. *Nat. Med.* **2008**, *14*, 501–503. [CrossRef]
30. Misra, V.; Lal, A.; El Khoury, R.; Chen, P.R.; Savitz, S.I. Intra-arterial delivery of cell therapies for stroke. *Stem Cells Dev.* **2012**, *21*, 1007–1015. [CrossRef]
31. Reynolds, B.A.; Tetzlaff, W.; Weiss, S. A multipotent EGF-responsive striatal embryonic progenitor cell produces neurons and astrocytes. *J. Neurosci.* **1992**, *12*, 4565–4574. [CrossRef] [PubMed]

32. Ray, J.; Peterson, D.A.; Schinstine, M.; Gage, F.H. Proliferation, differentiation, and long-term culture of primary hippocampal neurons. *Proc. Natl. Acad. Sci. USA* **2006**, *90*, 3602–3606. [CrossRef] [PubMed]
33. Balenci, L.; Inoue, T.; Karpowicz, P.; Willaime-Morawek, S.; DeVeale, B.; van der Kooy, D. E-Cadherin regulates neural stem cell self-renewal. *J. Neurosci.* **2009**, *29*, 3885–3896.
34. Götz, M.; Sirko, S.; Beckers, J.; Irmler, M. Reactive astrocytes as neural stem or progenitor cells: In vivo lineage, In vitro potential, and Genome-wide expression analysis. *Glia* **2015**, *63*, 1452–1468. [CrossRef] [PubMed]
35. Jeong, H.-K.; Ji, K.; Min, K.; Joe, E.-H. Brain inflammation and microglia: Facts and misconceptions. *Exp. Neurobiol.* **2013**, *22*, 59–67. [CrossRef]
36. Lotia, M.; Jankovic, J. New and emerging medical therapies in Parkinson's disease. *Expert Opin. Pharmacother.* **2016**, *17*, 895–909. [CrossRef]
37. Georgiev, D.; Hamberg, K.; Hariz, M.; Forsgren, L.; Hariz, G.M. Gender differences in Parkinson's disease: A clinical perspective. *Acta Neurol. Scand.* **2017**, *136*, 570–584. [CrossRef]
38. Chao, Y.X.; He, B.P.; Tay, S.S.W. Mesenchymal stem cell transplantation attenuates blood brain barrier damage and neuroinflammation and protects dopaminergic neurons against MPTP toxicity in the substantia nigra in a model of Parkinson's disease. *J. Neuroimmunol.* **2009**, *216*, 39–50. [CrossRef]
39. Kesslak, J.P. Transplantation of embryonic dopamine neurons for severe parkinson's disease. *N Engl J Med* **2001**, *344*, 710–719. [CrossRef]
40. Chen, D.; Fu, W.; Zhuang, W.; Lv, C.; Li, F.; Wang, X. Therapeutic effects of intranigral transplantation of mesenchymal stem cells in rat models of Parkinson's disease. *J. Neurosci. Res.* **2017**, *95*, 907–917. [CrossRef]
41. Zawada, W.M.; Banninger, G.P.; Thornton, J.; Marriott, B.; Cantu, D.; Rachubinski, A.L.; Das, M.; Griffin, W.S.T.; Jones, S.M. Generation of reactive oxygen species in 1-methyl-4-phenylpyridinium (MPP+) treated dopaminergic neurons occurs as an NADPH oxidase-dependent two-wave cascade. *J. Neuroinflammation* **2011**, *8*, 129. [CrossRef] [PubMed]
42. Zhou, Y.; Wu, J.; Sheng, R.; Li, M.; Wang, Y.; Han, R.; Han, F.; Chen, Z.; Qin, Z.H. Reduced Nicotinamide Adenine Dinucleotide Phosphate inhibits MPTP-induced neuroinflammation and neurotoxicity. *Neuroscience* **2018**, *391*, 140–153. [CrossRef] [PubMed]
43. Kordower, J.H.; Emborg, M.E.; Bloch, J.; Ma, S.Y.; Chu, Y.; Leventhal, L.; McBride, J.; Chen, E.Y.; Palfi, S.; Roitberg, B.Z.; et al. Neurodegeneration prevented by lentiviral vector delivery of GDNF in primate models of Parkinson's disease. *Science* **2000**, *290*, 767–773. [CrossRef] [PubMed]
44. Patel, N.K.; Bunnage, M.; Plaha, P.; Svendsen, C.N.; Heywood, P.; Gill, S.S. Intraputamenal infusion of glial cell line-derived neurotrophic factor in PD: A two-year outcome study. *Ann. Neurol.* **2005**, *57*, 298–302. [CrossRef] [PubMed]
45. Onorati, M.; Camnasio, S.; Binetti, M.; Jung, C.B.; Moretti, A.; Cattaneo, E. Neuropotent self-renewing neural stem (NS) cells derived from mouse induced pluripotent stem (iPS) cells. *Mol. Cell. Neurosci.* **2010**, *43*, 287–295. [CrossRef] [PubMed]
46. Ogawa, N.; Hirose, Y.; Ohara, S.; Ono, T.; Watanabe, Y. A simple quantitative bradykinesia test in MPTP-treated mice. *Res. Commun. Chem. Pathol. Pharmacol.* **1985**, *50*, 435–441. [PubMed]
47. van Putten, M.; Aartsma-Rus, A.; Louvain-la-Neuve, L. The use of hanging wire tests to monitor muscle strength and condition over time. *Treat. NMD Neuromuscul. Netw.* **2012**, *4*, 1–12.
48. Colotla, V.A.; Flores, E.; Oscos, A.; Meneses, A.; Tapia, R. Effects of MPTP on locomotor activity in mice. *Neurotoxicol. Teratol.* **1990**, *12*, 405–407. [CrossRef]
49. Moreno-Gonzalez, I.; Baglietto-Vargas, D.; Sanchez-Varo, R.; Jimenez, S.; Trujillo-Estrada, L.; Sanchez-Mejias, E.; Del Rio, J.C.; Torres, M.; Romero-Acebal, M.; Ruano, D.; et al. Extracellular amyloid-β and cytotoxic glial activation induce significant entorhinal neuron loss in young $PS1^{M146L}/APP^{751SL}$ mice. *J. Alzheimer's Dis.* **2009**, *18*, 755–776. [CrossRef]
50. Jimenez, S.; Baglietto-Vargas, D.; Caballero, C.; Moreno-Gonzalez, I.; Torres, M.; Sanchez-Varo, R.; Ruano, D.; Vizuete, M.; Gutierrez, A.; Vitorica, J. Inflammatory response in the Hippocampus of $PS1_{M146L}/APP_{751SL}$ mouse model of Alzheimer's disease: Age-dependent switch in the microglial phenotype from alternative to classic. *J. Neurosci.* **2008**, *28*, 11650–11661. [CrossRef]

3

The Effect of Sertoli Cells on Xenotransplantation and Allotransplantation of Ventral Mesencephalic Tissue in a Rat Model of Parkinson's Disease

Yun-Ting Jhao [1], Chuang-Hsin Chiu [2], Chien-Fu F. Chen [3], Ta-Kai Chou [2], Yi-Wen Lin [4], Yu-Ten Ju [5], Shinn-Chih Wu [5], Ruoh-Fang Yan [6], Chyng-Yann Shiue [6], Sheau-Huei Chueh [7], Christer Halldin [8], Cheng-Yi Cheng [2,*,†] and Kuo-Hsing Ma [9,*,†]

1 Graduate Institute of Medical Sciences, National Defense Medical Center, Taipei 114, Taiwan; k6520319@yahoo.com.tw
2 Department of Nuclear Medicine, Tri-Service General Hospital, Taipei 114, Taiwan; treasure316@gmail.com (C.-H.C.); dakaichou@gmail.com (T.-K.C.)
3 Graduate Institute of Life Sciences, National Defense Medical Center, Taipei 114, Taiwan; t70cyy@gmail.com
4 Institute of Oral Biology, National Yang-Ming University, Taipei 112, Taiwan; ywlin7@yahoo.com.tw
5 Department of Animal Science and Technology, National Taiwan University, Taipei 106, Taiwan; ytju@ntu.edu.tw (Y.-T.J.); scw01@ntu.edu.tw (S.-C.W.)
6 Department of Nuclear Medicine, National Taiwan University Hospital, Taipei 100, Taiwan; rfyen@ntu.edu.tw (R.-F.Y.); shiue@ntuh.gov.tw (C.-Y.S.)
7 Department of Biochemistry, National Defense Medical Center, Taipei 114, Taiwan; shch@ndmctsgh.edu.tw
8 Department of Clinical Neuroscience, Karolinska Institutet, 171 77 Stockholm, Sweden; christer.halldin@ki.se
9 Department of Biology and Anatomy, National Defense Medical Center, Taipei 114, Taiwan
* Correspondence: chengcy60@gmail.com (C.-Y.C.); kuohsing91@yahoo.com.tw (K.-H.M.)
† These authors contributed equally to this work.

Abstract: Intra-striatal transplantation of fetal ventral mesencephalic (VM) tissue has a therapeutic effect on patients with Parkinson's disease (PD). Sertoli cells (SCs) possess immune-modulatory properties that benefit transplantation. We hypothesized that co-graft of SCs with VM tissue can attenuate rejection. Hemi-parkinsonian rats were generated by injecting 6-hydroxydopamine into the right medial forebrain bundle of Sprague Dawley (SD) rats. The rats were then intrastriatally transplanted with VM tissue from rats or pigs (rVM or pVM), with/without a co-graft of SCs (rVM+SCs or pVM+SCs). Recovery of dopaminergic function and survival of the grafts were evaluated using the apomorphine-induced rotation test and small animal-positron emission tomography (PET) coupled with [^{18}F] DOPA or [^{18}F] FE-PE2I, respectively. Immunohistochemistry (IHC) examination was used to determine the survival of the grafted dopaminergic neurons in the striatum and to investigate immune-modulatory effects of SCs. The results showed that the rVM+SCs and pVM+SCs groups had significantly improved drug-induced rotational behavior compared with the VM alone groups. PET revealed a significant increase in specific uptake ratios (SURs) of [^{18}F] DOPA and [^{18}F] FE-PE2I in the grafted striatum of the rVM+SCs and pVM+SCs groups as compared to that of the rVM and pVM groups. SC and VM tissue co-graft led to better dopaminergic (DA) cell survival. The co-grafted groups exhibited lower populations of T-cells and activated microglia compared to the groups without SCs. Our results suggest that co-graft of SCs benefit both xeno- and allo-transplantation of VM tissue in a PD rat model. Use of SCs enhanced the survival of the grafted dopaminergic neurons and improved functional recovery. The enhancement may in part be attributable to the immune-modulatory properties of SCs. In addition, [^{18}F]DOPA and [^{18}F]FE-PE2I coupled with PET may provide a feasible method for in vivo evaluation of the functional integrity of the grafted DA cell in parkinsonian rats.

Keywords: sertoli cell; ventral mesencephalic tissue; transplantation; positron emission tomography; Parkinson's disease

1. Introduction

Parkinson's disease (PD) is typified by selective degeneration of dopaminergic (DA) neurons in substantia nigra pars compacta and is a prevalent neurodegenerative disorder [1]. Symptoms of PD include bradykinesia, resting tremor, postural instability, and ultimately rigidity [1]. Pharmacological and surgical treatment, such as deep brain stimulation, can partially alleviate these symptoms [2] but do not stop disease progression. Furthermore, the effectiveness of medications can decrease with time, and side effects such as paresthesia, depression, and dyskinesia may occur after long-term treatment [3,4].

Cell replacement therapy is a promising therapeutic strategy for PD and aims to re-establish DA function via re-introduction of exogenic DA cells into the affected brain circuit. Intra-striatal transplantation of cells harvested from human fetal ventral mesencephalic (hVM) tissue have been shown to restore striatal dopamine concentration in PD patients and provide long-lasting relief of the symptoms [5–7]. However, hVM tissues are unlikely to become a major PD therapy in that the supply of human fetal tissues is limited and there are serious ethical issues to consider. Alternatively, DA cells derived from human embryonic stem cells (hESCs) and induced pluripotent stem cells (iPSCs) can, in theory, resolve the supply issue and thus have been heavily tested in PD animal models [8–10]. Although hESCs and iPSCs have a strong potential for PD therapy, safety concerns with the use of these cells, including incomplete and unsynchronized differentiation, tumorigenesis, and neural overgrowth, are obstacles that have prevented their widespread use [11,12].

Fetal ventral mesencephalic (VM) tissue have been shown to have multiple benefits for PD cell therapy. Grafted DA cells from hVM tissue can re-innervate the striatum, and also release dopamine at high rate with auto-modulation via dopamine receptors [13–15]. Moreover, grafted hVM tissue can survive in patients for as long as 8–16 years [16,17]. Patients treated with this method had striatal dopamine concentration restored and also exhibited improvement of behavior. Despite all of the benefits of hVM tissue, its limitations have made the search for alternative cell sources, especially from non-human animals, an essential aim of PD cell therapy.

Porcine VM (pVM) tissue has become a promising cell source for PD therapy for a number of reasons. The supply of porcine embryos is abundant, and they are available whenever required. Pigs for transplantation can be raised in a specific pathogen–free environment [18,19]. DA cell grafts from pVM have shown no tumorigenesis in PD animal models and human patients [20,21]. Moreover, DA cells derived from pVM grafts can re-innervate the denervated striatum of patients, leading to recovery of motor function [22].

However, as in all xenotransplantation, immune suppression is required for pVM transplantation. Tacrolimus, neural, and cyclosporine are commonly used immunosuppressants for xenotransplantation, but side effects such as hepatotoxicity, nephrotoxicity, neurotoxicity, osteoporosis, diabetes, and hypertension are not uncommon [23–25]. Furthermore, studies have shown that immunosuppression drugs do not provide sufficient protection against immune rejection in the grafted area when treating PD [26–28]. To overcome this obstacle, we sought to identify a cell resource that could attenuate the immune response and provide benefits for DA cell survival.

Sertoli cells (SCs) are located in testis, which is an immune-privileged site that prevents germ cells from triggering an immune response [29]. SCs secrete multiple immune modulating molecules, such as the Fas ligand and transforming growth factor (TGF)-β [30]. Use of SCs in transplantation of other tissues and/or organ leads to a significant decrease in the levels of pro-inflammatory mediators, corresponding with increased anti-inflammatory cytokines at the grafted area [31,32]. Co-transplantation of SCs and tissues has resulted in prolonged survival of allograft and xenograft tissues [31,33,34]. Currently, SCs have been applied to a wide range of transplantation procedures, such as skin transplantation [31], cardiac grafts [34], and hepatocyte xenografts [33]. In addition, SCs hold great therapeutic potential for the treatment of neurodegeneration diseases. For example, rat ventral mesencephalic (rVM) tissue co-cultured with rat or porcine SCs results in enhanced neuronal survival and neuronal differentiation in vitro [35]. Study has also investigated the immune-modulatory effect in a hemiparkinsonian mouse model after co-grafting with rVM and rat SCs or mouse VM and rat SCs [36]. Another study showed that the neurite outgrowth was significantly increased in neurons co-cultured with SCs [37]. Taken together, these results suggest that SCs might be beneficial in cell replacement therapy for the treatment of PD.

The purpose of this study was to determine whether SCs are capable of enhancing the survival of VM allografts [18] and/or xenografts (porcine) in a PD model rat, and if so to reveal the underlying mechanisms of this enhancement. In order to determine DA neuron survival, we used small animal positron emission tomography (PET) with 2 [^{18}F]-labeled radioligands, [^{18}F]DOPA and [^{18}F]FE-PE2I, to study dopamine synthesis and the dopamine transporter, respectively. [^{18}F]FE-PE2I has been successfully used as an imaging agent to examine disease progression in PD patients [38]. We observed the immune response and DA cell survival of different transplantation groups via PET imaging and immunohistochemistry (IHC) studies. In addition, we used an apomorphine-induced rotation test to evaluate recovery of dopamine neuronal function.

2. Material and Methods

2.1. Animals

The Institutional Animal Care and Use Committee (IACUC) of the National Defense Medical Center (NDMC, Taipei, Taiwan) approved all of the animal experiment protocols (project identification code: IACUC-16-095 and IACUC-17-114; date of approval: August 1, 2016 and August 1, 2017). All animal experiments were conducted in accord with the Guidelines for the Use and Care of Laboratory Animals in Experimental Studies. Male Sprague–Dawley (SD) rats (eight weeks old, 280–300 g) were purchased from BioLASCO Ltd. (Taipei, Taiwan). Animals were housed in the NDMC animal facility under a 12 h light/dark cycle (light from 07:00 to 19:00), at a constant temperature of 23 ± 2 °C. The animals were given ad libitum access to a complete pellet diet and tap water. Lee–Sung pigs were obtained from National Taiwan University (Taipei, Taiwan), and raised in a pig house with an ISO9001-2015 certification by the British Standards Institution (certification number: FS686535). The experimental schedule is shown in Figure 1.

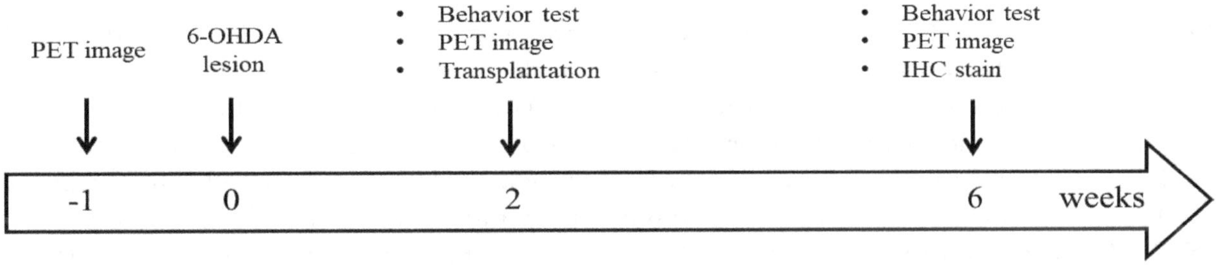

Figure 1. Experimental flowchart. [^{18}F]-DOPA and [^{18}F] FE-PE2I PET scans were performed at three time points (once before and twice after a unilateral 6-hydroxydopamine (6-OHDA) lesion was made to medial forebrain bundle of the animals). Behavior test refers to the apomorphine-induced rotation test. Transplantation was performed two weeks after the 6-OHDA lesion was made. At six weeks after the lesion, the rats were sacrificed for immunohistochemistry (IHC) studies.

2.2. *Hemiparkinsonian Rat Model*

A hemiparkinsonian rat model was generated as previously described [39]. Briefly, eight-week-old male SD rats were deeply anesthetized with Zolitel (20 mg/kg, i.p.), and placed in a stereotactic apparatus. Then, 6-OHDA (20 μg in 4 μL of 0.02% ascorbic acid–holding saline) (Sigma-Aldrich, Saint Louis, MO, USA) was delivered unilaterally to the medial forebrain bundle. The coordinates were 7.8, 1.2, and 4.4 mm below the dura, lateral to the midline, and posterior to the bregma, respectively.

2.3. *Behavioral Test*

Severity of DA pathway damage in rats was evaluated by the apomorphine-induced rotational behavior test two weeks after the 6-OHDA lesion was created [40]. Rats with lesions were administered apomorphine subcutaneously at a dose of 0.5 mg/kg in 0.2% ascorbic acid–holding saline (Sigma-Aldrich, Saint Louis, MO, USA). Subsequently, drug-induced rotational responses were recorded by a rotometer system (MED Associates, Inc., St. Albans, VT, USA) [41]. A hemiparkinsonian disorder was defined as rats that rotated over four turns/min away from the lesion side. These rats were selected for transplantation experiments. Four weeks after transplantation, the rotation test was performed again to evaluate recovery of DA function.

2.4. *SC Isolation*

Rat SCs were isolated as previously described, with minor modifications [42]. Briefly, testes were surgically removed from 21-day-old (d21) SD rats. The seminiferous tubules were dispersed into 40 mL 1X HBSS, and washed two times with 1X HBSS followed by incubation with 25 mL of 0.05% collagenase solution in 1X HBSS (C2674, Sigma-Aldrich, Saint Louis, MO, USA) at 37 °C for 15 min, with shaking at 80 oscillations/min. The tubules were subsequently washed three times with 50 mL of 1X HBSS and thereafter incubated with 25 mL of 0.05% trypsin (T5266, Sigma-Aldrich, Saint Louis, MO, USA) in 1X HBSS at 37 °C for 10 min. Next, the tubules were washed with a 0.03% trypsin inhibitor (T6522, Sigma-Aldrich, Saint Louis, MO, USA) solution (20 mL 1X HBSS, 5–10 min) and then incubated in a multi-enzyme solution (37 °C, 40 min, shaking at 80 oscillations/min). The multi-enzyme solution contained 0.03% trypsin inhibitor, 25 mL of 1X HBSS, 0.1% collagenase, 0.04% DNase I (DN25, Sigma-Aldrich, Saint Louis, MO, USA), and 0.2% hyaluronidase (H6254, Sigma-Aldrich, Saint Louis, MO, USA). The derived supernatant was sieved via a 70 mm–pore size nylon mesh and centrifuged at 200× *g* for 10 min to derive a pellet of SCs. Finally, the pellet was washed three times with 1X HBSS and used for the experiments.

After SC isolation, IHC staining was used to confirm that the cells in pellet were indeed SCs, as numerous cells are stained with both a nuclear biomarker (nuclear red) and SC biomarker (follicle stimulating hormone receptor; FSHr) (Figure 2a–c). The cells were first stained with rabbit-anti FSHr (1:250; Aviva Systems Biology Corporation, San Diego, CA, USA), and then incubated with Alexa488-conjugated donkey anti-rabbit IgG (1:250; Jackson ImmunoResearch Laboratories, West Grove, PA, USA). Finally, the cells were stained with nuclear red (1:1000; AAT Bioquest, Inc., Sunnyvale, CA, USA). SCs were identified as being double-positive ($FSHr^+$/nuclear red^+). Flow cytometry was then used to isolate SCs from the cell pellet and to estimate the purity of SCs by calculating the percentage of FSHr positive cells (Figure 2d,e). The results indicated that approximately 80% of the cells isolated from the testis were SCs.

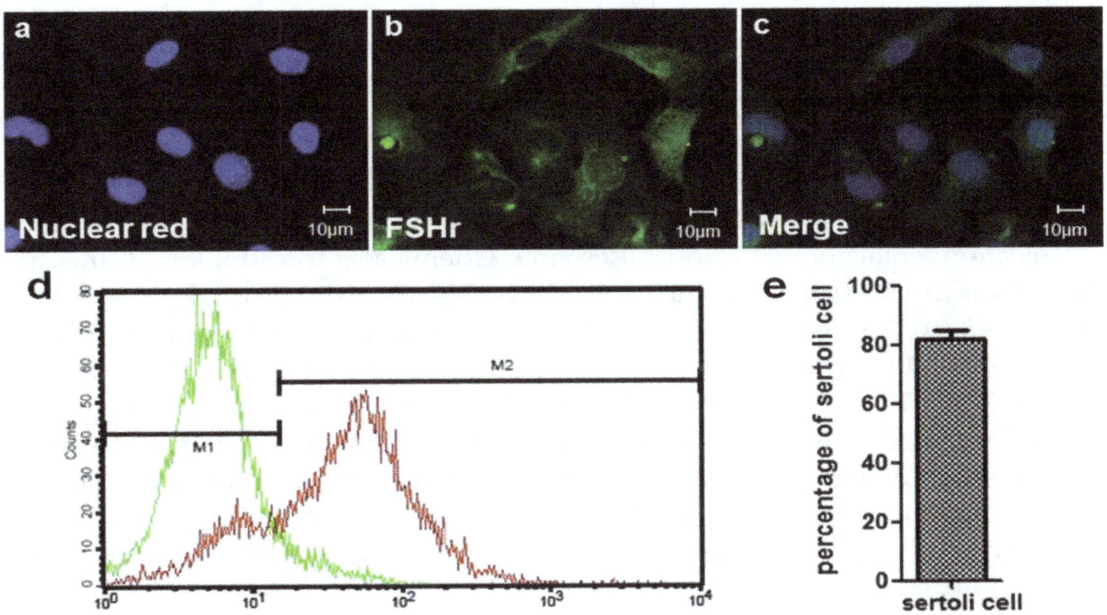

Figure 2. Isolation of Sertoli cells (SCs). IHC staining was used to identify SCs isolated from the testis. Staining included (**a**) nuclear red staining (biomarker of nucleus) and (**b**) immunostaining for FSHr (biomarker of SCs). (**c**) SCs were identified as double-labeled cells. (**d**) Flow cytometry showed different fluorescence intensity in M1 (cell suspension only stained with florescent secondary antibody) and M2 (cell suspension stained with FSHr primary antibody and florescent secondary antibody). The SCs (M2) exhibited a tremendously shifted peak as compared to the control (M1). (**e**) The purity of the SCs was calculated by flow cytometry.

2.5. Mesencephalic Tissue Preparation and Transplantation

VM tissues used to establish allotransplantation and xenotransplantation models were obtained from embryonic day 14 SD rats and embryonic day 27 Lee-Sung pigs [39,43,44]. Dissection areas were selected according to a previous study, with some modifications [40,45]. The dissected tissues containing abundant DA cell bodies were kept in 1X HBSS. VM tissue was cut to small sections and subsequently grafted into the lesioned striatum using glass micropipettes, with the coordinates 2.5, 0.5, and 5.5 mm in length lateral to the midline, posterior to the bregma, and below the dura, respectively. Thirty-three hemiparkinsonian rats were divided into six groups, and different combinations of tissues were grafted into the striatum. (1) The sham group (n = 3) was injected with 4 μL 1X HBSS. (2) The SCs group (n = 6) received ~1.25×10^5 SCs. (3) The rVM group (n = 6) was transplanted with rVM tissue. (4) The pVM group (n = 6) was transplanted with pVM tissue. (5) The rVM + SCs group (n = 6) was co-grafted rVM tissue and SCs (~1.25×10^5 cells). (6) The pVM + SCs group (n = 6) was co-grafted pVM tissue and SCs (~1.25×10^5 cells).

2.6. Radiopharmaceuticals

[^{18}F] DOPA was synthesized and provided by the Department of Nuclear Medicine affiliated with National Taiwan University Hospital. [^{18}F] FE-PE2I was synthesized as previously reported, with some modifications [46]. Briefly, nucleophilic fluorination of a tosyl precursor was performed in dimethyl sulfoxide with dried K [1 8F]/$K_{2.2.2}$, followed by modified HPLC purification (without a pre-purified cartridge). The desired compound was obtained after solid phase extraction and formulation in phosphate buffered saline.

The non-decay corrected radiochemical yield for [^{18}F] FE-PE2I was 4.98% ± 1.73% (n = 15), with a radiochemical purity > 96%. The specific activity was 179 ± 52 GBq/μmol at the time of end-of-synthesis (EOS). The injected mass of [^{18}F] FE-PE2I was about 94.2 ng/kg in subsequent imaging studies, with intravenous injection of 18–22 MBq of [^{18}F] FE-PE2I.

2.7. Small Animal PET Imaging

The protocol for small animal PET imaging was adapted from previous research [47]. PET imaging was performed with a small animal PET scanner (BIOPET 105, BIOSCAN, Santa Clara, CA, USA) at one week before the 6-OHDA lesion was created, two weeks after the 6-OHDA lesion was created, and four weeks after transplantation. [^{18}F] DOPA (22.2–25.9 MBq; 0.6–0.7 mCi) and [^{18}F] FE-PE2I (14.8–18.5 MBq; 0.4–0.5 mCi) was administrated via the tail vein of the rats. Thirty minutes before [^{18}F] DOPA injection, entacapone (Toronto Research Chemicals, Toronto, ON, Canada; 10 mg/kg) and carbidopa (Toronto Research Chemicals, Toronto, ON, Canada; 10 mg/kg) were administrated intraperitoneally [48,49]. The SD rats were anesthetized by passive inhalation of an isoflurane–oxygen mixture (2% and 5% isoflurane for maintenance and induction, respectively). PET image collection was executed 50 min ([^{18}F] DOPA) and 20 min ([^{18}F] FE-PE2I) after radioligand injection. The collection process was performed at an energy window of 250–700 keV. Image reconstruction was performed using a 2D filtered back-projection (ramp filter, with the cutoff being determined at Nyquist frequency) and the Fourier re-binning algorithm. Amide software (stanford university, Santa Clara, CA, USA) was used for analysis of the PET images. The rat brain atlas and magnetic resonance imaging (MRI) results were used for brain region confirmation [50]. The atlas and MRI results were also used to delineate volumes of interest of the cerebellum and striatum as observed in the PET images that had been reconstructed and summated. The specific uptake ratio (SUR) was expressed as (striatum − cerebellum)/cerebellum.

2.8. IHC Staining

Four weeks after transplantation, rats from each group were sacrificed for IHC experiments [43]. Rat brains were removed and post-fixed with 4% paraformaldehyde overnight at 4 °C. Brains were then soaked in 20% sucrose in 0.1M PBS for two days and then 30% sucrose in 0.1M PBS for two days. Brains were sliced into a series of coronal sections (30 μm) on a Cryostat Microtome (Leica CM 3050; Leica Microsystem, Wetzlar, Germany). All brain slices (approximately 90 slices per animal) that contained grafted region were divided into four sets, and prepared for tyrosine hydroxylase (TH), dopamine transporter (DAT) [51], CD3, and major histocompatibility complex (MHC) class II (OX-6)/ionized calcium binding adaptor molecule 1 (Iba1) staining. The survival of DA neurons in grafted regions was evaluated by immunostaining with anti-TH and anti-DAT antibodies. Anti-CD3 antibody was used to identify infiltrating T-cells in the grafted region. Anti-Iba1 and anti-OX-6 antibodies were used to calculate the total and activated microglia in the grafted striatum, respectively. 3,3-diaminobenzidine (DAB) was used in IHC staining of TH, DAT, and CD3; a positive reaction was identified as a brown color in brain slices.

First, brain slices were rinsed with blocking solution containing 3% normal goat serum (Vector, Burlingame, CA, USA) in PBS, and 0.5% Triton X-100 (Sigma-Aldrich, Saint Louis, MO, USA). The rinsed slices were incubated overnight at 4 °C with rabbit anti-CD3 antibody (1:200; Abcam, Cambridge, UK), rabbit anti-TH antibody (1:2000 dilution; Millipore Corporation, Billerica, MA, USA), or rabbit anti-DAT antibody (1:500; Abcam, Cambridge, UK). They were then incubated for 1 h with goat anti-rabbit biotinylated IgG (1:200; Vector). After washing three times, the slices were incubated for 1 h with avidin–biotin complex at a 1:200 dilutions (Vectastain ABC kit, Vector, Burlingame, CA, USA). For visualization, 0.05% DAB was applied for 6 min, and then the slices were mounted on glass slides.

The VM tissue distribution in the transplantation side striatum served as the basis for the definition of the grafted area. The optical density (OD) of the grafted areas was measured on all IHC sections. Images of the striatum stained with different antibodies were acquired, and examined using a slide scanner (Axio Scan.Z1; ZEISS, Oberkochen, Germany) and ImageJ software (version 1.8.0, NIH, Bethesda, MD, USA, 2014). The OD ratio was calculated as follows: OD ratio = $(OD_{striatum}-OD_{corpus\ callosum})/OD_{corpus\ callosum}$. The numbers of TH immunoreactive (TH-ir), DAT-ir, and CD3-ir cells of grafted regions were calculated and then divided by the areas, and they were further divided into 30 μm. Therefore, data of cell number was expressed as the number of cells per cubic millimeter.

Microglia activation was determined using immunofluorescence staining. Brain slices were double-stained with rabbit anti-Iba1 (1:500; Wako, Osaka, Japan) and mouse anti-OX-6 (1:150; Abcam, Cambridge, UK) and then stained with Alexa488-conjugated donkey anti-rabbit IgG (1:250; Jackson ImmunoResearch Laboratories, West Grove, PA, USA) and cy3-conjugated donkey anti-mouse IgG (1:250; Jackson ImmunoResearch Laboratories, West Grove, PA, USA). Finally, the brain sections were stained with nuclear red (1:1000; AAT Bioquest, Inc., Sunnyvale, CA, USA) for nuclear quantitation. The fluorescence image was captured by confocal microscopy (LSM880; Zeiss, Oberkochen, Germany). The quantification of microglia or SCs was performed as previously described [51]. Six consecutive brain sections containing graft areas were chosen, and the number of microglia in the whole graft was calculated for each brain section. The size of the grafted areas was 850.2 μm^2. The total number of microglia and activated microglia was divided by 850.2 μm^2 and expressed as the number per square millimeter.

2.9. Statistical Analysis

Data were presented as mean ± standard deviation. PET image SURs, behavioral test results, OD ratios and cell counts in the grafted areas derived through IHC staining were compared between groups. ANOVA with Bonferroni post-test was used for multiple comparisons, and Student's *t*-test was used to compare the results of two independent groups. Values of $p < 0.05$ were considered to indicate statistical significance.

3. Results

3.1. SCs Enhanced the Effect of VM Allotransplantation on DA Functional Recovery in Hemiparkinsonian Rats

The apomorphine-induced rotation test was used to evaluate DA function in hemiparkinsonian rats before and after transplantation (Figure 3). An increase in net rotation was noted in hemiparkinsonian rats, whereas no behavioral improvement was observed in the sham (n = 3) and SCs groups (n = 6). The hemiparkinsonian rats that received VM tissue from both rats (allotransplantation, rVM) (n = 6) and pigs (xenotransplantation, pVM) (n = 6) exhibited significant DA recovery. Co-transplantation of SCs in allo- and xenotransplantation of VM tissue led to improved DA function in the rats (n = 6/group). Moreover, rats that received rVM+SCs exhibited less rotations compared to rVM-grafted rats, indicating SCs as a co-graft material could enhance the effect of VM allotransplantation. However, the same effect was not observed with SCs in xenotransplantation.

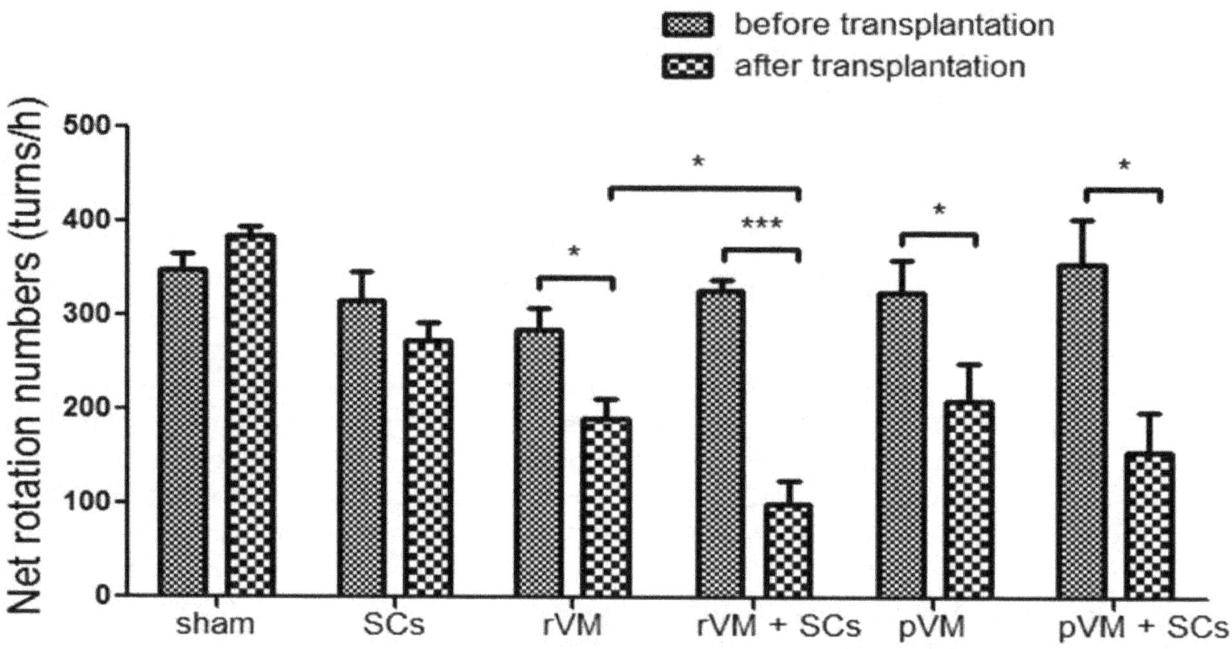

Figure 3. The apomorphine-induced rotational behavior test was used to evaluate dopaminergic (DA) function of hemiparkinsonian rats that received different neural grafts. The test was performed before and after transplantation. * $p < 0.05$; *** $p < 0.001$.

3.2. SCs Enhanced the Functional Result of DA Allografts and Xenografts in the Striatum of Hemiparkinsonian Rats as Determined by [^{18}F]DOPA Coupled with Small Animal PET

The functional results of DA allo- and xeno-transplantation in hemiparkinsonian rats were evaluated by [^{18}F] DOPA coupled with small animal PET. The PET images were acquired before and after creation of the 6-OHDA lesion and after transplantation (Figure 4a). A 6-OHDA lesion was created in the right median forebrain bundle of male SD rats to establish hemiparkinsonian rats, whose striatal [^{18}F] DOPA uptake level on the lesioned side (left side of the brain sections, Figure 4a) was decreased (middle column of each group, Figure 4a). Weak recovery of uptake was found in the rVM (n = 6), pVM (n = 6), SCs+rVM (n = 6), and SCs+pVM groups (n = 6) (right column of each group, Figure 4a). To measure the effects of transplantation and 6-OHDA lesion, the SUR of [^{18}F] DOPA on the lesioned side of striatum was used for further analysis (Figure 4b). After 6-OHDA injection, the SUR of [^{18}F] DOPA of the lesioned striatum dropped 85–95% from the baseline value. In sham (n = 3) and SC group (n = 6), there was no difference is the SUR of [^{18}F] DOPA, suggesting SCs alone do not induce recovery of DA cells. However, the uptake of [^{18}F] DOPA significantly increased in the rat (allograft) and porcine (xenograft) VM tissue transplantation groups. SCs co-grafted with rat or porcine VM tissue led greater uptake of [^{18}F] DOPA than when SCs were not co-grafted (both, $p < 0.05$). These results suggest that SCs can enhance the outcome intra-striatal allo- and xeno-transplantation of VM tissue grafts in hemiparkinsonian rats.

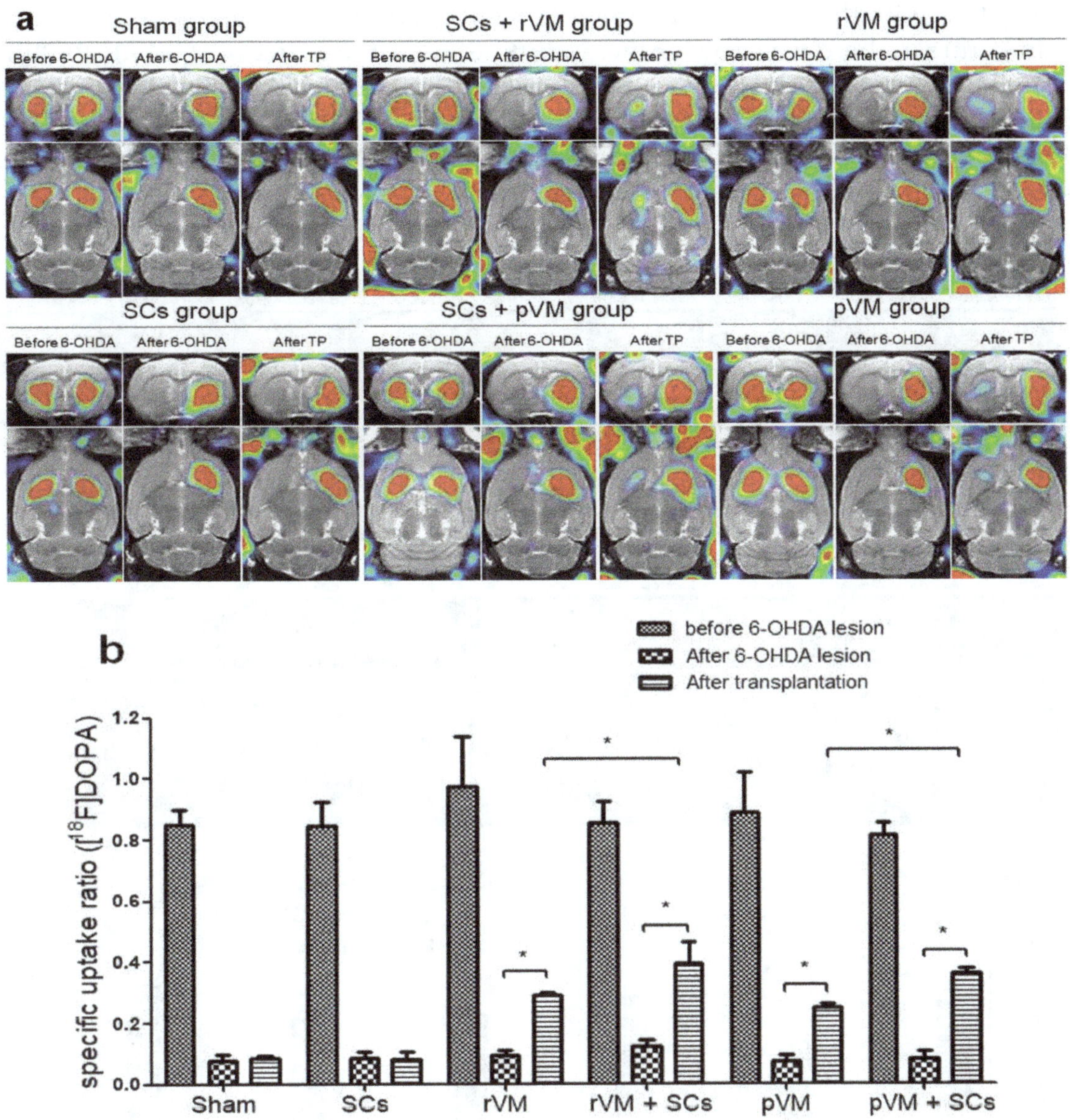

Figure 4. PET images of ^{18}F-DOPA uptake distribution in rat brains. (**a**) Coronal (upper panel) and horizontal (lower panel) sections of each group were acquired 50–80 min after injection of ^{18}F-DOPA. Left, middle, and right columns of each group represent the ^{18}F-DOPA uptake before the 6-OHDA lesion, after the 6-OHDA lesion, and after transplantation with different grafting materials, respectively. (**b**) Specific uptake ratios (SURs) of ^{18}F-DOPA of the grafted striatum at different time points and under different grafting conditions (* $p < 0.05$).

3.3. SCs Enhance Maturation of DA Allografts and Xenografts in the Striatum of Hemiparkinsonian Rats as Determined by [^{18}F]FE-PE2I Coupled with Small Animal PET

[^{18}F] FE-PE2I coupled with small animal PET is a good method for the in vivo measurement of dopamine transporter expression and was used to evaluate the maturation of the DA neural grafts in the rat striatum [52]. Similar to the [^{18}F]DOPA imaging results, striatal [^{18}F] FE-PE2I uptake levels on the lesioned side were decreased after the lesion was created (middle column of each group, Figure 5a), and a slight recovery of the uptake was observed in the rVM (n = 6), pVM (n = 6), SCs+rVM (n = 6), and SCs+pVM groups (n = 6) after transplantation (right column of each group, Figure 5a). Results of the quantitative analysis of the SURs are shown in Figure 5b. The SURs of [^{18}F] FE-PE2I were decreased after intracerebral injection of 6-OHDA in all groups. Transplantation of HBSS (n = 3) or SCs (n = 6) in the lesioned striatum did not induce a significant change in [^{18}F] FE-PE2I uptake. This result is consistent with the [^{18}F] DOPA imaging results, i.e., SCs are unable to induce recovery of DA cells. In contrast, there was no significant difference in the SURs of [^{18}F] FE-PE2I in the rVM group and pVM group after transplantation. In addition, co-grafts of SCs with VM tissue in allo- and xeno-transplantation groups resulted in significant uptake of [^{18}F] FE-PE2I in the grafted striatum (Figure 5a). This result suggests that SCs co-grafted with VM tissue can enhance maturation of transplanted DA cells in allo- and xeno-transplantation.

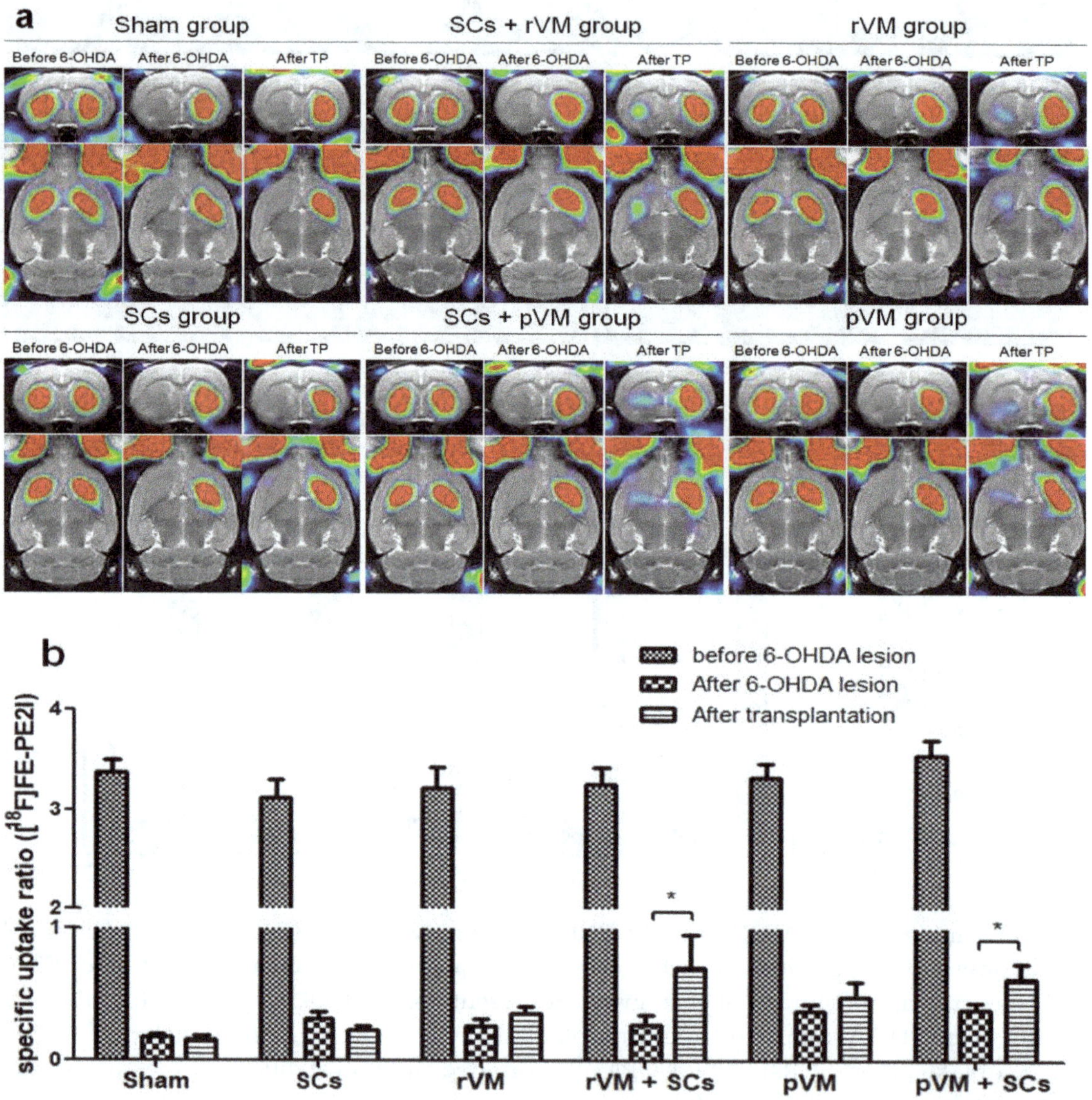

Figure 5. PET images of [^{18}F] FE-PE2I uptake distribution in rat brains. (**a**) Coronal (upper panel) and horizontal (lower panel) sections of each group were acquired 20–40 min after injection of [^{18}F]FE-PE2I. Left, middle, and right columns of each group represent [^{18}F]FE-PE2I uptake before the 6-OHDA lesion, after the 6-OHDA lesion, and after transplantation with different grafting materials, respectively. (**b**) SURs of [^{18}F] FE-PE2I in the grafted striatum at different time points and under different grafting conditions (* $p < 0.05$).

3.4. SCs Enhance the Survival of DA Cells in Allo- and Xeno-Transplantation

To determine whether the higher SURs of [^{18}F] DOPA and [^{18}F] FE-PE2I were due to higher DA cell survival in the grafted striatum, we performed TH staining of the grafted striatum. In photomicrographs of brain sections of the sham (n = 3) and SCs (n = 6) groups, TH-ir (DA) cells or fibers in the grafted striatum (right side of the brain sections, Figure 6a,b) were hardly visible. In contrast,

DA cell bodies and fibers were visible in the grafted striatum of the rVM (n = 6) (Figure 6c), rVM+SCs (Figure 6d) (n = 6), pVM (n = 6) (Figure 6e), and pVM+SCs (n = 6) (Figure 6f) transplantation groups.

We then quantified these IHC results with two different methods. First, we counted the number of TH-ir cell bodies in the grafted striatum (Figure 6g). The DA cell body densities of the rVM, pVM, rVM+SCs, and pVM+SCs groups were 1341 ± 269/mm^3, 1050 ± 187/mm^3, 2130 ± 385/mm^3, and 1598 ± 398/mm^3, respectively. Using the sham group (22 ± 2/mm^3) as the standard, we found that all four groups contained abundant DA cell bodies over the grafted striatum after transplantation. Moreover, the co-grafted groups (rVM+SCs and pVM+SCs) had higher DA cell body densities in the grafted striatum than the single grafted groups (rVM or pVM). Second, we used OD ratio to evaluate the density of the DA cells (cell bodies + fibers) in the grafted striatum (Figure 6h). Analysis of the OD ratios was consistent with the cell body density results; the rVM+SCs and pVM+SCs groups had significantly higher DA cell density than the rVM and pVM group. These results suggested that SCs co-grafted with VM tissue enhanced the survival and neuritis growth of DA cell grafts. Furthermore, we found that grafted SCs coexisted with DA cells in the brains of the rVM+SCs and pVM+SCs groups at the fourth week after the transplantation (Figure 7). This observation indicated that SCs themselves can survive in the host brain and might be able to provide long-term beneficial effects to the grafts.

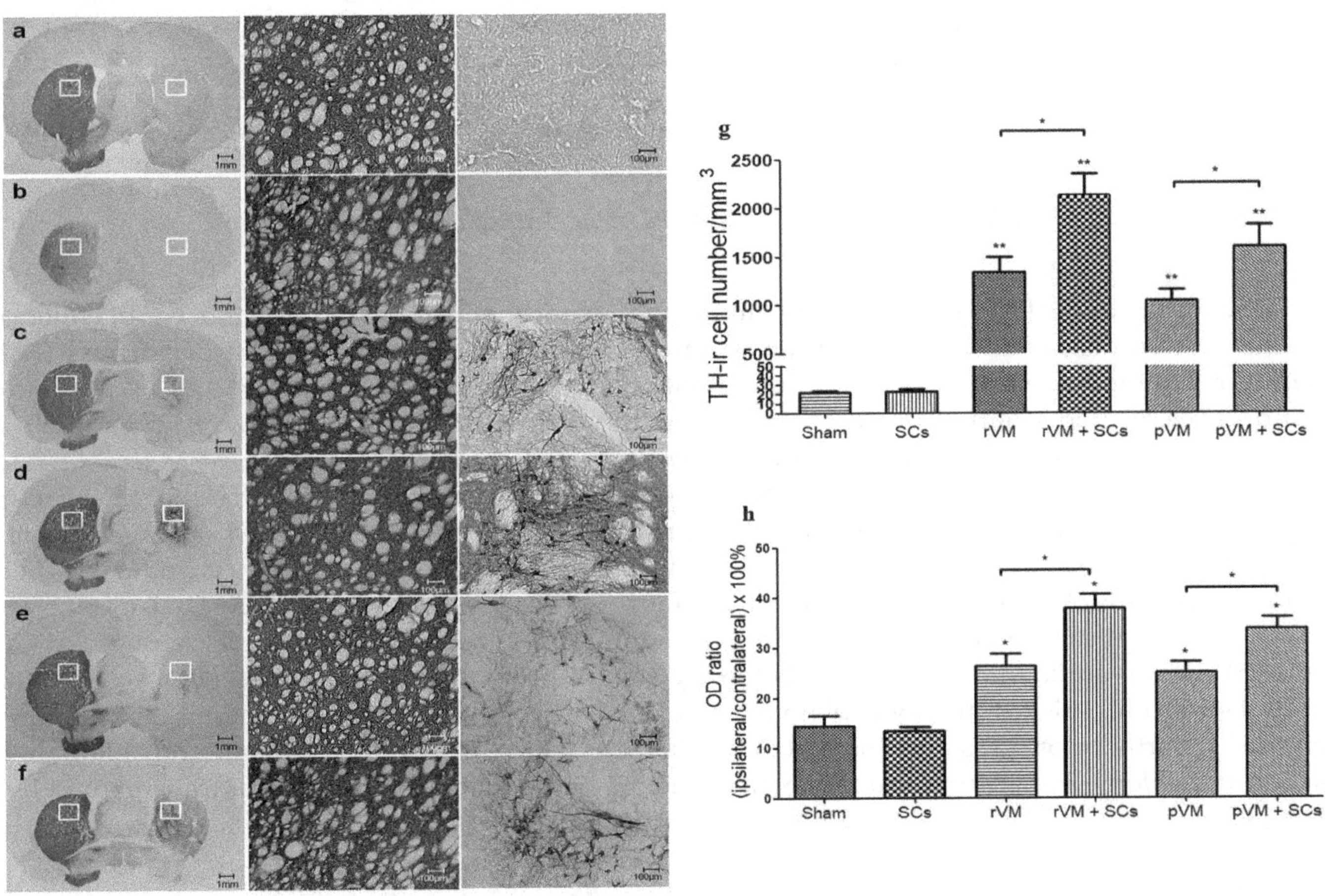

Figure 6. Photomicrographs of tyrosine hydroxylase immunoreactive (TH-ir) cells, and quantification results in hemiparkinsonian rat striatum at four weeks following transplantation. TH-ir cell bodies and fibers in the grafted side (right side of the coronal brain sections) of the rat ventral mesencephalic (rVM), rVM+SCs, pig ventral mesencephalic (pVM), and pVM+SCs groups. Greater densities of TH-ir cell bodies and fibers were noted in the rVM+SCs and pVM+SCs groups as compared to the rVM and pVM groups. Quantification data were consistent with the IHC results. (**a**) Sham group. (**b**) SC group. (**c**) rVM group. (**d**) rVM+SCs group. (**e**) pVM group. (**f**) pVM+SCs group. (**g**) Quantification of TH-ir cell density. (**h**) Optical density (OD) ratio in grafted striatum (* $p < 0.05$; ** $p < 0.01$).

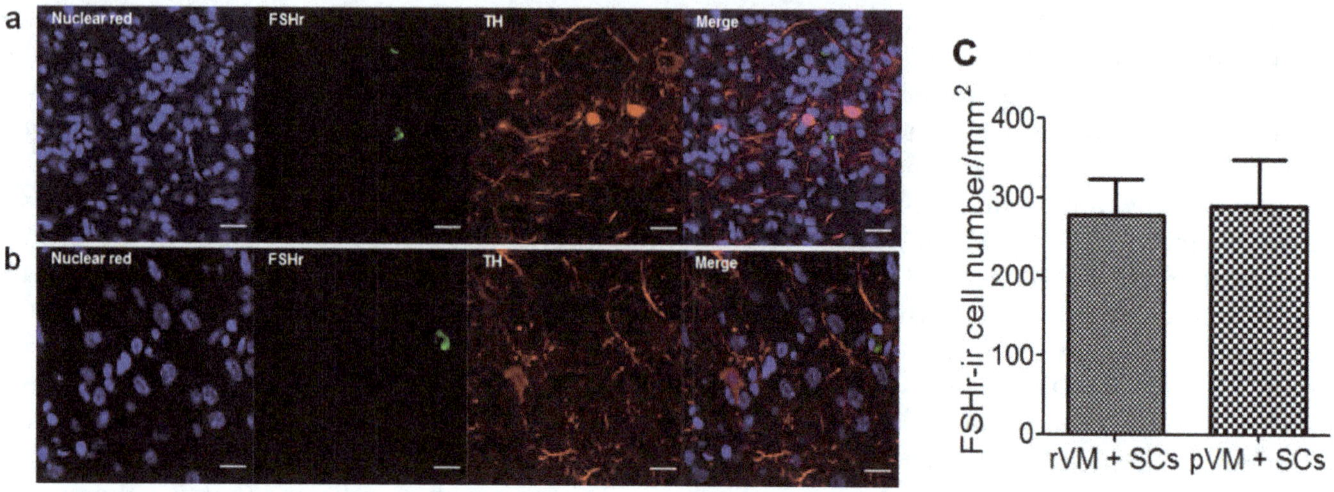

Figure 7. SCs and DA cells co-existed in the grafted striatum of the rVM+SCs and pVM+SCs groups four weeks after transplantation. Brain sections were stained with a dopaminergic neuron marker (TH) and SCs marker (FSHr). (**a**) rVM+SCs group. (**b**) pVM+SCs group. (**c**) Quantification data of FSHr-ir cell numbers. Scale bar = 20 μm.

3.5. SCs Enhance DA Cell Maturation in Both Allo- and Xeno-Transplantation

The results of the [^{18}F] FE-PE2I small animal PET experiment suggested that SCs could enhance the maturation of grafted DA cells in the striatum. To confirm this finding, we performed DAT immunostaining of the brain sections. Consistent with the PET results, very few DAT-ir cells were found in the grafted striatum of the sham (n = 3) and SCs groups (n = 6) (Figure 8a,b,g). DAT-ir cells were observed in the grafted striatum of the rVM (n = 6), rVM+SCs (n = 6), pVM (n = 6), and pVM+SCs groups (n = 6) (Figure 8c–f). Quantification of the IHC results showed that the sham group only contained 12 ± 2/mm^3 DAT-ir cells in the grafted striatum (Figure 8g). The rVM and pVM groups contained 331 ± 81/mm^3 and 279 ± 44/mm^3 DAT-ir cells, respectively. The rVM+SCs and pVM+SCs groups contained 1392 ± 145/mm^3 and 1122 ± 144/mm^3 DAT-ir cells, respectively. The DAT-ir cell body density in grafted striatum of the four groups was significantly higher than that of the sham group (all, $p < 0.05$). The co-grafts were consistently better than single grafts in generating mature DA cells in the grafted striatum. This result may explain the results of the [^{18}F] FE-PE2I small animal PET imaging (Figure 5). Furthermore, it is possible that SCs, when co-grafted with VM tissues (allotransplantation or xenotransplantation), can produce beneficial factors that enhance maturation of transplanted DA cells [35].

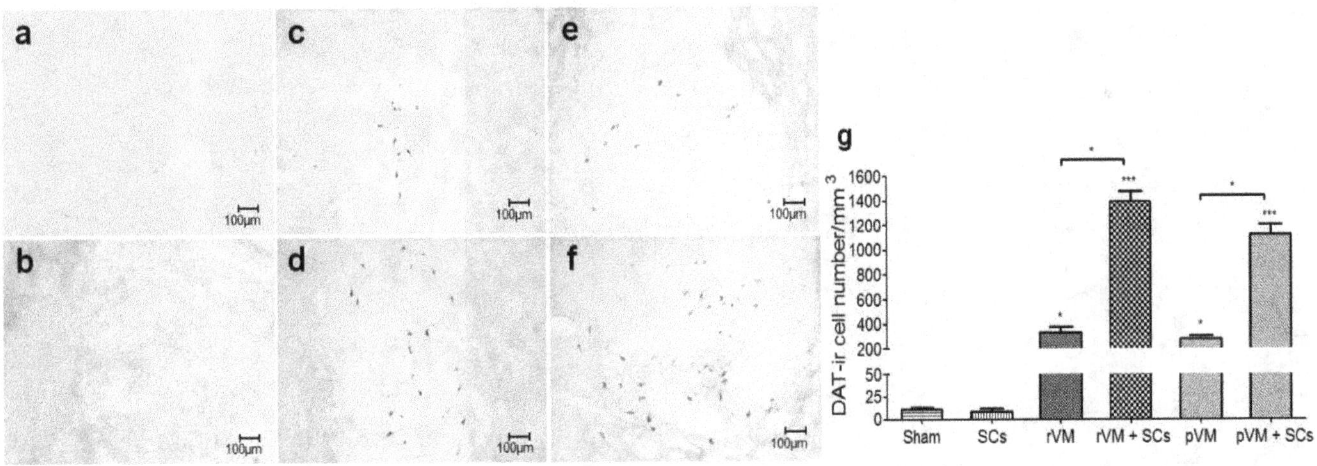

Figure 8. Photomicrographs of DAT-ir and quantification results at four weeks after transplantation. DAT-ir cell bodies in the grafted sides of the rVM, rVM+SCs, pVM, and pVM+SCs groups. The densities of DAT-ir cell bodies in the rVM+SCs and pVM+SCs groups were higher than those of the rVM and pVM groups. Quantification data were consistent with the IHC results. (**a**) Sham group. (**b**) SC group. (**c**) rVM group. (**d**) rVM+SCs group. (**e**) pVM group. (**f**) pVM+SCs group. (**g**) Quantification of DAT-ir cell density. (**h**) OD ratio in grafted striatum (* $p < 0.05$; *** $p < 0.01$). Scale bar = 100 μm.

3.6. SCs Attenuate the Immune Response of Microglia in Grafted Striatum after Allo- and Xeno-Transplantation

The enhancement effect of SCs on DA cell survival is likely due to immune-modulatory properties of SCs. We evaluated the immune response in grafted striatum via immunostaining brain slices with Iba1 (total microglia) and OX6 (activated microglia). The microglia cell numbers were higher after transplantation in all groups, except the sham group (n = 3) and the SCs group (n = 6) (Figure 9c–f). For quantification, we calculated the density of total microglia and activated microglia per mm^2 in the grafted striatum and contralateral striatum (Figure 9g). In the contralateral striatum, the density of total microglia and activated microglia were approximately 150/mm^2 and 5/mm^2, respectively. In hemiparkinsonian rats that received rVM (n = 6) or pVM tissue (n = 6), the total microglia in the grafted striatum were 967 ± 467/mm^2 and 1171 ± 650/mm^2, and the activated microglia were 630 ± 277/mm^2 and 772 ± 379/mm^2, respectively. In rats that received rVM+SCs (n = 6) or pVM+SCs (n = 6), the total microglia in the grafted striatum were 339 ± 71/mm^2 and 456 ± 88/mm^2, and the activated microglia were 107 ± 21/mm^2 and 219 ± 73/mm^2, respectively. Moreover, rats that received rVM and pVM co-grafted with SCs had a lower OX6+/Iba1+ ratio than rats in the rVM group and pVM group (Figure 9h). These results indicated that co-grafting with SCs can provide immune-modulatory effects via attenuating activated microglia in allo- and xeno-transplantation.

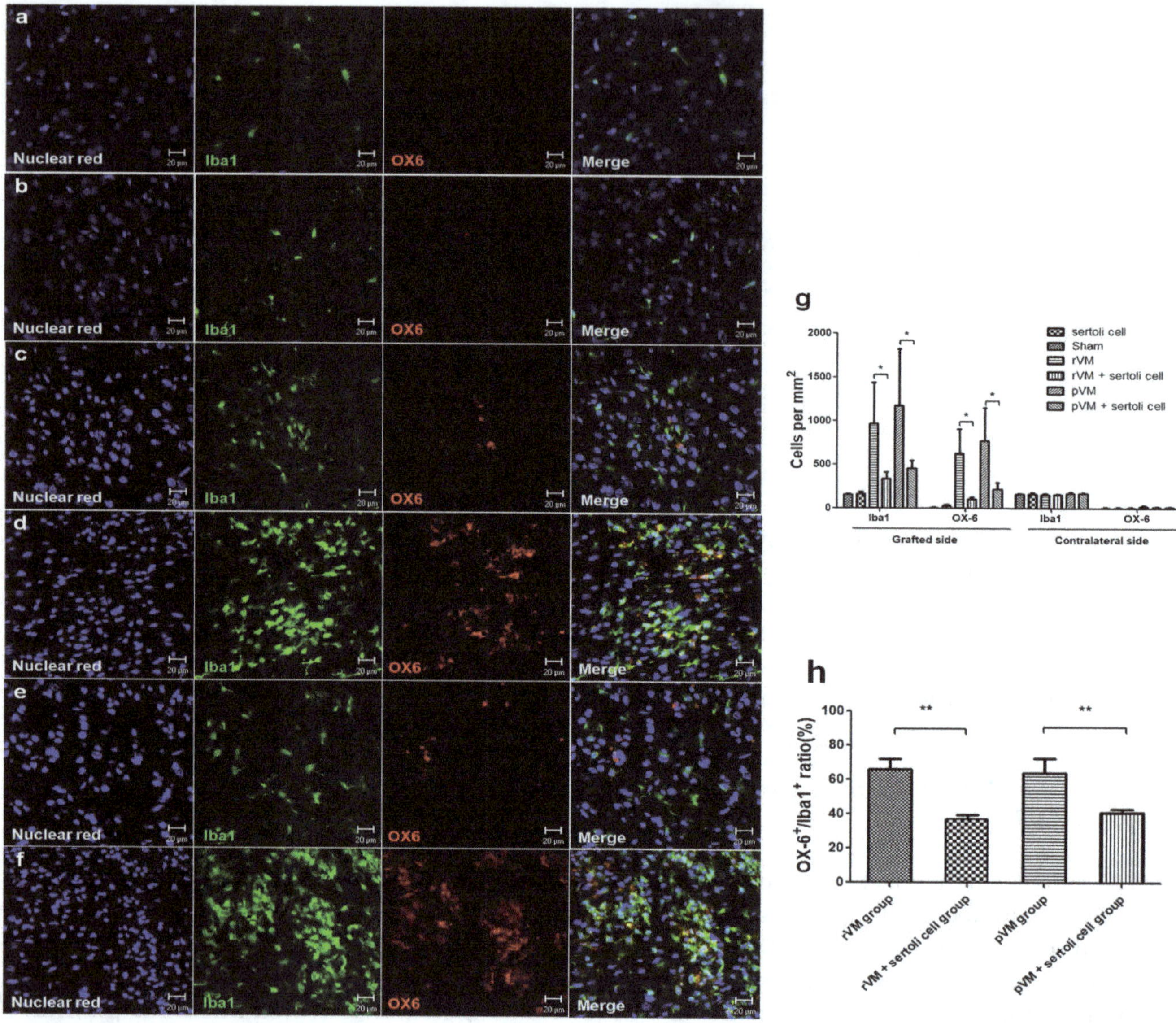

Figure 9. Immunofluorescence staining with anti-Iba1 and anti-OX6 antibody and quantification results at four weeks after transplantation. A severe immune response in grafted striatum was noted in the rVM group and pVM group. Co-grafted groups showed lower microglia cell numbers compared with the rVM and pVM groups. Quantification results showed that Iba1-ir and OX6-ir were obviously increased in the rVM and pVM groups. The OX-6+/Iba1+ ratio was lower in the rVM+SCs and pVM+SCs groups than in the rVM and pVM groups. (**a**) Sham group. (**b**) SCs group. (**c**) rVM group. (**d**) rVM+SCs group. (**e**) pVM group. (**f**) pVM+SCs group. (**g**) Quantification of Iba1-ir and OX-6-ir cell numbers. (**h**) OX-6+/Iba1+ ratio. Scale bar = 20 μm.

3.7. SCs Attenuate the T-Cell Infiltration Immune Response of after Allo- and Xeno-Transplantation

To investigate whether the immune-modulatory effect of SCs was via attenuating T-cell infiltration in the grafted striatum, we calculated the density of T-cells using the CD3 biomarker. The CD3-ir cell number was counted in the grafted striatum (Figure 10a–f) and the contralateral striatum (Figure 10g–l). The results indicated the T-cell number was higher in the rVM (n = 6), pVM (n = 6), and pVM+SCs groups (n = 6). For quantification, the T-cell densities of the contralateral side were used as the standard. The T-cell density in contralateral striatum was approximately 200/mm^3 in all of the experimental groups, and was similar to that of the sham group (207 ± 26/mm^3). The T-cell densities were 809 ± 133/mm^3, 322 ± 39/mm^3, 1707 ± 445/mm^3, and 446 ± 38/mm^3 in the rVM, rVM+SCs (n = 6), pVM, and pVM+SCs group, respectively. The density was markedly increased in the allotransplantation (rVM) group and the xenotransplantation (pVM) group, compared with the density of the contralateral striatum. These results showed that SCs co-grafted with rVM and pVM tissues significantly attenuated the T-cell number in grafted striatum compared with the rVM and pVM groups. This suggested that SCs co-grafted with rVM or pVM tissues provide an immune-modulatory effect that markedly reduced T-cell infiltration in the grafted striatum.

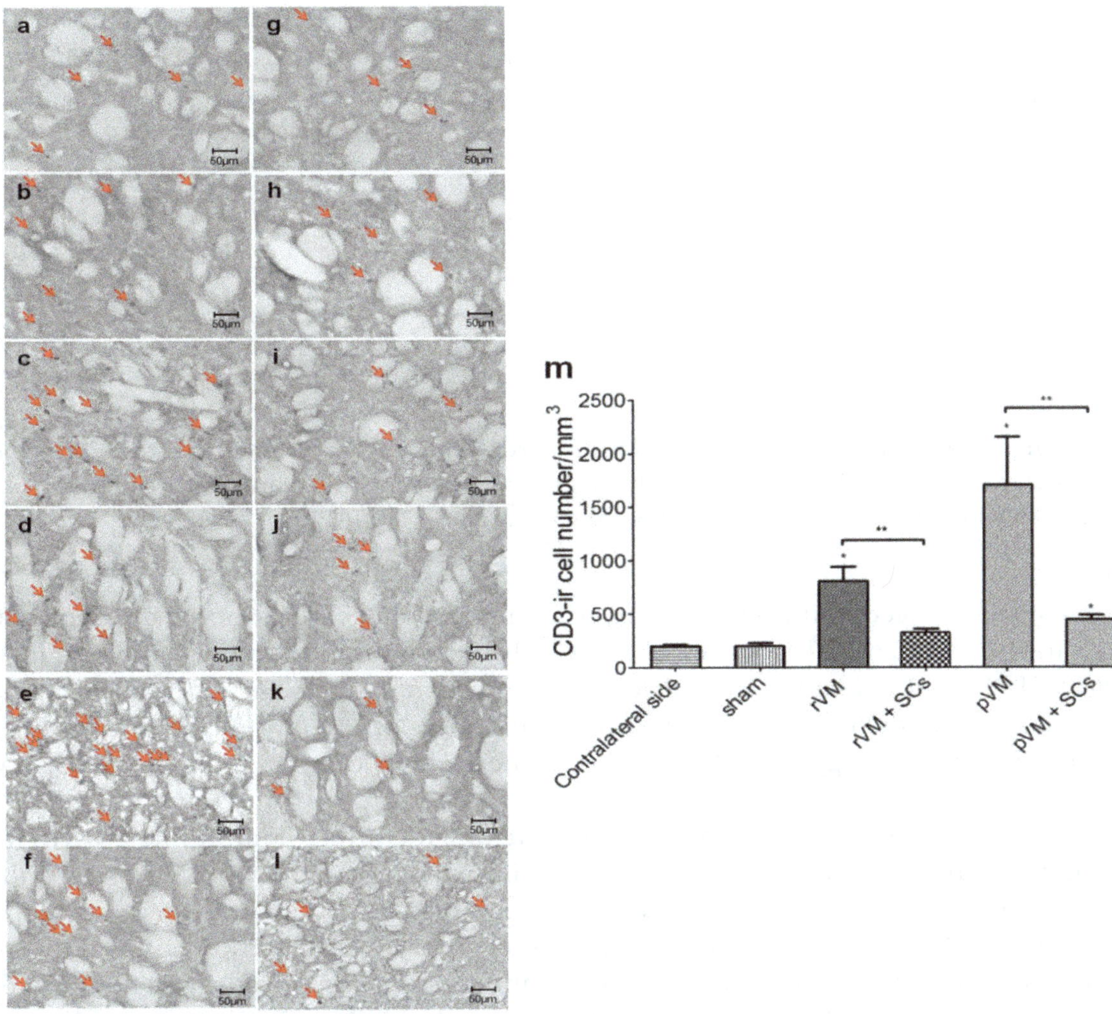

Figure 10. Grafted striatum and contralateral striatum stained with anti-CD3 antibody. Photomicrographs of grafted striatum: (**a**) Sham group, (**b**) SCs group, (**c**) rVM group, (**d**) rVM+SCs group, (**e**) pVM group and (**f**) pVM+SCs group. Photomicrographs of contralateral striatum: (**g**) Sham group, (**h**) SCs group, (**i**) rVM group, (**j**) rVM+SCs group, (**k**) pVM group and (**l**) pVM+SCs group. The number of CD3-ir cells in the rVM group and pVM group were greater than in the rVM+SCs group and pVM+SCs. (**m**) Quantification of CD3-positive cells in striatal grafts at four weeks after transplantation. The number of CD3-ir cells were increased in the grafted striatum of the rVM, pVM, and pVM+SCs groups at four weeks after transplantation (** $p < 0.01$, the grafted side compared to each contralateral side). The number of CD3-ir cell in the grafted striatum of the rVM+SCs and pVM+SCs groups was markedly less than in the rVM group and pVM group (* $p < 0.05$; ** $p < 0.01$). Scale bar = 50 μm.

3.8. The SURs of [^{18}F]DOPA and [^{18}F]FE-PE2I Were Highly Correlated

Pearson's correlation was used to investigate the relations between the SURs of [^{18}F]DOPA and [^{18}F]FE-PE2I in the grafted striatum of the rVM (n = 3), rVM+SCs (n = 4), pVM (n = 4), and pVM+SCs (n = 4) groups (Figure 11). The SURs of each group at all time points were included in the analysis. The Pearson's R square results were rVM group, R square = 0.9679; rVM+SCs group, R square = 0.9449;

pVM group, R square = 0.9626; pVM+SCs group, R square = 0.9280. These results indicated the SURs of [^{18}F] DOPA and [^{18}F] FE-PE2I were highly correlated in all of the animal groups tested in this study.

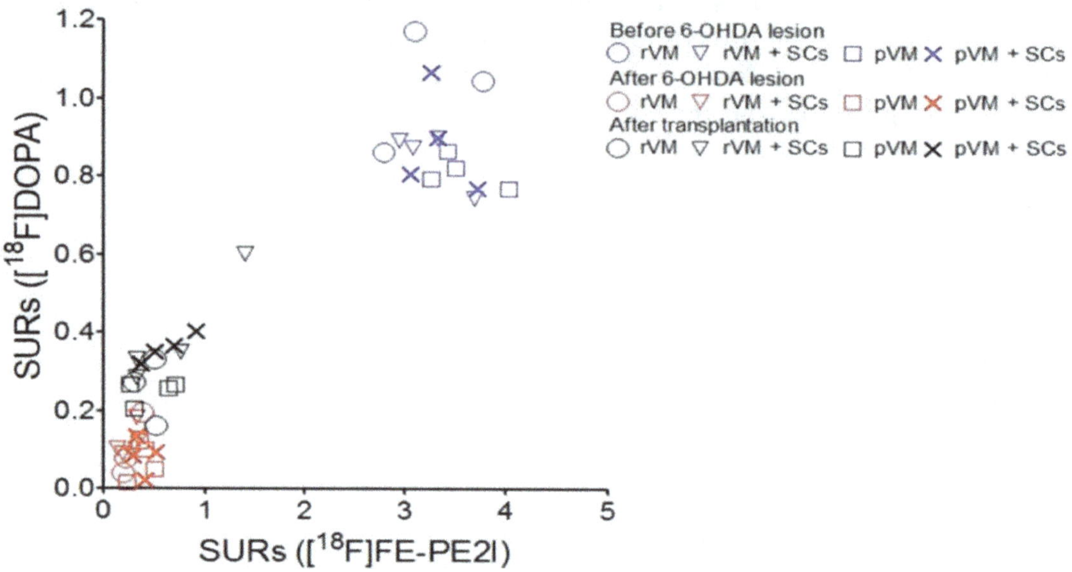

Figure 11. Pearson's correlation analysis of ^{18}F-DOPA and [^{18}F]FE-PE2I in the grafted striatum of the rVM, rVM+SCs, pVM, and pVM+SCs groups. In all groups, Pearson's R square was greater than 0.9, indicating a high correlation between the SUR of ^{18}F-DOPA and [^{18}F]FE-PE2I.

4. Discussion

The present study showed that transplanting SCs with rat (allograft) or porcine (xenograft) VM tissues resulted in better functional recovery of drug-induced rotational behavior and better striatal uptake recovery of [^{18}F] FE-PE2I and [^{18}F] DOPA in hemiparkinsonian rats. In addition, a co-graft SCs and VM tissues (rat or porcine) led to an increase in the numbers of TH-ir and DAT-ir neurons in the grafted striatum. These results suggest that SCs enhance the survival of grafted dopaminergic neurons in vivo.

SCs have been shown to secret numerous trophic factors, including insulin-like growth factor-I (IGF-I), basic fibroblast growth factor (bFGF), and glial cell line-derived neurotrophic factor (GDNF) [53]. The beneficial effect of SCs observed in this study may be in part due to the release of GDNF, which enhances TH-positive cell survival and nerve fiber formation after transplantation of VM tissues in the striatum [54]. Previous study showed that grafted dopaminergic precursor cells may not differentiate completely after transplantation [54] and may require several neurotrophic factors, such as GDNF and bFGF, to differentiate into mature cells [55]. As mentioned above, SCs may be beneficial for the differentiation of dopaminergic neurons due to the release of GDNF and bFGF.

The transplantation-induced significant recovery in the striatal uptake of [^{18}F] FE-PE2I was only found in the rVM+SCs and pVM+SCs groups, whereas significant recovery in the uptake of [^{18}F] DOPA was found in the rVM+SCs, pVM+SCs, rVM alone, and pVM alone groups. This discrepancy may be because the 2 radioligands target different biomarkers. [^{18}F] DOPA is a dopamine analogue, whereas [^{18}F] FE-PE2I is a DAT imaging agent [56–58]. In the rat brain, DAT is regarded as a biomarkers of DA neuron maturation, whereas dopamine synthesis may be present as early as embryonic day 12.5 [59].

The numbers of OX6-positive and CD3-positve cells were increased in lesioned striatum of the hemiparkinsonian rats. Grafting rat or porcine VM tissue with SCs led to a significant decrease in the number of OX6- and CD3-positive cells, whereas this phenomenon was absent in VM alone grafted group. These findings suggested that SCs exert immune-modulatory effects in the grafted striatum of hemiparkinsonian rats. SCs are known to secrete several immune-stimulators that generate an immune-privileged microenvironment [29]. These cells modulate the T-cell response through transforming growth factor B (TGFB) and consequently influence Th1 and Th2 responses [29]. SCs have been reported to modulate the expression of pro-inflammatory cytokines, including IL-1 and IL-6, and to inhibit macrophage migration via Activin A [60].

Mesenchymal Stem cells (MSCs) is another promising cellular material for PD therapy. Similar to SCs, MSCs could secret neurotrophic factors, such as nerve growth factor (NGF), brain-derived neurotrophic factor (BDNF), and GDNF [61,62]. These factors could slow down degenerative progression of dopaminergic neurons and induce proliferation of neural stem cells [63]. MSCs have been grafted bilaterally into striatum of early stage PD patients, leading to improvement in Unified Parkinson's Disease Rating Scale (UPDRS) scores of the patients. [64]. These neuroprotective effects of MSCs may result from their anti-inflammatory properties. The effects of MSCs on lipopolysaccharides (LPS)-induced microglial activation have been evaluated in an in vitro study, indicating that MSCs could inhibit activation of microglia, reduce production of TNF-α and inducible nitric oxide synthase (iNOS), and increase production of anti-inflammatory cytokine IL-10 and transforming growth factor β (TGF-β) [65]. These results suggest that MSCs might be used as a co-graft with the VM tissue in the cell therapy for late-stage PD patients.

5. Conclusions

The use of SCs as a co-graft material improves survival rates of VM allografts and xenografts in a parkinsonian rat model. This method circumvents the toxicity of immunosuppressants that may be used for inflammation-associated rejection and thus represents an advantageous approach. We also confirmed that [^{18}F] DOPA and [^{18}F] FE-PE2I coupled with PET is capable of monitoring the functional integrity of grafted DA cells. The co-graft of SCs and non-human VM tissues might provide a new therapeutic approach for the treatment of human PD.

Author Contributions: Conceptualization, Y.-T.J., K.-H.M., and C.-Y.C.; methodology, C.-H.C. and C.H.; software, C.-H.C. and T.-K.C.; validation, Y.-T.J., C.-F.F.C., and Y.-W.L.; formal analysis, Y.-T.J.; investigation, R.-F.Y. and C.-Y.S.; resources, Y.-T.J., S.-C.W., and C.H.; data curation, K.-H.M. and S.-H.C.; writing—original draft preparation, Y.-T.J.; writing—review and editing, C.-F.F.C., K.-H.M., and C.-Y.C.; visualization, Y.-T.J.; supervision, K.-H.M. and C.-Y.C.; project administration, Y.-T.J.; funding acquisition, K.-H.M.

Acknowledgments: The authors acknowledge the technical services provided by the Instrument Center of the National Defense Medical Center.

References

1. Dauer, W.; Przedborski, S. Parkinson's disease: Mechanisms and models. *Neuron* **2003**, *39*, 889–909. [CrossRef]
2. Kringelbach, M.L.; Jenkinson, N.; Owen, S.L.; Aziz, T.Z. Translational principles of deep brain stimulation. *Nat. Rev. Neurosci.* **2007**, *8*, 623–635. [CrossRef] [PubMed]
3. Jenner, P. Molecular mechanisms of L-DOPA-induced dyskinesia. *Nat. Rev. Neurosci.* **2008**, *9*, 665–677. [CrossRef] [PubMed]
4. Bejjani, B.; Houeto, J.; Hariz, M.; Yelnik, J.; Mesnage, V.; Bonnet, A.; Pidoux, B.; Dormont, D.; Cornu, P.; Agid, Y. Aggressive behavior induced by intraoperative stimulation in the triangle of Sano. *Neurology* **2002**, *59*, 1425–1427. [CrossRef] [PubMed]
5. Barker, R.A.; Drouin-Ouellet, J.; Parmar, M. Cell-based therapies for Parkinson disease-past insights and future potential. *Nat. Rev. Neurol.* **2015**, *11*, 492–503. [CrossRef] [PubMed]
6. Brundin, P.; Barker, R.A.; Parmar, M. Neural grafting in Parkinson's disease: Problems and possibilities. *Prog. Brain Res.* **2010**, *184*, 265–294. [PubMed]
7. Kefalopoulou, Z.; Politis, M.; Piccini, P.; Mencacci, N.; Bhatia, K.; Jahanshahi, M.; Widner, H.; Rehncrona, S.; Brundin, P.; Björklund, A.; et al. Long-term clinical outcome of fetal cell transplantation for Parkinson disease: Two case reports. *JAMA Neurol.* **2014**, *71*, 83–87. [CrossRef] [PubMed]

8. Ambasudhan, R.; Dolatabadi, N.; Nutter, A.; Masliah, E.; McKercher, S.R.; Lipton, S.A. Potential for cell therapy in Parkinson's disease using genetically programmed human embryonic stem cell-derived neural progenitor cells. *J. Comp. Neurol.* **2014**, *522*, 2845–2856. [CrossRef] [PubMed]
9. Brederlau, A.; Correia, A.S.; Anisimov, S.V.; Elmi, M.; Paul, G.; Roybon, L.; Morizane, A.; Bergquist, F.; Riebe, I.; Nannmark, U.; et al. Transplantation of Human Embryonic Stem Cell-Derived Cells to a Rat Model of Parkinson's Disease: Effect of In Vitro Differentiation on Graft Survival and Teratoma Formation. *Stem Cells* **2006**, *24*, 1433–1440. [CrossRef] [PubMed]
10. Kikuchi, T.; Morizane, A.; Doi, D.; Magotani, H.; Onoe, H.; Hayashi, T.; Mizuma, H.; Takara, S.; Takahashi, R.; Inoue, H.; et al. Human iPS cell-derived dopaminergic neurons function in a primate Parkinson's disease model. *Nature* **2017**, *548*, 592–596. [CrossRef] [PubMed]
11. Park, C.H.; Minn, Y.K.; Lee, J.Y.; Choi, D.H.; Chang, M.Y.; Shim, J.W.; Ko, J.Y.; Koh, H.C.; Kang, M.J.; Kang, J.S.; et al. In vitro and in vivo analyses of human embryonic stem cell-derived dopamine neurons. *J. Neurochem.* **2005**, *92*, 1265–1276. [CrossRef] [PubMed]
12. de Melo-Martín, I.; Hellmers, N.; Henchcliffe, C. First-in-human cell transplant trials in Parkinson's disease: The need for an improved informed consent process. *Parkinsonism Relat. Disord.* **2015**, *21*, 829–832. [CrossRef] [PubMed]
13. Brundin, P.; Strecker, R.E.; Lindvall, O.; Isacson, O.; Nilsson, O.G.; Barbin, G.; Prochiantz, A.; Forni, C.; Nieoullon, A.; Widner, H.; et al. Intracerebral Grafting of Dopamine Neurons. *Ann. N. Y. Acad. Sci.* **1987**, *495*, 473–495. [CrossRef] [PubMed]
14. Henderson, B.T.H.; Clough, C.G.; Hughes, R.C.; Hitchcock, E.R.; Kenny, B.G. Implantation of Human Fetal Ventral Mesencephalon to the Right Caudate Nucleus in Advanced Parkinson's Disease. *Arch. Neurol.* **1991**, *48*, 822–827. [CrossRef] [PubMed]
15. Kordower, J.H.; Freeman, T.B.; Snow, B.J.; Vingerhoets, F.J.; Mufson, E.J.; Sanberg, P.R.; Hauser, R.A.; Smith, D.A.; Nauert, G.M.; Perl, D.P.; et al. Neuropathological Evidence of Graft Survival and Striatal Reinnervation after the Transplantation of Fetal Mesencephalic Tissue in a Patient with Parkinson's Disease. *N. Engl. J. Med.* **1995**, *332*, 1118–1124. [CrossRef] [PubMed]
16. Pogarell, O.; Koch, W.; Gildehaus, F.J.; Kupsch, A.; Lindvall, O.; Oertel, W.H.; Tatsch, K. Long-term assessment of striatal dopamine transporters in parkinsonian patients with intrastriatal embryonic mesencephalic grafts. *Eur. J. Nucl. Med. Mol. Imaging* **2006**, *33*, 407–411. [CrossRef] [PubMed]
17. Politis, M.; Lindvall, O. Clinical application of stem cell therapy in Parkinson's disease. *BMC Med.* **2012**, *10*, 1. [CrossRef] [PubMed]
18. Fink, J.S.; Schumacher, J.M.; Ellias, S.L.; Palmer, E.P.; Saint-Hilaire, M.; Shannon, K.; Penn, R.; Starr, P.; Vanhorne, C.; Kott, H.S.; et al. Porcine xenografts in Parkinson's disease and Huntington's disease patients: Preliminary results. *Cell Transplant.* **2000**, *9*, 273–278. [CrossRef] [PubMed]
19. Cooper, D.K.C. A brief history of cross-species organ transplantation. *Bayl. Univ. Med Cent. Proc.* **2012**, *25*, 49–57. [CrossRef] [PubMed]
20. Parmar, M.; Torper, O.; Drouin-Ouellet, J. Cell-based therapy for Parkinson's disease: A journey through decades toward the light side of the Force. *Eur. J. Neurosci.* **2018**, *49*, 463–471. [CrossRef] [PubMed]
21. Chatterjee, D.; Wakeman, D.R.; Kordower, J.H. Cell Replacement Strategies for Parkinson's Disease. In *Cell Therapy*; Emerich, D.F., Orive, G., Eds.; Humana Press: Totowa, NJ, USA, 2017; pp. 73–83.
22. Busauschina, A.; Schnuelle, P.; van der Woude, F.J. Cyclosporine nephrotoxicity. *Transplant. Proc.* **2004**, *36*, S229–S233. [CrossRef] [PubMed]
23. Schumacher, J.M.; Ellias, S.; Palmer, E.P.; Kott, H.S.; Dinsmore, J.; Dempsey, P.K.; Fischman, A.J.; Thomas, C.; Feldman, R.G.; Kassissieh, S.; et al. Transplantation of embryonic porcine mesencephalic tissue in patients with PD. *Neurology* **2000**, *54*, 1042–1050. [CrossRef] [PubMed]
24. Danovitch, G.M. Immunosuppressant-induced metabolic toxicities. *Transplant. Rev.* **2000**, *14*, 65–81. [CrossRef]
25. Abouljoud, M.S.; Kumar, M.S.A.; Brayman, K.L.; Emre, S.; Bynon, J.S. For the OLN-452 Study Group Neoral rescue therapy in transplant patients with intolerance to tacrolimus. *Clin. Transplant.* **2002**, *16*, 168–172. [CrossRef] [PubMed]
26. Larsson, L.C.; Czech, K.A.; Brundin, P.; Widner, H. Intrastriatal ventral mesencephalic xenografts of porcine tissue in rats: Immune responses and functional effects. *Cell Transplant.* **2000**, *9*, 261–272. [CrossRef] [PubMed]

27. Wenker, S.D.; Leal, M.C.; Farías, M.I.; Zeng, X.; Pitossi, F.J. Cell therapy for Parkinson's disease: Functional role of the host immune response on survival and differentiation of dopaminergic neuroblasts. *Brain Res.* **2016**, *1638*, 15–29. [CrossRef] [PubMed]
28. Larsson, L.C.; Frielingsdorf, H.; Mirza, B.; Sophia, J.; Anderson, P.; Czech, K.A.; Strandberg, M.; Widner, H. Porcine neural xenografts in rats and mice: Donor tissue development and characteristics of rejection. *Exp. Neurol.* **2001**, *172*, 100–114. [CrossRef] [PubMed]
29. Mital, P.; Kaur, G.; Dufour, J.M. Immunoprotective Sertoli cells: Making allogeneic and xenogeneic transplantation feasible. *Reproduction* **2010**, *139*, 495–504. [CrossRef] [PubMed]
30. Loftis, J.M. Sertoli cell therapy: A novel possible treatment strategy for treatment-resistant major depressive disorder. *Med Hypotheses.* **2011**, *77*, 35–42. [CrossRef] [PubMed]
31. Shamekh, R.; El-Badri, N.S.; Saporta, S.; Pascual, C.; Sanberg, P.R.; Cameron, D.F. Sertoli cells induce systemic donor-specific tolerance in xenogenic transplantation model. *Cell Transplant.* **2006**, *15*, 45–53. [CrossRef] [PubMed]
32. Suarez-Pinzon, W.; Korbutt, G.S.; Power, R.; Hooton, J.; Rajotte, R.V.; Rabinovitch, A. Testicular sertoli cells protect islet beta-cells from autoimmune destruction in NOD mice by a transforming growth factor-beta1-dependent mechanism. *Diabetes* **2000**, *49*, 1810–1818. [CrossRef] [PubMed]
33. Rahman, T.M.; Diakanov, I.; Selden, C.; Hodgson, H. Co-transplantation of encapsulated HepG2 and rat Sertoli cells improves outcome in a thioacetamide induced rat model of acute hepatic failure. *Transpl. Int.* **2005**, *18*, 1001–1009. [CrossRef] [PubMed]
34. Lim, H.-G.; Lee, H.-M.; Oh, B.C.; Lee, J.R. Cell-mediated Immunomodulation of Chemokine Receptor 7–expressing Porcine Sertoli Cells in Murine Heterotopic Heart Transplantation. *J. Hear. Lung Transplant.* **2009**, *28*, 72–78. [CrossRef] [PubMed]
35. Shamekh, R.; Mallery, J.; Newcomb, J.; Hushen, J.; Saporta, S.; Cameron, D.F.; Sanberg, C.D.; Sanberg, P.R.; Willing, A.E. Enhancing tyrosine hydroxylase expression and survival of fetal ventral mesencephalon neurons with rat or porcine Sertoli cells in vitro. *Brain Res.* **2006**, *1096*, 1–10. [CrossRef] [PubMed]
36. Shamekh, R.; Newcomb, J.; Mallery, J.; Cassady, C.J.; Saporta, S.; Cameron, D.F.; Sanberg, P.R.; Willing, A.E. Survival of Rat or Mouse Ventral Mesencephalon Neurons after Cotransplantation with Rat Sertoli Cells in the Mouse Striatum. *Cell Transplant.* **2005**, *14*, 551–564. [CrossRef] [PubMed]
37. Deng, L.; Shi, B.; Zhuang, Y.; Chu, J.; Shi, X.; Zhang, S.; Guo, M. Performance and Mechanism of Neuroleukin in the Growth and Survival of Sertoli Cell-Induced Neurons in a Coculture System. *Cell Transplant.* **2014**, *23*, 381–394. [CrossRef] [PubMed]
38. Fazio, P.; Svenningsson, P.; Cselenyi, Z.; Halldin, C.; Farde, L.; Varrone, A. Nigrostriatal dopamine transporter availability in early Parkinson's disease. *Mov. Disord.* **2018**, *33*, 592–599. [CrossRef] [PubMed]
39. Weng, S.J.; Li, I.H.; Huang, Y.S.; Chueh, S.H.; Chou, T.K.; Huang, S.Y.; Shiue, C.Y.; Cheng, C.Y.; Ma, K.H. KA-bridged transplantation of mesencephalic tissue and olfactory ensheathing cells in a Parkinsonian rat model. *J. Tissue Eng. Regen. Med.* **2017**, *11*, 2024–2033. [CrossRef] [PubMed]
40. Weng, S.-J.; Shiue, C.-Y.; Huang, W.-S.; Cheng, C.-Y.; Huang, S.-Y.; Li, I.-H.; Tao, C.-C.; Chou, T.-K.; Liao, M.-H.; Chang, Y.-P.; et al. PET Imaging of Serotonin Transporters with 4-[^{18}F]-ADAM in a Parkinsonian Rat Model. *Cell Transplant.* **2013**, *22*, 1295–1305. [CrossRef] [PubMed]
41. Truong, L.; Allbutt, H.; Kassiou, M.; Henderson, J. Developing a preclinical model of Parkinson's disease: A study of behaviour in rats with graded 6-OHDA lesions. *Behav. Brain Res.* **2006**, *169*, 1–9. [CrossRef] [PubMed]
42. Anway, M.D.; Folmer, J.; Wright, W.W.; Zirkin, B.R. Isolation of sertoli cells from adult rat testes: An approach to ex vivo studies of Sertoli cell function. *Biol. Reprod.* **2003**, *68*, 996–1002. [CrossRef] [PubMed]
43. Takeuchi, Y.; Sawada, T.; Blunt, S.; Jenner, P.; Marsden, C. Effects of 6-hydroxydopamine lesions of the nigrostriatal pathway on striatal serotonin innervation in adult rats. *Brain Res.* **1991**, *562*, 301–305. [CrossRef]
44. Galpern, W.R.; Burns, L.H.; Deacon, T.W.; Dinsmore, J.; Isacson, O. Xenotransplantation of Porcine Fetal Ventral Mesencephalon in a Rat Model of Parkinson's Disease: Functional Recovery and Graft Morphology. *Exp. Neurol.* **1996**, *140*, 1–13. [CrossRef] [PubMed]
45. Dunnett, S.B.; Björklund, A. Dissecting embryonic neural tissues for transplantation. In *Neural Transplantation Methods*; Dunnett, S.B., Boulton, A.A., Baker, G.B., Eds.; Humana Press: Totowa, NJ, USA, 2000; pp. 3–25.

46. Bang, J.-I.; Jung, I.S.; Song, Y.S.; Park, H.S.; Moon, B.S.; Lee, B.C.; Kim, S.E. PET imaging of dopamine transporters with [18 F]FE-PE2I: Effects of anti-Parkinsonian drugs. *Nucl. Med. Biol.* **2016**, *43*, 158–164. [CrossRef] [PubMed]
47. Ma, K.-H.; Huang, W.-S.; Kuo, Y.-Y.; Peng, C.-J.; Liou, N.-H.; Liu, R.-S.; Hwang, J.-J.; Liu, J.-C.; Chen, H.-J.; Shiue, C.-Y. Validation of 4-[^{18}F]-ADAM as a SERT imaging agent using micro-PET and autoradiography. *NeuroImage* **2009**, *45*, 687–693. [CrossRef] [PubMed]
48. Kyono, K.; Takashima, T.; Katayama, Y.; Kawasaki, T.; Zochi, R.; Gouda, M.; Kuwahara, Y.; Takahashi, K.; Wada, Y.; Onoe, H.; et al. Use of [^{18}F] FDOPA-PET for in vivo evaluation of dopaminergic dysfunction in unilaterally 6-OHDA-lesioned rats. *EJNMMI Res.* **2011**, *1*, 25. [CrossRef] [PubMed]
49. Forsback, S.; Niemi, R.; Marjamäki, P.; Eskola, O.; Bergman, J.; Grönroos, T.; Haaparanta, M.; Haapalinna, A.; Rinne, J.; Solin, O. Uptake of 6-[^{18}F]fluoro-L-dopa and [^{18}F]CFT reflect nigral neuronal loss in a rat model of Parkinson's disease. *Synapse* **2004**, *51*, 119–127. [CrossRef] [PubMed]
50. Paxinos, G.; Watson, C. *The Rat Brain in Stereotaxic Coordinates, Hard cover Edition*; Academic Press: Cambridge, MA, USA, 2006.
51. Tay, T.L.; Mai, D.; Dautzenberg, J.; Fernández-Klett, F.; Lin, G.; Sagar; Datta, M.; Drougard, A.; Stempfl, T.; Ardura-Fabregat, A.; et al. A new fate mapping system reveals context-dependent random or clonal expansion of microglia. *Nat. Neurosci.* **2017**, *20*, 793–803. [CrossRef] [PubMed]
52. Inaji, M.; Yoshizaki, T.; Okauchi, T.; Maeda, J.; Nagai, Y.; Nariai, T.; Ohno, K.; Ando, K.; Okano, H.; Obayashi, S.; et al. In vivo PET measurements with [11C]PE2I to evaluate fetal mesencephalic transplantations to unilateral 6-OHDA-lesioned rats. *Cell Transplant.* **2005**, *14*, 655–663. [CrossRef] [PubMed]
53. Willing, A.E.; Othberg, A.I.; Saporta, S.; Anton, A.; Sinibaldi, S.; Poulos, S.G.; Cameron, D.F.; Freeman, T.B.; Sanberg, P.R. Sertoli cells enhance the survival of co-transplanted dopamine neurons. *Brain Res.* **1999**, *822*, 246–250. [CrossRef]
54. Wakeman, D.R.; Redmond, D.E.; Dodiya, H.B.; Sladek, J.R.; Leranth, C.; Teng, Y.D.; Samulski, R.J.; Snyder, E.Y. Human neural stem cells survive long term in the midbrain of dopamine-depleted monkeys after GDNF overexpression and project neurites toward an appropriate target. *Stem Cells Transl. Med.* **2014**, *3*, 692–701. [CrossRef] [PubMed]
55. Yang, F.; Liu, Y.; Tu, J.; Wan, J.; Zhang, J.; Wu, B.; Chen, S.; Zhou, J.; Mu, Y.; Wang, L. Activated astrocytes enhance the dopaminergic differentiation of stem cells and promote brain repair through bFGF. *Nat. Commun.* **2014**, *5*, 5627. [CrossRef] [PubMed]
56. Garnett, S.; Firnau, G.; Nahmias, C.; Chirakal, R. Striatal dopamine metabolism in living monkeys examined by positron emission tomography. *Brain Res.* **1983**, *280*, 169–171. [CrossRef]
57. Varrone, A.; Steiger, C.; Schou, M.; Takano, A.; Finnema, S.J.; Guilloteau, D.; Gulyás, B.; Halldin, C. In vitro autoradiography and in vivo evaluation in cynomolgus monkey of [^{18}F]FE-PE2I, a new dopamine transporter PET radioligand. *Synapse* **2009**, *63*, 871–880. [CrossRef] [PubMed]
58. Schou, M.; Steiger, C.; Varrone, A.; Guilloteau, D.; Halldin, C. Synthesis, radiolabeling and preliminary in vivo evaluation of [^{18}F]FE-PE2I, a new probe for the dopamine transporter. *Bioorg. Med. Chem. Lett.* **2009**, *19*, 4843–4845. [CrossRef] [PubMed]
59. Fiszman, M.L.; Zuddas, A.; Masana, M.I.; Barker, J.L.; Porzio, U.; Di Porzio, U. Dopamine Synthesis Precedes Dopamine Uptake in Embryonic Rat Mesencephalic Neurons. *J. Neurochem.* **1991**, *56*, 392–399. [CrossRef] [PubMed]
60. Zhao, S.; Zhu, W.; Xue, S.; Han, D. Testicular defense systems: Immune privilege and innate immunity. *Cell. Mol. Immunol.* **2014**, *11*, 428–437. [CrossRef] [PubMed]
61. García, R.; Aguiar, J.; Alberti, E.; de la Cuétara, K.; Pavón, N. Bone marrow stromal cells produce nerve growth factor and glial cell line-derived neurotrophic factors. *Biochem. Biophys. Res. Commun.* **2004**, *316*, 753–754. [CrossRef] [PubMed]
62. Mahmood, A.; Lu, D.; Chopp, M. Intravenous Administration of Marrow Stromal Cells (MSCs) Increases the Expression of Growth Factors in Rat Brain after Traumatic Brain Injury. *J. Neurotrauma* **2004**, *21*, 33–39. [CrossRef] [PubMed]
63. Bonuccelli, U.; Del Dotto, P. New pharmacologic horizons in the treatment of Parkinson disease. *Neurology* **2006**, *67*, S30–S38. [CrossRef] [PubMed]

64. Venkataramana, N.K.; Pal, R.; Rao, S.A.; Naik, A.L.; Jan, M.; Nair, R.; Sanjeev, C.C.; Kamble, R.B.; Murthy, D.P.; Chaitanya, K. Bilateral transplantation of allogenic adult human bone marrow-derived mesenchymal stem cells into the subventricular zone of Parkinson's disease: A pilot clinical study. *Stem Cells Int.* **2012**, *2012*. [CrossRef] [PubMed]
65. Lee, P.H.; Park, H.J. Bone Marrow-Derived Mesenchymal Stem Cell Therapy as a Candidate Disease-Modifying Strategy in Parkinson's Disease and Multiple System Atrophy. *J. Clin. Neurol.* **2009**, *5*, 1–10. [CrossRef] [PubMed]

4

Epigenetic Study in Parkinson's Disease: A Pilot Analysis of DNA Methylation in Candidate Genes in Brain

Luis Navarro-Sánchez [1], Beatriz Águeda-Gómez [1], Silvia Aparicio [1] and Jordi Pérez-Tur [1,2,3,*]

1 Unitat de Genètica Molecular, Instituto de Biomedicina de Valencia, CSIC, 46010 València, Spain; sannalu84@hotmail.com (L.N.-S.); beatrizaguedagomez@gmail.com (B.Á.-G.); silapdo@gmail.com (S.A.)
2 Centro de Investigación Biomédica en Red en Enfermedades Neurodegenerativas (CIBERNED), 46010 València, Spain
3 Unidad Mixta de Genética y Neurología, Instituto de Investigación Sanitaria La Fe, 46026 València, Spain
* Correspondence: jpereztur@ibv.csic.es

Abstract: Efforts have been made to understand the pathophysiology of Parkinson's disease (PD). A significant number of studies have focused on genetics, despite the fact that the described pathogenic mutations have been observed only in around 10% of patients; this observation supports the fact that PD is a multifactorial disorder. Lately, differences in miRNA expression, histone modification, and DNA methylation levels have been described, highlighting the importance of epigenetic factors in PD etiology. Taking all this into consideration, we hypothesized that an alteration in the level of methylation in PD-related genes could be related to disease pathogenesis, possibly due to alterations in gene expression. After analysing promoter regions of five PD-related genes in three brain regions by pyrosequencing, we observed some differences in DNA methylation levels (hypo and hypermethylation) in *substantia nigra* in some CpG dinucleotides that, possibly through an alteration in Sp1 binding, could alter their expression.

Keywords: epigenetics; Parkinson's disease; brain; DNA methylation

1. Introduction

Parkinson's disease (PD) is a progressive neurodegenerative disorder characterized by neural loss in *substantia nigra pars compacta* and the presence of numerous Lewy bodies in surviving neurons [1]. Affected individuals suffer from postural instability, tremor at rest, bradykinesia, and rigidity, as well as other clinical signs such as secondary motor symptoms and non-motor symptoms involving abnormalities in cognition/behavior/sensory system, sleep disorders, and autonomic dysfunctions [2].

Despite being the second most common neurodegenerative disorder, with a prevalence of 4–5% by the age of 85 [3], the mechanism that triggers the neurodegenerative process is still unknown, as is whether the biological processes which are altered in PD (oxidative stress, mitochondrial dysfunction, protein aggregation, inflammation, altered Ca^{2+} homeostasis) are causes or consequences.

Pathogenic mutations have been described only in around 10% of patients, mainly in monogenic or familial forms of PD [4–7]. They are found in 6 genes: *SNCA* [8], *PRKN* [9], *PINK1* [10], *DJ-1* [11], *LRRK2* [12], and *VSP35* [13]. Nevertheless, the remaining 90% of PD patients suffer from idiopathic or sporadic forms of PD with unknown cause. According to this, PD is considered a prototypic multifactorial, or complex, disorder.

Recently, some studies have also focused on epigenetics: mitotically and/or meiotically heritable changes that cannot be explained by changes in DNA sequence, i.e., stable and long-term alterations not present in the DNA sequence [14].

Epigenetics include RNA-mediated processes, histone modifications, and DNA methylation. They are flexible and dynamic processes that change depending on time and environment. Therefore, one person carries one genome, but hundreds or even thousands of epigenomes that could explain different phenotypes [15].

Differences in miRNA expression, histone modifications, and DNA methylation levels have been described in PD patients. With regard to this last epigenetic mechanism, no differences in DNA methylation levels in the brain have been observed between patients and controls in *UCHL1*, *MAPT* promoter, and *PRKN* promoter [16,17] whereas contradictory results have been reported for CpG islands in intron 1 of *SNCA* [18–20]. In PD patients and animal models, it has been demonstrated that α-synuclein could sequester DNMT1, which maintains DNA methylation, in the cytoplasm leading to global DNA hypomethylation [21]. These results highlight the possible influence that DNA methylation could have on PD origin by regulating genes linked to PD. Although more studies are needed to confirm this hypothesis, these reports point out that not only genetic variability should be considered in the future when looking for pathogenic factors in Parkinson's disease.

Singleton et al. first described a triplication in *SNCA* causing familial PD [22], pointing towards the fact that an increase in the expression of this protein may have the same consequences as a pathological mutation. As expression levels can also be regulated by altering DNA methylation in the promoter region, we wanted to analyze the possible relation between epigenetics and PD, more specifically, whether DNA methylation levels were altered in the brain of PD cases, and thus, would affect their expression levels. We focused on 5 extensively studied genes (*SNCA*, *PRKN*, *PINK1*, *DJ-1* and *LRRK2*) which are responsible for familial PD.

2. Material and Methods

2.1. Subjects

Frozen *substantia nigra*, parietal cortex, and occipital cortex were obtained from the Biobanc HCB-IDIBAPS, Barcelona (Catalonia, Spain) for 5 controls and 5 PD patients. All subjects gave their informed consent for inclusion before participating in the study. The study was conducted in accordance with the Declaration of Helsinki, and the protocol was approved by the Institutional Review Board from the Biobanc. Their clinical status was confirmed by *post-mortem* brain analysis. Controls, unlike PD patients, did not present Lewy bodies. However, 4 had other neurological injuries such as vascular encephalopathy and/or AD related pathology; in addition, 3 of them had suffered from vascular dementia.

There were 3 men and 2 women in each group. The mean age at the moment of death for the controls was 77.80 ± 6.80 years, whereas for the PD cases, was 81.00 ± 3.81 years. The mean post-mortem delay was 6 h 2 min for the controls and 10 h 14 min for the PD cases.

2.2. DNA Extraction

We followed the Maxwell 16 Mouse Tail DNA Purification Kit (Promega, Madison, WI, USA) instructions to extract DNA from each individual brain region: around 30 mg of tissue were dissected and then introduced into the Maxwell 16 Instrument (Promega, Madison, WI, USA). Finally, the sample was quantified using the Qubit dsDNA BR Assay kit and the Qubit Fluorometer (Life Technologies, Carlsbad, CA, USA).

For one of the controls, we could not obtain DNA from *substantia nigra*, so for this brain region our results are based on the values of 4 individuals in the control group.

2.3. Bisulfite Treatment

We treated the DNA with bisulfite, as in this process unmethylated C converts into U, whereas 5 mC remains unaltered. It is the gold standard technique to analyze DNA methylation levels. We opted for the EZ DNA Methylation-Gold kit (Zymo Research, Irvine, CA, USA) to treat 1 μg of DNA per individual brain region, following the supplier instructions. Finally, it was quantified using NanoDrop 2000 (Thermo Scientific, Waltham, MA, USA), taking into account that bisulfite-treated DNA has an absorption coefficient at 260 nm, resembling that of RNA.

2.4. CpG Island Prediction

We focused on the 5 most extensively studied genes in PD. All of them are responsible for familial forms of Parkinson's disease: *SNCA, PRKN, PINK1, DJ-1,* and *LRRK2*. As their expression is ubiquitous, although with some differences between regions, CpG islands in their promoters should exhibit low methylation levels. Therefore, gene transcription should not be repressed, thus allowing their expression in all the cells [23,24].

In order to determine the location of the CpG island, for each gene we considered around 3000 bp upstream and 3500 bp downstream from the transcription start site (TSS) to include the promoter and the first exons, where CpG islands are frequently present. We predicted the presence of CpG islands using 5 prediction programs: Softberry (http://www.softberry.com/berry.phtml?topic=cpgfinder&group=programs&subgroup=promoter), CpG cluster (http://bioinfo2.ugr.es/CpGcluster/) [25], Bioinformatics (http://www.bioinformatics.org/sms2/cpg_islands.html) and Emboss (http://www.ebi.ac.uk/Tools/seqstats/emboss_cpgplot/). Furthermore, we annotated the CpG islands predicted at the UCSC genome browser [26]. The CpG cluster is based on the physical distance between neighboring CGs, and not in the search of CpG islands [27] as others are; therefore, it can find shorter CpG islands. Also, we predicted the position of putative promoters for those genes using FirstEF (http://rulai.cshl.edu/tools/FirstEF/) and WWW Promoter Scan (http://www-bimas.cit.nih.gov/molbio/proscan/).

As results differed between predictions, we chose for our analysis those areas where all or most of the predictions agreed (Figure S1, Supplementary Material).

2.5. Primer Design and Pyrosequencing

The methylation analysis was carried out by pyrosequencing. Ten pyrosequencing assays were designed using the PyroMark Assay Design Software 2.0 (Qiagen, Hilden, Germany), avoiding homopolymers longer than 4 residues, with a maximum difference in primers T_m of 2 °C, an amplicon length between 100–500 bp, no more than two CG dinucleotides in the primer sequence, and with some T bases derived from unmethylated C to ensure only amplification of bisulfite modified DNA [28]. In the target sequences, we analyzed the presence of frequent SNPs using the Single Nucleotide Polymorphism database, dbSNP (http://www.ncbi.nlm.nih.gov/projects/SNP/), because their presence could alter the assay. There was none in any of the regions analyzed by our assays.

In the Supplementary Material, Table S1, all the pyrosequencing assays conducted are described, as are the primer sequences, the size of the fragment amplified, and the PCR conditions.

For each assay we performed three independent replicates. All pyrosequencing reactions were carried out in a PyroMark MD sequencer using NDTS (nucleotide dispensing tips) at the Servei d'Anàlisi d'ADN of the Instituto de Biomedicina de Valencia-CSIC. All reagents were from Qiagen. Results were analyzed by the program PyroMark Q-CpG 1.0.9, which calculated the percentage of DNA methylation per position.

Furthermore, we checked for PCR bias [29]. For this purpose, PCRs were carried out with DNA with known methylation percentages (0, 50 and 100) which were previously bisulfite-treated. Then, DNA methylation levels were analyzed by pyrosequencing. We confirmed that the observed and the expected methylation levels matched. There was an efficiency of conversion of unmethylated cytosines into uracils of nearly 100%, and, thus there was no bias. We used the EpiTect PCR Control DNA Set from Qiagen.

2.6. Statistical Analysis

Due to the low number of samples, a non-parametric test (the Mann-Whitney test) was used to compare means. The statistical analysis was conducted using SPSS (IBM Analytics, Armonk, NY, USA), version 20.

Results are shown as "overall", representing the arithmetic mean of the methylation levels at all CpG dinucleotides in an assay, as well as for the individual CpG dinucleotides. The arithmetic mean is based on the values obtained for the 5 individuals that compose each group (controls and PD patients), except for the results from the control group in *substantia nigra*, which are based on only 4 individuals, as we could not obtain DNA for one of the controls.

2.7. Transcription Factor Binding Site Prediction

We predicted in silico whether CpG dinucleotides differentially methylated could be affecting transcription factor binding sites using TFSEARCH (http://www.cbrc.jp/research/db/TFSEARCH.html) [30], JASPAR CORE (http://jaspar.genereg.net/) [31] and AliBaBa 2.1 and PATCH (http://www.gene-regulation.com/pub/programs.html). The expression of each transcription factor in brain was taken from The Human Protein Atlas (http://www.proteinatlas.org) [32] and the Allen Human Brain Atlas (2010), from The Allen Institute for Brain Science (http://human.brain-map.org) [33].

3. Results

In Table 1, we summarize the significant results of those CpG dinucleotides differentially methylated in patients compared to controls. The results obtained for individual CpG dinucleotides, and overall for the 10 pyrosequencing assays, are shown in Figures S2–S6 (Supplementary Material).

Table 1. Statistically significant results for the comparison of DNA methylation levels between controls and PD patients ($p < 0.05$).

Assay [1]	Position [2]	Region [3]	Location [4]	*p*-Value	Methylation Ratio [5]	Sp1 Binding [6]
SNCA #1	3	PC	−1586	0.008	2.571	NB
SNCA #1	6	PC	−1551	0.016	2.958	−strand
SNCA #1	6	SN	−1551	0.016	0.000	−strand
SNCA #2	2	OC	−1458	0.032	0.326	NB
SNCA #2	7	OC	−1442	0.008	0.526	+strand
SNCA #2	8	SN	−1440	0.016	0.033	+strand
PRKN #1	3	SN	−187	0.016	0.193	NB
PRKN #2	2	OC	+44	0.032	0.114	+strand
PRKN #2	6	PC	+69	0.032	0.128	+strand
PINK1 #1	2	SN	+355	0.016	0.143	+strand

[1]: Each assay is named with the gene followed by the number of the fragment analyzed, according to notation in Figure S1 and Table S1, supplementary material. [2]: Position indicates the CpG pair within the assay, as shown in supplementary Figures S2–S6. [3]: PC: parietal cortex; OC: occipital cortex; SN: substantia nigra. [4]: +1 was assigned to the A from the first translated codon. The position is taken using the following references: NM_001146054 (SNCA), NM_013988 (PRKN) and NM_032409 (PINK1). [5]: Ratio obtained by dividing the methylation level in PD patients with respect to its equivalent in healthy individuals. [6]: When a Sp1 site is predicted to exist in the region where there is a CpG pair showing statistical significant differences between cases and controls, the strand where this binding may exist is shown. NB: not binding predicted for Sp1 nor any other transcription factor.

No single island showed differences in its level of methylation when the information for all CpG dinucleotides was combined (Supplementary Figures S2–S6), although some specific positions in *SNCA*, *PRKN* and *PINK1* showed differences between cases and controls in *substantia nigra*. Also, some significant differences were observed in individual dinucleotides in both parietal and occipital cortices.

For these sites, we checked if they could be part of transcription factor binding sites, and thus, whether changes in their methylation level could affect gene expression.

Some candidates were proposed by sequence-based prediction programs, but due to their expression pattern that did not include *substantia nigra*, and/or their targets that did not include our genes, only Sp1 could be considered as a possible candidate; it is ubiquitous and binds to GC-rich sequences. Moreover, it is involved in many cellular processes, including cell differentiation, cell growth, apoptosis, immune responses, response to DNA damage, and chromatin remodeling. It can be an activator or a repressor, and its activity is highly regulated by post-translational modifications. It interacts with HDAC1 and DNMT1 [34–36], amongst others. The precise location of the Sp1 binding sites can be found in Table 1 and supplementary Figure S1. Most of the Sp1 putative binding sites are found within the predicted promoters, also affecting in some instances (*SNCA* #2, *PRKN* #2 and *PINK1* #1) exonic and intronic sequences that are included in the prediction.

4. Discussion

DNA methylation has been implicated in a diverse range of cellular functions and pathologies. It is generally associated with a repressed chromatin state and inhibition of promoter activity, i.e., transcriptional repression [37,38]. We analyzed whether the five genes responsible for the familial forms of Parkinson's disease could be related to PD pathogenesis by having differential DNA methylation patterns. Our focus was on DNA methylation levels around their transcription start site, because variations of this epigenetic mark in this area can influence gene expression.

Overall, no major differences were observed, although for a few positions, the level of methylation of specific CpG dinucleotides differed significantly between cases and controls. Nevertheless, these changes stem from a very low level of methylation, i.e., small variations in the methylation level could have a large numerical impact. Due to those two characteristics, no statistically significant results would have been obtained if corrections for multiple tests were applied.

Most statistically significant islands share one common, and suggestive, characteristic: Sp1 is the only transcription factor that could be mediating the effect of the differential methylation seen here. Nevertheless, more work is needed to confirm our results prior to exploring the role of Sp1 in PD.

For *SNCA*, overexpression is pathogenic [22], and thus, lower DNA methylation levels that involve higher transcription and gene expression could be pathogenic too. However, for *PRKN*, *PINK1*, *DJ-1*, and *LRRK2*, the pathogenic factor in PD is the lack of enough active protein; therefore, higher DNA methylation levels that involve transcription silencing could lead to decreased gene expression, and thus, could be pathogenic [9–12]. As expected for housekeeping genes, the promoters of *SNCA*, *PRKN*, *PINK1*, *DJ-1*, and *LRRK2* were poorly methylated in all regions analyzed, allowing their ubiquitous expression. Our few statistically significant results were obtained for specific CpG sites mainly located in *SNCA* and *PRKN*. The majority of them showed that DNA methylation levels were higher in controls than in cases in those specific CpG dinucleotides. Remarkably, a higher methylation level was observed only for two specific CpG dinucleotides in *SNCA* with respect to controls (Table 1).

We analyzed *substantia nigra* and parietal and occipital cortices, with a special focus on the first region, as it is extensively affected in PD [39], unlike the parietal cortex and occipital cortex, that could be considered "control brain regions". We did not compare the values between brain regions because, as Ladd-Acosta et al. concluded, the DNA methylation pattern correlates much more strongly within a brain region across individuals than within an individual across brain regions [40]. Nevertheless, we found CpG sites with statistically significant differences in all three tissues (Table 1), and in some instances with the same CpG dinucleotide being altered in more than one region.

Previous studies also addressed the question of whether methylation levels in PD-related genes could be linked to disease pathogenesis, although with controversial results possibly related to, in part, methodological and design differences and to study limitations. In those studies, when statistical significance is found, the results point towards a hypomethylation in PD cases.

In brain, with regard to *SNCA*, Matsumoto et al., Jowaed et al. and de Boni et al. analyzed, in a similar number of individuals to our study, a region in *SNCA* that was not included in our study, although close and downstream to the area in intron 1 that we studied [18–20]. Matsumoto et al. [18] and Jowaed et al. [19] reported hypomethylation across *substantia nigra* in PD cases when compared to controls (and Jowaed et al. even in cortex and *putamen* [19]), whereas de Boni et al. [20] didn't observe any difference in any of the five brain regions analyzed, *substantia nigra* included, and not even for global values on the region analyzed or single CpG dinucleotides. Our results are similar to those of de Boni et al. [20], although we could see some hypo and hypermethylation in some specific CpG dinucleotides. In addition, our technical procedure is similar, using pyrosequencing instead of cloning and sequencing, as Matsumoto et al. and Jowaed et al. did, which could result in a technically-biased evaluation [18,19]. We all analyzed *substantia nigra* from PD cases and controls, despite the fact that dopaminergic neurons are largely eliminated in this region in PD cases as a consequence of the evolution of the disease, and are almost intact in healthy subjects. Nevertheless, the range of our values was similar to those observed by Jowaed et al. and de Boni et al., i.e., very low levels, which would not be affected by such cellular heterogeneity [19,20]. In the opposite direction, the methylation levels observed by Matsumoto et al. were completely different, and this difference could be related to their methodological approach [18].

For *PRKN*, de Mena et al. observed that there was almost no methylation in any of the 3 brain regions analyzed in cases and controls in the region of the analyzed promoter, which does not overlap with our assays [17]. There are similarities between that study and ours, in terms of the number of individuals and regions analyzed (we studied parietal cortex instead of cerebellum); although, again, we used pyrosequencing instead of cloning and sequencing. This could explain the different results observed between that study and ours. We reported some differences in specific CpG dinucleotides for this gene (Table 1).

In addition, our methodological approach does not differentiate between 5-hydroxy-methylC (5 hmC) and 5-methylC. Although 5 hmC has its highest levels in the central nervous system, and its amount increases with age, Munzel et al. and Kriaucionis et al. observed that this effect on methylation percentages is expected to be almost imperceptible [41,42].

Although there are no studies that have analyzed the influence of the post-mortem delay on DNA methylation, previous reports did not find any effect on pH, RNA quality [43], or protein concentration (post-synaptic proteins [44] and primary microglia [45]) in brain. Therefore, we considered that the difference in post-mortem delay between both groups, healthy controls, and PD cases would not affect our results.

Masliah et al. concluded, after a genome-wide study of methylation levels in brain and blood, that there were similar methylation patterns between them [46]. Unfortunately, no blood was available from the individuals we have analyzed here, so we could not test the hypothesis that epigenetic marks could be potential biomarkers for non-invasive predictive testing for PD. In blood, some studies have been carried out where no differences between cases and controls for SNCA [47,48] or for *PRKN*, *DJ-1* [49,50] were observed. However hypomethylation has been described in other studies for sporadic cases [51–53]. Tan et al. also observed no differences in other regions of *SNCA* and *LRRK2*, both overlapping with our assays [53]. The main difference between these studies relies on the number of individuals analyzed, as different technical approaches have given similar results.

We conclude that imbalances in the methylation of specific CpG islands in PD-related genes do not seem to be related to the disease process. We studied three different tissues in five different genes in triplicates, which increases the confidence of our results. However, due to the small size of our population and the low level of methylation observed, our results could be considered as trends that should be replicated in a larger study, maybe using genome-wide approaches to account for the effect of genes not related to familial PD. Moreover, functional validation of these results is also needed. This is especially significant, as doubling or halving the amount of methylation with such a low baseline is unlikely to have a major impact on the pathophysiology of the disease.

Other epigenetic mechanisms, i.e., small RNAs or histone modification, could also be relevant in the dysregulation of the expression of some genes important in the pathogenesis of Parkinson's disease. For example, miR-7 [54] and miR-153 [55] repress α-synuclein expression, and miR-205 expression is down regulated in sporadic PD patients with enhanced LRRK2 expression [56]. In addition, it has been reported that the expression of some miRNAs is altered in in vivo PD models [57,58] and in PD patients [59]. On the other hand, it has also been observed that α-synuclein associates with histones and inhibits their acetylation [60]. Therefore, histone deacetylase (HDAC) inhibitors are the most recent emerging therapeutic targets in the treatment of PD. It has been demonstrated in cellular and animal models that the neurotoxicity of α-synuclein in the nucleus could be rescued by their administration [61,62]. Thus, it is important to study the epigenetic regulation of PD genes in order to obtain a clearer picture of the etiology of this disorder.

Supplementary Materials:
Figure S1: Schematic representation of the regions of interest in SNCA (A, B), LRRK2 (C, D), PRKN (E, F), PINK1 (G, H) and DJ-1 (I, J), Figure S2: Results for all three assays in SNCA in the parietal and occipital cortices and Substantia nigra from PD patients and controls, Figure S3: Results for the two assays in LRRK2 in the parietal and occipital cortices and Substantia nigra from PD patients and controls, Figure S4: Results for the two assays in PRKN in the parietal and occipital cortices and Substantia nigra from PD patients and controls, Figure S5: Results for the assay in PINK1 in the parietal and occipital cortices and Substantia nigra from PD patients and controls, Figure S6: Results for the two assays in DJ-1 in the parietal and occipital cortices and Substantia nigra from PD patients and controls, Table S1. Description of the parameters used for the epigenetic analysis of DNA methylation levels by pyrosequencing, Table S2. Characteristics of the CpG islands predicted by the software employed in this work.

Author Contributions: L.N.-S.: Research organization and execution. Design, execution, review and critique of the statistical analysis. Writing of the first draft. J.P.T.: Research conception and organization. Design, review and critique of the statistical analysis. Review and critique of the manuscript. B.Á.-G.: Research execution and Review and critique of the manuscript. S.A.: Research execution and Review and critique of the manuscript.

Acknowledgments: Patients and voluntary controls are warmly acknowledged for their contribution to this research. We acknowledge support of the publication fee by the CSIC Open Access Publication Support Initiative through its Unit of Information Resources for Research (URICI).

References

1. Jellinger, K.A. Formation and development of Lewy pathology: A critical update. *J. Neurol.* **2009**, *256* (Suppl. 3), 270–279. [CrossRef] [PubMed]
2. Jankovic, J. Parkinson's disease: Clinical features and diagnosis. *J. Neurol. Neurosurg. Psychiatry* **2008**, *79*, 368–376. [CrossRef] [PubMed]
3. De Lau, L.M.; Breteler, M.M. Epidemiology of Parkinson's disease. *Lancet Neurol.* **2006**, *5*, 525–535. [CrossRef]
4. Thomas, B.; Beal, M.F. Parkinson's disease. *Hum. Mol. Genet.* **2007**, *16*, R183–R194. [CrossRef] [PubMed]
5. Coppede, F. Genetics and epigenetics of Parkinson's disease. *Sci. World J.* **2012**, *2012*, 489830. [CrossRef] [PubMed]
6. Bras, J.; Guerreiro, R.; Hardy, J. SnapShot: Genetics of Parkinson's disease. *Cell* **2015**, *160*, 570–570.e1. [CrossRef] [PubMed]
7. Kalinderi, K.; Bostantjopoulou, S.; Fidani, L. The genetic background of Parkinson's disease: Current progress and future prospects. *Acta Neurol. Scand.* **2016**, *134*, 314–326. [CrossRef] [PubMed]
8. Polymeropoulos, M.H.; Lavedan, C.; Leroy, E.; Ide, S.E.; Dehejia, A.; Dutra, A.; Pike, B.; Root, H.; Rubenstein, J.; Boyer, R.; et al. Mutation in the alpha-synuclein gene identified in families with Parkinson's disease. *Science* **1997**, *276*, 2045–2047. [CrossRef] [PubMed]
9. Kitada, T.; Asakawa, S.; Hattori, N.; Matsumine, H.; Yamamura, Y.; Minoshima, S.; Yokochi, M.; Mizuno, Y.; Shimizu, N. Mutations in the parkin gene cause autosomal recessive juvenile parkinsonism. *Nature* **1998**, *392*, 605–608. [CrossRef] [PubMed]
10. Valente, E.M.; Abou-Sleiman, P.M.; Caputo, V.; Muqit, M.M.; Harvey, K.; Gispert, S.; Ali, Z.; Del Turco, D.; Bentivoglio, A.R.; Healy, D.G.; et al. Hereditary early-onset Parkinson's disease caused by mutations in PINK1. *Science* **2004**, *304*, 1158–1160. [CrossRef] [PubMed]
11. Bonifati, V.; Rizzu, P.; van Baren, M.J.; Schaap, O.; Breedveld, G.J.; Krieger, E.; Dekker, M.C.; Squitieri, F.; Ibanez, P.; Joosse, M.; et al. Mutations in the DJ-1 gene associated with autosomal recessive early-onset parkinsonism. *Science* **2003**, *299*, 256–259. [CrossRef] [PubMed]

12. Paisan-Ruiz, C.; Jain, S.; Evans, E.W.; Gilks, W.P.; Simon, J.; van der Brug, M.; Lopez de Munain, A.; Aparicio, S.; Gil, A.M.; Khan, N.; et al. Cloning of the gene containing mutations that cause PARK8-linked Parkinson's disease. *Neuron* **2004**, *44*, 595–600. [CrossRef] [PubMed]
13. Vilarino-Guell, C.; Wider, C.; Ross, O.A.; Dachsel, J.C.; Kachergus, J.M.; Lincoln, S.J.; Soto-Ortolaza, A.I.; Cobb, S.A.; Wilhoite, G.J.; Bacon, J.A.; et al. VPS35 mutations in Parkinson disease. *Am. J. Hum. Genet.* **2011**, *89*, 162–167. [CrossRef] [PubMed]
14. Riggs, A.D.; Martienssen, R.A.; Russo, V.E.A. *Epigenetic Mechanisms of Gene Regulation*; Cold Spring Harbor Laboratory Press: Cold Spring Harbor, NY, USA, 1996; p. 4.
15. Weber, M.; Schubeler, D. Genomic patterns of DNA methylation: Targets and function of an epigenetic mark. *Curr. Opin. Cell Biol.* **2007**, *19*, 273–280. [CrossRef] [PubMed]
16. Barrachina, M.; Ferrer, I. DNA methylation of Alzheimer disease and tauopathy-related genes in postmortem brain. *J. Neuropathol. Exp. Neurol.* **2009**, *68*, 880–891. [CrossRef] [PubMed]
17. De Mena, L.; Cardo, L.F.; Coto, E.; Alvarez, V. No differential DNA methylation of PARK2 in brain of Parkinson's disease patients and healthy controls. *Mov. Disord.* **2013**, *28*, 2032–2033. [CrossRef] [PubMed]
18. Matsumoto, L.; Takuma, H.; Tamaoka, A.; Kurisaki, H.; Date, H.; Tsuji, S.; Iwata, A. CpG demethylation enhances alpha-synuclein expression and affects the pathogenesis of Parkinson's disease. *PLoS ONE* **2010**, *5*, e15522. [CrossRef] [PubMed]
19. Jowaed, A.; Schmitt, I.; Kaut, O.; Wullner, U. Methylation regulates alpha-synuclein expression and is decreased in Parkinson's disease patients' brains. *J. Neurosci.* **2010**, *30*, 6355–6359. [CrossRef] [PubMed]
20. De Boni, L.; Tierling, S.; Roeber, S.; Walter, J.; Giese, A.; Kretzschmar, H.A. Next-generation sequencing reveals regional differences of the alpha-synuclein methylation state independent of Lewy body disease. *Neuromol. Med.* **2011**, *13*, 310–320. [CrossRef] [PubMed]
21. Desplats, P.; Spencer, B.; Coffee, E.; Patel, P.; Michael, S.; Patrick, C.; Adame, A.; Rockenstein, E.; Masliah, E. Alpha-synuclein sequesters Dnmt1 from the nucleus: A novel mechanism for epigenetic alterations in Lewy body diseases. *J. Biol. Chem.* **2011**, *286*, 9031–9037. [CrossRef] [PubMed]
22. Singleton, A.B.; Farrer, M.; Johnson, J.; Singleton, A.; Hague, S.; Kachergus, J.; Hulihan, M.; Peuralinna, T.; Dutra, A.; Nussbaum, R.; et al. alpha-Synuclein locus triplication causes Parkinson's disease. *Science* **2003**, *302*, 841. [CrossRef] [PubMed]
23. Bird, A. The essentials of DNA methylation. *Cell* **1992**, *70*, 5–8. [CrossRef]
24. Antequera, F. Structure, function and evolution of CpG island promoters. *Cell. Mol. Life Sci.* **2003**, *60*, 1647–1658. [CrossRef] [PubMed]
25. Hackenberg, M.; Previti, C.; Luque-Escamilla, P.L.; Carpena, P.; Martinez-Aroza, J.; Oliver, J.L. CpGcluster: A distance-based algorithm for CpG-island detection. *BMC Bioinform.* **2006**, *7*, 446. [CrossRef] [PubMed]
26. Kent, W.J.; Sugnet, C.W.; Furey, T.S.; Roskin, K.M.; Pringle, T.H.; Zahler, A.M.; Haussler, D. The human genome browser at UCSC. *Genome Res.* **2002**, *12*, 996–1006. [CrossRef] [PubMed]
27. Gardiner-Garden, M.; Frommer, M. CpG islands in vertebrate genomes. *J. Mol. Biol.* **1987**, *196*, 261–282. [CrossRef]
28. Wojdacz, T.K.; Hansen, L.L.; Dobrovic, A. A new approach to primer design for the control of PCR bias in methylation studies. *BMC Res. Notes* **2008**, *1*, 54. [CrossRef] [PubMed]
29. Warnecke, P.M.; Stirzaker, C.; Melki, J.R.; Millar, D.S.; Paul, C.L.; Clark, S.J. Detection and measurement of PCR bias in quantitative methylation analysis of bisulphite-treated DNA. *Nucleic Acids Res.* **1997**, *25*, 4422–4426. [CrossRef] [PubMed]
30. Heinemeyer, T.; Wingender, E.; Reuter, I.; Hermjakob, H.; Kel, A.E.; Kel, O.V.; Ignatieva, E.V.; Ananko, E.A.; Podkolodnaya, O.A.; Kolpakov, F.A.; et al. Databases on transcriptional regulation: TRANSFAC, TRRD and COMPEL. *Nucleic Acids Res.* **1998**, *26*, 362–367. [CrossRef] [PubMed]
31. Bryne, J.C.; Valen, E.; Tang, M.H.; Marstrand, T.; Winther, O.; da Piedade, I.; Krogh, A.; Lenhard, B.; Sandelin, A. JASPAR, the open access database of transcription factor-binding profiles: New content and tools in the 2008 update. *Nucleic Acids Res.* **2008**, *36*, D102–D106. [CrossRef] [PubMed]
32. Uhlen, M.; Fagerberg, L.; Hallstrom, B.M.; Lindskog, C.; Oksvold, P.; Mardinoglu, A.; Sivertsson, A.; Kampf, C.; Sjostedt, E.; Asplund, A.; et al. Proteomics. Tissue-based map of the human proteome. *Science* **2015**, *347*, 1260419. [CrossRef] [PubMed]

33. Hawrylycz, M.J.; Lein, E.S.; Guillozet-Bongaarts, A.L.; Shen, E.H.; Ng, L.; Miller, J.A.; van de Lagemaat, L.N.; Smith, K.A.; Ebbert, A.; Riley, Z.L.; et al. An anatomically comprehensive atlas of the adult human brain transcriptome. *Nature* **2012**, *489*, 391–399. [CrossRef] [PubMed]
34. Hung, J.J.; Wang, Y.T.; Chang, W.C. Sp1 deacetylation induced by phorbol ester recruits p300 to activate 12(S)-lipoxygenase gene transcription. *Mol. Cell. Biol.* **2006**, *26*, 1770–1785. [CrossRef] [PubMed]
35. Olofsson, B.A.; Kelly, C.M.; Kim, J.; Hornsby, S.M.; Azizkhan-Clifford, J. Phosphorylation of Sp1 in response to DNA damage by ataxia telangiectasia-mutated kinase. *Mol. Cancer Res.* **2007**, *5*, 1319–1330. [CrossRef] [PubMed]
36. Spengler, M.L.; Guo, L.W.; Brattain, M.G. Phosphorylation mediates Sp1 coupled activities of proteolytic processing, desumoylation and degradation. *Cell Cycle* **2008**, *7*, 623–630. [CrossRef] [PubMed]
37. Bernstein, B.E.; Meissner, A.; Lander, E.S. The mammalian epigenome. *Cell* **2007**, *128*, 669–681. [CrossRef] [PubMed]
38. Klose, R.J.; Bird, A.P. Genomic DNA methylation: The mark and its mediators. *Trends Biochem. Sci.* **2006**, *31*, 89–97. [CrossRef] [PubMed]
39. Braak, H.; Del Tredici, K.; Rub, U.; de Vos, R.A.; Jansen Steur, E.N.; Braak, E. Staging of brain pathology related to sporadic Parkinson's disease. *Neurobiol. Aging* **2003**, *24*, 197–211. [CrossRef]
40. Ladd-Acosta, C.; Pevsner, J.; Sabunciyan, S.; Yolken, R.H.; Webster, M.J.; Dinkins, T.; Callinan, P.A.; Fan, J.B.; Potash, J.B.; Feinberg, A.P. DNA methylation signatures within the human brain. *Am. J. Hum. Genet.* **2007**, *81*, 1304–1315. [CrossRef] [PubMed]
41. Munzel, M.; Globisch, D.; Carell, T. 5-Hydroxymethylcytosine, the sixth base of the genome. *Angew. Chem. Int. Ed. Engl.* **2011**, *50*, 6460–6468. [CrossRef] [PubMed]
42. Kriaucionis, S.; Heintz, N. The nuclear DNA base 5-hydroxymethylcytosine is present in Purkinje neurons and the brain. *Science* **2009**, *324*, 929–930. [CrossRef] [PubMed]
43. Robinson, A.C.; Palmer, L.; Love, S.; Hamard, M.; Esiri, M.; Ansorge, O.; Lett, D.; Attems, J.; Morris, C.; Troakes, C.; et al. Extended post-mortem delay times should not be viewed as a deterrent to the scientific investigation of human brain tissue: A study from the Brains for Dementia Research Network Neuropathology Study Group, UK. *Acta Neuropathol.* **2016**, *132*, 753–755. [CrossRef] [PubMed]
44. Siew, L.K.; Love, S.; Dawbarn, D.; Wilcock, G.K.; Allen, S.J. Measurement of pre- and post-synaptic proteins in cerebral cortex: Effects of post-mortem delay. *J. Neurosci. Methods* **2004**, *139*, 153–159. [CrossRef] [PubMed]
45. Mizee, M.R.; Miedema, S.S.; van der Poel, M.; Schuurman, K.G.; van Strien, M.E.; Melief, J.; Smolders, J.; Hendrickx, D.A.; Heutinck, K.M.; Hamann, J.; et al. Isolation of primary microglia from the human post-mortem brain: Effects of ante- and post-mortem variables. *Acta Neuropathol. Commun.* **2017**, *5*, 16. [CrossRef] [PubMed]
46. Masliah, E.; Dumaop, W.; Galasko, D.; Desplats, P. Distinctive patterns of DNA methylation associated with Parkinson disease: Identification of concordant epigenetic changes in brain and peripheral blood leukocytes. *Epigenetics* **2013**, *8*, 1030–1038. [CrossRef] [PubMed]
47. Richter, J.; Appenzeller, S.; Ammerpohl, O.; Deuschl, G.; Paschen, S.; Bruggemann, N.; Klein, C.; Kuhlenbaumer, G. No evidence for differential methylation of alpha-synuclein in leukocyte DNA of Parkinson's disease patients. *Mov. Disord.* **2012**, *27*, 590–591. [CrossRef] [PubMed]
48. Song, Y.; Ding, H.; Yang, J.; Lin, Q.; Xue, J.; Zhang, Y.; Chan, P.; Cai, Y. Pyrosequencing analysis of SNCA methylation levels in leukocytes from Parkinson's disease patients. *Neurosci. Lett.* **2014**, *569*, 85–88. [CrossRef] [PubMed]
49. Cai, M.; Tian, J.; Zhao, G.H.; Luo, W.; Zhang, B.R. Study of methylation levels of parkin gene promoter in Parkinson's disease patients. *Int. J. Neurosci.* **2011**, *121*, 497–502. [CrossRef] [PubMed]
50. Tan, Y.; Wu, L.; Li, D.; Liu, X.; Ding, J.; Chen, S. Methylation status of DJ-1 in leukocyte DNA of Parkinson's disease patients. *Transl. Neurodegener.* **2016**, *5*, 5. [CrossRef] [PubMed]
51. Ai, S.X.; Xu, Q.; Hu, Y.C.; Song, C.Y.; Guo, J.F.; Shen, L.; Wang, C.R.; Yu, R.L.; Yan, X.X.; Tang, B.S. Hypomethylation of SNCA in blood of patients with sporadic Parkinson's disease. *J. Neurol. Sci.* **2014**, *337*, 123–128. [CrossRef] [PubMed]
52. Schmitt, I.; Kaut, O.; Khazneh, H.; deBoni, L.; Ahmad, A.; Berg, D.; Klein, C.; Frohlich, H.; Wullner, U. L-dopa increases alpha-synuclein DNA methylation in Parkinson's disease patients in vivo and in vitro. *Mov. Disord.* **2015**, *30*, 1794–1801. [CrossRef] [PubMed]

53. Tan, Y.Y.; Wu, L.; Zhao, Z.B.; Wang, Y.; Xiao, Q.; Liu, J.; Wang, G.; Ma, J.F.; Chen, S.D. Methylation of alpha-synuclein and leucine-rich repeat kinase 2 in leukocyte DNA of Parkinson's disease patients. *Parkinsonism Relat. Disord.* **2014**, *20*, 308–313. [CrossRef] [PubMed]
54. Junn, E.; Lee, K.W.; Jeong, B.S.; Chan, T.W.; Im, J.Y.; Mouradian, M.M. Repression of alpha-synuclein expression and toxicity by microRNA-7. *Proc. Natl. Acad. Sci. USA* **2009**, *106*, 13052–13057. [CrossRef] [PubMed]
55. Doxakis, E. Post-transcriptional regulation of alpha-synuclein expression by mir-7 and mir-153. *J. Biol. Chem.* **2010**, *285*, 12726–12734. [CrossRef] [PubMed]
56. Cho, H.J.; Liu, G.; Jin, S.M.; Parisiadou, L.; Xie, C.; Yu, J.; Sun, L.; Ma, B.; Ding, J.; Vancraenenbroeck, R.; et al. MicroRNA-205 regulates the expression of Parkinson's disease-related leucine-rich repeat kinase 2 protein. *Hum. Mol. Genet.* **2013**, *22*, 608–620. [CrossRef] [PubMed]
57. Gillardon, F.; Mack, M.; Rist, W.; Schnack, C.; Lenter, M.; Hildebrandt, T.; Hengerer, B. MicroRNA and proteome expression profiling in early-symptomatic alpha-synuclein(A30P)-transgenic mice. *Proteom. Clin. Appl.* **2008**, *2*, 697–705. [CrossRef] [PubMed]
58. Asikainen, S.; Rudgalvyte, M.; Heikkinen, L.; Louhiranta, K.; Lakso, M.; Wong, G.; Nass, R. Global microRNA expression profiling of Caenorhabditis elegans Parkinson's disease models. *J. Mol. Neurosci.* **2010**, *41*, 210–218. [CrossRef] [PubMed]
59. Martins, M.; Rosa, A.; Guedes, L.C.; Fonseca, B.V.; Gotovac, K.; Violante, S.; Mestre, T.; Coelho, M.; Rosa, M.M.; Martin, E.R.; et al. Convergence of miRNA expression profiling, alpha-synuclein interacton and GWAS in Parkinson's disease. *PLoS ONE* **2011**, *6*, e25443. [CrossRef] [PubMed]
60. Urdinguio, R.G.; Sanchez-Mut, J.V.; Esteller, M. Epigenetic mechanisms in neurological diseases: Genes, syndromes, and therapies. *Lancet Neurol.* **2009**, *8*, 1056–1072. [CrossRef]
61. Kontopoulos, E.; Parvin, J.D.; Feany, M.B. Alpha-synuclein acts in the nucleus to inhibit histone acetylation and promote neurotoxicity. *Hum. Mol. Genet.* **2006**, *15*, 3012–3023. [CrossRef] [PubMed]
62. Kidd, S.K.; Schneider, J.S. Protective effects of valproic acid on the nigrostriatal dopamine system in a 1-methyl-4-phenyl-1,2,3,6-tetrahydropyridine mouse model of Parkinson's disease. *Neuroscience* **2011**, *194*, 189–194. [CrossRef] [PubMed]

5

Novel Immunotherapeutic Approaches to Target Alpha-Synuclein and Related Neuroinflammation in Parkinson's Disease

Maria Angela Samis Zella [1,2], Judith Metzdorf [1], Friederike Ostendorf [1], Fabian Maass [3], Siegfried Muhlack [1], Ralf Gold [1,4], Aiden Haghikia [1,4] and Lars Tönges [1,4,*]

[1] Department of Neurology, St. Josef-Hospital, Ruhr-University Bochum, 44791 Bochum, Germany; maria.zella@rub.de (M.A.S.Z.); judith.metzdorf@rub.de (J.M.); friederike.ostendorf@googlemail.com (F.O.); siegfried.muhlack@rub.de (S.M.); ralf.gold@rub.de (R.G.); aiden.haghikia@rub.de (A.H.)

[2] Department of Neurology, St. Josef- Hospital, Katholische Kliniken Ruhrhalbinsel, Contilia Gruppe, 45257 Essen, Germany

[3] Department of Neurology, University Medical Center, 37075 Goettingen, Germany; fabian.maass@med.uni-goettingen.de

[4] Neurodegeneration Research, Protein Research Unit Ruhr (PURE), Ruhr University Bochum, 44801 Bochum, Germany

* Correspondence: lars.toenges@rub.de

Abstract: The etiology of Parkinson's disease (PD) is significantly influenced by disease-causing changes in the protein alpha-Synuclein (aSyn). It can trigger and promote intracellular stress and thereby impair the function of dopaminergic neurons. However, these damage mechanisms do not only extend to neuronal cells, but also affect most glial cell populations, such as astroglia and microglia, but also T lymphocytes, which can no longer maintain the homeostatic CNS milieu because they produce neuroinflammatory responses to aSyn pathology. Through precise neuropathological examination, molecular characterization of biomaterials, and the use of PET technology, it has been clearly demonstrated that neuroinflammation is involved in human PD. In this review, we provide an in-depth overview of the pathomechanisms that aSyn elicits in models of disease and focus on the affected glial cell and lymphocyte populations and their interaction with pathogenic aSyn species. The interplay between aSyn and glial cells is analyzed both in the basic research setting and in the context of human neuropathology. Ultimately, a strong rationale builds up to therapeutically reduce the burden of pathological aSyn in the CNS. The current antibody-based approaches to lower the amount of aSyn and thereby alleviate neuroinflammatory responses is finally discussed as novel therapeutic strategies for PD.

Keywords: Parkinson's disease; neuroinflammation; alpha-Synuclein; immunotherapy

1. Introduction

Parkinson's disease (PD) is the second-most common neurodegenerative disorder and affects 2–3% of the general population with an age of more than 65 years. It is characterized by a multitude of pathophysiological alterations that are linked to the dysfunctional state of the protein alpha-Synuclein (aSyn). In their typical clinical presentation, PD patients exhibit deficits in motor function with bradykinesia, rigidity, tremor, and possibly postural instability, reflecting the degeneration of dopamine processing neurons in the substantia nigra pars compacta and alterations of dopaminergic neurotransmission to basal ganglia motor circuits [1–3]. Apart from the motor symptoms, there are substantial impairments in non-motor domains, including signs of cognitive deficits, psychiatric comorbidities, daytime sleepiness, or autonomic failure [4]. The wide variety of symptoms clearly indicates that many different brain regions and neurotransmitter systems are affected [5]. As a causal pathological agent, aSyn has been identified, but the entirety of disease processes cannot be attributed to the dysfunction of a single protein [6]. After intensive research over the last decades, it is now generally accepted that aSyn pathology is tightly associated with CNS neuroinflammation in PD. The evidence for this phenomenon extends over basic science studies in cell culture or animal models to increasing data by human examinations, which have been collected post mortem but also in vivo [7]. However, the importance of neuroinflammation for the development and progression of the PD is still controversial [8–10]. In order to weigh the scientific findings, clearly identified pathomechanisms in humans should play a central role in further analysis. This is of particular importance considering the new immunological therapeutic approaches currently being tested in human clinical trials. In vitro and in vivo animal models of PD serve to evaluate and differentiate human pathophysiological findings. In addition, they enable us to design targeted immunological therapies for human use.

In this review article, we present the current state of basic research on the role of aSyn and its direct lesioning effects on neuronal cells. The impact of aSyn as an aggravator but also a potential facilitator [11] for neuroinflammation is a strong focus of this review, and will be analyzed based on cell types involved in neuroinflammation. The interactions of aSyn with astroglia, monocytes and microglia, and T lymphocytes are depicted and resulting cellular pathologies and their associated pathomechanisms are discussed. These findings are related to human neuropathological data and analyzed for their significance, but also plausibility. Finally, a clear rationale emerges that reducing the burden of aSyn pathology is a promising approach for a disease modifying therapy of PD. The preclinical results and designs of current human studies employing anti-aSyn directed antibodies as immunotherapy for PD will be presented and evaluated in a look-ahead for future developments.

2. Alpha-Synuclein as Direct Stressor for CNS Neurons

ASyn is a soluble cytoplasmic protein with a length of 140 aa (amino acids) being encoded in the *SNCA* gene. Its primary structure can be subclassified into three regions: an amphipathic N-terminal domain (1–61 aa), a hydrophobic domain (61–95 aa) with the non-amyloid-beta component (NAC), and the N-terminal domain (96–140 aa), which is highly acidic [12–14]. ASyn is expressed throughout the CNS, but can also be found in the PNS and other tissues like red blood cells [15]. ASyn aggregates that are localized in so-called Lewy bodies are a neuropathological hallmark of human PD [1] and point mutations in the *SNCA* gene cause familial forms [14].

The physiological role of aSyn is only incompletely understood, but diverse studies have shown that it plays an important role in synaptic plasticity and neurotransmitter release [12,15,16]. ASyn is stored in the presynaptic terminal, where it presumably interacts with the soluble N-ethylmaleimide sensitive factor (NSF) attachment receptor (SNARE) complex, whose circle of assembly and disassembly is important for the continuing release of neurotransmitters. A direct interaction with synaptobrevin-2/vesicle-associated membrane protein 2 and phospholipids seems to promote the assembly of the SNARE-complexes [13,17], and over-expression of aSyn leads to an impairment of different synaptic functions, like vesicle trafficking and recycling [18,19].

Under physiological conditions, native aSyn appears as a soluble monomer, but in cases of oxidative stress or a change in pH level, it aggregates to insoluble fibrils, which have a β-sheet conformation. It is able to organize itself in oligomers and amyloid fibrils as aggregates that can harm different cell organelles, like the Golgi apparatus or mitochondria, and then affect mechanisms like synaptic vesicle release [14]. An interesting feature of aSyn is its propagation mechanism, as there is increasing evidence that aSyn is transmitted between neurons. After transport of aSyn assemblies along the axons, it is released in the extracellular space and can then be uptaken by another neuron [16]. Because of the demonstrated spread of aSyn, it is considered by some authors to have a prion-like behavior [20], although this hypothesis is heavily debated [21,22]. Interestingly, Yamada et al. demonstrated recently that the release of aSyn in vivo is also regulated by neuronal activity, as well as by extrinsic mechanisms and stressors [23].

Extracellular aSyn seems to have deleterious effects on neighboring neurons and glia. Various in vitro studies have been performed to demonstrate direct effects on neuronal cells. Exogenous fibrils, which were internalized by primary hippocampal neurons, induced pathological misfolding in endogenous aSyn. This led to phosphorylation, ubiquitination, and finally aSyn aggregation. Interestingly, there was no need for an overexpression of wild-type aSyn because the presence of endogenous aSyn was sufficient [24,25]. Hassink et al. used primary cortical neurons from rats to evaluate the effect of exogenous aSyn. After exposition to extracellular aSyn, they observed an increased uptake of aSyn, followed by intracellular formations of aSyn assemblies. At the same time, a spontaneous initial increase of synaptic activity was detected, which later reversed and resulted in overall activity reduction [26]. Importantly, the impairment of cellular functions by aggregated aSyn is not limited to dopaminergic neuronal cells. Long-term potentiation in hippocampal brain slices is negatively impacted by extracellular aSyn, supposedly by a temporary inhibition of NMDA receptors [14,27]. Direct injections of disaggregated aSyn in the substantia nigra of wild-type rats caused neuronal cell death and behavioral motor deficits [28]. Even after intravenous injection, aSyn crosses the blood brain barrier [28], or is found in the contralateral hemisphere of wild type mice following previous injection of aSyn into the ipsilateral striatum [29]. Virus-mediated overexpression of wildtype or transgenic aSyn injected into the mouse substantia nigra can lead to loss of nigral dopaminergic cells and of their axons, and additionally impairs the intrinsic regenerative capacity of dopaminergic axons [30].

These data clearly demonstrate direct and often deleterious effects of both soluble and aggregated aSyn on neuronal cell populations, including dopaminergic neurons. However, extracellular aSyn also activates glial cells and provokes a multitude of immunological reactions, which can be classified under the term neuroinflammation.

3. Alpha-Synuclein as Promotor of Neuroinflammation

With more than 50% of all cells in the brain, glial cells form the largest cell population and thus comprise a number of approximately 1 trillion [31]. In the CNS, they mainly differentiate into astroglia, microglia, and oligodendroglia, which are all susceptible to aSyn and may react with a neuroinflammatory response. Whether a glial cell activation is deleterious or even beneficial for the CNS milieu and its neurons is often hard to say and strongly depends on the specific context. However, in order to develop an interfering strategy to alleviate aSyn-triggered mechanisms, the cellular responses of glia must be well known. However, with respect to neuroinflammation, T lymphocytes and their interaction with microglia also seem to play an important role in PD. Here, we present the most relevant findings from cell-based and animal studies, differentiated according to the three most important cell populations involved.

3.1. Astroglia

Astroglia are essential for the proper functioning of CNS neurons. They provide structural and metabolic cerebral homeostasis, regulate water transport, blood flow and synaptic transmission, and produce neurotrophic factors. Thus, their role of a more passive cell population, initially assumed to be rather insignificant, was significantly expanded. In addition, there is a close interaction with microglial and oligodendroglial cells [32].

There is only little experimental data on aSyn-mediated stimulation of astroglia available. In cell culture, aSyn was a strong TLR-4 mediated stimulation factor for astrocytes and was taken up by these via endocytosis [33], so that it can be assumed that astrocytes contribute significantly to the clearance and degradation of extracellular aSyn. In the case of astrocyte-selective overexpression of the aSyn A53T mutant, it accumulated significantly and led to reactive astrogliosis associated with cerebral microhemorrhages and reduction of astroglial glutamate transporters [34]. On the other hand, transgenic model systems were used and also demonstrated an important neuroprotective potential of astrocytes. Selective overexpression of the transcription factor Nrf2 in astrocytes in transgenic aSyn-A53T mutant mice delayed the onset of clinical disease and significantly prolonged survival. Furthermore, the amount of aSyn-A53T protein aggregates in mouse brain was significantly reduced and dysfunctional protein degradation, including chaperone-mediated autophagy and macroautophagy, was reduced. Interestingly, the survival of motoneurons was also improved and gliosis and oxidative stress in the spinal cord were reduced, so that non-dopaminergic cells and the CNS tissue structure in general benefited from the modified astrocyte function [35].

Astrocytes can absorb extracellular aSyn from neurons and degrade it into lysosomal compartments [36]. Studies on human astrocytes derived from embryonic stem cells have shown that aggregated aSyn can be intercellularly exchanged between astrocytes via small tubular "tunneling nanotubes" (TNT) to avoid overloading of individual cells. In return, the endangered astrocytes received energy-providing mitochondria from the healthier cells [37].

It can be concluded that astrocytes play an important modulating role in the progression of PD neuropathology, not least due to their very high abundance and their buffering effect, e.g., for spreading of aSyn. In the case of overloading with aSyn, there is a risk of decompensation with consecutive pro-inflammatory effects, activation of microglia, and collapse of the homeostasis in the CNS.

3.2. Microglia

Microglial cells are a central part of the innate immune system of the CNS and account for up to 20% of the total glial population [38]. The physiological role of resident microglia in the CNS comprises the preservation of homeostasis by continuous monitoring of the local milieu, and if necessary, phagocytose or pinocytose degradation products or cellular residues [39,40]. In this case, the microglia present morphologically with strongly ramified and highly mobile cell processes that are in close contact with other glial and neuronal cells [41,42]. Different stimuli, such as neuronal cell death, mechanical stress, infectious agents, or cellular toxins, can drive microglia to a transformation to an altered morphological phenotype. This induces a polarization of microglia into a proinflammatory neurotoxic or anti-inflammatory neuroprotective phenotype, traditionally referred to as M1 or M2 type. For example, proinflammatory M1 microglia are characterized by the secretion of factors, such as NO, ROS, IL-1b, IL-6, or TNF-α, whereas anti-inflammatory M2 microglia secrete IL-4, IL-10, IL-13, or TGF-β [43,44]. The functional phenotype of the reactive microglia strongly depends on distinct stimulating factors at the beginning or during the progression of PD, and on the respective localization in the CNS [45,46], so that the M1/M2 concept for the differentiated function of the microglia seems to be rather simplified [47].

Constitutive transgenic overexpression or local viral-induced overexpression of aSyn results in an increased number of microglial cells in the SN [48,49]. Transgenic mice overexpressing aSyn also showed a distinct microglial activation in the striatum from the first month of life, and only then in the SN from the fifth month of life persisting until the 14th month of life. This was accompanied by increased levels of TNF-α in the striatum and increased levels of TNF-α, TLR1, TLR4, and TLR8 in the SN [50]. Extracellular aSyn released from neuronal cells was demonstrated in an elegant study using in silico, in vitro, and in vivo approaches to be an endogenous agonist for Toll-like receptor 2 (TLR2), which activates inflammatory responses in microglia [51].

In vitro, numerous experiments have shown that the aSyn wildtype or mutant variants activate microglia and can pathologically increase microglial inflammatory signals in an autocrine or paracrine manner [7]. In addition to activation of ROS, matrix metalloproteinases or activation of the NFB/AP-1/Nrf2 pathway play an important role [52,53]. Williams et al. investigated the involvement of microglial MHCII complex in alpha-synuclein-induced neurodegeneration. By knock-out of the class II transactivator (CIITA), they inhibited MHCII expression on primary microglia. Stimulation with fibrillar aSyn in vitro then led only to a reduced inflammatory response with less iNOS expression and antigen processing. In vivo, RNA silencing of CIITA led to a reduced MHCII expression, less peripheral immune cell infiltration into the CNS, and reduced aSyn-induced neurodegeneration [54].

In a recent study, a possible pathogenetic contribution of the intestinal microbiome was investigated in mice overexpressing aSyn. It was shown that the microbiome is essential for the development of motor deficits, aSyn aggregation, and microglia activation. Short-chain fatty acids were identified as modulators of microglial activation [55].

Beside this inflammatory response, microglia cells are directly involved in the clearance mechanism of aSyn [56]. In investigating the phagocytic activity of BV2 microglial cells after incubation with aSyn in different aggregation states, an increased internalization of fibrillar aSyn was observed [57]. Another study showed high uptake of aSyn by primary microglia, independent of its aggregation, whereas the C-terminally truncated form of aSyn had the strongest effect on inflammation and oxidative stress, with the TLR4 pathway being critically involved [58]. Besides the aggregation state, PD related mutations of aSyn influence the phagocytic activity of microglia. For the wildtype and A53T mutant form, an enhanced phagocytosis by primary microglia was observed, whilst incubation with the A30P or E46K mutant showed no effect [59]. Interestingly, concentration levels of aSyn also affect the cellular clearance capacity. In an elegant study with induced Pluripotent Stem Cell (PSC) lines generated from early onset PD patients with *SNCA* A53T and *SNCA* triplication mutations, it was shown that PSC-macrophages have a significantly reduced phagocytosis capability with the *SNCA* triplication. This could be phenocopied by adding monomeric aSyn to the cell culture medium. While the PSC-macrophages were, in principle, capable of clearing aSyn, their ability was compromised by high levels of exogenous or endogenous aSyn [60].

Overall, it can be stated that the microglia detect damaging events in the CNS very sensitively and quickly react with an adaptive response. To what extent this response may have a neuroprotective anti-inflammatory effect, or if rather the progression of the disease is amplified in a harmful pro-inflammatory manner, can often only be analyzed incompletely with standard methods. More precise molecular methods must be used to map the entire spectrum of microglial phenotypes and their functional consequences.

3.3. Lymphocyte Cells

T lymphocytes establish the connection from the peripheral immune system to the immune system of the CNS, and are of importance for the modulation of neuroinflammation in conjunction with CNS microglia. Together with B lymphocytes, T lymphocytes build the adaptive immune system. After maturing in the thymus, T lymphocytes circulate continuously in the blood until they are activated by antigen-presenting cells, such as dendritic cells, B cells, or microglia. A rough distinction is made between two classes, the CD8$^+$ T cells and the CD4$^+$ T cells. CD8$^+$ T cells, also known as cytotoxic T cells, are activated by the MHCI complex and predominantly induce apoptosis of damaged or infected cells. CD4$^+$ T-helper cells, which are activated by the MHCII complex, ensure a regulation of the immune response. The proteins of the MHC complex are a central element in immune regulation and present antigenic structures that can activate T lymphocytes [61].

In the midbrain of mice overexpressing aSyn, CD3$^+$ T cells were detected in significant numbers. In an additional analysis of the entire brain, similar amounts of CD4$^+$ and CD8$^+$ cells were detected [62]. Local overexpression of aSyn in midbrain neurons by viral vectors led to dopaminergic neurodegeneration [30], which was shown to be associated with a locally significantly increased proliferation of T and B lymphocytes being accompanied with microglial activation [49]. The maturation of T lymphocytes similarly seems to depend on the expression of alpha-synuclein, because aSyn knockout mice showed significantly reduced numbers of CD4$^+$, CD8$^+$, and even double CD4$^+$/CD8$^+$ negative cells compared to controls [63].

Looking at these data, it can be stated that the lymphocyte cell population is importantly involved in the modulation of neuroinflammation in PD models through its interaction with the innate immune system.

4. Alpha-Synuclein Pathology and Neuroinflammation in Human PD

In the human PD brain, perikaryal Lewy bodies represent the neuropathological hallmark and contain aggregated aSyn as a main component. Lewy neurites represent filamentous neuritic inclusions of degenerating dopaminergic neurons [64–66]. In familial PD, various point mutations in aSyn (A30P, E46K, H50Q, G51D, A53E, and A53T) have been identified, but also rare gene duplications or triplications exist which cause additional autosomal-dominant or -recessive forms of PD [67]. The pathological findings on neuroinflammation in human PD are extensive but also heterogeneous. This may not least be due to the fact that the clinical diagnosis of PD continues to be associated with uncertainties [68]. For this reason, it has to be assumed that the human data collected so far contain a proportion of patients who have been incorrectly diagnosed, and therefore findings on neuroinflammation may be more diverse. Another ongoing challenge is the pre-analytical processing of biosamples, as well as the analytical and clinical procedures, which need to be standardized in the best possible way and should be performed with a harmonized protocol in multicentric cohorts [69,70].

Initial analyzes of the neuroinflammation of PD were already performed 100 years ago on brain post mortem tissue [71]. In the neuropathological studies, the degeneration of dopaminergic nigral neurons and Lewy-body pathology were in focus first [72], but soon after the involvement of extra-nigral brain regions [73] and glial pathology was appreciated [74].

4.1. Astroglia

There are indications of astrocytic involvement in PD since very early investigations [74,75], although their impact varied in severity with different analytical techniques [76]. The topographical distribution pattern of aSyn immunoreactive astrocytes appears to parallel the formation of neuronal aSyn inclusion bodies in cortical neurons. This could depend on the proximity to terminal axons of affected cortico-striatal or cortico-thalamic neurons [77]. Compared to other brain regions, the astrocyte density in the SN is rather low, which could also explain the increased vulnerability of dopaminergic neurons to cellular stressors [78]. Astrocytes appear in two morphological phenotypes as protoplasmic astrocytes in the gray matter or as fibrous astrocytes in the white matter. Protoplasmic astrocytes surround the neuronal cell bodies and synapses, while fibrous astrocytes surround Ranvier's rings and oligodendroglia [32].

The function of astroglia is strongly influenced by microglia due to their close (patho)physiological interactions. Recently, in analogy to the classification of microglial cells into an M1 and M2 phenotype, a subdivision of the astrocytes into a destructive A1 and protective A2 phenotype has been postulated [79]. An astrocytic A1 phenotype was shown to be induced by classically activated microglia via secretion of the cytokines IL-1α, TNF-α, and complement component 1q (C1q), thereby losing the ability to promote neuronal survival and outgrowth or to support synaptogenesis. Rather, in the presence of this cell type, the probability for cell death of neurons and oligodendrocytes is increased. The histological analysis of astrocytes in the SN in human PD showed a significantly increased population of destructive A1 cells, so that a close connection with dopaminergic cell pathology could be established [79]. In knowledge of the strong activating stimulus of aSyn for microglia, it seems very likely that such a context promotes the A1 phenotype. In an own analysis, we detected increases of astrocytic cell numbers in the striatum of PD patients, which was also associated with increased expression of the protein ROCK2, a regeneration-inhibiting protein. This alteration was not found to the same extent in the SN. The striatal expression of regeneration inhibitory factors may thus indicate another cause for why dopaminergic axonal pathology develops [80].

The expression of aSyn in astrocytes in PD human brains is rather low, but it has been found in various aggregation states [77,81]. ASyn can accumulate in astrocytes, predominantly in non-fibrillar forms. In human neuropathological investigations, mostly fibrous astrocytes were shown to be affected [82]. They were located not only in the SN, but also in regions such as the striatum and dorsal thalamus, where dopaminergic terminals reside [77]. This indicates again a possibly harmful constellation for dopaminergic neurotransmission.

4.2. Microglia

The microglial cell population probably plays the most important role in neuroinflammation in human PD, but also in other neurodegenerative diseases [83]. It is found in an altered state both in the early stages and in the progressive course of the disease. Compared to healthy controls, a microglial cell number increase was shown in several post mortem studies on PD brains. Microglia were primarily located to the SN as identified by immunohistochemical detection of the complement receptor CR3/43, ferritin, or HLA-DR [76,84]. In the case of MHCII-positive microglia, an increased aSyn deposition was additionally found [85]. Other brain regions, such as putamen, hippocampus, transentorhinal cortex, or cingulus gyrus, also exhibited increased microglial cell populations [86]. In our own study, we detected a striatal increase in the number of ED1-positive microglial cells compared to age-matched healthy controls. This was accompanied with an increased expression of the regeneration-inhibitory protein ROCK2 and astrogliosis, indicating that there may be a microglial-astroglial interplay that impairs dopaminergic regenerative function [80]. The morphological phenotype of human microglia ranges from the ramified putative "resting state" to the "activated" or "phagocytotic" variant [87]. However, in our opinion this does not permit drawing a definite inference to their (patho)physiological function. Activated microglia express enzymes, such as iNOS, NADPH oxidase, COX, or myeloperoxidase, which are oxidative stressors that can potentiate inflammatory processes. An increased expression of these enzymes can be found in post mortem analyzes in the SN of PD patients [88,89]. Interestingly, microglia-associated pro-inflammatory cytokines, such as TNF-α, IL-1β, and IL-2, IL-10, or CRP, may be significantly increased in PD peripheral blood analyzes [90].

Much attention is currently being paid to nuclear medicine methods that label microglia in PD patients in vivo with positron emission tomography (PET) technology. The application of the marker [^{11}C]*(R)*-PK11195 for activated microglia to drug naïve early disease-stage PD patients showed a significantly higher density of microglia in the midbrain and putamen on the clinically more affected side. This also correlated with a reduced activity of dopamine transporter ligand [^{11}C]CFT and motor function [91,92]. Increased accumulation of microglia, also labelled with [^{11}C]*(R)*-PK11195, has recently been demonstrated, even in patients with REM sleep behavior disorder (RBD)—a high-risk prodromal feature of PD. In this study, a left-sided increased nigral accumulation of [^{11}C]*(R)*-PK11195 correlated with a decreased left-sided putaminal signal of the dopaminergic marker [^{18}F]-DOPA [93].

However, these findings have to be interpreted with caution, because [^{11}C]*(R)*-PK11195 or other TSPO ligands are not exclusively specific for microglial cells, as recent studies have demonstrated. It is known that microglial activation in response to various stimuli is associated with an upregulation of its mitochondrial translocator protein-18 kDa (TSPO) [94], which is also the case in various neurodegenerative diseases, including PD [95]. But TSPO is also involved in several physiological processes, such as steroidogenesis, mitochondrial oxidation, cell growth, and differentiation, and thus is not a discriminating factor to differentiate microglial phenotypes. Furthermore, a common single nucleotide polymorphism (rs6971) in exon 4 of the *TSPO* gene affects the ligand-binding affinity [95]. Importantly, TSPO can also be expressed on activated astroglia [96]. Novel second-generation TSPO ligands labeled with the short-lived positron-emitter isotopes ^{11}C and ^{18}F like [^{11}C]-PBR06 or [^{18}F]-DPA-714 enable the visualization of TSPO expression with improved signal to nose ratios and are currently extensively studied in human clinical studies. Here, quantification of data is a complex task, because microglia are distributed ubiquitously throughout the entire brain, and premorbid disease may occur. Advanced methods using cluster analysis could be performed with dynamic PET scans of each individual to determine suitable reference regions [95]. The identification of new targets that specify either pro- inflammatory or anti-inflammatory microglial phenotypes will provide further information on the role of these immune cells and provide a mechanistic rationale for the development of more effective neuroprotective drugs, including antibody-based therapies. A novel microglial marker candidate is the macrophage colony-stimulating factor 1 receptor (CSF1R), which was successfully targeted with a positron-emitting, high-affinity ligand [^{11}C]-CPPC. It showed specific and elevated uptake in a murine model of AD and in post mortem brain tissue of AD patients [97].

4.3. Lymphocyte Cells

The appearance of lymphocyte cells has been primarily demonstrated in post mortem SN analyzes of PD patients [84,85,98,99]. Interestingly, in the peripheral blood of PD patients, significant differences were found in the abundance of lymphocyte subpopulations, which indicated altered adaptive immunity and were associated with motor dysfunction [100]. In other studies, the ratio of $CD4^+$ to $CD8^+$ cells decreased, the number of regulatory $CD4^+/CD25^+$ T cells was reduced, or the total number of $CD4^+$ T cells and $CD19^+$ B cells were decreased [101,102]. This indicates that in PD patients, possibly due to aSyn pathology, maturation deficits or dysregulated populations of T lymphocytes develop, as has also been observed in animal studies [63].

In several genetic examinations, an association of PD with the individual configuration of alleles of the MHC complex was found [103,104]. To clarify the question of whether PD is associated with T-cell recognition of aSyn epitopes presented by specific MHC alleles, Sulzer et al. examined PD patients and age-matched controls. After stimulation of human PBMC with IFN-γ or IL-5 in one or two-thirds of the patients, respectively, a reaction to aSyn peptides of the Y39-near N-terminal or S129-related C-terminal region was found. This response was mediated by $CD4^+$ or $CD8^+$ T cells and was especially robust with defined subsets of MHC alleles [105]. Approximately 40% of the PD patients in the study exhibited immune responses to aSyn epitopes, which could reflect variations in disease progression or environmental factors. In classic autoimmune disorders, such as type-1 diabetes or rheumatoid arthritis, a similar percentage of patients display these responses [106]. Here, the MHC class II response may precede MHC class I, and it was noted in the Sulzer manuscript that exposing microglia to aSyn activated MHC class I expression by dopamine neurons [105].

These data demonstrate that aSyn protein fragments can induce T cell-mediated immune responses in PD patients and raise the suspicion that PD could be an autoimmune disorder. To what extent the risk of progression of disease in PD individuals with certain MHC allele configurations may be increased must be investigated in further studies.

From the overall point of view, it becomes clear that there is a relevant neuroinflammatory component also in human PD neuropathology, which is documented in post mortem analyzes as well as in in vivo investigations at different time points during the disease course [74,93,107]. Unfortunately, longitudinal interindividual or even intraindividual analyzes are missing or are only available for small patient numbers and periods of time in which no significant alterations have been observed [107].

5. Immunotherapies to Target Alpha-Synuclein

As shown in the previous sections, aSyn pathology is a central hallmark of PD. Both in animal experiments and in human studies, it has been shown that accompanying neuroinflammatory processes involving astroglia, microglia, and T cells play a role for disease development. Therapeutic strategies with direct targeting of single or multiple aspects of cellular neuroinflammation are currently under investigation and are described elsewhere [44,108]. Another therapeutic approach is to primarily focus on aSyn as the causative substrate of this pathology. The main goal here is to lower the burden of extracellular aSyn in the brain, and as a consequence reduce neuroinflammation. Until now, passive immunization with administration of antibodies directed against aSyn, seems to be one of the most promising strategies to accomplish this objective [109]. In the following section, an overview of the most important recent results with the passive aSyn immunotherapies are presented.

5.1. Preclinical Data

Passive immunization in PD is performed by an anti-aSyn antibody transfer into the CNS. The two primary targets for monoclonal antibodies are the aSyn C-terminal region [110–112] and the aSyn N-terminal region [113,114]. In the following section, only data with aSyn transgenic mice will be discussed because toxin-based PD mouse models have not been examined for this therapeutic approach.

In 2011, a monoclonal antibody that targets the C-terminus of aSyn (9E4) was administered for a period of 6 months with once weekly intravenous injections of a dose of 10 mg/kg body weight to a transgenic mouse model of PD. The mice exhibited an improved behavioral response in the water maze and showed less aggregation of calpain-cleaved aSyn in synapses and axons of both cortical but also hippocampal neurons [111]. One year later, a second group applied a different C-terminal antibody (AB274) to the identical mouse model with once weekly injections intraperitoneally. In their study, behavioral motor deficits improved too, and reduced neurodegeneration in the passively immunized group could be demonstrated after 4 weeks. Interestingly, an enhanced localization of aSyn and AB274 in microglial cells was found. It could be shown on a microglial cell line that the uptake of the AB274/aSyn complex through microglial cells was mediated via the Fcγ receptor and directed into lysosomal compartments [110]. In a third study with anti-aSyn antibodies directed against the C-terminally, the compounds 1H7, 5C1, and 5D12 were administered once weekly for a total of six weeks in another transgenic PD mouse model. Here, the behavioral performance was improved and the expression of aSyn in mouse cortex and striatum after application of 1H7 and 5C1was significantly reduced. This was accompanied by a less severe axonal pathology [112].

With regards to anti-aSyn antibodies that are N-terminally directed, Weihofen et al. reported recently on the identification, binding characteristics, and efficacy in mouse models of PD with the human-derived aSyn antibody (BIIB054) [115]. In three different experiments with intracerebrally inoculated preformed aSyn fibrils, BIIB054 treatment for 3 months or longer was shown to be able to attenuate the spreading of aSyn pathology, rescue motor impairments, and reduce loss of dopamine transporter density in dopaminergic striatum terminals [115]. In another study, intraperitoneal administration of an anti-aSyn antibody (AB1) two weekly for a total of three months resulted in a moderate but insignificant improvement in behavioral parameters. Nigral dopaminergic neurons were robustly protected against cellular loss and the general expression of aSyn, as well as the activation of microglia was found to be decreased in this brain region [113]. A treatment of mice that had been intrastriatally injected with aSyn pre-synthetic fibrils with an anti-aSyn N-terminal antibody (Syn303) was undertaken in order to find a more specific antibody design for misfolded aSyn. This antibody was applied weekly for a period of up to half a year and successfully inhibited the spread of aSyn aggregates, as well as improved behavioral deficits and loss of dopaminergic cells [114].

In additional studies on transgenic PD mice, these were either weekly immunized for more than three months with antibodies (Syn-F1, Syn-O1, and Syn-O4) directed specifically against oligomeric and fibrillary aSyn but not against monomeric aSyn [116], or received weekly passive immunizations for a total of 14 weeks with a monoclonal antibody that targeted oligomeric/protofibrillary forms of aSyn (mAb47 antibody) [117]. Both approaches led to significant decreases in the accumulation of pathogen aSyn fibrils and improved behavioral performance measures. Passive immunization of mice overexpressing human aSyn in oligodendrocytes (PLP-aSyn mice) resulted in a reduction of total aSyn in the hippocampus and less intracellular accumulation of aggregated aSyn, particularly in the spinal cord. This was accompanied by a reduced amount of activated microglial cells [118]. If non-transgenic, aSyn knock-out, or aSyn transgenic mice received unilateral intra-cerebral injections with an aSyn-expressing viral vector construct, the axonal accumulation of aSyn in the contralateral side could be reduced with a three-month systemic administration of a monoclonal anti-aSyn antibody (1H7, recognizing amino acids 91–99) [119].

The evaluation of neuroinflammatory changes in the above-mentioned animal studies after passive immunization differs greatly in the scope of the analyzes. Overall, the regulation of microglia is a common parameter, which is often interpreted to have been shifted to a more beneficial level. The function of astrocytes is only analyzed in a few studies and T cell responses are not described. An overview is shown in Table 1.

Table 1. Passive immunization studies in Parkinson's disease animal models and therapeutic responses.

Antibody	Target Site of aSyn	Behavioural Alterations	Cellular Responses of				Reference
			Neurons	Astroglia	Microglia	T Cells	
9E4	C-terminal	Cylinder Test ↑	DA cell loss (30%) ↓	-	-	-	[111]
Ab274	C-terminal	Pole test and total activity ↑	NeuN cell loss ↓	Astroglial aSyn accumulation ↓	Microgliosis ↓ Increased aSyn clearance by microglia	-	[110]
1H7 5C1	C-terminal	Probe/transversal round beam test ↑	DA cell loss (35%) ↓	Astrogliosis ↓	Microgliosis ↓	-	[112]
AB1 AB2	N-terminal Central region	Cylinder test ↑	DA and NeuN cell loss ↓	-	Microgliosis ↓	-	[113]
Syn303	N-terminal	Wirehang ↑	Blocked aSyn spreading DA cell loss ↓	-	-	-	[114]
BIIB054	N-terminal	Wirehang ↑	Loss of dopamine transporter density in the SN ↓	No change	No change	-	[115]
mAB47	Protofibrils	-	aSyn protofibrils in spinal cord ↓	No change	No change	-	[117]
Syn-O1 Syn-O4 Syn-F1	Oligomers Oligomers Fibrils	Total activity in open field test ↑	aSyn accumulation ↓ NeuN cell loss ↓	Astrogliosis in hippocampus ↓ (O1,O4)	Microgliosis in hippocampus ↓ (O1,O4)	-	[116]

Apart from antibody-based therapeutic strategies to alleviate aSyn toxicity and promote its clearance, another option is to employ small molecules. Nilotinib is a c-Abl tyrosine kinase inhibitor that is used in the treatment of chronic myelogenous leukemia. It has recently been shown to be able to both alleviate c-Abl associated aSyn aggregation and impaired autophagy, as well as protect dopaminergic neurons in mouse models of PD [120]. Importantly, the level and activity of c-Abl can be elevated in human PD in both the substantia nigra and striatum [121], so that the potential of c-Abl inhibitors as a symptomatic and causative treatment is currently under evaluation in a drug repurposing approach [122].

In view of these data, it can be stated that the anti-aSyn-directed immunotherapies are effective in animal models—whether directed against the C-terminus or the N-terminus of aSyn—and clearly demonstrate "target engagement", reduce the lesion burden of aSyn (mostly in a specific manner), can to some extent probably alleviate "αactivation" of cellular inflammatory responses, and achieve improvements in motor behavior. These promising findings were strong arguments to proceed to clinical trials in humans.

5.2. Human Clinical Data

Various human clinical trials using anti-aSyn antibodies have been performed in a phase I clinical design with healthy volunteers. Overall, they have shown a good safety profile and tolerability. More recently, large phase II clinical trials have been launched and will soon provide data on safety, tolerability and to some extent on treatment effects in PD patients.

PRX002 was the first passive therapeutic immunization using an anti-aSyn directed antibody that was developed by Prothena. It is a C-terminally directed humanized IgG1 monoclonal antibody and has been studied in animal models before, as mentioned above [111]. In a phase I study with a single ascending dose, PRX002 show good tolerability and safety with intravenous infusions of 0.3, 1.0, 3.0, 10, or 30 mg/kg. After a single infusion of 30 mg/kg, serum PRX002 antibody levels were shown to be increased up to 578 μg/mL. Within one hour from PRX002 administration, there was a significant reduction of free aSyn in serum, which was dose-dependent. Interestingly, levels of total aSyn that comprise free and bound aSyn increased in a dose-dependent way. This was presumably due to an expected change in kinetics after the binding of antibodies. The average terminal half-life across all doses was 18.2 days [123].

This trial was followed by another phase I study with multiple ascending-doses in patients with mild or moderate idiopathic PD (HY stages 1–3). These received three intravenous infusions of PRX002 (0.3 mg/kg, 1.0 mg/kg, 3.0 mg/kg, 10 mg/kg, 30 mg/kg, or 60 mg/kg) or placebo at intervals of 4 weeks. Data on tolerability and safety were favorable. Serum levels of PRX002 increased dose proportionally, whereas serum free-to-total serum levels were strongly reduced. The antibody mean terminal elimination half-life was comparable for all doses, approximating 10.2 days. The mean CSF concentration of the study compound increased dose dependently and was approximately 0.3% relative to serum across all dose cohorts. However, it was not possible to demonstrate any alteration in the CSF aSyn level, probably due to the relatively weak affinity of the antibody for the monomeric forms, as opposed to aSyn aggregates, which are much less prominent in the CSF [124]. In 2017, Prothena and Roche launched a multinational phase II study of PRX002/RO7046015 in patients with newly diagnosed PD (PASADENA Study, ClinicalTrials identifier NCT03100149).

BIIB054 is a fully human anti-aSyn IgG1 monoclonal antibody directed at the N-terminus of aSyn. It has been developed by Neurimmune, a Swiss biotech company, and was originally isolated from B cell lines that had been generated from neurologically healthy individuals. It is very selective for the aggregated form of human aSyn that has been detected in tissue sections from PD and DLB patients [125]. A good safety and tolerance were recently demonstrated at single doses with up to 90 mg/kg in a phase I study in healthy volunteers. The BIIB054 serum half-life was 28 days, its CSF concentrations obtained were 0.2% of those observed in plasma. In PD patients, single doses of BIIB054 up to 45 mg/kg were well tolerated, the CSF:plasma ratio was 0.4%, and the pharmacokinetic profile was comparable to that observed in healthy volunteers [126]. Currently, BIIB054 is being evaluated in a multinational phase II study with patients with newly diagnosed PD (SPARK Study, ClinicalTrials identifier NCT03318523).

The MEDI1341 monoclonal and the BAN0805 monoclonal anti-aSyn antibodies are two more compounds which are currently being planned to be tested in single-center studies of single ascending intravenous doses in phase I clinical trials. MEDI1341 appears to have only a reduced immune effector function and is now developed in collaboration of AstraZeneca and Takeda (ClinicalTrials.gov identifier NCT03272165). BAN0805 is more specifically targeting aSyn oligomers as well as protofibrils, and is being developed in a collaboration between Abbvie and Bioarctic [127].

These clinical studies are all carried out at a very high pace of development and take place at a high medical technology level. Unfortunately, no data on the modulation of neuroinflammatory parameters, e.g., in blood, serum, or other compartments, are publicly available.

6. Conclusions

It is undisputed that neuroinflammatory processes are significantly associated with the neuropathological progression and clinical course of PD. Through very elaborate studies on patients, but also in animal models of the disease, core pathomechanisms have been identified, which apart from the damage to neuronal cells also affect glial cells and are regulated by neuroimmunological interactions. This opens up a large window for neuroimmunological interventions with the ultimate goal to modify the course of disease.

Although more focused interventions with targeted regulation of the function of astrocytes, microglia, or T lymphocytes will become possible in the future, the knowledge of the multicellular spreading PD neuropathology seems to make a rather broad and systemic approach meaningful. Therefore, the reduction of CNS aSyn by targeted antibody-based technologies is a valid tool. In future human clinical studies, the target engagement and effectiveness of anti-aSyn antibodies to reduce aSyn burden in the CNS should be very precisely monitored. This could be achieved by detection of CSF aSyn with enzyme-linked sandwich-type immunosorbent assays (ELISA) or even more precise analyzes for its aggregated states with Protein Misfolding Cyclic Amplification (PMCA) or Real-Time Quaking-Induced Conversion (RT-QuIC) [128]. A non-invasive in vivo analysis of the brain aSyn load would be possible via nuclear medicine PET or SPECT methodology, but this is currently not established [129].

In contrast, the PET technology for specific labeling of microglial cells has improved and could be integrated with the use of second-generation microglial TSPO tracers [130]. First-generation tracers have been shown to perform well already in prodromal stages of PD [93] and have also been used as cellular readout in microglia-targeted therapeutic approaches [131]. One challenge remains to quantify the extent of CNS neuroinflammation as precisely as possible. In addition to the most specific markers possible, a comprehensive assessment must map the various cell populations and also represent individual phenotypes. With regard to anti-aSyn immunotherapies, the characterization and quantification of aSyn-dependent immune effects should be a main goal, especially during the course of therapy. Without such measures, the assessment of a change in neuroinflammation under passive immunotherapy remains limited.

From a practical point of view, an improved applicability of antibodies with subcutaneous injections would facilitate their use in clinical studies even more. Safety and tolerability issues will always remain primary goals for the development of novel agents. If these requirements can be met, it will only be a matter of time before an effective method of anti-aSyn antibody-based therapy is developed that effectively modulates cellular neuroinflammation and also achieves clinically relevant improvements in PD patients.

Author Contributions: Conceptualization: M.A.S.Z., L.T.; literature search: M.A.S.Z., J.M., F.O., F.M., S.M., R.G., A.H., L.T.; writing—original draft preparation: M.A.S.Z., L.T.; writing—review and editing: M.A.S.Z., J.M., F.O., F.M., S.M., R.G., A.H., L.T.; project administration: M.A.S.Z., L.T.

Abbreviations

AP	Activator protein
aSyn	Alpha-Synuclein
CD	Cluster of differentiation
CIITA	Class II transactivator
CNS	Central nervous system
COX	Cyclooxygenase
CR	Complement receptor
CRP	C-reactive protein
CSF	Cerebrospinal fluid
DA	Dopamine
DOPA	Dihydroxyphenylalanine
ELISA	Enzyme-linked sandwich-type immunosorbent assays
HLA-DR	Human leukocyte antigen—DR isotype
IFN	Interferon
Ig	Immunoglobulin
IL	Interleukin
iNOS	Inducible nitric oxide synthase
MHC	Major histocompatibility complex
NAC	Non-amyloid-beta component
NADPH	Nicotinamide adenine dinucleotide phosphate
NFκB	Nuclear factor 'kappa-light-chain-enhancer' of activated B-cells
NMDA receptor	N-methyl-D-aspartate receptor
NO	Nitrogen monoxide
Nrf2	Nuclear factor erythroid 2-related factor 2
NSF	N-ethylmaleimide sensitive factor
PBMC	Peripheral blood mononuclear cell
PD	Parkinson's disease
PET	Positron emission tomography
PMCA	Protein misfolding cyclic amplification
PNS	Peripheral nervous system
RNA	Ribonucleic acid
ROCK2	Rho associated coiled-coil containing protein kinase 2
ROS	Reactive oxygen species
RT-QuIC	Real-time quaking-induced conversion
SN	Substantia nigra
SNARE	Soluble N-ethylmaleimide sensitive factor attachment receptor
SPECT	Single photon emission computed tomography
TGF	Transforming growth factor
TLR	Toll-like receptor
TNF	Tumor necrosis factor
TNT	Tunneling nanotubes
TSPO	Translocator protein-18 kDa

References

1. Poewe, W.; Seppi, K.; Tanner, C.M.; Halliday, G.M.; Brundin, P.; Volkmann, J.; Schrag, A.E.; Lang, A.E. Parkinson disease. *Nat. Rev. Dis. Primers* **2017**, *3*, 17013. [CrossRef]
2. Przedborski, S. The two-century journey of Parkinson disease research. *Nat. Rev. Neurosci.* **2017**, *18*, 251–259. [CrossRef]
3. Kalia, L.V.; Lang, A.E. Parkinson's disease. *Lancet* **2015**, *386*, 896–912. [CrossRef]
4. Schapira, A.H.V.; Chaudhuri, K.R.; Jenner, P. Non-motor features of Parkinson disease. *Nat. Rev. Neurosci.* **2017**, *18*, 435–450. [CrossRef]
5. Titova, N.; Chaudhuri, K.R. Non-motor Parkinson disease: New concepts and personalised management. *Med. J. Aust.* **2018**, *208*, 404–409. [CrossRef]
6. Zeng, X.S.; Geng, W.S.; Jia, J.J.; Chen, L.; Zhang, P.P. Cellular and Molecular Basis of Neurodegeneration in Parkinson Disease. *Front. Aging Neurosci.* **2018**, *10*, 109. [CrossRef]
7. Rocha, E.M.; De Miranda, B.; Sanders, L.H. α-synuclein: Pathology, mitochondrial dysfunction and neuroinflammation in Parkinson's disease. *Neurobiol. Dis.* **2018**, *109*, 249–257. [CrossRef]
8. Obeso, J.A.; Stamelou, M.; Goetz, C.G.; Poewe, W.; Lang, A.E.; Weintraub, D.; Burn, D.; Halliday, G.M.; Bezard, E.; Przedborski, S.; et al. Past, present, and future of Parkinson's disease: A special essay on the 200th Anniversary of the Shaking Palsy. *Mov. Disord.* **2017**, *32*, 1264–1310. [CrossRef]
9. Troncoso-Escudero, P.; Parra, A.; Nassif, M.; Vidal, R.L. Outside in: Unraveling the Role of Neuroinflammation in the Progression of Parkinson's Disease. *Front. Neurol.* **2018**, *9*, 860. [CrossRef]
10. Tonges, L.; Metzdorf, J.; Zella, S. [Parkinson's disease and neuroinflammation—Cellular pathology, mechanisms and therapeutic options]. *Fortschr. Neurol. Psychiatr.* **2018**. [CrossRef]
11. Johnson, M.E.; Stecher, B.; Labrie, V.; Brundin, L.; Brundin, P. Triggers, Facilitators, and Aggravators: Redefining Parkinson's Disease Pathogenesis. *Trends Neurosci.* **2019**, *42*, 4–13. [CrossRef]
12. Bendor, J.T.; Logan, T.P.; Edwards, R.H. The function of α-synuclein. *Neuron* **2013**, *79*, 1044–1066. [CrossRef]
13. Emamzadeh, F.N. α-synuclein structure, functions, and interactions. *J. Res. Med. Sci.* **2016**, *21*, 29. [CrossRef]
14. Bengoa-Vergniory, N.; Roberts, R.F.; Wade-Martins, R.; Alegre-Abarrategui, J. α-synuclein oligomers: A new hope. *Acta Neuropathol.* **2017**, *134*, 819–838. [CrossRef]
15. Burre, J. The Synaptic Function of α-Synuclein. *J. Parkinson's Dis.* **2015**, *5*, 699–713. [CrossRef]
16. Marques, O.; Outeiro, T.F. α-synuclein: From secretion to dysfunction and death. *Cell Death Dis.* **2012**, *3*, e350. [CrossRef]
17. Burré, J.; Sharma, M.; Tsetsenis, T.; Buchman, V.; Etherton, M.R.; Südhof, T.C. α-synuclein promotes SNARE-complex assembly in vivo and in vitro. *Science* **2010**, *329*, 1663–1667. [CrossRef]
18. Bridi, J.C.; Hirth, F. Mechanisms of α-Synuclein Induced Synaptopathy in Parkinson's Disease. *Front. Neurosci.* **2018**, *12*, 80. [CrossRef]
19. Burre, J.; Sharma, M.; Sudhof, T.C. Definition of a molecular pathway mediating α-synuclein neurotoxicity. *J. Neurosci.* **2015**, *35*, 5221–5232. [CrossRef]
20. Brundin, P.; Melki, R. Prying into the Prion Hypothesis for Parkinson's Disease. *J. Neurosci.* **2017**, *37*, 9808–9818. [CrossRef]
21. Walsh, D.M.; Selkoe, D.J. A critical appraisal of the pathogenic protein spread hypothesis of neurodegeneration. *Nat. Rev. Neurosci.* **2016**, *17*, 251–260. [CrossRef]
22. Surmeier, D.J.; Obeso, J.A.; Halliday, G.M. Parkinson's Disease Is Not Simply a Prion Disorder. *J. Neurosci.* **2017**, *37*, 9799–9807. [CrossRef] [PubMed]
23. Yamada, K.; Iwatsubo, T. Extracellular α-synuclein levels are regulated by neuronal activity. *Mol. Neurodegener.* **2018**, *13*, 9. [CrossRef] [PubMed]
24. Volpicelli-Daley, L.A.; Luk, K.C.; Lee, V.M. Addition of exogenous α-synuclein preformed fibrils to primary neuronal cultures to seed recruitment of endogenous α-synuclein to Lewy body and Lewy neurite-like aggregates. *Nat. Protoc.* **2014**, *9*, 2135–2146. [CrossRef] [PubMed]
25. Volpicelli-Daley, L.A.; Luk, K.C.; Patel, T.P.; Tanik, S.A.; Riddle, D.M.; Stieber, A.; Meaney, D.F.; Trojanowski, J.Q.; Lee, V.M. Exogenous α-synuclein fibrils induce Lewy body pathology leading to synaptic dysfunction and neuron death. *Neuron* **2011**, *72*, 57–71. [CrossRef]

26. Hassink, G.C.; Raiss, C.C.; Segers-Nolten, I.M.J.; van Wezel, R.J.A.; Subramaniam, V.; le Feber, J.; Claessens, M. Exogenous α-synuclein hinders synaptic communication in cultured cortical primary rat neurons. *PLoS ONE* **2018**, *13*, e0193763. [CrossRef] [PubMed]
27. Diogenes, M.J.; Dias, R.B.; Rombo, D.M.; Vicente Miranda, H.; Maiolino, F.; Guerreiro, P.; Nasstrom, T.; Franquelim, H.G.; Oliveira, L.M.; Castanho, M.A.; et al. Extracellular α-synuclein oligomers modulate synaptic transmission and impair LTP via NMDA-receptor activation. *J. Neurosci.* **2012**, *32*, 11750–11762. [CrossRef]
28. Peelaerts, W.; Bousset, L.; Van der Perren, A.; Moskalyuk, A.; Pulizzi, R.; Giugliano, M.; Van den Haute, C.; Melki, R.; Baekelandt, V. α-Synuclein strains cause distinct synucleinopathies after local and systemic administration. *Nature* **2015**, *522*, 340–344. [CrossRef]
29. Luk, K.C.; Kehm, V.; Carroll, J.; Zhang, B.; O'Brien, P.; Trojanowski, J.Q.; Lee, V.M. Pathological α-synuclein transmission initiates Parkinson-like neurodegeneration in nontransgenic mice. *Science* **2012**, *338*, 949–953. [CrossRef]
30. Tonges, L.; Szego, E.M.; Hause, P.; Saal, K.A.; Tatenhorst, L.; Koch, J.C.; Hedouville, Z.D.; Dambeck, V.; Kugler, S.; Dohm, C.P.; et al. α-synuclein mutations impair axonal regeneration in models of Parkinson's disease. *Front. Aging Neurosci.* **2014**, *6*, 239. [CrossRef]
31. Herculano-Houzel, S. The human brain in numbers: A linearly scaled-up primate brain. *Front. Hum. Neurosci.* **2009**, *3*, 31. [CrossRef] [PubMed]
32. Sofroniew, M.V.; Vinters, H.V. Astrocytes: Biology and pathology. *Acta Neuropathol.* **2010**, *119*, 7–35. [CrossRef]
33. Lee, H.J.; Suk, J.E.; Patrick, C.; Bae, E.J.; Cho, J.H.; Rho, S.; Hwang, D.; Masliah, E.; Lee, S.J. Direct transfer of α-synuclein from neuron to astroglia causes inflammatory responses in synucleinopathies. *J. Biol. Chem.* **2010**, *285*, 9262–9272. [CrossRef] [PubMed]
34. Gu, X.L.; Long, C.X.; Sun, L.; Xie, C.; Lin, X.; Cai, H. Astrocytic expression of Parkinson's disease-related A53T α-synuclein causes neurodegeneration in mice. *Mol. Brain* **2010**, *3*, 12. [CrossRef] [PubMed]
35. Gan, L.; Vargas, M.R.; Johnson, D.A.; Johnson, J.A. Astrocyte-specific overexpression of Nrf2 delays motor pathology and synuclein aggregation throughout the CNS in the α-synuclein mutant (A53T) mouse model. *J. Neurosci.* **2012**, *32*, 17775–17787. [CrossRef] [PubMed]
36. Loria, F.; Vargas, J.Y.; Bousset, L.; Syan, S.; Salles, A.; Melki, R.; Zurzolo, C. α-Synuclein transfer between neurons and astrocytes indicates that astrocytes play a role in degradation rather than in spreading. *Acta Neuropathol.* **2017**, *134*, 789–808. [CrossRef] [PubMed]
37. Rostami, J.; Holmqvist, S.; Lindstrom, V.; Sigvardson, J.; Westermark, G.T.; Ingelsson, M.; Bergstrom, J.; Roybon, L.; Erlandsson, A. Human Astrocytes Transfer Aggregated α-Synuclein via Tunneling Nanotubes. *J. Neurosci.* **2017**, *37*, 11835–11853. [CrossRef]
38. Soulet, D.; Rivest, S. Microglia. *Curr. Biol.* **2008**, *18*, R506–R508. [CrossRef]
39. Hanisch, U.K.; Kettenmann, H. Microglia: Active sensor and versatile effector cells in the normal and pathologic brain. *Nat. Neurosci.* **2007**, *10*, 1387–1394. [CrossRef]
40. Wolf, S.A.; Boddeke, H.W.; Kettenmann, H. Microglia in Physiology and Disease. *Annu. Rev. Physiol.* **2017**, *79*, 619–643. [CrossRef]
41. Nimmerjahn, A.; Kirchhoff, F.; Helmchen, F. Resting microglial cells are highly dynamic surveillants of brain parenchyma in vivo. *Science* **2005**, *308*, 1314–1318. [CrossRef]
42. Wake, H.; Moorhouse, A.J.; Jinno, S.; Kohsaka, S.; Nabekura, J. Resting microglia directly monitor the functional state of synapses in vivo and determine the fate of ischemic terminals. *J. Neurosci.* **2009**, *29*, 3974–3980. [CrossRef] [PubMed]
43. Tang, Y.; Le, W. Differential Roles of M1 and M2 Microglia in Neurodegenerative Diseases. *Mol. Neurobiol.* **2016**, *53*, 1181–1194. [CrossRef] [PubMed]
44. Subramaniam, S.R.; Federoff, H.J. Targeting Microglial Activation States as a Therapeutic Avenue in Parkinson's Disease. *Front. Aging Neurosci.* **2017**, *9*, 176. [CrossRef] [PubMed]
45. Graeber, M.B. Changing face of microglia. *Science* **2010**, *330*, 783–788. [CrossRef] [PubMed]
46. Hu, X.; Leak, R.K.; Shi, Y.; Suenaga, J.; Gao, Y.; Zheng, P.; Chen, J. Microglial and macrophage polarization-new prospects for brain repair. *Nat. Rev. Neurol.* **2015**, *11*, 56–64. [CrossRef]
47. Ransohoff, R.M. A polarizing question: Do M1 and M2 microglia exist? *Nat. Neurosci.* **2016**, *19*, 987–991. [CrossRef] [PubMed]

48. Tofaris, G.K.; Garcia Reitböck, P.; Humby, T.; Lambourne, S.L.; O'Connell, M.; Ghetti, B.; Gossage, H.; Emson, P.C.; Wilkinson, L.S.; Goedert, M.; et al. Pathological changes in dopaminergic nerve cells of the substantia nigra and olfactory bulb in mice transgenic for truncated human α-synuclein(1-120): Implications for Lewy body disorders. *J. Neurosci.* **2006**, *26*, 3942–3950. [CrossRef]
49. Theodore, S.; Cao, S.; McLean, P.J.; Standaert, D.G. Targeted overexpression of human α-synuclein triggers microglial activation and an adaptive immune response in a mouse model of Parkinson disease. *J. Neuropathol. Exp. Neurol.* **2008**, *67*, 1149–1158. [CrossRef]
50. Watson, M.B.; Richter, F.; Lee, S.K.; Gabby, L.; Wu, J.; Masliah, E.; Effros, R.B.; Chesselet, M.F. Regionally-specific microglial activation in young mice over-expressing human wildtype α-synuclein. *Exp. Neurol.* **2012**, *237*, 318–334. [CrossRef]
51. Kim, C.; Ho, D.H.; Suk, J.E.; You, S.; Michael, S.; Kang, J.; Joong Lee, S.; Masliah, E.; Hwang, D.; Lee, H.J.; et al. Neuron-released oligomeric α-synuclein is an endogenous agonist of TLR2 for paracrine activation of microglia. *Nat. Commun.* **2013**, *4*, 1562. [CrossRef] [PubMed]
52. Lee, E.J.; Woo, M.S.; Moon, P.G.; Baek, M.C.; Choi, I.Y.; Kim, W.K.; Junn, E.; Kim, H.S. α-synuclein activates microglia by inducing the expressions of matrix metalloproteinases and the subsequent activation of protease-activated receptor-1. *J. Immunol.* **2010**, *185*, 615–623. [CrossRef] [PubMed]
53. Hoenen, C.; Gustin, A.; Birck, C.; Kirchmeyer, M.; Beaume, N.; Felten, P.; Grandbarbe, L.; Heuschling, P.; Heurtaux, T. α-Synuclein Proteins Promote Pro-Inflammatory Cascades in Microglia: Stronger Effects of the A53T Mutant. *PLoS ONE* **2016**, *11*, e0162717. [CrossRef] [PubMed]
54. Williams, G.P.; Schonhoff, A.M.; Jurkuvenaite, A.; Thome, A.D.; Standaert, D.G.; Harms, A.S. Targeting of the class II transactivator attenuates inflammation and neurodegeneration in an α-synuclein model of Parkinson's disease. *J. Neuroinflamm.* **2018**, *15*, 244. [CrossRef] [PubMed]
55. Sampson, T.R.; Debelius, J.W.; Thron, T.; Janssen, S.; Shastri, G.G.; Ilhan, Z.E.; Challis, C.; Schretter, C.E.; Rocha, S.; Gradinaru, V.; et al. Gut Microbiota Regulate Motor Deficits and Neuroinflammation in a Model of Parkinson's Disease. *Cell* **2016**, *167*, 1469–1480.e12. [CrossRef] [PubMed]
56. Lee, H.J.; Suk, J.E.; Bae, E.J.; Lee, S.J. Clearance and deposition of extracellular α-synuclein aggregates in microglia. *Biochem. Biophys. Res. Commun.* **2008**, *372*, 423–428. [CrossRef] [PubMed]
57. Hoffmann, A.; Ettle, B.; Bruno, A.; Kulinich, A.; Hoffmann, A.C.; von Wittgenstein, J.; Winkler, J.; Xiang, W.; Schlachetzki, J.C. α-synuclein activates BV2 microglia dependent on its aggregation state. *Biochem. Biophys. Res. Commun.* **2016**, *479*, 881–886. [CrossRef]
58. Fellner, L.; Irschick, R.; Schanda, K.; Reindl, M.; Klimaschewski, L.; Poewe, W.; Wenning, G.K.; Stefanova, N. Toll-like receptor 4 is required for α-synuclein dependent activation of microglia and astroglia. *Glia* **2013**, *61*, 349–360. [CrossRef]
59. Roodveldt, C.; Labrador-Garrido, A.; Gonzalez-Rey, E.; Fernandez-Montesinos, R.; Caro, M.; Lachaud, C.C.; Waudby, C.A.; Delgado, M.; Dobson, C.M.; Pozo, D. Glial innate immunity generated by non-aggregated α-synuclein in mouse: Differences between wild-type and Parkinson's disease-linked mutants. *PLoS ONE* **2010**, *5*, e13481. [CrossRef]
60. Haenseler, W.; Zambon, F.; Lee, H.; Vowles, J.; Rinaldi, F.; Duggal, G.; Houlden, H.; Gwinn, K.; Wray, S.; Luk, K.C.; et al. Excess α-synuclein compromises phagocytosis in iPSC-derived macrophages. *Sci. Rep.* **2017**, *7*, 9003. [CrossRef]
61. Fakhoury, M. Immune-mediated processes in neurodegeneration: Where do we stand? *J. Neurol.* **2016**, *263*, 1683–1701. [CrossRef] [PubMed]
62. Sommer, A.; Fadler, T.; Dorfmeister, E.; Hoffmann, A.C.; Xiang, W.; Winner, B.; Prots, I. Infiltrating T lymphocytes reduce myeloid phagocytosis activity in synucleinopathy model. *J. Neuroinflamm.* **2016**, *13*, 174. [CrossRef]
63. Shameli, A.; Xiao, W.; Zheng, Y.; Shyu, S.; Sumodi, J.; Meyerson, H.J.; Harding, C.V.; Maitta, R.W. A critical role for α-synuclein in development and function of T lymphocytes. *Immunobiology* **2016**, *221*, 333–340. [CrossRef] [PubMed]
64. Baba, M.; Nakajo, S.; Tu, P.H.; Tomita, T.; Nakaya, K.; Lee, V.M.; Trojanowski, J.Q.; Iwatsubo, T. Aggregation of α-synuclein in Lewy bodies of sporadic Parkinson's disease and dementia with Lewy bodies. *Am. J. Pathol.* **1998**, *152*, 879–884. [PubMed]
65. Braak, H.; Del Tredici, K. Neuropathological Staging of Brain Pathology in Sporadic Parkinson's disease: Separating the Wheat from the Chaff. *J. Parkinson's Dis.* **2017**, *7*, S71–S85. [CrossRef]

66. Dauer, W.; Przedborski, S. Parkinson's disease: Mechanisms and models. *Neuron* **2003**, *39*, 889–909. [CrossRef]
67. Lunati, A.; Lesage, S.; Brice, A. The genetic landscape of Parkinson's disease. *Rev. Neurol. (Paris)* **2018**, *174*, 628–643. [CrossRef]
68. Rizzo, G.; Copetti, M.; Arcuti, S.; Martino, D.; Fontana, A.; Logroscino, G. Accuracy of clinical diagnosis of Parkinson disease: A systematic review and meta-analysis. *Neurology* **2016**, *86*, 566–576. [CrossRef]
69. Lerche, S.; Heinzel, S.; Alves, G.W.; Barone, P.; Behnke, S.; Ben-Shlomo, Y.; Berendse, H.; Bloem, B.R.; Burn, D.; Dodel, R.; et al. Aiming for Study Comparability in Parkinson's Disease: Proposal for a Modular Set of Biomarker Assessments to be Used in Longitudinal Studies. *Front. Aging Neurosci.* **2016**, *8*, 121. [CrossRef]
70. Teunissen, C.E.; Tumani, H.; Engelborghs, S.; Mollenhauer, B. Biobanking of CSF: International standardization to optimize biomarker development. *Clin. Biochem.* **2014**, *47*, 288–292. [CrossRef]
71. Foix, C.; Nicolesco, J. *Cérébrale: Les Noyauz Gris Centraux Et La Région Mésencephalo-Soue-Optique. SuiviD'Un Appendice Sur L'Anatomic Pathologique De La Maladie De Parkinson*; Masson et Cie.: Paris, France, 1925.
72. Lewy, F. Zur pathologischen Anatomie der Paralysis agitans. *Dtsch. Z. Nervenheilkd.* **1914**, *50*, 50–55.
73. Braak, H.; Braak, E.; Yilmazer, D.; Schultz, C.; de Vos, R.A.; Jansen, E.N. Nigral and extranigral pathology in Parkinson's disease. *J. Neural Transm. Suppl.* **1995**, *46*, 15–31. [PubMed]
74. McGeer, P.L.; McGeer, E.G. Glial reactions in Parkinson's disease. *Mov. Disord.* **2008**, *23*, 474–483. [CrossRef] [PubMed]
75. Forno, L.S.; DeLanney, L.E.; Irwin, I.; Di Monte, D.; Langston, J.W. Astrocytes and Parkinson's disease. *Prog. Brain Res.* **1992**, *94*, 429–436.
76. Mirza, B.; Hadberg, H.; Thomsen, P.; Moos, T. The absence of reactive astrocytosis is indicative of a unique inflammatory process in Parkinson's disease. *Neuroscience* **2000**, *95*, 425–432. [CrossRef]
77. Braak, H.; Sastre, M.; Del Tredici, K. Development of α-synuclein immunoreactive astrocytes in the forebrain parallels stages of intraneuronal pathology in sporadic Parkinson's disease. *Acta Neuropathol.* **2007**, *114*, 231–241. [CrossRef] [PubMed]
78. Mena, M.A.; Garcia de Yebenes, J. Glial cells as players in parkinsonism: The "good," the "bad," and the "mysterious" glia. *Neuroscientist* **2008**, *14*, 544–560. [CrossRef]
79. Liddelow, S.A.; Guttenplan, K.A.; Clarke, L.E.; Bennett, F.C.; Bohlen, C.J.; Schirmer, L.; Bennett, M.L.; Munch, A.E.; Chung, W.S.; Peterson, T.C.; et al. Neurotoxic reactive astrocytes are induced by activated microglia. *Nature* **2017**, *541*, 481–487. [CrossRef] [PubMed]
80. Saal, K.A.; Galter, D.; Roeber, S.; Bahr, M.; Tonges, L.; Lingor, P. Altered Expression of Growth Associated Protein-43 and Rho Kinase in Human Patients with Parkinson's Disease. *Brain Pathol.* **2017**, *27*, 13–25. [CrossRef]
81. Solano, S.M.; Miller, D.W.; Augood, S.J.; Young, A.B.; Penney, J.B., Jr. Expression of α-synuclein, parkin, and ubiquitin carboxy-terminal hydrolase L1 mRNA in human brain: Genes associated with familial Parkinson's disease. *Ann. Neurol.* **2000**, *47*, 201–210. [CrossRef]
82. Song, Y.J.; Halliday, G.M.; Holton, J.L.; Lashley, T.; O'Sullivan, S.S.; McCann, H.; Lees, A.J.; Ozawa, T.; Williams, D.R.; Lockhart, P.J.; et al. Degeneration in different parkinsonian syndromes relates to astrocyte type and astrocyte protein expression. *J. Neuropathol. Exp. Neurol.* **2009**, *68*, 1073–1083. [CrossRef] [PubMed]
83. Salter, M.W.; Stevens, B. Microglia emerge as central players in brain disease. *Nat. Med.* **2017**, *23*, 1018–1027. [CrossRef] [PubMed]
84. McGeer, P.L.; Itagaki, S.; Boyes, B.E.; McGeer, E.G. Reactive microglia are positive for HLA-DR in the substantia nigra of Parkinson's and Alzheimer's disease brains. *Neurology* **1988**, *38*, 1285–1291. [CrossRef]
85. Croisier, E.; Moran, L.B.; Dexter, D.T.; Pearce, R.K.; Graeber, M.B. Microglial inflammation in the parkinsonian substantia nigra: Relationship to α-synuclein deposition. *J. Neuroinflamm.* **2005**, *2*, 14. [CrossRef] [PubMed]
86. Imamura, K.; Hishikawa, N.; Sawada, M.; Nagatsu, T.; Yoshida, M.; Hashizume, Y. Distribution of major histocompatibility complex class II-positive microglia and cytokine profile of Parkinson's disease brains. *Acta Neuropathol.* **2003**, *106*, 518–526. [CrossRef] [PubMed]
87. Halliday, G.M.; Stevens, C.H. Glia: Initiators and progressors of pathology in Parkinson's disease. *Mov. Disord.* **2011**, *26*, 6–17. [CrossRef]
88. Hunot, S.; Boissière, F.; Faucheux, B.; Brugg, B.; Mouatt-Prigent, A.; Agid, Y.; Hirsch, E.C. Nitric oxide synthase and neuronal vulnerability in Parkinson's disease. *Neuroscience* **1996**, *72*, 355–363. [CrossRef]

89. Knott, C.; Stern, G.; Wilkin, G.P. Inflammatory regulators in Parkinson's disease: iNOS, lipocortin-1, and cyclooxygenases-1 and -2. *Mol. Cell. Neurosci.* **2000**, *16*, 724–739. [CrossRef]
90. Qin, X.Y.; Zhang, S.P.; Cao, C.; Loh, Y.P.; Cheng, Y. Aberrations in Peripheral Inflammatory Cytokine Levels in Parkinson Disease: A Systematic Review and Meta-analysis. *JAMA Neurol.* **2016**, *73*, 1316–1324. [CrossRef]
91. Ouchi, Y.; Yagi, S.; Yokokura, M.; Sakamoto, M. Neuroinflammation in the living brain of Parkinson's disease. *Parkinsonism Relat. Disord.* **2009**, *15* (Suppl. 3), S200–S204. [CrossRef]
92. Ouchi, Y.; Yoshikawa, E.; Sekine, Y.; Futatsubashi, M.; Kanno, T.; Ogusu, T.; Torizuka, T. Microglial activation and dopamine terminal loss in early Parkinson's disease. *Ann. Neurol.* **2005**, *57*, 168–175. [CrossRef] [PubMed]
93. Stokholm, M.G.; Iranzo, A.; Ostergaard, K.; Serradell, M.; Otto, M.; Svendsen, K.B.; Garrido, A.; Vilas, D.; Borghammer, P.; Santamaria, J.; et al. Assessment of neuroinflammation in patients with idiopathic rapid-eye-movement sleep behaviour disorder: A case-control study. *Lancet Neurol.* **2017**, *16*, 789–796. [CrossRef]
94. Trapani, A.; Palazzo, C.; de Candia, M.; Lasorsa, F.M.; Trapani, G. Targeting of the translocator protein 18 kDa (TSPO): A valuable approach for nuclear and optical imaging of activated microglia. *Bioconjug. Chem.* **2013**, *24*, 1415–1428. [CrossRef] [PubMed]
95. Alam, M.M.; Lee, J.; Lee, S.Y. Recent Progress in the Development of TSPO PET Ligands for Neuroinflammation Imaging in Neurological Diseases. *Nucl. Med. Mol. Imaging* **2017**, *51*, 283–296. [CrossRef] [PubMed]
96. Lavisse, S.; Guillermier, M.; Herard, A.S.; Petit, F.; Delahaye, M.; Van Camp, N.; Ben Haim, L.; Lebon, V.; Remy, P.; Dolle, F.; et al. Reactive astrocytes overexpress TSPO and are detected by TSPO positron emission tomography imaging. *J. Neurosci.* **2012**, *32*, 10809–10818. [CrossRef] [PubMed]
97. Horti, A.G.; Naik, R.; Foss, C.A.; Minn, I.; Misheneva, V.; Du, Y.; Wang, Y.; Mathews, W.B.; Wu, Y.; Hall, A.; et al. PET imaging of microglia by targeting macrophage colony-stimulating factor 1 receptor (CSF1R). *Proc. Natl. Acad. Sci. USA* **2019**. [CrossRef] [PubMed]
98. McGeer, P.L.; Itagaki, S.; Akiyama, H.; McGeer, E.G. Rate of cell death in parkinsonism indicates active neuropathological process. *Ann. Neurol.* **1988**, *24*, 574–576. [CrossRef] [PubMed]
99. Miklossy, J.; Doudet, D.D.; Schwab, C.; Yu, S.; McGeer, E.G.; McGeer, P.L. Role of ICAM-1 in persisting inflammation in Parkinson disease and MPTP monkeys. *Exp. Neurol.* **2006**, *197*, 275–283. [CrossRef]
100. Saunders, J.A.; Estes, K.A.; Kosloski, L.M.; Allen, H.E.; Dempsey, K.M.; Torres-Russotto, D.R.; Meza, J.L.; Santamaria, P.M.; Bertoni, J.M.; Murman, D.L.; et al. CD4+ regulatory and effector/memory T cell subsets profile motor dysfunction in Parkinson's disease. *J. Neuroimmune Pharmacol.* **2012**, *7*, 927–938. [CrossRef]
101. Baba, Y.; Kuroiwa, A.; Uitti, R.J.; Wszolek, Z.K.; Yamada, T. Alterations of T-lymphocyte populations in Parkinson disease. *Parkinsonism Relat. Disord.* **2005**, *11*, 493–498. [CrossRef]
102. Stevens, C.H.; Rowe, D.; Morel-Kopp, M.C.; Orr, C.; Russell, T.; Ranola, M.; Ward, C.; Halliday, G.M. Reduced T helper and B lymphocytes in Parkinson's disease. *J. Neuroimmunol.* **2012**, *252*, 95–99. [CrossRef] [PubMed]
103. Hamza, T.H.; Zabetian, C.P.; Tenesa, A.; Laederach, A.; Montimurro, J.; Yearout, D.; Kay, D.M.; Doheny, K.F.; Paschall, J.; Pugh, E.; et al. Common genetic variation in the HLA region is associated with late-onset sporadic Parkinson's disease. *Nat. Genet.* **2010**, *42*, 781–785. [CrossRef]
104. Kannarkat, G.T.; Cook, D.A.; Lee, J.K.; Chang, J.; Chung, J.; Sandy, E.; Paul, K.C.; Ritz, B.; Bronstein, J.; Factor, S.A.; et al. Common Genetic Variant Association with Altered HLA Expression, Synergy with Pyrethroid Exposure, and Risk for Parkinson's Disease: An Observational and Case-Control Study. *NPJ Parkinson's Dis.* **2015**, *1*. [CrossRef]
105. Sulzer, D.; Alcalay, R.N.; Garretti, F.; Cote, L.; Kanter, E.; Agin-Liebes, J.; Liong, C.; McMurtrey, C.; Hildebrand, W.H.; Mao, X.; et al. T cells from patients with Parkinson's disease recognize α-synuclein peptides. *Nature* **2017**, *546*, 656–661. [CrossRef] [PubMed]
106. Petrich de Marquesini, L.G.; Fu, J.; Connor, K.J.; Bishop, A.J.; McLintock, N.E.; Pope, C.; Wong, F.S.; Dayan, C.M. IFN-γ and IL-10 islet-antigen-specific T cell responses in autoantibody-negative first-degree relatives of patients with type 1 diabetes. *Diabetologia* **2010**, *53*, 1451–1460. [CrossRef]
107. Gerhard, A.; Pavese, N.; Hotton, G.; Turkheimer, F.; Es, M.; Hammers, A.; Eggert, K.; Oertel, W.; Banati, R.B.; Brooks, D.J. In vivo imaging of microglial activation with [11C](R)-PK11195 PET in idiopathic Parkinson's disease. *Neurobiol. Dis.* **2006**, *21*, 404–412. [CrossRef] [PubMed]

108. Zella, S.M.A.; Metzdorf, J.; Ciftci, E.; Ostendorf, F.; Muhlack, S.; Gold, R.; Tonges, L. Emerging Immunotherapies for Parkinson Disease. *Neurol. Ther.* **2018**. [CrossRef] [PubMed]
109. Sardi, S.P.; Cedarbaum, J.M.; Brundin, P. Targeted Therapies for Parkinson's Disease: From Genetics to the Clinic. *Mov. Disord.* **2018**, *33*, 684–696. [CrossRef]
110. Bae, E.J.; Lee, H.J.; Rockenstein, E.; Ho, D.H.; Park, E.B.; Yang, N.Y.; Desplats, P.; Masliah, E.; Lee, S.J. Antibody-aided clearance of extracellular α-synuclein prevents cell-to-cell aggregate transmission. *J. Neurosci.* **2012**, *32*, 13454–13469. [CrossRef] [PubMed]
111. Masliah, E.; Rockenstein, E.; Mante, M.; Crews, L.; Spencer, B.; Adame, A.; Patrick, C.; Trejo, M.; Ubhi, K.; Rohn, T.T.; et al. Passive immunization reduces behavioral and neuropathological deficits in an α-synuclein transgenic model of Lewy body disease. *PLoS ONE* **2011**, *6*, e19338. [CrossRef]
112. Games, D.; Valera, E.; Spencer, B.; Rockenstein, E.; Mante, M.; Adame, A.; Patrick, C.; Ubhi, K.; Nuber, S.; Sacayon, P.; et al. Reducing C-terminal-truncated α-synuclein by immunotherapy attenuates neurodegeneration and propagation in Parkinson's disease-like models. *J. Neurosci.* **2014**, *34*, 9441–9454. [CrossRef] [PubMed]
113. Shahaduzzaman, M.; Nash, K.; Hudson, C.; Sharif, M.; Grimmig, B.; Lin, X.; Bai, G.; Liu, H.; Ugen, K.E.; Cao, C.; et al. Anti-human α-synuclein N-terminal peptide antibody protects against dopaminergic cell death and ameliorates behavioral deficits in an AAV-α-synuclein rat model of Parkinson's disease. *PLoS ONE* **2015**, *10*, e0116841. [CrossRef] [PubMed]
114. Tran, H.T.; Chung, C.H.; Iba, M.; Zhang, B.; Trojanowski, J.Q.; Luk, K.C.; Lee, V.M. A-synuclein immunotherapy blocks uptake and templated propagation of misfolded α-synuclein and neurodegeneration. *Cell Rep.* **2014**, *7*, 2054–2065. [CrossRef] [PubMed]
115. Weihofen, A.; Liu, Y.; Arndt, J.W.; Huy, C.; Quan, C.; Smith, B.A.; Baeriswyl, J.L.; Cavegn, N.; Senn, L.; Su, L.; et al. Development of an aggregate-selective, human-derived α-synuclein antibody BIIB054 that ameliorates disease phenotypes in Parkinson's disease models. *Neurobiol. Dis.* **2018**, *124*, 276–288. [CrossRef] [PubMed]
116. El-Agnaf, O.; Overk, C.; Rockenstein, E.; Mante, M.; Florio, J.; Adame, A.; Vaikath, N.; Majbour, N.; Lee, S.J.; Kim, C.; et al. Differential effects of immunotherapy with antibodies targeting α-synuclein oligomers and fibrils in a transgenic model of synucleinopathy. *Neurobiol. Dis.* **2017**, *104*, 85–96. [CrossRef] [PubMed]
117. Lindström, V.; Fagerqvist, T.; Nordström, E.; Eriksson, F.; Lord, A.; Tucker, S.; Andersson, J.; Johannesson, M.; Schell, H.; Kahle, P.J.; et al. Immunotherapy targeting α-synuclein protofibrils reduced pathology in (Thy-1)-h[A30P] α-synuclein mice. *Neurobiol. Dis.* **2014**, *69*, 134–143. [CrossRef]
118. Kallab, M.; Herrera-Vaquero, M.; Johannesson, M.; Eriksson, F.; Sigvardson, J.; Poewe, W.; Wenning, G.K.; Nordstrom, E.; Stefanova, N. Region-Specific Effects of Immunotherapy With Antibodies Targeting α-synuclein in a Transgenic Model of Synucleinopathy. *Front. Neurosci.* **2018**, *12*, 452. [CrossRef]
119. Spencer, B.; Valera, E.; Rockenstein, E.; Overk, C.; Mante, M.; Adame, A.; Zago, W.; Seubert, P.; Barbour, R.; Schenk, D.; et al. Anti-α-synuclein immunotherapy reduces α-synuclein propagation in the axon and degeneration in a combined viral vector and transgenic model of synucleinopathy. *Acta Neuropathol. Commun.* **2017**, *5*, 7. [CrossRef]
120. Hebron, M.L.; Lonskaya, I.; Moussa, C.E. Nilotinib reverses loss of dopamine neurons and improves motor behavior via autophagic degradation of α-synuclein in Parkinson's disease models. *Hum. Mol. Genet.* **2013**, *22*, 3315–3328. [CrossRef]
121. Ko, H.S.; Lee, Y.; Shin, J.H.; Karuppagounder, S.S.; Gadad, B.S.; Koleske, A.J.; Pletnikova, O.; Troncoso, J.C.; Dawson, V.L.; Dawson, T.M. Phosphorylation by the c-Abl protein tyrosine kinase inhibits parkin's ubiquitination and protective function. *Proc. Natl. Acad. Sci. USA* **2010**, *107*, 16691–16696. [CrossRef]
122. Stoker, T.B.; Torsney, K.M.; Barker, R.A. Emerging Treatment Approaches for Parkinson's Disease. *Front. Neurosci.* **2018**, *12*, 693. [CrossRef] [PubMed]
123. Schenk, D.B.; Koller, M.; Ness, D.K.; Griffith, S.G.; Grundman, M.; Zago, W.; Soto, J.; Atiee, G.; Ostrowitzki, S.; Kinney, G.G. First-in-human assessment of PRX002, an anti-α-synuclein monoclonal antibody, in healthy volunteers. *Mov. Disord.* **2017**, *32*, 211–218. [CrossRef] [PubMed]
124. Jankovic, J.; Goodman, I.; Safirstein, B.; Marmon, T.K.; Schenk, D.B.; Koller, M.; Zago, W.; Ness, D.K.; Griffith, S.G.; Grundman, M.; et al. Safety and Tolerability of Multiple Ascending Doses of PRX002/RG7935, an Anti-α-Synuclein Monoclonal Antibody, in Patients With Parkinson Disease: A Randomized Clinical Trial. *JAMA Neurol.* **2018**, *75*, 1206–1214. [CrossRef]

125. Weihofen, A.; Patel, H.; Huy, C.; Liu, C.; Combaluzier, I.; Mueller-Steiner, S.; Cavegn, N.; Strobel, L.; Kuznetsov, G.; Engber, T.M.; et al. Binding and functional characterization of human-derived anti-α-Synuclein antibody BIIB054. *Neurodegener. Dis.* **2017**, *17* (Suppl. 1), 59.

126. Brys, M.; Hung, S.; Fanning, L.; Penner, N.; Yang, M.; Welch, M.; Koenig, E.; David, E.; Fox, T.; Makh, S.; et al. Randomized, double-blind, placebo-controlled, single ascending dose study of antiα-synuclein antibody BIIB054 in patients with Parkinson's disease. *Neurology* **2018**, *90*, S26.001.

127. BioArctic. *BioArctic Enters into Collaboration with AbbVie for Parkinson Disease Research*; BioArctic: Stockholm, Sweden, 2018.

128. Maass, F.; Schulz, I.; Lingor, P.; Mollenhauer, B.; Bahr, M. Cerebrospinal fluid biomarker for Parkinson's disease: An overview. *Mol. Cell. Neurosci.* **2018**. [CrossRef] [PubMed]

129. Verdurand, M.; Levigoureux, E.; Zeinyeh, W.; Berthier, L.; Mendjel-Herda, M.; Cadarossanesaib, F.; Bouillot, C.; Iecker, T.; Terreux, R.; Lancelot, S.; et al. In Silico, in Vitro, and in Vivo Evaluation of New Candidates for α-Synuclein PET Imaging. *Mol. Pharm.* **2018**, *15*, 3153–3166. [CrossRef] [PubMed]

130. Dupont, A.C.; Largeau, B.; Santiago Ribeiro, M.J.; Guilloteau, D.; Tronel, C.; Arlicot, N. Translocator Protein-18 kDa (TSPO) Positron Emission Tomography (PET) Imaging and Its Clinical Impact in Neurodegenerative Diseases. *Int. J. Mol. Sci.* **2017**, *18*, 785. [CrossRef] [PubMed]

131. Jucaite, A.; Svenningsson, P.; Rinne, J.O.; Cselenyi, Z.; Varnas, K.; Johnstrom, P.; Amini, N.; Kirjavainen, A.; Helin, S.; Minkwitz, M.; et al. Effect of the myeloperoxidase inhibitor AZD3241 on microglia: A PET study in Parkinson's disease. *Brain* **2015**, *138*, 2687–2700. [CrossRef]

6

Impairment of Motor Function Correlates with Neurometabolite and Brain Iron Alterations in Parkinson's Disease

Beate Pesch [1,†], Swaantje Casjens [1,†], Dirk Woitalla [2,3], Shalmali Dharmadhikari [4,5,6], David A. Edmondson [4,5], Maria Angela Samis Zella [3], Martin Lehnert [1], Anne Lotz [1], Lennard Herrmann [2], Siegfried Muhlack [2], Peter Kraus [2], Chien-Lin Yeh [4,5], Benjamin Glaubitz [7], Tobias Schmidt-Wilcke [8,9], Ralf Gold [2], Christoph van Thriel [10], Thomas Brüning [1], Lars Tönges [2,†] and Ulrike Dydak [4,5,*,†]

1 Institute for Prevention and Occupational Medicine of the German Social Accident Insurance, Institute of the Ruhr University Bochum (IPA), 44789 Bochum, Germany; pesch@ipa-dguv.de (B.P.); casjens@ipa-dguv.de (S.C.); lehnert@ipa-dguv.de (M.L.); lotz@ipa-dguv.de (A.L.); bruening@ipa-dguv.de (T.B.)
2 Department of Neurology, St. Josef-Hospital, Ruhr-University Bochum, 44791 Bochum, Germany; d.woitalla@contilia.de (D.W.); lennard.herrmann@rub.de (L.H.); siegfried.muhlack@rub.de (S.M.); peter.kraus@rub.de (P.K.); ralf.gold@rub.de (R.G.); lars.toenges@rub.de (L.T.)
3 Department of Neurology, St. Josef-Hospital, Katholische Kliniken Ruhrhalbinsel, Contilia Gruppe, 45257 Essen, Germany; maria.zella@rub.de
4 School of Health Sciences, Purdue University, West Lafayette, IN 47907, USA; stdharm@emory.edu (S.D.); edmondsd@purdue.edu (D.A.E.); ln7511@gmail.com (C.-L.Y.)
5 Department of Radiology and Imaging Sciences, Indiana University School of Medicine, Indianapolis, IN 46202, USA
6 Department of Radiology and Imaging Sciences, Emory University, Atlanta, GA 30322, USA
7 Department of Neurology, BG University Hospital Bergmannsheil, Ruhr University Bochum, 44789 Bochum, Germany; benjamin.glaubitz@rub.de
8 Department of Neurology, St. Mauritius Therapieklinik, 40670 Meerbusch, Germany; tobias-schmidt-wilcke@t-online.de
9 Institute of Clinical Neuroscience and Medical Psychology, Universitätsklinikum Düsseldorf, 40225 Düsseldorf, Germany
10 Leibniz Research Centre for Working Environment and Human Factors (IfADo), TU Dortmund, 44139 Dortmund, Germany; thriel@ifado.de
* Correspondence: udydak@purdue.edu
† These authors contributed equally to this work.

Abstract: We took advantage of magnetic resonance imaging (MRI) and spectroscopy (MRS) as non-invasive methods to quantify brain iron and neurometabolites, which were analyzed along with other predictors of motor dysfunction in Parkinson's disease (PD). Tapping hits, tremor amplitude, and the scores derived from part III of the Movement Disorder Society-Sponsored Revision of the Unified Parkinson Disease Rating Scale (MDS-UPDRS3 scores) were determined in 35 male PD patients and 35 controls. The iron-sensitive MRI relaxation rate R2* was measured in the globus pallidus and substantia nigra. γ-aminobutyric acid (GABA)-edited and short echo-time MRS was used for the quantification of neurometabolites in the striatum and thalamus. Associations of R2*, neurometabolites, and other factors with motor function were estimated with Spearman correlations and mixed regression models to account for repeated measurements (hands, hemispheres). In PD patients, R2* and striatal GABA correlated with MDS-UPDRS3 scores if not adjusted for age. Patients with akinetic-rigid PD subtype (N = 19) presented with lower creatine and striatal glutamate and glutamine (Glx) but elevated thalamic GABA compared to controls or mixed PD subtype. In PD patients, Glx correlated with an impaired dexterity when adjusted for covariates.

Elevated myo-inositol was associated with more tapping hits and lower MDS-UPDRS3 scores. Our neuroimaging study provides evidence that motor dysfunction in PD correlates with alterations in brain iron and neurometabolites.

Keywords: Parkinson's disease; brain iron; motor dysfunction; neurometabolites; magnetic resonance imaging; magnetic resonance spectroscopy; GABA; spectroscopy

---

1. Introduction

Parkinson's disease (PD) is one of the most common movement disorders and is characterized by dopaminergic neurodegeneration primarily in the iron (Fe)-rich substantia nigra (SN) and α-synuclein pathology [1]. There is strong evidence that α-synuclein may act as cellular ferrireductase involved in the generation of bioavailable iron (Fe) [2]. Fe is the most abundant redox-active metal in the brain and subject to chelation in neuromelanin to avoid oxidative damage [3]. Brain Fe accumulation occurs in areas primarily associated with motor activity [4] and correlates with age and neurometabolite levels [5]. Studies in PD patients frequently found associations of nigral Fe with disease progression, including an impairment of gross motor functions [6–9]. A study in rhesus monkeys demonstrated that brain Fe accumulation was associated with an impaired fine motor function such as dexterity and motor speed [10]. A large prospective study showed that elevated systemic Fe can impair dexterity in healthy elderly men [11]. In PD patients, systemic Fe correlated with brain Fe [12], but less is known about the effects of brain Fe on fine motor functions.

With the advent of neuroimaging, magnetic resonance spectroscopy (MRS) became available as non-invasive method to quantify neurometabolites, including neurotransmitters, as potential biomarkers of PD [13]. MRS allows for the estimation of low-molecular weight chemicals, such as *N*-acteylaspartate (NAA; a marker of neuronal function), total creatine (tCr; involved in energy metabolism), myo-inositol (mI; a glial cell marker), glutamate (Glu; an excitatory neurotransmitter), and glutamine (Gln; precursor of Glu and γ-aminobutyric acid (GABA)) [14]. Several MRS studies on PD exist, but due to differences in resolution and sensitivity, small sample sizes, and heterogeneous patient populations across the studies, results are rather heterogeneous [13,14].

Advanced MRS techniques, such as spectral editing, also enable the in vivo study of GABA, the primary inhibitory neurotransmitter in the central nervous system [15]. The pathophysiology of PD involves the indirect and direct pathways of motor control in the basal ganglia, which are neuronal circuits facilitating the initiation and execution of voluntary movement. These pathways depend on excitatory and inhibitory signaling and well-balanced regulation of the neurotransmitters dopamine, GABA, and Glu. Thus, exploring GABA and Glu neurotransmitters by MRS is of great interest to understand the disruption of motor pathways in PD, and their implications on gross and fine motor function. Recently, an association of brain neurotransmitters on gross motor function was demonstrated in PD patients [16]. Fine motor performance was predicted by thalamic GABA levels in a study on manganese-exposed workers, with manganese toxicity leading to a particular form of parkinsonism [17]. Less is known about the influence of neurometabolites, including GABA, on dexterity in PD.

The aim of this analysis was to explore associations between brain Fe, GABA, and neurometabolites involved in the energy metabolism, i.e., tCr, mI and the combined signal of Glu and Gln (Glx), with motor function in PD patients compared to controls within the framework of the study on WELDing and Oxidative damage (WELDOX II). We hypothesized that PD patients may also show associations of these neuroimaging variables with fine motor function. Furthermore, we evaluated differences in Parkinson clinical phenotypes.

2. Materials and Methods

2.1. Ethics Statement

This study protocol was approved by the ethics committee of Ruhr University Bochum (registration number 4762-13) and conducted according to the principles expressed in the Declaration of Helsinki. All participants provided written informed consent.

2.2. Study Groups

For this analysis, we investigated 35 male PD patients and 35 male controls of similar age with complete neuroimaging data and variables from various motor performance tests. Subjects were recruited for the neuroimaging study WELDOX II from 2013 to 2016, as previously described [5,18,19]. Eligible for the present analysis were right-handed men, who did not suffer from alcohol abuse or claustrophobia.

Education, smoking habits, and other characteristics were assessed by questionnaires in face-to-face interviews. The analysis of blood samples was performed with methods as formerly described [5,18,20]. Carbohydrate-deficient transferrin >2.6% was presumptive for alcohol abuse, which may affect Fe metabolism and motor functions.

2.3. Diagnosis of PD, the Hoehn and Yahr Scale, and MDS-UPDRS 3 Scores

PD patients were enrolled at the Department of Neurology, St. Josef-Hospital, Bochum, Germany, and were diagnosed according to the criteria of the UK Parkinson Disease Society Brain Bank [21]. Disease severity was assessed by a neurologist (DW, SM, LH) using the Movement Disorder Society-Sponsored Revision of the Unified Parkinson Disease Rating Scale (MDS-UPDRS, part I–III) [22] and the Hoehn and Yahr scale [23]. All assessments reported in this study were performed with patients being on their regular medication. Patients on GABAergic medication were ineligible for participation.

The MDS-UPDRS part III scores (further referred to as MDS-UPDRS3 total scores) were additionally verified—except for the rigidity subscore—by an independent rater based on video documentation of motor functions in all participants with the permission of the subject and blinded for disease status (DW, MZ). An MDS-UPDRS3 motor rigidity score (further referred to as MDS-UPDRS3 rigidity subscore) was defined as the sum of rigidity measures of the neck, the right and left upper extremity and the right and left lower extremity. Each motor function was rated with 0–4 points, resulting in a range from 0 to 20 for the summary subscore. In the present analysis, we included only patients with an akinetic-rigid ($N = 19$) or mixed subtype ($N = 16$) [16].

2.4. Fine Motor Tests

All fine motor tests were performed while the patients were on medication. Dexterity was tested for the right and left hand with a tapping test. Tapping hits were determined with the Motor Performance Series (Schuhfried, Mödling, Austria) as previously described [11]. The number of hits, a measure of motor speed, was acquired by tapping a stylus within 32 s as often as possible on a 1600 mm^2 plate. Average tremor amplitude was calculated from drawn spirals (participants were asked to track a given spiral) and gave a measure of kinetic tremor [24].

2.5. MRI and MRS Data Acquisition and Processing

MRI scans were performed on a 3 T Philips Achieva X-series whole-body clinical scanner (Philips Healthcare, Best, The Netherlands) with a 32-channel head coil as previously described [5,18]. In brief, R2* (= 1/T2*) relaxation rates were measured using a high-resolution (isotropic 1.5 mm^3) 3D fast field echo (FFE) sequence with multiple flip angles (repetition time (TR) = 24.3 ms, echo time (TE) = 3.7 ms, ΔTE = 4.4 ms). R2* regions of interest (ROIs) were manually placed bilaterally in the substantia nigra (SN) and the globus pallidus (GP).

For the quantification of neurometabolites, 30 mm × 30 mm × 25 mm voxels of interest (VOIs) were centered on the thalamus and the striatum (head of caudate nucleus, putamen and part of GP interna) of both hemispheres. Both short echo time point resolved spectroscopy (PRESS) spectra (TE/TR = 30/2000 ms, 32 averages) and MEGA-PRESS edited GABA spectra (TE/TR = 68/2000 ms, edit ON acquisitions = 128, edit OFF acquisitions = 128) were acquired from each VOI [15]. In addition, reference spectra without water suppression were obtained for phase, frequency, and eddy current correction. Brain tissue segmentation into gray matter (GM), white matter (WM), and cerebrospinal fluid (CSF) was performed using the partial volume correction tool by Nia Goulden and Paul Mullins (https://www.bangor.ac.uk/psychology/biu/Wiki.php.en) with SPM8 as formerly described [25].

Post-processing and quantification of MRS data was done using LCModel (v 6.2-0R) [26] and was focused on GABA, Glx, tCr, and mI. Neurometabolite concentrations were referenced to the unsuppressed water signal. Tissue correction of neurometabolite content is applied to adjust for CSF in VOIs, assuming no metabolic activity in CSF. Only concentrations that were estimated with a relative standard deviation (%SD) <20%, as reported by LCModel, were used for further statistical analysis.

The placement of the MRI ROIs, MRS VOIs, and representative spectra are shown in Figure 1.

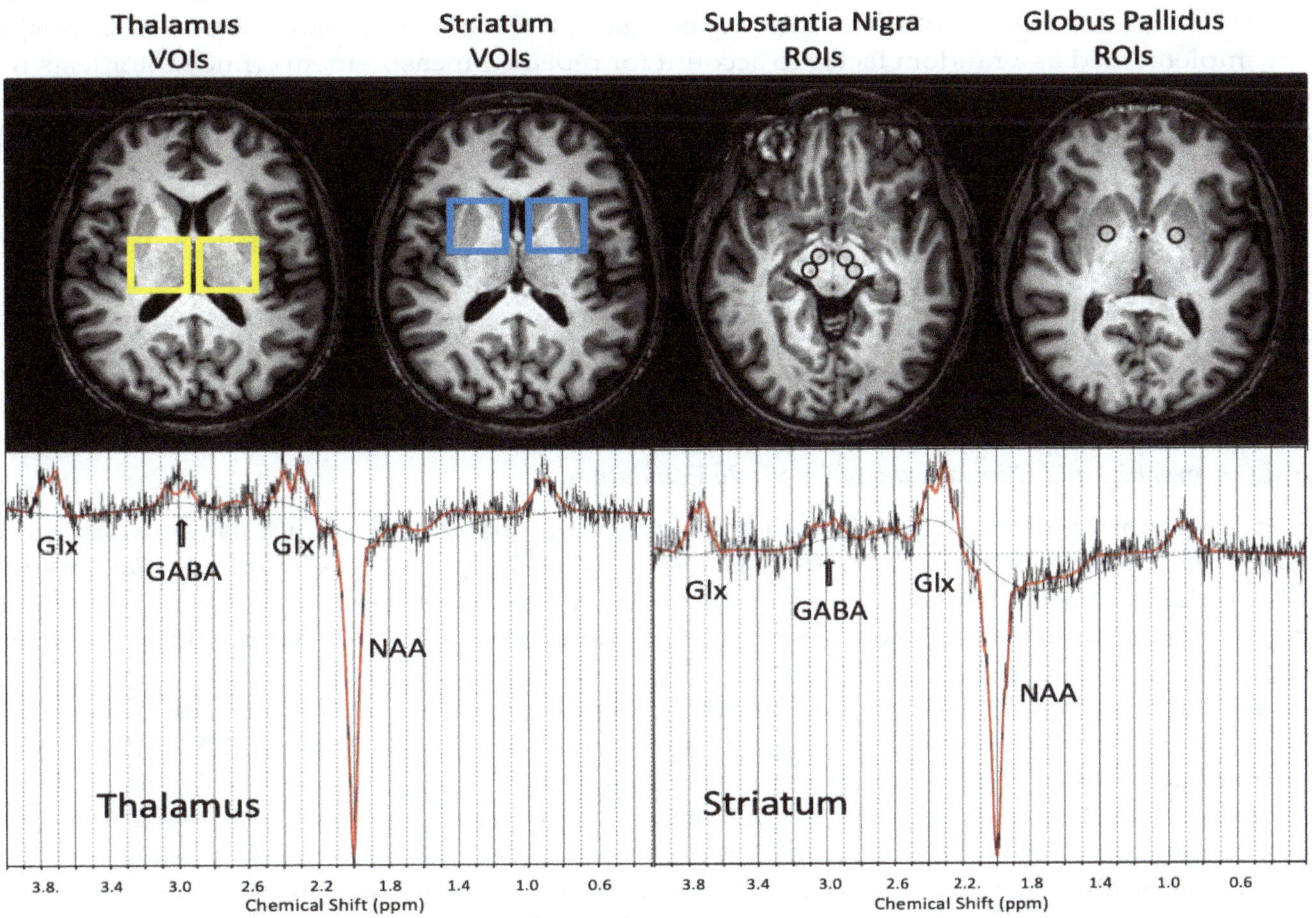

Figure 1. T1-weighted MRI images showing the placement of MRS volumes of interest (VOIs) centered on thalamus and striatum, and placement of regions of interest (ROIs) in the substantia nigra and globus pallidus for R2* analysis. Representative edited GABA spectra from the right thalamus and right striatum, indicating the LCModel fit (red solid line) with GABA, Glx, and NAA peaks shown below the MRI images. Abbreviations: T1: transverse relaxation time, MRI: magnetic resonance imaging, MRS: magnetic resonance spectroscopy, VOI: voxel of interest, ROI: region of interest, NAA: N-acetylaspartate, GABA: γ-aminobutyric acid, Glx: combined signal of glutamate and glutamine.

2.6. Statistics

For describing the distribution of neurometabolites, CSF-corrected concentrations (mM) were presented. For the analysis of associations, water-scaled values in institutional units (i.u.) were adjusted for the CSF content of the voxels by using partial correlation coefficients or by implementing CSF as covariate in the regression models. Neuroimaging data were presented with arithmetic means of measurements in both hemispheres for descriptive purposes, whereas the hemisphere was implemented as covariate in the regression models.

Median and inter-quartile range (IQR) were used to describe the distribution of continuous variables. Group differences of continuous variables were tested with the Kruskal–Wallis or Wilcoxon rank-sum test and of categorical variables with the χ^2 test. Spearman correlation coefficients (r_s) were presented with 95% confidence intervals (CIs).

We applied mixed models to fine motor tests (Poisson respectively loglinear regression) in all men and linear regression to MDS-UPDRS3 motor scores in PD patients for the association of neuroimaging data with motor dysfunctions, adjusted for CSF, and with PD subtype, age (in decades), education (high for university entrance level vs. lower levels), and the more affected hand as covariates. Education was a significant predictor of tapping and other fine motor tests that involve cognition or reaction time [11]. In addition, we also ran the models using CSF-corrected neurometabolites. Subjects were implemented as a random factor to account for repeated measurements (fine motor tests with both hands, neuroimaging in both hemispheres). Point estimates of these potential predictors were presented with β (additive in linear models) or $\exp(\beta)$ (factor in loglinear models) along with 95% CIs. Effects were considered significant when 0 (linear models) or 1 (loglinear models) was not included in the CIs.

Two-sided p-values ≤ 0.05 are shown in bold, one-sided p-values ≤ 0.05 in italic, not corrected for multiple tests.

The calculations were performed with the statistical software SAS, version 9.4 (SAS Institute Inc., Cary, NC, USA).

3. Results

3.1. Demographics and Clinical Data of the Study Population

Characteristics of this right-handed male study population are shown by the study group in Table 1. Median age at PD diagnosis was 54 (IQR 49–60) years for all patients. Patients with akinetic-rigid PD were diagnosed at a slightly younger age than patients with a mixed PD clinical phenotype (52 vs. 55 years), had longer disease duration (6.0 vs. 3.5 years) and a lower level of education.

PD patients presented with a median MDS-UPDRS3 total score of 34. Men with akinetic-rigid PD had similar total scores but higher rigidity subscores than those with mixed PD (median 7 vs. 4). Controls had MDS-UPDRS3 scores of 2 or lower (median 1).

As compared with controls, PD patients showed larger tremor amplitudes (1.1 vs. 0.9 mm) and fewer tapping hits (164 vs. 186) with their left or more affected hand. Patients with akinetic-rigid PD performed slightly fewer tapping hits and smaller tremor amplitudes than those with mixed PD.

Table S1 presents the motor variables for the right and left hand by potentially influencing factors, such as disease status, age, and education, which were implemented as predictors of motor function into the statistical models. In all strata, tapping with the right hand resulted in more hits than with the left hand (e.g., 212 vs. 186 in controls and 178 vs. 164 in all PD patients). This effect is less pronounced in PD patients where the right hand is the more affected side (176 vs. 171). A similar but weaker pattern for the results by hand can be observed for the amplitudes in spiral drawing. Subjects with a high school diploma performed more tapping hits, smaller tremor amplitudes and higher rigidity subscores than men with a lower education.

Table 1. Demographics and clinical data of male patients with Parkinson disease and of controls.

Participants Characteristics	Controls N = 35	PD N = 35	p-Value	Akinetic-Rigid PD N = 19	Mixed PD N = 16	p-Value
	Median (IQR)	Median (IQR)		Median (IQR)	Median (IQR)	
Age [years]	55 (50;66)	59 (54; 66)	0.14	60 (54; 67)	59 (55; 66)	0.64
Age at diagnosis [years]		54 (49; 60)		52 (47; 60)	55 (53; 59)	0.37
PD duration [years]		4.7 (2.5; 7.7)		**6.0 (3.6; 8.8)**	**3.5 (1.5; 5.4)**	**0.03**
MDS-UPDRS3 total score; range (0–132)	**1 (0; 2)**	**34 (24; 43)**	**<0.0001**	34 (24; 43)	34.5 (25.5; 45.5)	0.95
MDS-UPDRS3 rigidity subscore; range (0–20)	**0 (0; 0)**	**7 (3; 8)**	**<0.0001**	*7 (5; 8)*	*4 (2.5; 8)*	*0.10*
Tapping hits (left or more affected hand)	**186 (180; 205)**	**164 (134; 184)**	**0.0001**	163 (143; 178)	176 (126; 190)	0.80
Tremor amplitude (mm) (left or more affected hand)	**0.9 (0.7; 1.1)**	**1.1 (0.8; 1.3)**	**0.019**	1.1 (0.8; 1.2)	1.2 (0.8; 1.9)	0.26
	N (%)	N (%)		N (%)	N %	
Education						
Low	*8 (22.9)*	*17 (48.6)*	*0.08*	*12 (63.2)*	*5 (31.3)*	*0.06*
Medium	8 (22.9)	6 (17.1)		4 (21.1)	2 (12.5)	
High	19 (54.3)	12 (34.3)		3 (15.8)	9 (56.3)	
Clinically more affected side						
Left		12 (34.3)		5 (26.3)	7 (43.8)	0.24
Right		14 (40.0)		7 (36.8)	7 (43.8	
No preference		9 (25.7)		7 (36.8)	2 (12.5)	

Abbreviations: IQR: inter-quartile range (25th; 75th percentile), PD: Parkinson's disease, MDS-UPDRS3: Movement Disorder Society-Sponsored Revision of the Unified Parkinson Disease Rating Scale part III, p-value obtained by Kruskal–Wallis test or χ^2 test (significant effects marked in bold and marginal effects marked in italic).

Table S2 shows that there were no obvious correlations between gross and fine motor dysfunctions in PD patients (e.g., between MDS-UPDRS3 rigidity subscore and tapping $r_s = 0.14$, 95% CI −0.20–0.45) but depicts a weak negative association between tapping hits and tremor amplitude ($r_s = -0.32$, 95% CI −0.59–0.02) and a strong positive correlation between MDS-UPDRS3 total and rigidity subscores ($r_s = 0.78$, 95% CI 0.61–0.89).

3.2. Distribution of Brain Iron and Neurometabolites

Table 2 presents the distribution of R2* in SN and GP, as well as the distribution of the CSF-corrected neurometabolites in the striatum or the thalamus with median and IQR values. There were no obvious group differences of R2* values in these ROIs. PD patients showed lower striatal tCr than controls (7.2 vs. 7.5 mM), and akinetic PD patients had lower striatal tCr than mixed PD patients (6.7 vs. 7.4 mM). Akinetic PD patients also displayed lower striatal Glx levels compared to mixed PD (9.6 vs. 11.4 mM), with overall similar concentrations of 10.5 mM in all PD patients and controls. No group difference was found for local GABA concentrations. Box plots of the distributions of striatal and thalamic GABA, Glx, mI and tCr are displayed in Figure S1.

Table 2. Distribution of brain iron and neurometabolites in male Parkinson patients and controls.

Neuroimaging Data	Controls N = 35	PD N = 35	p-Value	Akinetic-Rigid PD N = 19	Mixed PD N = 16	p-Value
	Median (IQR)	Median (IQR)		Median (IQR)	Median (IQR)	
R2* (1/s), SN (N = 70)	45.6 (40.8; 51.1)	48.0 (39.6; 61.8)	0.38	45.5 (40.4; 56)	53.4 (36.1; 62.4)	0.84
R2* (1/s), GP (N = 70)	43.4 (40.4; 47.7)	44.1 (37.9; 47.2)	0.43	39.9 (37.9; 46.2)	46.0 (38.2; 47.8)	0.41
Striatum						
GABA (mM) (N = 68)	1.9 (1.8; 2.1)	2.0 (1.8; 2.2)	0.30	2.0 (1.6; 2.2)	2.1 (1.9; 2.4)	0.13
Glx (mM) (N = 69)	10.5 (9.1; 11.3)	10.5 (8.9; 11.7)	0.88	**9.6 (8.8; 11.0)**	**11.4 (10.1; 12.3)**	**0.04**
Myo-inositol (mM) (N = 69)	4.3 (3.9; 4.7)	4.4 (3.9; 5.1)	0.37	4.3 (3.9; 4.8)	4.6 (4.0; 5.2)	0.30
Total creatine (mM) (N = 69)	**7.5 (7.1; 7.8)**	**7.2 (6.5; 7.6)**	**0.04**	**6.7 (6.4; 7.4)**	**7.4 (6.9; 7.8)**	**0.03**

Table 2. *Cont.*

Neuroimaging Data	Controls N = 35	PD N = 35	p-Value	Akinetic-Rigid PD N = 19	Mixed PD N = 16	p-Value
	Median (IQR)	Median (IQR)		Median (IQR)	Median (IQR)	
Thalamus						
GABA (mM) (N = 69)	1.9 (1.7; 2.4)	2.1 (1.9; 2.4)	0.21	*2.2 (1.9; 2.4)*	*2.0 (1.9; 2.1)*	*0.06*
Glx (mM) (N = 69)	7.2 (6.6; 8.3)	7.7 (6.8; 8.5)	0.15	7.5 (6.8; 8.3)	7.8 (7.2; 8.9)	0.22
Myo-inositol (mM) (N = 69)	4.6 (4.2; 4.8)	4.9 (4.3; 5.4)	0.19	4.7 (3.9; 5.3)	5.0 (4.5; 5.5)	0.64
Total creatine (mM) (N = 69)	6.3 (6.1; 6.8)	6.3 (6.0; 6.8)	0.77	*6.2 (6.0; 6.5)*	*6.5 (6.2; 6.8)*	*0.06*

Abbreviations: PD: Parkinson's disease, IQR: inter-quartile range (25th; 75th percentile) of the distribution using the arithmetic mean of both hemispheres, R: relaxation rate, SN: substantia nigra, GP: globus pallidus, GABA: γ-aminobutyric acid, Glx: glutamate and glutamine, p-value obtained by the Wilcoxon rank-sum test (significant effects marked in bold and marginal effects marked in italic).

3.3. Correlations Between Neuroimaging Data and Motor Dysfunctions

The correlations between neuroimaging data and motor dysfunctions in PD patients are depicted in Table 3. Nigral and pallidal R2* correlated with both the MDS-UPDRS3 total scores and the rigidity sub-scores (e.g., R2* in SN with total score: $r_s = 0.39$, 95% CI 0.07–0.64). Likewise, increased striatal GABA correlated with worse motor variables (e.g., with MDS-UPDRS3 total score: $r_s = 0.37$, 95% CI −0.08–0.69), although with wide confidence limits. Higher striatal mI was associated with more tapping hits ($r_s = 0.49$, 95% CI 0.06–0.77). Scatter plots for these associations are displayed in Figure S2.

Table 3. Spearman correlation coefficients with 95% confidence interval between age, neuroimaging data, and motor variables in 35 Parkinson patients.

r_s (95% CI)	Tapping Hits Non-dominant or More Affected Hand	Tremor Amplitude (mM) Non-Dominant Hand or More Affected Hand	MDS-UPDRS3 Total Score	MDS-UPDRS3 Rigidity Subscore
Age (years)	−0.10 (−0.42, 0.24)	0.09 (−0.25, 0.41)	**0.34 (0.01, 0.61)**	0.19 (−0.16, 0.49)
Age at diagnosis (years)	−0.11 (−0.43, 0.23)	0.17 (−0.18, 0.47)	0.26 (−0.08, 0.54)	0.07 (−0.27, 0.39)
R2* (1/s) SN	0.11 (−0.23, 0.43)	0.02 (−0.32, 0.35)	**0.39 (0.07, 0.64)**	*0.33 (−0.01, 0.59)*
R2* (1/s) GP	−0.04 (−0.37, 0.29)	0.00 (−0.34, 0.33)	*0.32 (−0.01, 0.59)*	*0.32 (−0.01, 0.59)*
Striatum				
GABA	0.31 (−0.14, 0.65)	0.13 (−0.32, 0.53)	*0.37 (−0.08, 0.69)*	0.30 (−0.15, 0.65)
Glx	−0.30 (−0.66, 0.16)	0.03 (−0.42, 0.47)	−0.14 (−0.55, 0.32)	−0.23 (−0.61, 0.24)
Myo-inositol	**0.49 (0.06, 0.77)**	*−0.34 (−0.68, 0.12)*	0.30 (−0.16., 0.66)	0.03 (−0.42, 0.47)
Total creatine	0.06 (−0.39, 0.49)	−0.29 (−0.65, 0.18)	0.14 (−0.32, 0.55)	0.01 (−0.44, 0.45)
Thalamus				
GABA	−0.11 (−0.55, 0.38)	−0.20 (−0.61, 0.30)	−0.01 (−0.47, 0.46)	−0.15 (−0.58, 0.34)
Glx	0.16 (−0.30, 0.56)	−0.03 (−0.47, 0.42)	0.15 (−0.31, 0.56)	0.11 (−0.35, 0.52)
Myo-inositol	−0.10 (−0.52, 0.36)	0.03 (−0.42, 0.47)	−0.19 (−0.58, 0.28)	−0.30 (−0.65, 0.17)
Total creatine	0.26 (−0.20, 0.63)	−0.01 (−0.45, 0.43)	0.01 (−0.44, 0.45)	−0.12 (−0.54, 0.34)

Abbreviations: MDS-UPDRS3: Movement Disorder Society-Sponsored Revision of the Unified Parkinson Disease Rating Scale part III, R: relaxation rate, SN: substantia nigra, GP: globus pallidus, GABA: γ-aminobutyric acid, Glx: glutamate and glutamine, rs: Spearman correlation coefficients (for neurometabolites partial correlation coefficients adjusted for CSF content in the voxels using data from the hemisphere contralateral to the non-dominant hand for tapping hits and tremor amplitude), CI: confidence interval (significant effects marked in bold and marginal effects marked in italic).

3.4. Predictors of Impaired Fine Motor Functions

Predictors of fine motor tests were estimated in all subjects as shown in Table 4 displaying impaired dexterity, especially in PD patients with mixed subtype. After adjustment for CSF and other covariates, mI was positively associated with tapping hits (exp(β) = 1.27, 95% CI 1.06–1.53) in PD patients. Higher regional Glx levels, primarily in the thalamus, were associated with larger amplitudes in PD (exp(β) = 1.22, 95% CI 1.02–1.47). Similar associations were observed for the reported neurometabolites with direct CSF correction (data not shown).

Table 4. Brain iron, neurometabolites, and other potential predictors of fine motor skills determined by a mixed model in 35 Parkinson patients and 35 controls.

	Tapping Hits			Tremor Amplitude (mm)		
	Exp(β)	95% CI		Exp(β)	95% CI	
Study groups ($N = 70$)						
Intercept	227.30	190.62	271.04	0.54	0.30	1.00
Akinetic-rigid PD vs. controls	**0.91**	**0.85**	**0.97**	1.06	0.87	1.30
Mixed PD vs. controls	**0.87**	**0.81**	**0.93**	**1.32**	**1.08**	**1.62**
High education vs. lower levels	**1.05**	**1.00**	**1.10**	0.95	0.80	1.12
Age [per 10 years]	*0.98*	*0.96*	*1.01*	1.06	0.96	1.18
Non-dominant hand	**0.90**	**0.86**	**0.95**	**1.18**	**1.12**	**1.25**
More affected side	**0.92**	**0.86**	**0.99**	**1.11**	**1.02**	**1.22**
Brain iron ($N = 35$ PD patients)						
R2* [log 1/s]	0.96	0.84	1.10	0.87	0.70	1.08
Globus pallidus vs. substantia nigra	0.99	0.94	1.06	0.98	0.92	1.04
Neurometabolites ($N = 35$ PD patients)						
GABA	1.01	0.85	1.22	1.03	0.84	1.25
Striatum vs. thalamus	0.99	0.91	1.07	0.97	0.89	1.05
Glutamate and glutamine	0.89	0.75	1.07	**1.22**	**1.02**	**1.47**
Striatum vs. thalamus	1.02	0.94	1.12	*0.93*	*0.85*	*1.01*
Myo-inositol	**1.27**	**1.06**	**1.53**	1.09	0.87	1.36
Striatum vs. thalamus	1.02	0.95	1.10	0.97	0.90	1.05
Total creatine	1.01	0.69	1.49	1.16	0.73	1.84
Striatum vs. thalamus	1.00	0.92	1.08	0.96	0.88	1.04

Abbreviations: PD: Parkinson's disease, R: relaxation rate, GABA: γ-aminobutyric acid, exp(β): estimate of the regression coefficients (as factor of change per unit) of potential predictors of motor dysfunction in all subjects and for neuroimaging data in PD patients in separate models for each influencing factor adjusted for CSF content of the voxel (neurometabolites only), age (per 10 years), education (high, lower educational levels), hand (dominant, non-dominant hand), and affected side (more, less affected or no preference), CI: confidence interval (significant effects marked in bold and marginal effects marked in italic).

3.5. Predictors of Gross-Motor Symptoms

Finally, potential predictors of MDS-UPDRS3 total scores were estimated in PD patients, with age as strongest factor ($\beta = 7.22$, 95% CI 0.38–14.06) as demonstrated in Table 5. Low thalamic mI showed a notable association with higher scores ($\beta = -29.00$, 95% CI −58.21–0.21). The distribution of the rigidity subscores did not allow parametric regression modeling.

Table 5. Potential predictors of the MDS-UPDRS3 total scores in 35 Parkinson patients.

	B	95% CI	
Intercept	−4.10	−46.71	38.51
Akinetic-rigid PD vs. mixed PD	−4.13	−14.40	6.14
High education	−4.71	−15.53	6.11
Age [per 10 years]	**7.22**	**0.38**	**14.06**
Brain iron			
R2*, globus pallidus [log 1/s]	19.72	−8.70	48.15
R2*, substantia nigra [log 1/s]	10.19	−7.41	27.78
Neurometabolites			
Striatum			
GABA	7.71	−21.54	36.95
Glutamate and glutamine	−5.48	−33.40	22.45
Myo-inositol	0.06	−31.11	31.23
Total creatine	14.59	−38.66	67.84
Thalamus			
GABA	−9.83	−46.20	26.54
Glutamate and glutamine	1.18	−28.92	31.28
Myo-inositol	*−29.00*	*−58.21*	*0.21*
Total creatine	−5.49	−75.39	64.42

Abbreviations: PD: Parkinson's disease, GABA: γ-aminobutyric acid, R: relaxation rate, MDS-UPDRS3: Movement Disorder Society-Sponsored Revision of the Unified Parkinson Disease Rating Scale part III, β: estimate of regression coefficient (linear model) for potential predictors of motor dysfunction, in separate models for each neuroimaging variable adjusted for CSF content of the voxel (neurometabolites only), age (per 10 years), education (high, lower educational levels), CI: confidence interval (significant effects marked in bold and marginal effect marked in italic).

4. Discussion

Impairments in motor function are key clinical manifestations of PD and are hypothesized to be associated with changes in brain iron and neurometabolites, including those involved in energy metabolism and damage to dopaminergic neurons. The present analysis provides evidence of a positive correlation of nigral and pallidal R2* as well as striatal GABA with MDS-UPDRS3 total scores and rigidity subscores in all PD patients. After adjustment for covariates, lower mI levels predicted impaired gross and fine motor functions in PD, primarily in the thalamus. We further observed a positive association of thalamic Glx with larger tremor amplitudes.

The present study found higher R2* values, representative of increased iron content, in the SN of PD patients and a correlation with gross motor function as evaluated with the MDS-UPDRS3, in line with results reported in recent studies [6,7] and several previous studies subjected to a narrative review on nigral Fe as biomarker of PD [9]. Using quantitative susceptibility mapping in addition to R2*-based iron mapping, the study by Langkammer et al. showed iron-based pathologic alterations also in GP and thalamus, as well as a correlation with PD severity. At higher resolution with 7T MRI, an association between R2*-weighted signals in clusters of dopaminergic cells within the SN and disease severity was found [27]. Importantly, our study shows that the correlation between R2* and MDS-UPDRS3 motor scales becomes weaker when adjusting for age because, in general, brain iron is known to increase with age [28,29]. Adjustment for age may therefore capture variance of an underlying causal association between iron accumulation and disease severity.

There is substantial variability of reported neurometabolite levels between MRS studies in Parkinson's disease [13,14]. A major reason is the small size of these complex neuroimaging studies, which hinders the adjustment for pertinent covariates. Creatine plays a role in energy homeostasis by maintaining ATP levels constant in cells with high and fluctuating energy demands, primarily in muscles and glia [30]. In our study, patients with akinetic-rigid PD, an often clinically more severe phenotype of PD than the mixed or tremor-dominant phenotype, presented with lower striatal and thalamic tCr compared to the other phenotypes. Another analysis found basal ganglia tCr to be decreased in all PD phenotypes [31]. Therefore, using tCr as an internal reference for metabolite concentrations was inappropriate for this study, and internal water scaling was used instead, with adjusting for CSF fractions of the MRS VOI.

MDS-UPDRS3 scores strongly discriminated PD patients from controls. In contrast, differences in fine motor tests were less pronounced although PD patients displayed larger amplitudes and lower tapping speed than controls in line with other investigations [32,33]. In a large cohort of healthy subjects, we have demonstrated that age is a strong determinant of dexterity [11]. This was not as obvious in this study population, where the mixed PD phenotype and the left or more affected hand were stronger predictors. We observed lower tCr in PD patients as compared to controls, but could not observe an association with motor dysfunction. As already discussed for brain iron, implementing age into regression models may overadjust two parallel age-related processes, for example a putative correlation between tCr and motor dysfunction. Increasing age is correlated with both tCr levels [5,34] and motor dysfunction [11]. This may explain why we could not discern an association between tCr and fine motor function in our PD patients after adjustment for age.

Alterations in neurotransmission in PD are not restricted to the dopaminergic systems but importantly also involve the glutamatergic (glutamate and its precursor glutamine) and GABAergic networks [16]. Glx is an important precursor in the synthesis of GABA [35]. Furthermore, the cycling between Gln and Glu accounts for more than 80% of cerebral glucose consumption and is important for brain energetic metabolism [36]. In our study, akinetic-rigid PD patients exhibited significantly lower striatal Glx levels than the mixed PD phenotype, while there were no significant differences between the combined groups of PD patients and controls. However, better fine motor performance of PD patients was associated with lower Glx levels. O'Gorman et al. have shown that prefrontal Glx levels are associated with axial symptoms in PD, thus indicating that Glx seems to play a role in more complex motor function [16]. Furthermore, in our previous analysis, we had observed an age-related decline of Glx [5].

Non-MRS, invasive and ex-vivo techniques in animal models and postmortem human studies have established that the loss of dopaminergic striatal neurons in PD is accompanied by increased striatal GABA content [37–40]. In vivo GABA levels remain challenging to quantify due to the small MRS signal; however, several MRS studies at 3 T and higher magnetic fields have so far reported increased GABA levels in basal ganglia structures both in PD [16,41,42] as well as in manganese-exposed workers, who are at risk developing a form of parkinsonism [17,43]. Several of these studies also reported an association of basal ganglia GABA levels with gross motor function [16,42] or fine motor function as measured by a pegboard test [17]. The present study used statistical modeling, accounting for several confounding factors such as age. We found thalamic GABA levels to be slightly increased in the akinetic-rigid PD phenotype when compared to the mixed PD form. However, a correlation of striatal GABA levels with MDS-UPDRS3 scores showed wide confidence intervals and disappeared in the full prediction model. This is in line with a recent MRS study on manganese-exposed welders, that did not find any correlation of thalamic or striatal GABA levels with MDS-UPDRS3 scores, or with rigidity or tremor subscores, but showed that only age remained as predictor for the MDS-UPDRS3 scores in the statistical model [44].

Elevated levels of mI, a suggested marker for gliosis, in the brainstem and several cortical areas have been associated with neurodegenerative diseases [45], and have been shown to correlate negatively with gross motor function in ataxias and amyotrophic lateral sclerosis [46,47]. Notably, low systemic mI can increase blood glucose levels due to insulin-mimetic properties [48]. We found that lower mI levels in PD patients predicted motor dysfunctions, associated with less tapping hits and higher MDS-UPDRS3 total scores. This finding is in line with a previous report of decreased thalamic mI in manganese toxicity [49], which can lead to a Parkinsonian syndrome. In our previous analysis of mI in a mixed study population with healthy persons and PD or hematochromatosis patients, brain iron assessed with R2* in GP was a predictor of lower mI in these VOIs [5].

Overall, confidence limits of most estimates for associations between the study variables were rather wide, indicating additional influencing factors, such as potential effects of alcohol intake or body mass index on motor function in PD [50]. With recent reports on medication effecting both the on neurometabolites NAA and tCr [51] as well as on brain Fe levels and symptom scores [7], it is well possible that the medication status can modify the effects of neurometabolites or brain Fe on MDS-UPDRS3 scores. While patients receiving any GABAergic medication were ineligible for participation, nearly all patients were on dopaminergic treatment at the time of motor testing and neuroimaging. However, detailed information on medication, including a precise quantitative estimation of the levodopa equivalent dose and hours to time of investigation, were not available for all patients of this study. A further limitation of this and other costly neuroimaging studies is the small study size, which impairs a sound statistical modeling of the association between brain iron or neurometabolites with motor dysfunction allowing for the inclusion of pertinent confounders, effect modifiers, and interaction terms between study variables.

5. Conclusions

The present study provides evidence that brain iron and neurometabolite concentrations involved in energy metabolism and neurotransmission are associated with motor dysfunction in PD. To which extent neurometabolite alterations can be attributed to clinical PD phenotypes remains to be elucidated.

Supplementary Materials:
Figure S1: Box plots of the distributions of striatal and thalamic GABA, Glx, mI and tCr in 35 Parkinson patients and 35 controls; Figure S2: Correlations between brain iron (R2* placed in SN and GP) and GABA (in striatal and thalamic VOIs) with MDS-UPDRS3 total scores and of mI with tapping hits in 35 Parkinson patients; Table S1: Fine motor test results for both hands presented with medians for tapping hits and tremor amplitude in Parkinson patients and controls; Table S2: Spearman correlation coefficients with 95% confidence interval between motor functions in 35 Parkinson patients.

Author Contributions: Conceptualization, B.P., D.W., C.v.T., T.S.-W., and U.D.; methodology, S.C., U.D., A.L., B.P., and P.K.; software: U.D., B.G., T.S.-W., and P.K.; validation: A.L., B.P., B.G., T.S.-W., and U.D.; data curation: A.L., S.C., P.K., B.G., U.D., T.S.-W., S.D., C.-L.Y., and D.A.E.; formal analysis: S.C., A.L., and B.P.; investigation, B.G., T.S.-W., M.L., D.W., M.A.S.Z., L.H., and S.M.; Writing—original draft preparation: B.P., S.C., U.D., and L.T.; Writing—review and editing: B.P., S.C., U.D., L.T., D.W., A.L., M.L. T.B., C.v.T., M.A.S.Z., L.H., P.K., S.M., B.G., T.S.W., S.D., C.-L.Y., R.G., and D.A.E.; Project administration: B.P., D.W., T.B., R.G., and U.D.; Funding acquisition: B.P., D.W., T.B., C.v.T., T.S.-W., and U.D.

Acknowledgments: We appreciate the scientific support of Burkhard Mädler from PHILIPS, Germany, and the contribution of colleagues from the WELDOX II team. Further we thank Richard Edden for providing his MEGA-PRESS sequence for the present study.

References

1. Poewe, W.; Seppi, K.; Tanner, C.M.; Halliday, G.M.; Brundin, P.; Volkmann, J.; Schrag, A.-E.; Lang, A.E. Parkinson disease. *Nat. Rev. Dis. Primers* **2017**, *3*, 17013. [CrossRef] [PubMed]
2. Davies, P.; Moualla, D.; Brown, D.R. Alpha-synuclein is a cellular ferrireductase. *PLoS ONE* **2011**, *6*, e15814. [CrossRef]
3. Zucca, F.A.; Segura-Aguilar, J.; Ferrari, E.; Muñoz, P.; Paris, I.; Sulzer, D.; Sarna, T.; Casella, L.; Zecca, L. Interactions of iron, dopamine and neuromelanin pathways in brain aging and Parkinson's disease. *Prog. Neurobiol.* **2017**, *155*, 96–119. [CrossRef] [PubMed]
4. Rouault, T.A. Iron metabolism in the CNS: Implications for neurodegenerative diseases. *Nat. Rev. Neurosci.* **2013**, *14*, 551–564. [CrossRef] [PubMed]
5. Casjens, S.; Dydak, U.; Dharmadhikari, S.; Lotz, A.; Lehnert, M.; Quetscher, C.; Stewig, C.; Glaubitz, B.; Schmidt-Wilcke, T.; Edmondson, D.; et al. Association of exposure to manganese and iron with striatal and thalamic GABA and other neurometabolites—Neuroimaging results from the WELDOX II study. *Neurotoxicology* **2018**, *64*, 60–67. [CrossRef] [PubMed]
6. Langkammer, C.; Pirpamer, L.; Seiler, S.; Deistung, A.; Schweser, F.; Franthal, S.; Homayoon, N.; Katschnig-Winter, P.; Koegl-Wallner, M.; Pendl, T.; et al. Quantitative Susceptibility Mapping in Parkinson's Disease. *PLoS ONE* **2016**, *11*, e0162460. [CrossRef]
7. Jin, L.; Wang, J.; Jin, H.; Fei, G.; Zhang, Y.; Chen, W.; Zhao, L.; Zhao, N.; Sun, X.; Zeng, M.; et al. Nigral iron deposition occurs across motor phenotypes of Parkinson's disease. *Eur. J. Neurol.* **2012**, *19*, 969–976. [CrossRef]
8. Ulla, M.; Bonny, J.M.; Ouchchane, L.; Rieu, I.; Claise, B.; Durif, F. Is R2* a new MRI biomarker for the progression of Parkinson's disease? A longitudinal follow-up. *PLoS ONE* **2013**, *8*, e57904. [CrossRef]
9. Guan, X.; Xu, X.; Zhang, M. Region-Specific Iron Measured by MRI as a Biomarker for Parkinson's Disease. *Neurosci. Bull.* **2017**, *33*, 561–567. [CrossRef] [PubMed]
10. Kastman, E.K.; Willette, A.A.; Coe, C.L.; Bendlin, B.B.; Kosmatka, K.J.; McLaren, D.G.; Xu, G.; Canu, E.; Field, A.S.; Alexander, A.L.; et al. A calorie-restricted diet decreases brain iron accumulation and preserves motor performance in old rhesus monkeys. *J. Neurosci.* **2012**, *32*, 11897–11904. [CrossRef]
11. Pesch, B.; Casjens, S.; Weiss, T.; Kendzia, B.; Arendt, M.; Eisele, L.; Behrens, T.; Ulrich, N.; Pundt, N.; Marr, A.; et al. Occupational Exposure to Manganese and Fine Motor Skills in Elderly Men: Results from the Heinz Nixdorf Recall Study. *Ann. Work Expo. Health* **2017**, *61*, 1118–1131. [CrossRef] [PubMed]
12. Yu, S.-Y.; Cao, C.-J.; Zuo, L.-J.; Chen, Z.-J.; Lian, T.-H.; Wang, F.; Hu, Y.; Piao, Y.-S.; Li, L.-X.; Guo, P.; et al. Clinical features and dysfunctions of iron metabolism in Parkinson disease patients with hyper echogenicity in substantia nigra: A cross-sectional study. *BMC Neurol.* **2018**, *18*, 9. [CrossRef]
13. Ciurleo, R.; Di Lorenzo, G.; Bramanti, P.; Marino, S. Magnetic resonance spectroscopy: An in vivo molecular imaging biomarker for Parkinson's disease? *BioMed. Res. Int.* **2014**, *2014*, 519816. [CrossRef]
14. Dydak, U.; Edmondson, D.; Zauber, S.E. MRS of Parkinsonian Disorders. In *Magnetic Resonance Spectroscopy of Degenerative Brain Diseases*; Oz, G., Ed.; Springer: Basel, Switzerland, 2016; pp. 71–102.

15. Mullins, P.G.; McGonigle, D.J.; O'Gorman, R.L.; Puts, N.A.J.; Vidyasagar, R.; Evans, C.J.; Edden, R.A.E. Current practice in the use of MEGA-PRESS spectroscopy for the detection of GABA. *NeuroImage* **2014**, *86*, 43–52. [CrossRef] [PubMed]
16. O'Gorman Tuura, R.L.; Baumann, C.R.; Baumann-Vogel, H. Beyond Dopamine: GABA, Glutamate, and the Axial Symptoms of Parkinson Disease. *Front. Neurol.* **2018**, *9*, 806. [CrossRef]
17. Long, Z.; Li, X.-R.; Xu, J.; Edden, R.A.E.; Qin, W.-P.; Long, L.-L.; Murdoch, J.B.; Zheng, W.; Jiang, Y.-M.; Dydak, U. Thalamic GABA predicts fine motor performance in manganese-exposed smelter workers. *PLoS ONE* **2014**, *9*, e88220. [CrossRef] [PubMed]
18. Pesch, B.; Dydak, U.; Lotz, A.; Casjens, S.; Quetscher, C.; Lehnert, M.; Abramowski, J.; Stewig, C.; Yeh, C.-L.; Weiss, T.; et al. Association of exposure to manganese and iron with relaxation rates R1 and R2*-magnetic resonance imaging results from the WELDOX II study. *Neurotoxicology* **2018**, *64*, 68–77. [CrossRef]
19. Van Thriel, C.; Quetscher, C.; Pesch, B.; Lotz, A.; Lehnert, M.; Casjens, S.; Weiss, T.; van Gelder, R.; Plitzke, K.; Brüning, T.; et al. Are multitasking abilities impaired in welders exposed to manganese? Translating cognitive neuroscience to neurotoxicology. *Arch. Toxicol.* **2017**, *91*, 2865–2877. [CrossRef]
20. Casjens, S.; Henry, J.; Rihs, H.-P.; Lehnert, M.; Raulf-Heimsoth, M.; Welge, P.; Lotz, A.; van Gelder, R.; Hahn, J.-U.; Stiegler, H.; et al. Influence of welding fume on systemic iron status. *Ann. Occup. Hyg.* **2014**, *58*, 1143–1154. [CrossRef]
21. Hughes, A.J.; Daniel, S.E.; Kilford, L.; Lees, A.J. Accuracy of clinical diagnosis of idiopathic Parkinson's disease: A clinico-pathological study of 100 cases. *J. Neurol. Neurosurg. Psychiatry* **1992**, *55*, 181–184. [CrossRef]
22. Goetz, C.G.; Fahn, S.; Martinez-Martin, P.; Poewe, W.; Sampaio, C.; Stebbins, G.T.; Stern, M.B.; Tilley, B.C.; Dodel, R.; Dubois, B. Movement Disorder Society-sponsored revision of the Unified Parkinson's Disease Rating Scale (MDS-UPDRS): Process, format, and clinimetric testing plan. *Mov. Disord.* **2007**, *22*, 41–47. [CrossRef] [PubMed]
23. Hoehn, M.M.; Yahr, M.D. Parkinsonism: Onset, progression and mortality. *Neurology* **1967**, *17*, 427–442. [CrossRef]
24. Kraus, P.H.; Hoffmann, A. Spiralometry: Computerized assessment of tremor amplitude on the basis of spiral drawing. *Mov. Disord.* **2010**, *25*, 2164–2170. [CrossRef] [PubMed]
25. Ashburner, J.; Friston, K.J. Unified segmentation. *NeuroImage* **2005**, *26*, 839–851. [CrossRef] [PubMed]
26. Provencher, S.W. Estimation of metabolite concentrations from localized in vivo proton NMR spectra. *Magn. Reson. Med.* **1993**, *30*, 672–679. [CrossRef]
27. Schwarz, S.T.; Mougin, O.; Xing, Y.; Blazejewska, A.; Bajaj, N.; Auer, D.P.; Gowland, P. Parkinson's disease related signal change in the nigrosomes 1-5 and the substantia nigra using T2* weighted 7T MRI. *NeuroImage* **2018**, *19*, 683–689. [CrossRef]
28. Schenker, C.; Meier, D.; Wichmann, W.; Boesiger, P.; Valavanis, A. Age distribution and iron dependency of the T2 relaxation time in the globus pallidus and putamen. *Neuroradiology* **1993**, *35*, 119–124. [CrossRef]
29. Hallgren, B.; Sourander, P. The effect of age on the non-haemin iron in the human brain. *J. Neurochem.* **1958**, *3*, 41–51. [CrossRef]
30. Xiao, Y.; Luo, M.; Luo, H.; Wang, J. Creatine for Parkinson's disease. *Cochrane Database Syst. Rev.* **2014**, CD009646. [CrossRef]
31. Gong, T.; Xiang, Y.; Saleh, M.G.; Gao, F.; Chen, W.; Edden, R.A.E.; Wang, G. Inhibitory motor dysfunction in parkinson's disease subtypes. *J. Magn. Reson. Imaging* **2018**, *47*, 1610–1615. [CrossRef]
32. Růžička, E.; Krupička, R.; Zárubová, K.; Rusz, J.; Jech, R.; Szabó, Z. Tests of manual dexterity and speed in Parkinson's disease: Not all measure the same. *Parkinsonism Relat. Disord.* **2016**, *28*, 118–123. [CrossRef] [PubMed]
33. Bologna, M.; Guerra, A.; Paparella, G.; Giordo, L.; Alunni Fegatelli, D.; Vestri, A.R.; Rothwell, J.C.; Berardelli, A. Neurophysiological correlates of bradykinesia in Parkinson's disease. *Brain* **2018**. [CrossRef] [PubMed]
34. Angelie, E.; Bonmartin, A.; Boudraa, A.; Gonnaud, P.M.; Mallet, J.J.; Sappey-Marinier, D. Regional differences and metabolic changes in normal aging of the human brain: Proton MR spectroscopic imaging study. *AJNR Am. J. Neuroradiol.* **2001**, *22*, 119–127. [PubMed]

35. Govindpani, K.; Calvo-Flores Guzmán, B.; Vinnakota, C.; Waldvogel, H.J.; Faull, R.L.; Kwakowsky, A. Towards a Better Understanding of GABAergic Remodeling in Alzheimer's Disease. *Int. J. Mol. Sci.* **2017**, *18*, 1813. [CrossRef] [PubMed]
36. Sibson, N.R.; Dhankhar, A.; Mason, G.F.; Rothman, D.L.; Behar, K.L.; Shulman, R.G. Stoichiometric coupling of brain glucose metabolism and glutamatergic neuronal activity. *Proc. Natl. Acad. Sci. USA* **1998**, *95*, 316–321. [CrossRef] [PubMed]
37. Galvan, A.; Wichmann, T. GABAergic circuits in the basal ganglia and movement disorders. *Progress Brain Res.* **2007**, *160*, 287–312. [CrossRef]
38. Kish, S.; Rajput, A.; Gilbert, J.; Rozdilsky, B.; Chang, L.J.; Shannak, K.; Hornykiewicz, O. GABA-dopamine relationship in Parkinson's disease striatum. *Adv. Neurol.* **1987**, *45*, 75–77. [PubMed]
39. Zecca, L.; Stroppolo, A.; Gatti, A.; Tampellini, D.; Toscani, M.; Gallorini, M.; Giaveri, G.; Arosio, P.; Santambrogio, P.; Fariello, R.G.; et al. The role of iron and copper molecules in the neuronal vulnerability of locus coeruleus and substantia nigra during aging. *Proc. Natl. Acad. Sci. USA* **2004**, *101*, 9843–9848. [CrossRef]
40. Hornykiewicz, O. Chemical neuroanatomy of the basal ganglia–normal and in Parkinson's disease. *J. Chem. Neuroanat.* **2001**, *22*, 3–12. [CrossRef]
41. Emir, U.E.; Tuite, P.J.; Öz, G. Elevated pontine and putamenal GABA levels in mild-moderate Parkinson disease detected by 7 tesla proton MRS. *PLoS ONE* **2012**, *7*, e30918. [CrossRef]
42. Dydak, U.; Dharmadhikari, S.; Snyder, S.; Zauber, S.E. Increased thalamic GABA levels correlate with Parkinson disease severity. In Proceedings of the AD/PD Conference, Nice, France, 18–21 March 2015.
43. Dydak, U.; Jiang, Y.M.; Long, L.L.; Zhu, H.; Chen, J.; Li, W.M.; Edden, R.A.; Hu, S.; Fu, X.; Long, Z.; et al. In vivo measurement of brain GABA concentrations by magnetic resonance spectroscopy in smelters occupationally exposed to manganese. *Environ. Health Perspect.* **2011**, *119*, 219–224. [CrossRef] [PubMed]
44. Ma, R.E.; Ward, E.J.; Yeh, C.-L.; Snyder, S.; Long, Z.; Gokalp Yavuz, F.; Zauber, S.E.; Dydak, U. Thalamic GABA levels and occupational manganese neurotoxicity: Association with exposure levels and brain MRI. *Neurotoxicology* **2018**, *64*, 30–42. [CrossRef] [PubMed]
45. Oz, G.; Alger, J.R.; Barker, P.B.; Bartha, R.; Bizzi, A.; Boesch, C.; Bolan, P.J.; Brindle, K.M.; Cudalbu, C.; Dinçer, A.; et al. Clinical proton MR spectroscopy in central nervous system disorders. *Radiology* **2014**, *270*, 658–679. [CrossRef]
46. Cheong, I.; Deelchand, D.K.; Eberly, L.E.; Marjańska, M.; Manousakis, G.; Guliani, G.; Walk, D.; Öz, G. Neurochemical correlates of functional decline in amyotrophic lateral sclerosis. *J. Neurol. Neurosurg. Psychiatry* **2018**. [CrossRef] [PubMed]
47. Oz, G.; Hutter, D.; Tkác, I.; Clark, H.B.; Gross, M.D.; Jiang, H.; Eberly, L.E.; Bushara, K.O.; Gomez, C.M. Neurochemical alterations in spinocerebellar ataxia type 1 and their correlations with clinical status. *Mov. Disord.* **2010**, *25*, 1253–1261. [CrossRef] [PubMed]
48. Croze, M.L.; Géloën, A.; Soulage, C.O. Abnormalities in myo-inositol metabolism associated with type 2 diabetes in mice fed a high-fat diet: Benefits of a dietary myo-inositol supplementation. *Brit. J. Nutr.* **2015**, *113*, 1862–1875. [CrossRef]
49. Long, Z.; Jiang, Y.-M.; Li, X.-R.; Fadel, W.; Xu, J.; Yeh, C.-L.; Long, L.-L.; Luo, H.-L.; Harezlak, J.; Murdoch, J.B.; et al. Vulnerability of welders to manganese exposure—A neuroimaging study. *Neurotoxicology* **2014**, *45*, 285–292. [CrossRef]
50. Abbas, M.M.; Xu, Z.; Tan, L.C.S. Epidemiology of Parkinson's Disease-East Versus West. *Mov. Disord. Clin. Pract.* **2018**, *5*, 14–28. [CrossRef]
51. Mazuel, L.; Chassain, C.; Jean, B.; Pereira, B.; Cladière, A.; Speziale, C.; Durif, F. Proton MR Spectroscopy for Diagnosis and Evaluation of Treatment Efficacy in Parkinson Disease. *Radiology* **2016**, *278*, 505–513. [CrossRef]

7

The Role of Lipids in Parkinson's Disease

Helena Xicoy [1,2], Bé Wieringa [1] and Gerard J. M. Martens [2,*]

1 Department of Cell Biology, Radboud Institute for Molecular Life Sciences (RIMLS), Radboudumc, 6525 GA Nijmegen, The Netherlands; Helena.xicoy@radboudumc.nl (H.X.); Be.Wieringa@radboudumc.nl (B.W.)

2 Department of Molecular Animal Physiology, Donders Institute for Brain, Cognition and Behaviour, Radboud University, 6525 GA Nijmegen, The Netherlands

* Correspondence: G.Martens@ncmls.ru.nl

Abstract: Parkinson's disease (PD) is a neurodegenerative disease characterized by a progressive loss of dopaminergic neurons from the nigrostriatal pathway, formation of Lewy bodies, and microgliosis. During the past decades multiple cellular pathways have been associated with PD pathology (i.e., oxidative stress, endosomal-lysosomal dysfunction, endoplasmic reticulum stress, and immune response), yet disease-modifying treatments are not available. We have recently used genetic data from familial and sporadic cases in an unbiased approach to build a molecular landscape for PD, revealing lipids as central players in this disease. Here we extensively review the current knowledge concerning the involvement of various subclasses of fatty acyls, glycerolipids, glycerophospholipids, sphingolipids, sterols, and lipoproteins in PD pathogenesis. Our review corroborates a central role for most lipid classes, but the available information is fragmented, not always reproducible, and sometimes differs by sex, age or PD etiology of the patients. This hinders drawing firm conclusions about causal or associative effects of dietary lipids or defects in specific steps of lipid metabolism in PD. Future technological advances in lipidomics and additional systematic studies on lipid species from PD patient material may improve this situation and lead to a better appreciation of the significance of lipids for this devastating disease.

Keywords: Parkinson's disease; fatty acyls; glycerolipids; glycerophospholipids; sphingolipids; sterol lipids; lipoproteins; α-synuclein-mediated pathology; disease-modifying effects; neuroprotection

---

1. Introduction

Parkinson's disease (PD) is the second most common neurodegenerative disease affecting 1% of the population above 60 years and up to 4% of individuals in the highest age groups [1]. Parkinson's disease is characterized by motor symptoms, such us tremor, rigidity, bradykinesia (slowed movement) and impaired balance [2], and non-motor manifestations, including sleep disorders, and autonomic, gastrointestinal, sensory, and neuropsychiatric symptoms [3]. These symptoms are associated with a progressive loss of dopaminergic (DA) neurons from the nigrostriatal pathway, formation of Lewy bodies (LB), and microgliosis [4]. In familial PD, which explains 5–10% of all cases, these abnormalities may be caused by a mutation in one of the thus far known 19 familial genes, including *SNCA*, *LRRK2*, *PRKN*, *PINK1* and *DJ-1*, among others [5]. The remaining 90–95% of PD cases are of sporadic nature with unknown etiology.

Despite a large number of studies on familial forms of PD or toxin-induced cell and animal PD models (e.g., use of 1-methyl-4-phenyl-1,2,3,6-tetrahydropyridine (MPTP), rotenone or 6-hydroxydopamine (6-OHDA)) [6–8], no disease-modifying treatment for PD has been developed yet. Thus, additional approaches are necessary to advance the field of PD. Previously, we used data from genome-wide association studies and other genetic studies of PD patients to build a molecular landscape [9]. This enabled us to identify, in an unbiased way, various processes and pathways that might be involved in PD. Interestingly, we found that lipids play a key role in most of the processes that have been (classically) associated with PD (i.e., oxidative stress, endosomal-lysosomal function, endoplasmic reticulum stress, and immune response), and thus in PD etiology. In agreement with this observation, not only mutations in the gene encoding the lipid-producing enzyme glucocerebroside (*GBA*) are associated with familial PD [10–12], but also multiple single-nucleotide polymorphisms (SNPs) located in other genes involved in lipid metabolism, e.g., *SREBF1* [13], *DGKQ* [14], *ASAH1* [15] or *SMPD1* [16], have been linked to sporadic PD.

Lipids are biomolecules soluble in nonpolar organic solvents, usually insoluble in water, and primarily known for their metabolic role in energy storage [17,18]. Furthermore, they are the main constituents of cellular membranes, part of membrane rafts and protein anchors, and signaling and transport molecules [19–23]. There are eight different classes of lipids, classified as fatty acyls, glycerolipids, glycerophospholipids, sphingolipids, sterols, prenols, saccharolipids, and polyketides [24]. Here we will review the current knowledge of the role of the first five lipid classes and of lipoproteins in PD (Figure 1). Certain aspects of the relationship between PD and lipids are beyond the scope of this review, including the complex interaction between (membrane) glycerophospholipids and α-synuclein, the interaction between lipid classes, and the role of cholesterol derivatives, such as bile acids, tocopherols, and tocotrienols (vitamin E), vitamin A and carotenoids, vitamin D, steroidal hormones (e.g., estrogen) and coenzyme Q10.

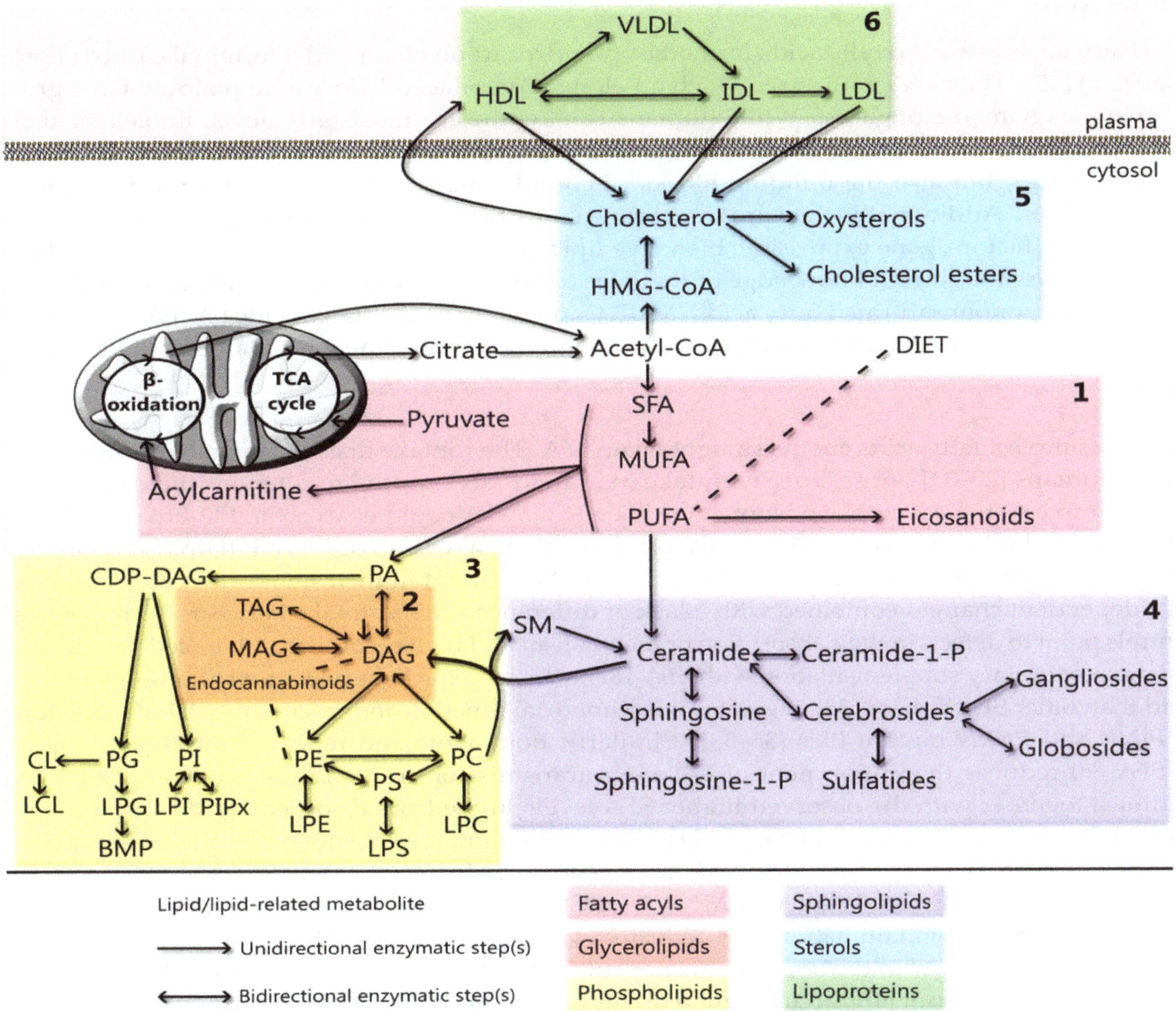

Figure 1. Cellular lipid metabolism and lipoprotein cycle. Schematic representation of lipid metabolism, whereby each colored box represents one lipid class: (**1**) fatty acyls, which include saturated (SFA), monounsaturated (MUFA), and polyunsaturated (PUFA) fatty acids, their mitochondrial-transporter, acylcarnitine, and the PUFA-derivatives eicosanoids; (**2**) glycyerolipids, including monoacylglycerol (MAG), diacylglycerol (DAG), and triacylglycerol (TAG), together with endocannabinoids (even though only some of them belong to this lipid class); (**3**) phospholipids, which include phosphatidic acid (PA), phosphatidylcholine (PC), phosphatidylserine (PS), phosphatidylethanolamine (PE), phosphatidylinositol (PI), phosphatidylglycerol (PG), cardiolipin (CL), and their lyso derivatives (lysoPC (LPC), lysoPS (LPS), lysoPE (LPE), lysoPI (LPI), lysoPG (LPG) and lysoCL (LCL)), and Bis(monoacylglycero)phosphate (BMP); (**4**) sphingolipids, including ceramide(-1-phosphate), sphingosine(-1-phosphate), sphingomyelin (SM), cerebrosides, sulfatides, gangliosides, and globosides; (**5**) sterols, which include the metabolites of cholesterol synthesis, such as β-hydroxy β-methylglutaryl-CoA (HMG-CoA), cholesterol, and its derivatives cholesterol esters and oxysterols; and (**6**) lipoproteins, including high-density lipoproteins (HDL), intermediate-density lipoproteins (IDL), low-density lipoproteins (LDL), and very low-density lipoproteins (VLDL). A depiction of the various lipid structures and of all the metabolic steps involved in their generation and interconversion(s) is given in Figures 2a,b–6a,b, respectively.

2. Fatty Acyls

Fatty acyls are carboxylic acids formed by a hydrocarbon chain and a terminal carboxyl group (Figure 2) [25]. They are synthesized by chain elongation of acetyl-CoA with malonyl-CoA groups by enzymes named elongases. While humans can synthesize most fatty acyls, linoleic acid (LA) and alpha-linoleic acid (ALA) need to be obtained through the diet [26]. Fatty acyls are not only energy sources, but also the building blocks of complex lipids and as such form a key category of metabolites. Additionally, they are membrane constituents and regulate intracellular signaling, transcription factors, gene expression, bioactive lipid production, and inflammation [27,28]. Below, we will discuss the current knowledge of the roles of fatty acyls, more specifically of saturated fatty acids (SFA), monounsaturated fatty acids (MUFA), polyunsaturatedfatty acids (PUFA), eicosanoids and (acyl)carnitine, in PD, and an overview can be found in Supplementary Materials Table S1.

2.1. SFA

The simplest fatty acids are the straight-chain SFA. Their intake does not seem to be linked to PD risk in humans [29–31] per se, but SFA intake in individuals exposed to rotenone increases PD risk, when compared to pesticide exposure alone [32]. Thus, SFA could exacerbate PD-linked pathology. Interestingly, higher levels of SFA (mainly 16:0 and 18:0) have been observed in lipid rafts from the frontal cortex of PD patients compared to controls [33], but not in their temporal cortex [34]. These area-dependent changes combined with a lack of differences in SFA intake between PD patients and controls point to defects in their absorption or metabolism and region-specific and/or cell-compartment differences. Dietary supplementation with SFA 18:0, which seems to be a less potent pro-inflammatory lipid than other SFA species [35], regulates mitochondrial function and rescues the PD-like phenotype of *PINK* and *PRKN* mutant flies [36–38]. Similarly, both acute and repeated intra-gastric gavage of SFA 8:0 reduces the impairment of DA neurotransmission in MPTP-treated mice [39]. These findings, together with the observed higher SFA levels in the frontal cortical lipid rafts, may point towards a compensatory mechanism in PD patients. In contrast, exposure of SH-SY5Y cells, primary neurons, and astrocytes to SFA 16:0 leads to apoptosis, reduces peroxisome proliferator-activated receptor gamma coactivator 1-alpha (PPARGC1A, PGC-1alpha) and estrogen receptor alpha (ER-alpha) expression, promotes inflammation, and activates cyclooxygenase-2 (COX-2) [35,40–42], features that have also been observed in the brains of PD patients [43–45]. Since α-synuclein modulates the uptake of SFA 16:0 into the brain [46], accumulation of this protein in PD brains might lead to increased levels of SFA 16:0, which in turn can trigger some of its neuropathological activities.

2.2. MUFA

Variants of SFA containing one double bond are known as MUFA. Higher MUFA intake has been variably associated with decreased PD risk [31], reduced risk only in women [29] or unchanged risk [30]. These discrepancies in findings could be due to variation in the ethnicity of the subjects, differences in the type of study (cohort or case-control), the number of participants, questionnaires employed to asses MUFA intake, the corrections used, or even PD etiology, since different MUFA levels in cerebrospinal fluid (CSF) have been described in PD patients carrying a GBA mutation and those that do not [47]. Of note, no abnormalities in MUFA level have been observed in the temporal cortex of PD patients [34]. Some MUFA, such as oleic acid and cis-vaccenic acid, trigger the production of dopamine in MN9D cells [48], and the amide of oleic acid and dopamine (*N*-oleoyl-dopamine) modulates the firing of nigrostriatal DA neurons [49]. Interestingly, α-synuclein has a motif homologous to a region in fatty acid-binding proteins, allowing it to bind to oleic acid [50], which facilitates the interaction of α-synuclein with lipid rafts [51]. Based on these suggestive but still tentative findings, the effect of MUFA intake on PD risk, and specifically, its effect on dopamine production and the intracellular location and function of α-synuclein, should be further examined.

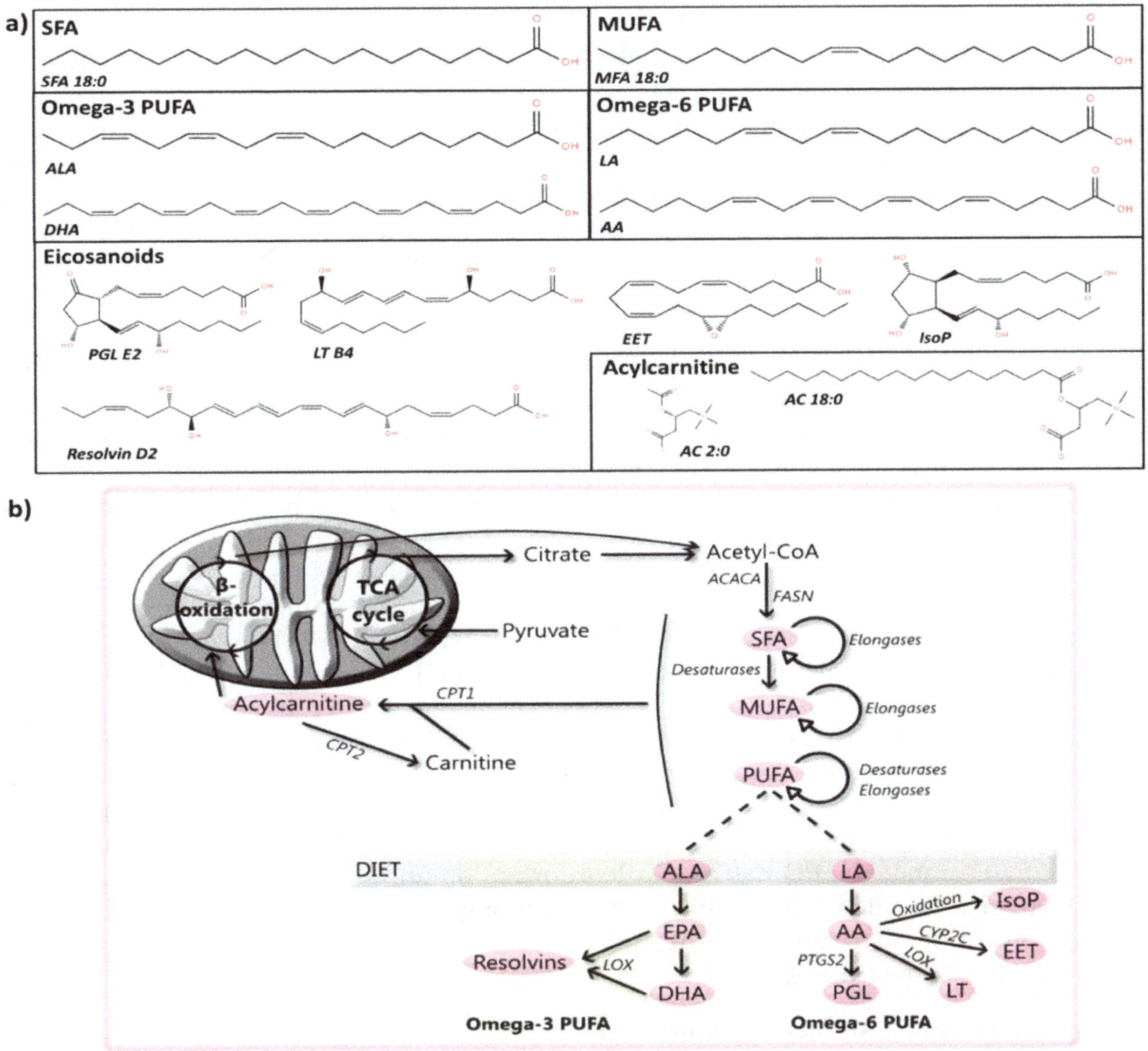

Figure 2. Fatty acyls: structures and metabolic steps involved. (**a**) Schematic representation of the chemical structures of fatty acyls, including saturated fatty acids (SFA 18:0), monounsaturated fatty acids (MUFA 18:1), omega-3 polyunsaturated fatty acids (PUFA, alpha-linoleic acid (ALA, top) and docosahexaenoic acid (DHA, bottom)), omega-6 PUFA (linoleic acid (LA, top) and arachidonic acid (AA, bottom)), eicosanoids (from left to right, prostaglandin E2 (PGL E2), leukotriene B4 (LT), 14,15-Epoxyeicosatrienoic acid (EET), 15-F2t-Isoprostane (IsoP), and resolvin D2 (bottom)), and acetylcarnitine (AC 2:0) and acylcarnitine (AC 18:0). Chemical structures are adapted from the LIPID MAPS structure database [25]. (**b**) Schematic overview of steps involved in the metabolism of fatty acyls, where fatty acids (FAs) can be obtained through the diet or by a multi-enzymatic reaction starting from acetyl-CoA and performed by enzymes such as acetyl-CoA carboxylase 1 (ACACA) and fatty acid synthase (FASN). Multiple steps of elongation, performed by elongases, and desaturation, carried out by desaturases, produce MUFA and PUFA. PUFA include, among others, omega-3 PUFA, such as ALA, which can be converted by a multistep reaction into eicosapentaenoic acid (EPA) and DHA, and omega-6 PUFA, including LA, which can be transformed by a multistep reaction to AA. PUFA can be further metabolized by enzymes such as lipoxygenase (LOX), prostaglandin-endoperoxide synthase 2 (PTGS2, also known as COX2), cytochrome p450 2C to various eicosanoids, including resolvins, PGL, LT, EET, or oxidized to isoP. Furthermore, transport of FA into mitochondria for their metabolism is preceded by their association with carnitine, which is catalyzed by the enzyme carnitine O-palmitoyltransferase 1 (CPT1) and reversed by carnitine O-palmitoyltransferase 2 (CPT2).

2.3. PUFA

Fatty acids containing two or more double bonds are known as PUFA and are usually classified according to the position of the first double bond counted from the tail (omega). The omega-3 family, for which ALA is the essential parent fatty acid, forms metabolic products that include eicosapentaenoic acid (EPA) and docosahexaenoic acid (DHA). Omega-6 PUFA, for which LA is the parent fatty acid, include arachidonic acid (AA), an intensely studied precursor of signaling lipids [52]. In general, PUFA are known to play a role in inflammation [53], epigenetics [54], and brain development [55] and function [56]. As such, they have been widely examined in PD patients, and animal and cellular PD models.

2.3.1. Human Studies on PUFA

Higher intake of omega-3 PUFA and ALA, but not other PUFA, such as omega-6 PUFA or LA, has been associated with reduced risk of PD [31,32], while other studies have reported a weak positive association between omega-6 PUFA and LA intake and PD risk [57], or have provided evidence against an association between PUFA intake and PD risk [30]. A link between AA intake and PD risk is also controversial: one study described a positive association [30], while another reported an inverse association [58]. Serum of PD patients has decreased concentrations of long-chain PUFA, including ALA, LA, and AA, compared to controls [59], while the CSF of PD patients has increased levels of 4-hydroxynonenal, a toxic product generated by AA peroxidation [60]. However, as described for MUFA, PUFA levels in CSF may depend on PD etiology [47]. The levels of PUFA in the anterior cingulate cortex and the occipital cortex of PD patients are increased or not changed, respectively [61]. Additionally, the levels of DHA and AA are decreased in frontal cortex lipid rafts from PD patients [33], while reduced LA, increased DHA and docosatetraenoic acid (an omega-6 PUFA), and no changes in AA have been reported for the cytosolic fraction of PD frontal cortex [62]. Moreover, no changes of PUFA were observed in the temporal cortex of PD patients [34]. Therefore, there is no agreement on the impact of PUFA intake on PD risk and little information on PUFA levels in the blood, CSF and brain of PD patients is available. The only consistent finding is the altered intracellular distribution of PUFA in neurons from the frontal cortex of PD patients, i.e., reduced levels of DHA in the lipid rafts and increased DHA in the cytosolic fraction.

2.3.2. Animal and Cellular Studies on PUFA

Omega-3 PUFA exert neuroprotective actions in MPTP-treated mice [63] by increasing the expression of brain-derived neurotrophic factor [64] and have also neuroprotective activity in 6-OHDA-treated rats [65]. A decrease in the level of this class of PUFA has been observed in the brains of an MPTP-induced goldfish PD model [66]. Furthermore, omega-3 PUFA deficiency leads to a reduced ability of the nigrostriatal system to maintain homeostasis under oxidative conditions, increasing the risk for PD [67]. Maternal omega-3 PUFA seem to partially protect a lipopolysaccharide (LPS)-model for PD [68]. Likewise, the omega-3 PUFA DHA protects DA neurons against MPTP–[69–71], paraquat—[72] or rotenone-induced toxicity [73] in rodent models and against effects of 6-OHDA-treatment in *Caenorhabditis elegans*, mice and rats [74–76], also when administered as TAG-DHA [77]. Moreover, DHA plays a crucial role in the differentiation of induced pluripotent stem cells (iPSCs) into functional DA neurons [78] and DHA supplementation protects DA neurons from the SN in MPTP-treated mice [79]. EPA and ethyl-EPA attenuate 1-methyl-4-phenylpyridinium (MPP+)-induced cell death in SH-SY5Y cells, primary mesencephalic neurons, and brain slices [80,81], and in vivo reduce MPTP/probecenid-induced dyskinesia and memory deficits (without preventing nigrostriatal DA loss) [82]. Thus, omega-3 PUFA appear to have a neuroprotective role in animal models for PD.

Additionally, pretreatment of rats with fish oil (which is rich in omega-3 PUFA) for 25 days before 6-OHDA treatment mitigates the loss of substantia nigra (SN) DA neurons [83]. In contrast, a chronic supplementation of fish oil in rats does not protect DA neurons but increases dopamine turnover [84]. These differential effects could be explained by the finding that the ethyl ester of DHA, a PUFA present in fish oil, enhances 6-OHDA-induced neuronal damage by triggering lipid peroxidation in mouse striatum [85]. Lipid peroxidation, which occurs frequently in PUFA, may lead to mitochondrial dysfunction [86] and α-synuclein oligomerization [87]. It is therefore not surprising that a number of studies have demonstrated beneficial effects when using deuterium-reinforced (deuterated) PUFA (which protects the PUFA sites susceptible for oxidation) [88–90], or PUFA in combination with antioxidants [91]. Therefore, omega-3 PUFA, probably in combination with the prevention of lipid peroxidation, should be further studied as a complementary therapy for PD.

Increased levels of omega-6 PUFA (LA and AA) have been reported in mice brain slices upon MPP+ treatment [81]. Similarly, upregulated AA signaling has been observed in the caudate-putamen and frontal cortex of 6-OHDA-treated rats [92], and the striatum and midbrain of MPTP-treated mice [93]. Both LA and AA are able to inhibit MPP+-induced toxicity in PC12 cells [94], while excess AA aggravates α-synuclein oligomerization in PC12 cells [95]. Interestingly, a mouse model with impaired incorporation of AA in the brain is resistant to MPTP treatment [96]. Hence, pharmacologically induced PD is linked to an increase in AA, the consequences of which are at present unclear and could be dose dependent.

2.3.3. Alpha-Synuclein and PUFA

Under physiological conditions, α-synuclein and PUFA are involved in endocytic mechanisms linked to synaptic vesicle recycling upon neuronal stimulation [97]. Moreover, α-synuclein and PUFA regulate each other, since α-synuclein increases endogenous levels of AA and DHA [62], and its oligomers control the ability of AA to stimulate SNARE-complex formation and endocytosis [98]. Reciprocally, PUFA strongly interact with the N-terminal region of α-synuclein [99], enhancing its oligomerization both in vivo and in vitro [100–102]. This might precede the formation of protective (LB-like) inclusions in DA cells [103].

Studies on specific PUFA species have shown that DHA induces α-synuclein oligomerization [104] by activating retinoic X receptor and PPAR-gamma 2 [105], effects that were prevented by co-administering aspirin [106]. The oligomers formed in the presence of DHA seem to be cytotoxic [107] and affect membrane integrity [108] and the physical properties of DHA itself (triggering formation of lipid droplets) [109]. Alpha-synuclein aggregation is also induced by AA [110], but the oligomers that are formed seem to be less toxic (more prone to disaggregation and enzymatic digestion) [111] and their formation is prevented or enhanced by low or high doses of dopamine, respectively [112]. The enhanced toxicity of dopamine might be related to its ability to form adducts with AA, which are able to trigger apoptosis [113].

Interestingly, a diet poor in omega-3 PUFA (with or without DHA supplementation) did not affect α-synuclein expression [114]. Accordingly, a DHA-rich diet had no effect on the DA system, motor impairments or α-synuclein levels in α-synuclein-overexpressing mice but increased the longevity of the mice [115]. This latter phenomenon might be related to the role of monomeric α-synuclein in sequestering early DHA peroxidation products and thus reducing oxidative stress [116]. The interaction between α-synuclein and (peroxidated) PUFA has recently been reviewed in more detail elsewhere [87,117].

2.4. Eicosanoids and Docosanoids

Eicosanoids and docosanoids constitute a family of bioactive fatty acyls mainly generated by AA, EPA, and DHA oxidation. They play a local role in infection and inflammation [118]. The family includes PGL, LT, EET, isoprostanes, HETE, isofurans, and resolvins, among others. Interestingly, one of the enzymes responsible for the formation of eicosanoids, COX-2, has been linked to PD pathology. Its role in the disease has been reviewed elsewhere [119,120].

2.4.1. PGL

No changes in or increased PGL E2 levels have been observed in the CSF and SN of PD patients, respectively [121,122]. In animal and cellular PD models, PGL E2 secretion is induced by LPS [123–125], 6-OHDA [126–128], rotenone [129,130], MPTP [131,132], and α-synuclein aggregation [133,134]. Nevertheless, PGL E2 levels in the striatum, hippocampus, and cortex of 6-OHDA-treated mice are decreased following a four-week exposure [135]. The eicosanoid PGL E2 mainly mediates its effects by binding to PGL E2 receptors (EP1-4), which trigger various intracellular pathways [136]. The EP1 receptor knock-out (KO) has neuroprotective effects on 6-OHDA-treated mice [137] and an EP1 antagonist protects embryonic rat mesencephalic primary cultures from 6-OHDA toxicity [138]. An agonist of EP2 protects primary neuronal cultures from 6-OHDA-induced toxicity [139] and an agonist of EP4 prevents DA loss in the SN of MPTP-treated mice [140]. Therefore, the effect of PGL E2 is also dependent on which receptor it binds. Interestingly, astrocytes KO for the familial PD gene *DJ-1* secrete less PGL E2 than WT astrocytes [141]. This could impair DA neuron survival mediated by EP2 [142]. Thus, the sparse data that are available regarding the effects of PGL suggest that increased PGL E2 levels may play a role in the pathology of animal and cellular PD models, but that this occurs in a time-, location-, phenotype- and receptor-dependent manner.

Both PGL A1 and lipocalin-type PGL D synthase (the enzyme that isomerizes PGL H2 to PGL D2) inhibit rotenone- and paraquat-induced apoptosis in SH-SY5Y cells, respectively [143,144]. Furthermore, enhanced prostacyclin synthesis seems to reduce glial activation and ameliorate motor dysfunction in 6-OHDA-treated rats [145]. Conversely, PGL J2 treatment of SK-N-SH cells leads to the formation of aggregates containing ubiquitinated α-synuclein [146], and infusion into the SN of mice induces a pathology that mimics the slow-onset cellular and behavioral pathology of PD, including loss of DA neurons in the SN, α-synuclein aggregation, posture impairment, and microgliosis [115,147,148]. Hence, PGL other than PGL E2 seem to play a role in PD pathology as well, with effects being protective or detrimental. To resolve this complexity and obtain deeper insight into the contributions of these oxidized PUFAs to PD pathology further research is needed.

2.4.2. LT

Increased plasma LT B3 has been suggested as a biomarker for PD [149]. However, the role of LT has only been tested in animal and cellular PD models, in which MPTP treatment upregulates arachidonate 5-lipoxygenase (5-LOX, the enzyme that synthesizes LT from AA). This work has demonstrated that 5-LOX inhibition has neuroprotective effects [150], a finding which would be in agreement with the observation that LT B4 enhances MPP+-induced neurotoxicity in midbrain cultures [150]. Moreover, inhibition of cysteinyl LT receptor 1 has neuroprotective effects in a rotenone-induced rat PD model [151,152]. Interestingly, 5-LOX KO in mice reduces striatal dopamine levels under normal conditions [153]. Thus, 5-LOX seems to be necessary for maintaining the DA tone but can become deleterious upon toxicant challenge.

2.4.3. EET

In PD patients, the SNP rs10889162 located in CYP2J2 (the enzyme that metabolizes AA into EET) is associated with age of diagnosis [154]. Interestingly, EETs are known to have cytoprotective effects in other diseases and may therefore also play a role in PD neuroinflammation [155]. In PD models, 14,15-EET, which is released from astrocytes, enhances cell viability against oxidative stress [156] and protects DA neuronal loss in MPTP-treated mice [157]. Inhibition or KO of the soluble epoxide hydrolase (sEH, inhibition of which elevates endogenous EET) protects MPTP-treated mice [157,158], and a double sEH and COX-2 inhibitor has protective effects on a rotenone-induced *Drosophila melanogaster* PD model [159]. Combined, these findings suggest that EET has widespread neuroprotective effects, not necessarily relevant for PD only.

2.4.4. Isoprostanes

The role of isoprostanes in PD is controversial. Both higher levels and no change in F2-isoprostane have been found in urine and plasma [160–162] of (early) PD patients. Moreover, no changes have been observed in CSF [163] or SN [164] of PD patients, but higher levels of F2-isoprostane have been described in anterior cingulate cortex of PD patients [61]. Higher F2-isoprostane levels have also been observed in rotenone-, but not manganese-treated DA neurons derived from healthy iPSC [165]. Thus, more research and proper stratification of findings need to be performed to understand the role of isoprostanes in PD.

2.4.5. Other Eicosanoids and Docosanoids

The classic and non-classic AA-derived eicosanoids, HETE and isofurans, are increased in plasma [161,162] and SN [164] of PD patients, respectively. The docosanoid resolvin D1 attenuates MPP+-induced PD by inhibiting inflammation in PC12 cells [166], and resolvin D2 seems to restore LPS-induced neural injury in a rat model, also by suppression of inflammation [167]. Thus, resolvins seem to have a protective role in PD.

2.5. Carnitine and Acylcarnitine

Carnitine is a trimethyllysine derivative that can associate with various fatty acids, forming acylcarnitine. This association facilitates their transport from the cytosol to the mitochondrial matrix, where fatty acids undergo β-oxidation. Both carnitine and acylcarnitines are involved in processes such as neurotransmission and apoptosis [168]. Decreased levels of carnitine and (long-chain) acylcarnitines have been detected in plasma from PD patients [149,169,170], while no changes in acylcarnitine levels have been found in either CSF or plasma from PD patients when compared to controls [171]. Acetylcarnitine (acylcarnitine 2:0) protects SK-N-MC cells from rotenone-induced toxicity [172], and carnitine reduces the effects of MPP+ on rat forebrain primary cultures [173] and LPS in SIM-A9 microglial cells [174]. Moreover, the neuroprotective properties of acetylcarnitine have been found in rats treated with 6-OHDA [175–177] and rotenone [178,179], and non-human primates treated with MPTP [180]. Additionally, increased levels of carnitine and acylcarnitine 16:0 and 18:0 have been detected in the striatum of 6-OHDA-treated rats and the mesencephalon of MPTP-treated mice, respectively [181,182], suggesting a compensatory mechanism against PD-associated toxicity. Thus, while it remains unclear whether the levels of acylcarnitine change in PD patients, mounting evidence points towards decreased plasma levels and a protective role of acylcarnitine in animal and cellular PD models.

3. Glycerolipids

The esterification of one, two or three fatty acyls with glycerol gives rise to the glycerolipids mono-, di-, and tri-substituted glycerol, known as monoacylglycerol (MAG), diacylglycerol (DAG), and triacylglycerol (TAG), respectively (Figure 3). There is little information available on the function of MAG, while DAG is a neutral lipid involved in the formation of membranes [55] and in the synaptic vesicle cycle [183]. Additionally, DAG fulfills a role as secondary lipid messenger [184]. The neutral lipid TAG is the main energy storage molecule [185]. Below, we will discuss the current knowledge of the roles of all glycerolipids with potential significance for PD, including MAG, endocannabinoids, DAG, and TAG, and an overview is given in Supplementary Materials Table S1.

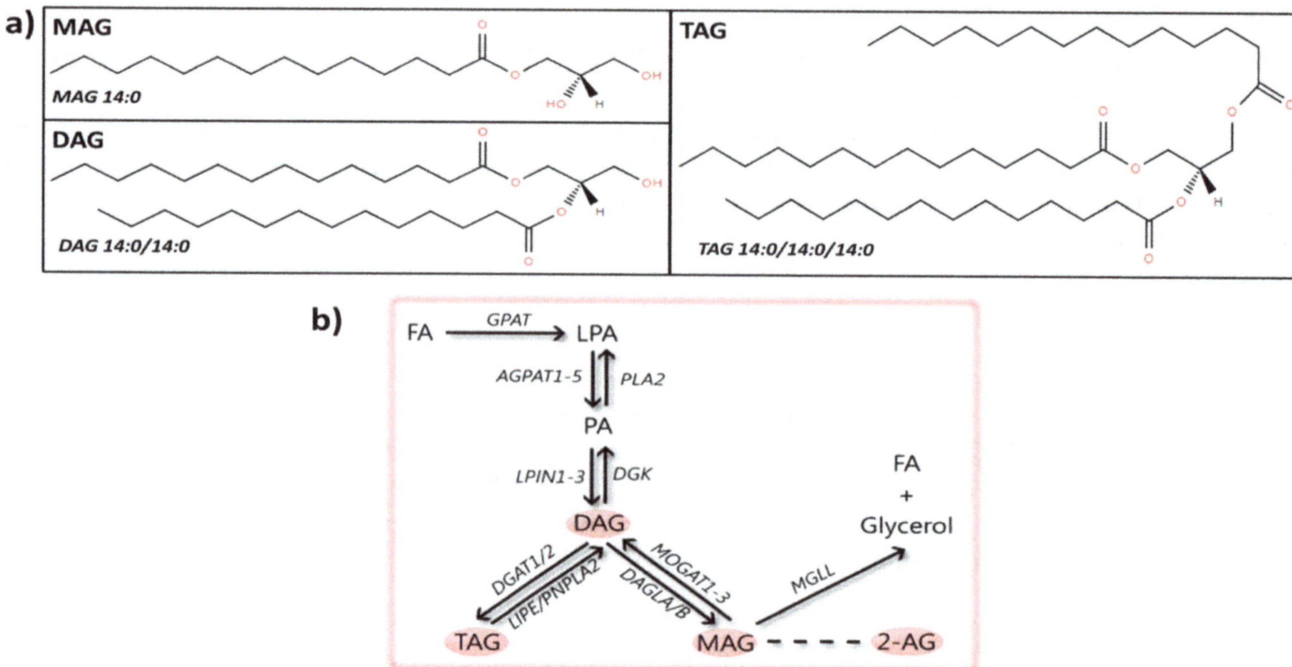

Figure 3. Glycerolipids: structures and metabolic steps involved. (**a**) Schematic representation of the chemical structures of glycerolipids, including monoacylglycerol (MAG 14:0), diacylglycerol (DAG 14:0/14:0), and triacylglycerol (TAG 14:0/14:0/14:0). Chemical structures are adapted from the LIPID MAPS structure database [25]. (**b**) Schematic overview of metabolic steps involved in the synthesis and conversion of glycerolipids: synthesis starts from fatty acids (FAs) by sequential conversion into LPA and PA, which are phospholipids (process described in Figure 4b). The enzyme phosphatide phosphatase (LPIN1-3) converts PA into DAG, a step that can be reversed by diacylglycerol kinase (DGK). From DAG, one FA can be added to the glycerol backbone by diacylglycerol O-acyltransferase 1/2 (DGAT1/2), creating TAG, a step that can be reversed by the hormone-sensitive lipase (LIPE) or patatin-like phospholipase domain-containing protein 2 (PNPLA2). Additionally, one FA can be removed from DAG by the enzyme sn1-specific diacylglycerol lipase alpha/beta (DAGLA/B), giving rise to MAG, which can be transformed back to DAG by 2-acylglycerol O-acyltransferase 1-3 (MOGAT1-3). One of the mostly studied MAG species is the endocannabinoid 2-arachidonoylglycerol (2-AG). MAG can be degraded to glycerol and a FA by monoglyceride lipase (MGLL).

3.1. MAG

Decreased and increased expression of MAG lipase, the enzyme that degrades MAG to glycerol and free fatty acids, has been observed in the SN and the putamen of PD patients, respectively [186]. These differential effects could be associated with the different mechanisms leading to the degeneration of these brain areas; in PD patients, the SN presents with cell loss [187], while the putamen has DA depletion [188]. In PD models, pharmacological inhibition of MAG lipase has neuroprotective effects in both SH-SY5Y cells treated with MPP+ [189] and chronic MPTP/probenecid mouse models [190,191]. Although there is no information on the levels of MAG in PD patients, the model studies suggest that MAG lipase inhibition, and thus higher levels of MAG, may be protective for PD.

Endocannabinoids

MAG include 2-arachidonoylglycerol (2-AG), which is classified as an endocannabinoid. Endocannabinoids are a heterogeneous and thus difficult to classify group of lipids, including not only 2-AG, but also fatty acyl amides, such as anandamide (AEA). They have been previously linked to PD [192].

Two studies have reported high AEA levels in the CSF of untreated PD patients, which were restored upon DA treatment [193,194]. Moreover, higher cannabinoid 1 receptor (CB1R) levels in the putamen, and higher and lower cannabinoid 2 receptor (CB2R) levels have been reported in the SN and putamen of PD patients, respectively [186]. Treatment of PD patients displaying no psychiatric comorbidities with cannabidiol, a naturally occurring cannabinoid constituent of cannabis which appears to lack psychoactive effects, improves quality of life measures, but does not improve Unified Parkinson's Disease Rating Scale scores [195]. The toxin 6-OHDA has been found to increase CB1R mRNA expression [196], downregulate CB1R protein density in multiple brain regions [197,198] or not produce any changes [199]. These differential effects may be explained by the fact that 6-OHDA seems to change the expression of CB1R protein in a region- and time-specific manner [200]. Decreased CB1R mRNA expression has been described in the striatum of reserpine-treated rats [201] and increased CB1R protein density has been observed in *PRKN* KO female mice [202]. Moreover, CB1R agonists fail to modulate spontaneous excitatory postsynaptic currents in cortical synapses of PINK1 KO mice [203], which points towards a CB1R dysfunction in these synapses. Hence, there is no agreement on the modulation of CB1R in different PD models and its correlation with the pathogenesis of PD in humans, making further studies necessary.

Drug-induced animal PD models (using rotenone, 6-OHDA or LPS) show increased CB2R mRNA expression [204,205], while a genetic model for PD (*LRRK2* KO) does not display changes in CB2R mRNA levels [206]. However, CB2R agonists appear to improve PD-linked impairments in both drug- and genetically-induced rodent PD models [206–208]. Thus, CB2R upregulation in animal models may reflect a compensatory mechanism, since the administration of CB2R agonists has positive effects on PD-linked pathology.

Reduced levels of the AEA precursor synthesizing enzyme, *N*-acyl-transferase [209], and reduced activities of the AEA membrane transporter and hydrolase [210,211] have been observed in the striatum of 6-OHDA-treated rats. Moreover, 6-OHDA treatment has been found to both decrease [212] and increase [210,211] striatal levels of AEA, while MPTP-lesions in monkeys increase striatal AEA levels [213]. Interestingly, an increase of AEA levels, by inhibition of the fatty acid amide hydrolase or administration of AM404 (an endogenous cannabinoid reuptake inhibitor), has neuroprotective effects [209,210,214,215]. Hence, AEA seems to have neuroprotective effects, which have been suggested to be mediated by activation of PI3K and inhibition of JNK signaling [216].

All MPTP, rotenone, and reserpine treatments lead to increased 2-AG levels in a time- and region-specific manner in various animal PD models [213,217–219], and 2-AG administration provides protection against MPTP-induced cell death [217]. The endocannabinoid *N*-arahidonoyl-dopamine has anti-inflammatory effects on both macrophages and activated BV-2 cells [96] and modulates the activity of SN neurons [220], together with the endocannabinoid-like *N*-oleoyl-dopamine [49]. This effect could be linked to the fact that, together with an inhibitor of endocannabinoid degradation, administration of a D2 receptor agonist improves motor performance in both a 6-OHDA-and a reserpine-model for PD [221]. Similar to AEA, different members of the endocannabinoid group of lipids may thus be neuroprotective.

3.2. DAG

Parkinson's disease patients have decreased plasma levels of DAG [222,223], and increased DAG levels in frontal cortex [224] and primary visual cortex [225]. Interestingly, SNPs from the chromosomal region that includes the gene encoding diacylglycerol kinase theta (*DGKQ*), which mediates the production of phosphatidic acid (PA) from DAG, are associated with PD susceptibility [14,226,227], and *DGKQ* is linked to increased PA 36:2 production and consequent α-synuclein aggregation [228]. The dysregulation of integral DAG metabolism in PD patients could be related to the observed genetic association between *DGKQ* and PD.

3.3. TAG

TAG levels are decreased in serum and plasma of (male) PD patients [222,223,229–233], even before diagnosis [234], and higher serum TAG is associated with reduced risk of idiopathic PD [235]. However, other studies have found no differences in blood TAG levels of PD patients and controls [236,237]. In the primary visual cortex of PD patients, the levels of TAG are decreased [225]. Thus, reduced levels of TAG seem to be linked to PD, although high heterogeneity has been described for TAG in PD patients [238]. Gender, ethnicity, or the technique used to measure TAG could bias the obtained results and contribute to the observed heterogeneity. Nevertheless, the trend from the majority of the results is in line with findings in animal models for PD, in which α-synuclein A53T overexpression leads to deceased serum TAG levels [239], and 6-OHDA treatment decreases TAG levels in retroperitoneal white adipose tissue [240]. Both rotenone and α-synuclein overexpression have been linked to intracellular deposition of TAG [241,242], which forms lipid droplets to which α-synuclein binds, and as such, the turnover of stored TAG is reduced and α-synuclein aggregation is enhanced [243]. In agreement, α-synuclein A53T overexpression in N27 cells leads to increased intracellular levels of TAG [244]. Thus, intracellular deposition and reduced turnover of TAG may explain the reduced levels of this acylglycerol in PD serum. Interestingly, *Saccharomyces cereviciae* that are unable to synthesize TAG are more tolerant to α-synuclein overexpression [242].

4. Glycerophospholipids

Glycerophospholipids, or phospholipids, have a glycerol backbone and a polar head group, which allows their classification into distinct subgroups, known as PA, phosphatidylethanolamine (PE), phosphatidylserine (PS), phosphatidylcholine (PC), phosphatidylinositol (PI), phosphatidyllycerol (PG) and cardiolipin (CL) (Figure 4). The hydrolysis of one acyl derivative gives rise to the lipid species known as lysophospholipids. Glycerophospholipids are key components of the lipid bilayers of cells, and as such play a role in organelle function [245] and processes like endocytosis [246] or mitophagy [247]. Moreover, they also act as signaling molecules [248–250] and regulate lipid metabolism-related gene expression [251]. Below, we will discuss the current research on glycerophospholipids in PD, more specifically PA, PE, PS, PC, PI, PG and CL, and an overview is given in Supplementary Materials Table S1.

4.1. PA

One of the best-known glycerophospholipid messengers is PA, which has a broad spectrum of functions, including intracellular vesicular trafficking, cell survival, cytoskeletal organization, neuronal development, and mitochondrial function [252–254]. Increased plasma PA (18:2/15:0) levels have been suggested as a biomarker for PD [149]. Additionally, PA is known to interact with residues 1–102 of α-synuclein [255,256], thus enhancing the formation of multimeric and protease-resistant α-synuclein aggregates [257,258]. ATP13A2, a lysosomal ATPase, which, when mutated, causes familial PD, constitutes another link between PA and PD. This ATPase requires the interaction with PA, and also PI(3,5)P2, to protect cells against rotenone-induced mitochondrial stress or other PD-related stress conditions, such as exposure to Fe(3+) [259,260]. Furthermore, overexpression of phospholipase D2 appears to induce DA neuronal cell loss via a mechanism involving PA signaling [261]. Given PA's role in the subcellular distribution and aggregation of α-synuclein, and in ATP13A2-mediated neuroprotection, it would be of interest to study PA levels and its partitioning in the brains of PD patients.

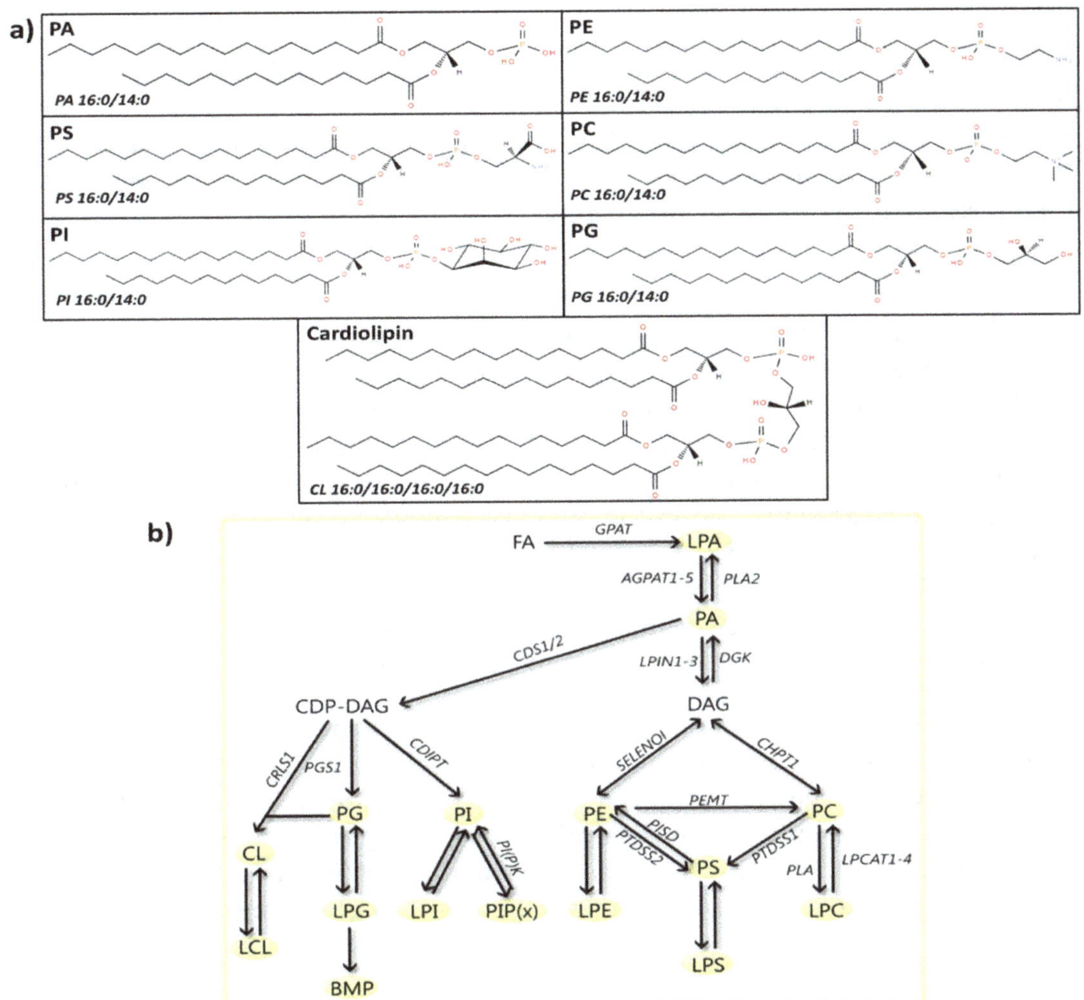

Figure 4. Phospholipids: structures and metabolic steps involved. (**a**) Schematic representation of the chemical structures of phospholipids, including phosphatidic acid (PA 16:0/14:0), phosphatidylethanolamine (PE 16:0/14:0), phosphatidylserine (PS 16:0/14:0), phosphatidylcholine (PC 16:0/14:0), phosphatidylinositol (PI 16:0/14:0), phosphatidylglycerol (PG 16:0/14:0), and cardiolipin (CL 16:0/16:0/16:0/16:0). Chemical structures are adapted from the LIPID MAPS structure database [25]. (**b**) Schematic overview of phospholipid metabolism. Synthesis starts with the conversion of fatty acids into lysophosphatidic acid (LPA) by glycerol-3-phosphate acyltransferase (GPAT). LPA is then metabolized to PA by 1-acyl-sn-glycerol-3-phosphate acyltransferase 1-5 (AGPAT1-5), a reaction that can be reversed by phospholipase A2 (PLA2). PA can then be metabolized to diacylglycerol (DAG) (process described in Figure 3b), which can be subsequently transformed to PE by ethanolaminephosphotransferase 1 (SELENOI) or PC by cholinephosphotransferase 1 (CHPT1). PE can also be converted into PC by the enzyme phosphatidylethanolamine *N*-methyltransferase (PEMT). Both compounds can be precursors for the synthesis of PS by the enzymes phosphatidylserine synthase1/2 (PTDSS1/2). The conversion of PE to PS can be reversed by phosphatidylserine decarboxylase proenzyme (PISD). Additionally, PA can be metabolized by phosphatidate cytidylyltransferase 1/2 (CDS1/2) to cytidine diphosphate DAG (CDP-DAG), which can then be transformed to either PI, PG or CL, by CDP-diacylglycerol–inositol 3-phosphatidyltransferase (CDIPT), CDP-diacylglycerol–glycerol-3-phosphate 3-phosphatidyltransferase (PGS1) and cardiolipin synthase (CRLS1), respectively. PI can be phosphorylated by PI (phosphate) kinases (PI(P)K), to produce PI phosphate (PIP(x)). Moreover, all phospholipids can be metabolized to their lyso-forms (LPC, LPS, LPE, LPI, LPG, and LCL) by PLA2, a reaction reversed by lysophospholipid acyltransferases (LPCATs). LPG can be further metabolized to Bis(monoacylglycero)phosphate (BMP).

LPA

Similar to PA, LPA is a lipid mediator in a wide range of biological actions, including cell proliferation, (nervous system) development, and cytokine secretion [262–264]. Furthermore, LPA is involved in neuronal (DA) differentiation [265]. The expression of LPA receptor 1 is reduced in the SN of a 6-OHDA rat PD model [265], and an LPA receptor ligand attenuates the MPTP mouse PD model [266]. Unfortunately, nothing is known about LPA in PD patients.

4.2. PE

The glycerophospholipid PE has a structural role in biological membranes, and it is a regulator of cell division, membrane fusion/fission, and hepatic secretion of very low-density lipoproteins (VLDL) [267]. Patients with PD show decreased plasma levels of PE 34:2 [222], and those carrying a *GBA* mutation have decreased serum levels of PE compared to non-*GBA* mutation carriers [268]. Decreased total PE levels have also been observed in the SN of PD patients before treatment [269], in males only after treatment [270], and in the primary visual cortex [225]. In contrast, increased PE has been found in frontal cortex lipid rafts from PD patients [33]. Of note, one of the enzymes linked to PE synthesis, phosphoethanolamine cytidylyltransferase, is elevated in the SN of PD patients [271]. All findings combined, most evidence points towards decreased levels of PE in PD patients, but the biological implications of the reduced levels need to be examined further.

In vitro, PE is necessary for the interaction between α-synuclein and biological membranes [272] and for the formation of stable, highly conductive channels by α-synuclein [273]. Both processes might have a role in the normal function of α-synuclein. Accordingly, in yeast and worm models, PE deficiency disrupts α-synuclein homeostasis and induces its aggregation [274,275]. This deficiency, also seen in PD patients, could be due to increased formation of LPC from PE, which occurs in MPP+ models [276]. Moreover, the inhibition of this metabolic step offers significant protection against cytotoxicity [277].

4.3. PS

The glycerophospholipid PE is involved in the triggering of both intracellular and extracellular cascades, such as the activation of kinases or the clearance of apoptotic cells [250,278]. It plays a role in neuronal survival and differentiation, and neurotransmitter release [279]. Plasma levels of PS 40:4 are decreased in PD patients [222], but higher levels of PS 36:1, PS 36:1, 36:2, and 38:3, or overall PS, have been found in parkin-mutant fibroblasts [280], frontal cortex [224], and primary visual cortex [225] of PD patients, respectively. This is in agreement with the increased PS synthase activity that has been observed in the SN of PD patients [271]. However, some groups reported contrasting findings and claimed that total PS levels in PD SN and frontal cortex lipid rafts are not significantly altered [33,270]. Yet another interesting finding is that parkin-mutant iPSC-derived neurons have a different subcellular distribution of PS [281], with increased and decreased PS in the mitochondrial and ER fractions, respectively.

The exposure of PS on the cellular surface, which acts as an "eat-me" signal for phagocytosis, is triggered by 6-OHDA [282], rotenone [283], paraquat [284], MPP+ [285], and WT, A53T and A30P α-synuclein [286]. Blockade by an antibody against PS is protective in a rotenone-induced neuronal/glial PD model [287], pointing towards a role of microglial-mediated phagocytosis in PD. This glyceroplipid is known to be associated with the N-terminal- and mid-region of α-synuclein [256,288], with some preference for acetylated α-synuclein [289,290]. This association correlates with membrane penetration [255], alpha-helix formation [256] and aggregation [291], and vesicle [256,292] and liposome [51] binding. Taken together, these findings suggest that PS is a modulator of apoptosis and α-synuclein-mediated pathology.

4.4. PC

The most-abundant glycerophospholipid in eukaryotic membranes, including mitochondrial membranes [293], where it plays a structural role, is PC. It is involved in anti-inflammation [294], cholesterol metabolism [295], and neuronal differentiation [296]. Decreased levels of PC 34:2 and 46:2, PC 34:5, 36:5, and 38:5, and total PC, have been observed in plasma and frontal cortex from PD patients [222,224], and in SN from only male PD patients [270], respectively. One of the enzymes involved in PC synthesis, PC cytidylyltransferase, is elevated in the SN of PD patients [271]. Interestingly, components of the pathway "PC biosynthesis", together with "PPAR signaling" components, allowed accurate classification of PD and control samples [297], highlighting altered PC metabolism as a consistent feature of PD.

Decreased PC levels have been found in the SN of a mouse model of early PD [298] and in brain tissue from MPTP-treated goldfish [66]. Interestingly, α-synuclein does not bind to but rather remodels pure PC membranes through weak interactions with this phospholipid [299,300], and α-synuclein E46K mutants form functionally distinct ion channels in PC membranes [301]. However, others observed binding of the physiologically relevant N-terminally acetylated α-synuclein to pure PC membranes, with preference for highly curved and ordered membranes [302]. Based on these multiple links, the significance of PC metabolism for PD pathology is an interesting and important topic for further study.

LPC

The most-abundant lysophospholipid in the blood is LPC. Its levels are critically related to major alterations in mitochondrial function (e.g., oxidation rate) and to minor defects in mitochondrial permeability [303,304]. Of note, saturated acyl LPCs have inflammatory properties, such as leukocyte extravasation and formation of pro-inflammatory mediators, which can be compensated by polyunsaturated acyl LPC, such as LPC 20:4 and LPC 22:6 [305]. Higher levels of LPC 16:0 and 18:1 have been found in the lipid profile of parkin-mutant fibroblasts compared to healthy controls [280]. Moreover, increased plasma LPC 18:2 has been suggested as a biomarker for PD [149].

Treatment with MPTP induces LPC formation, which leads to cytotoxic changes, dopamine release and inhibition of its uptake, a decreased mitochondrial potential, and increased reactive oxygen species (ROS) formation in PC12 cells [277]. The lysophospholipid LPC inhibits D1 and D2 receptor binding activities in the striatum of rats, inhibits the dopamine transporter, and decreases striatal dopamine turnover rate [306], leading to hypokinesia [307]. Interestingly, 6-OHDA treatment of rats gives rise to an overall decrease in LPC species, with the exception of LPC 16:0 and 18:1, which are increased in the SN [298]. Thus, LPC has negative effects on the DA system, but LPC levels in PD models seem to depend on the type of pharmacological treatment used and the LPC species involved.

4.5. PI

The glycerophospholipids PI and PI phosphates are part of intracellular signal transduction systems [308], but relatively little is known about their role in PD. In humans, higher levels of PI 34:1 and no changes of PI 36:1, 36:2, 38:4, 38:5, 40:5, and 40:6, or no changes in total PI have been observed in parkin-mutant skin fibroblasts [280], and in the lipid rafts of frontal cortex from PD patients [33], respectively. Decreased levels of overall PI have been observed in the SN of male PD patients [270]. In rodents, MPTP decreases the expression level of striatal PI-transfer protein [309], which is involved in the transfer of PI across membranes [310].

PI Phosphate (PIPx)

The role of PIPx species in PD is also poorly defined. However, PI and PIP2 effectively influence self-oligomerization of α-synuclein [311], while α-synuclein seems to prefer binding membranes containing PI(4,5)P2 [312]. Moreover, PIP3 is decreased in the nuclear fraction and whole-tissue homogenate, while PIP2 is increased in whole-tissue homogenate of SN from PD patients [313]. As mentioned above, ATP13A2 requires the interaction with PI(3,5)P2 to protect cells against PD-related stress conditions [259,260], an interaction that is able to reduce proteasomal inhibitor-induced accumulation of ubiquitin proteins [314]. Future systematic studies on the roles of the various PIPx species are required to understand their function in PD pathogenesis.

4.6. PG

Less than 1% of total glycerophospholipids in intracellular membranes is composed of PG and it is mainly localized to mitochondrial membranes, where it can be synthesized locally [315]. The levels of total PG are not changed in lipid rafts from PD frontal cortex [33], while increased PG 32:0 has been described in total extracts of the same brain area [224]. Alpha-synuclein is able to bind PG with various degrees of affinity depending on the variability in its structure (WT $\approx$ truncated > A53T > A30P) [316], and PG-containing membranes can promote α-synuclein aggregation [317–319]. Additionally, α-synuclein oligomers are able to induce PG clustering [319], connect PG-containing vesicles [320] and disrupt PG vesicles [321] through large membrane bilayer defects, rather than through a pore-like mechanism [322], leading to vesicle docking and fusion problems. Furthermore, low concentrations of α-synuclein inhibit and high concentrations stimulate lipid peroxidation of PG [323]. Unfortunately, information on the levels of PG in animal PD models is lacking.

4.7. CL

The glycerophospholipid specific for mitochondrial membranes is CL. Here, it plays both a structural and functional role [324–326]. No changes in total CL levels have been detected in the SN of PD patients [270]. However, *PINK1* KO mouse embryonic fibroblasts display decreased CL levels and supplementation with CL rescues mitochondrial dysfunction [327]. Moreover, rotenone induces oxidation of highly unsaturated CL in human peripheral blood lymphocytes [328], and increases levels of plasma PUFA CLs, but decreases oxidizable PUFA-containing CL levels and increases mono-oxygenated CL species in the SN of rats [329]. A proper CL content in the inner mitochondrial membrane and the presence of acyl side chains are crucial for α-synuclein localization [330,331], while CL content in the outer mitochondrial membrane buffers synucleinopathy [332]. Moreover, α-synuclein is able to disrupt artificial membranes containing CL [333], and its overexpression reduces CL content in MN9D cells [334] and in mouse brain [335]. Additionally, the formation of complexes between CL and α-synuclein, together with cytochrome c, may be a source of oxidative stress [336]. The role that CL plays in the interaction of α-synuclein with membranes and in mitophagy has been previously reviewed [337,338].

5. Sphingolipids

Sphingolipids constitute a family of lipids characterized by the presence of a sphingoid-base backbone. This complex family of compounds includes the sphingoid bases (e.g., sphingosine and sphingosine-1-phosphate), ceramides, phosphosphingolipids (e.g., sphingomyelin (SM)) and glycosphingolipids (e.g., cerebrosides, ganglisodes, and sulfatides) (Figure 5). Sphingolipids are not only structural components of cell membranes, but they also play a role in apoptosis, autophagy, and immune response [339]. Here, we will specifically focus on the involvement of sphingosine(-1-phosphate), ceramide, SM, cerebrosides, gangliosides, and sulfatides, and an overview is given in Supplementary Materials Table S1.

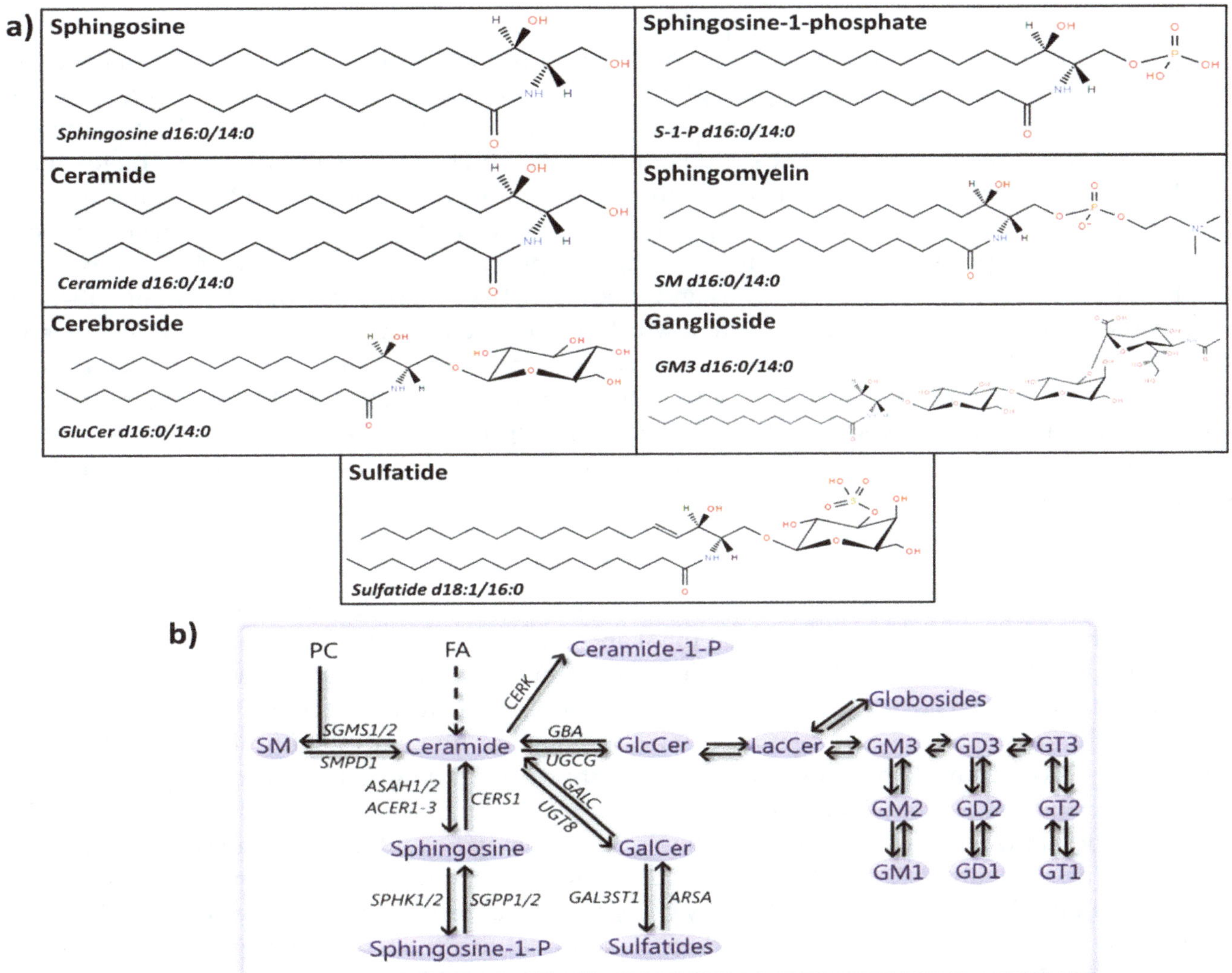

Figure 5. Sphingolipids: structures and metabolic steps involved. (**a**) Schematic representation of the chemical structures of sphingolipids, including sphingosine (d16:0/14:0), sphingosine-1-phosphate (S-1-P d16:0/14:0), ceramide (d16:0/14:0), sphingomyelin (SM d16:0/14:0), cerebroside (glucosylceramide, GluCer, d16:0/14:0), ganglioside (GM3 d16:0/14:0), and sulfatide (d18:1/16:0). Chemical structures are adapted from the LIPID MAPS structure database [25]. (**b**) Schematic overview of steps involved in the formation and metabolic conversion of sphingolipids. Synthesis of sphingolipids starts by a multistep process to convert fatty acids (FAs) into ceramide. Phosphatidylcholine can be fused to ceramide by phosphatidylcholine:ceramide cholinephosphotransferase 1/2 (SGMS1/2) to produce SM, which can be converted back into ceramide by sphingomyelin phosphodiesterase (SMPD1). Ceramide can also be phosphorylated by ceramide kinase (CERK) to ceramide-1-P and converted into sphingosine by acid or alkaline ceramidases (ASAH1/2 or ACER1-3), and phosphorylated by sphingosine kinase 1/2 (SPHK1/2) to sphingosine-1-phosphate. This process can be reversed by the sequential action of sphingosine-1-phosphate phosphatase 1/2 (SGPP1/2) and ceramide synthase 1 (CERS1). Furthermore, ceramide can be glycosylated via the addition of a galactose molecule by 2-hydroxyacylsphingosine 1-beta-galactosyltransferase (UGT8) to produce galactosylceramide (GalCer). Further addition of a sulfate group by galactosylceramide sulfotransferase (GAL3ST1) results in the formation of sulfatides. This process can be reversed by the sequential actions of arylsulfatase A (ARSA) and galactocerebrosidase (GALC). Finally, ceramide can also by glycosylated via the addition of a glucose molecule by ceramide glucosyltransferase (UGCG) to produce GlcCer, which can be further glycosylated to produce both globosides and gangliosides, such as GM3, GD2 and GT1, by multiple enzymes [340], or it can be converted back into ceramide by glucosylceramidase (GBA).

5.1. Sphingosine(-1-Phosphate)

Sphingosine is a bioactive lipid known to induce apoptosis and regulate endocytosis, while its phosphorylated form, sphingosine-1-phosphate (S1P), promotes cell survival and triggers diverse intracellular signaling pathways through G-protein-coupled receptors [339,341,342]. Sphingosine induces the formation of oligomeric α-synuclein species, which serve as template for the formation of endogenous α-synuclein aggregates in human and mammalian neurons [343]. Similarly, S1P accumulation, e.g., due to GBA deficiency, promotes α-synuclein aggregation [343]. Alpha-synuclein itself inhibits the expression and activity of sphingosine kinase 1, the enzyme that catalyzes the phosphorylation of sphingosine to S1P [344] and modulates S1P receptor-mediated signaling [345,346]. Sphingosine-1-phosphate supplementation of MPP+-treated cells is neuroprotective [347–349], and a selective S1P receptor agonist is protective in mouse and cellular models treated with 6-OHDA and rotenone [350]. Therefore, while S1P is protective in animal and cellular PD models, presumably through its pro-survival effects, it is clear that both sphingosine and S1P are linked to α-synuclein aggregation. Unfortunately, the lack of studies on human samples does not allow drawing a conclusion regarding the relevance of these lipids for PD pathogenesis.

5.2. Ceramide

Ceramide is involved in apoptosis, lipid raft formation, and regulation of the mitochondrial respiratory chain [340,351–353]. Both higher [354] and lower [222,355] plasma levels of ceramide have been reported in PD patients, while lower ceramide 18:0 and no differences in total ceramide levels are observed in their frontal cortex [224] and SN [270], respectively. Reduced levels of ceramide may be associated with α-synuclein accumulation [356,357]. This is in line with the finding that reduced and increased levels of ceramide have been observed in the anterior cingulate cortex and primary visual cortex of PD patients [225,356,358], which display and lack α-synuclein aggregation, respectively [359]. Thus, variation in the levels of ceramide in different tissues may be linked to α-synuclein accumulation.

Mimicking PD with *PLA2G6* KO, *LRRK2* KO, *PINK1* KO or rotenone treatment increases ceramide levels in fly brain, mouse brain, mouse olfactory bulb, and human erythrocytes, respectively [283,360–362]. C2-ceramide initiates a series of events leading to neuronal death, including an early inactivation of PI3K/AKT and ERK pathways, followed by activation of JNK, GSK3β activation and neuronal death [363]. Additionally, C2-ceramide induces cytotoxicity and ROS production in neuronal(-like) cells [364–366], which can be prevented by WT α-synuclein [367], *PINK1* [368,369] and *DJ-1* [370]. However, both in vivo and in vitro C2-ceramide seems to suppress microglial activation [371], protect neurons against α-synuclein-induced cell injury [372], and reverse rotenone-induced phosphorylation and aggregation of α-synuclein [373]. An increase in ceramide levels is thus commonly found in animal and cellular models for PD, but its effects are unclear and may be both beneficial and detrimental for different PD-related traits.

5.3. SM

The most abundant sphingolipid in eukaryotic cells and plasma is SM. It is one of the building blocks of the cellular membrane and a source of bioactive lipids, such as ceramide, ceramide-1-phosphate and S1P, which are involved in inflammation [374,375], cell death [376,377] and autophagy [378]. In the nervous system, SM is a major constituent of myelin. Mutations in sphingomyelinase-1, which lead to SM accumulation, are a risk factor for PD [16,379,380]. This feature may be linked to the increase in α-synuclein expression observed upon SM treatment [381] and the presence of SM in LB inclusions [382]. Parkinson's disease patients carrying GBA mutations have elevated levels of total plasma SM compared to PD patients not carrying the mutation [268]. Moreover, SM 18:1 and SM 26:1 are increased and decreased in the anterior cingulate cortex [358], respectively, while increased SM levels have been described in the primary visual cortex [225], and, in males only, in the SN of PD patients [270]. However, no changes have been found in the putamen or cerebellum of sporadic PD patients [383]. The role that SM accumulation appears to play in PD pathogenesis may thus be multifold, being linked to inflammation, autophagy dysfunction, and/or α-synuclein expression and aggregation.

5.4. Cerebrosides

Cerebrosides are lipids glycosylated via the addition of either glucose or galactose and known to be involved in intracellular membrane transport and cell survival [384]. In PD patients, cerebrosides are increased in plasma (of *GBA* mutation carriers) and, in males only, in the SN, whereas they are decreased in lipid rafts from the frontal cortex [33,268,270,385]. More specifically, PD patients have increased levels of glucosylceramide [223,354] in plasma but no changes in cerebroside levels in the temporal cortex [386], putamen or cerebellum [383], and decreased levels of galactosylceramide 24:1 and lactosylceramide 18:1 in the frontal cortex [224]. Thus, whereas a consistent coupling between PD and increased cerebrosides in plasma has been found, cerebroside changes in the brain are region dependent and their significance for PD needs to be determined.

Interestingly, mutations in the enzymes responsible for the degradation of cerebrosoides, namely GBA and galactocerebosides (GALC), which cause Gaucher's disease and Krabbe's disease, respectively, have been associated with α-synuclein aggregation and PD [387,388]. Glucosylceramide, a product that accumulates upon GBA deficiency, destabilizes α-synuclein tetramers and related multimers and frees α-synuclein monomers and leads to cellular toxicity [389]. These effects are caused by colocalization of glucosylceramide with α-synuclein and induction of a pathogenic conformational change of the protein [390]. This promotes aggregation of WT (but not mutated) α-synuclein into a β-sheeted conformation [343,391], and conversion of α-synuclein into a proteinase-resistant form [392]. Conversely, α-synuclein inhibits normal activity of GBA [393], which increases glucosylceramide, creating a feedback loop. Inhibition of glucosylceramide synthase, which decreases glucosylceramide levels, slows α-synuclein accumulation [394] and partially protects mice against MPTP-induced toxicity [395]. Thus, it is well established that glucosylceramide accumulation leads to α-synuclein aggregation and toxicity. Interestingly, aging of WT mice leads to brain accumulation of both glucosylceramide and lactosylceramide [396], suggesting that age-associated changes in its metabolism might be related to PD onset.

5.5. Gangliosides

Gangliosides are synthesized by the addition of carbohydrate moieties to lactosylceramide. One of the simplest and most widely distributed ganglioside is monosialodihexosylganglioside (GM3) that consists of lactosylceramide and sialic acid [397]. Gangliosides were initially discovered in the brain where they are involved in neurotransmission, receptor regulation, and stabilization of neural circuits, including the nigro-striatal DA pathway [398,399]. Parkinson's disease patients have higher plasma levels of gangliosides [385], GM3 gangliosides [223], and N-acetylneuraminic acid-3 (NANA-3) gangliosides [222] than controls. Likewise, higher GM2 and GM3 levels have been detected in parkin-mutant iPSCs compared to controls [280]. However, no accumulation of GM1, GM2 or GM3 has been observed in the putamen or cerebellum of sporadic (or heterozygous GBA-mutation) PD patients [383], nor in the SN of PD patients [270]. Even a GM1 deficiency, together with decreased expression of ganglioside biosynthetic enzymes (B3GALT4 and ST3GAL2), has been found in the SN from PD patients [400,401]. Hence, most publications point towards increased gangliosides in plasma of PD patients, but concomitant changes in ganglioside levels have not been observed in their brains.

Interestingly, GM1 supplementation seems to have a positive disease-modifying effect in PD patients [402–407]. Also, increased GM1 levels are neuroprotective in MPTP-treated animals [408]. For example, GM1 can partially protect against 6-OHDA treatment [409] and aging-related DA deficits [410] as well. However, studies on MPTP-treated non-human primates have shown that a short treatment with GM1 does not lead to any improvement [411], while a chronic treatment does have a positive effect [412], which might be restricted to the surviving DA neurons in the midbrain, rather than due to the prevention of cell death [413]. Mechanistically, GM1 treatment increases DA innervation, dopamine synthesis, and TH expression following an MPTP lesion [414–423]. Moreover, GM1 inhibits the inflammatory response triggered by 6-OHDA [424], protects against the toxic intracellular GPR37 aggregates observed in parkinsonism [425] and is involved in the internalization of α-synuclein into microglia [426]. Nonetheless, evidence for an α-synuclein-linked role of GM1 is controversial: in one study it was claimed that GM1 may accelerate α-synuclein aggregation [427] and the formation of proteinase-resistant α-synuclein [392], but other work demonstrated that it induces alpha-helical structure and inhibits or eliminates α-synuclein fibril formation (depending on the amount of GM1 present) [289,428]. It is also unclear whether membranes containing GM1 interact with α-synuclein [428,429]. Hence, GM1 is a promising candidate for PD treatment, but further clarification of its specific effects on α-synuclein is urgently needed.

Only a limited number of studies have analyzed the role of gangliosides other than GM1 in animal and cellular models. For instance, mice lacking GM2/GD2 synthase develop parkinsonism, which can be partially rescued by administration of GM1 [400,430]. However, GM2 accumulation, as seen in Tay Sachs and Sandhoff's diseases, leads to α-synuclein aggregation [431]. Thus, both deficiency and excess of GM2 may lead to PD-like pathology. Likewise, GM3 accelerates α-synuclein aggregation [427] and regulates α-synuclein-induced channel formation in PC-containing membranes [301]. Furthermore, deletion of GD3 synthase, which decreases production of the pro-apoptotic GD3 ganglioside, protects against MPTP treatment in mice [432]. In contrast, ganglioside GT1b is neurotoxic in nigral DA neurons by triggering nitric oxide release from activated microglia [433]. The gangliosides GD3 and GT1b are unchanged and decreased in the SN of (male) PD patients, respectively [270]. Together, these results indicate that GM3, GD3, and GT1b play aggravating roles in PD pathology. Finally, 1-phenyl-2-decanoylamino-3-morpholino-1-propanol (PDMP, an inhibitor of glycosylceramide synthase that decreases ganglioside content) enhances α-synuclein toxicity, which can be rescued by ganglioside addition [434].

5.6. Sulfatides

Sulfatides, which are sulfated galactocerebrosides, form a group of lipids involved in protein trafficking, immune responses and neural plasticity, among others [435]. Higher levels of sulfatides have been detected in the plasma [385] and visual cortex [225] of PD patients, and in the SN of male PD patients [270]. Arylsulfatase A, an enzyme that breaks down sulfatides, has been linked to PD recurrence [436,437]. However, no changes or reductions in sulfatide levels have been described in lipid rafts from the frontal cortex of PD patients [33] and in brain samples from PD patients [438], respectively. Thus, most evidence points towards increased sulfatide levels in PD, although a number of studies have not confirmed this finding, suggesting patient, technique and/or tissue-type differences among the various investigations.

6. Sterols

Sterols are amphipathic lipids synthesized from acetyl-CoA via the β-hydroxy β-methylglutaryl-CoA reductase pathway and containing a fused four-ring core structure (Figure 6). Sterols are known to play a role in immune cell function [439], influence membrane fluidity and permeability, and serve as signaling molecules and hormones [440], among others. Here we will review the current findings on sterols in PD, more specifically cholesterol, its precursors, CE, and oxysterols (Supplementary Materials Table S1).

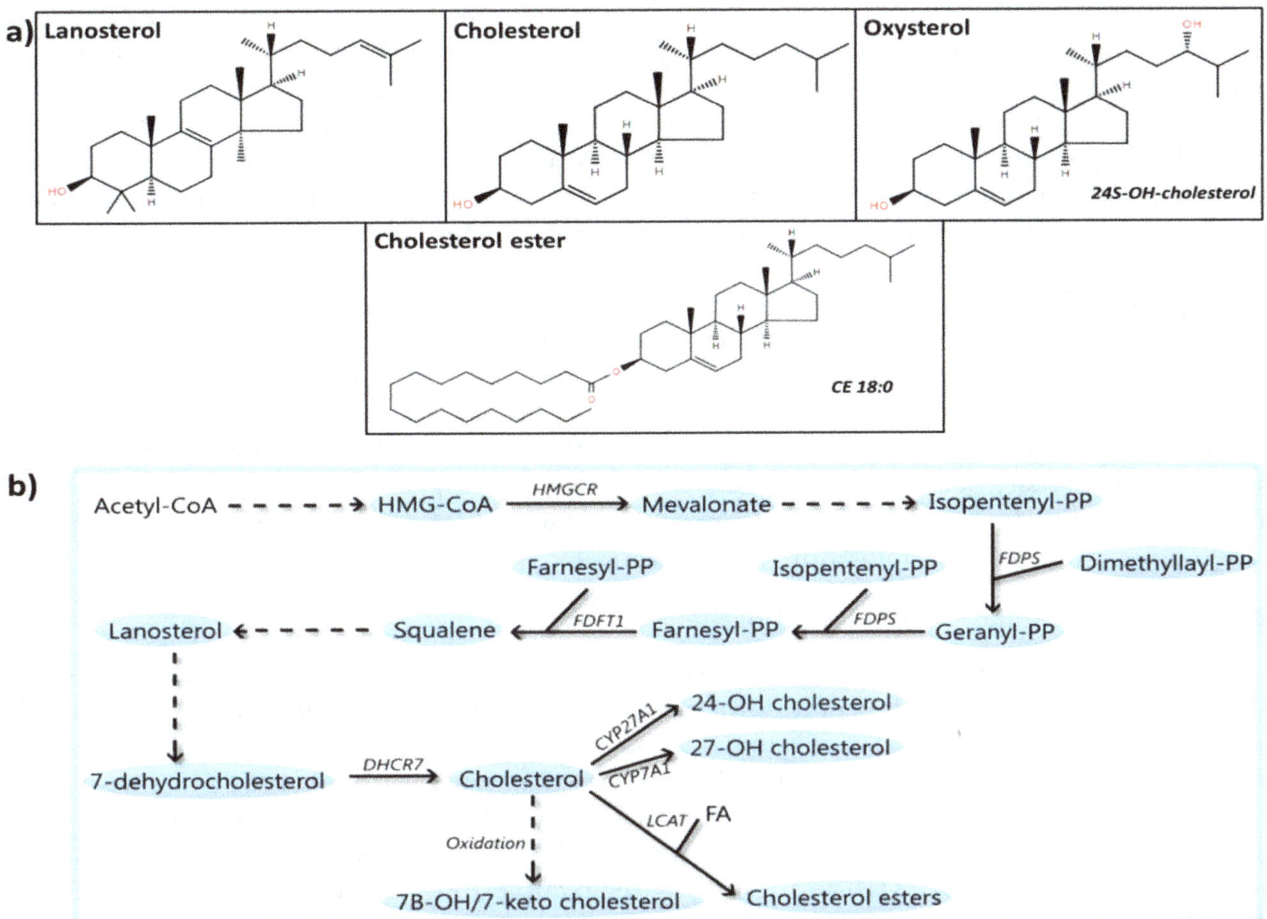

Figure 6. Sterols: structures and metabolic steps involved. (**a**) Schematic representation of the chemical structures of sterols, including lanosterol, cholesterol, oxysterols (24S-hydroxy-cholesterol), and cholesterol esters (CE 18:0). Chemical structures are adapted from the LIPID MAPS structure database [25]. (**b**) Schematic overview of steps involved in sterol metabolism. Acetyl-CoA is used to synthesize β-hydroxy β-methylglutaryl-CoA (HMG-CoA), which is converted into mevalonate by 3-hydroxy-3-methylglutaryl-coenzyme A reductase (HMGCR). Mevalonate is metabolized to isopentenyl-PP by a multistep process, followed by its conversion to geranyl-PP and farnesyl-PP by the enzyme farnesyl pyrophosphate synthase (FDPS). Two molecules of farnesyl-PP are condensed by squalene synthase (FDFT1) to create squalene, which is further metabolized to lanosterol and 7-dehydrocholesterol. Subsequently, cholesterol is synthesized from 7-hydrocholesterol by 7-dehydrocholesterol reductase (DHCR7). Finally, cholesterol can be oxidized to compounds such as 7-beta-hydroxycholesterol or 7-ketocholesterol. It can also be esterified to a fatty acid (FA) by phosphatidylcholine-sterol acyltransferase (LCAT) to create cholesterol esters or metabolized by the cytochrome p450 to produce compounds such as 24/27-hydroxycholesterol.

6.1. Cholesterol

6.1.1. Human Studies on Cholesterol

Cholesterol intake has been found to be negatively [29,441], positively [30,442], or not [31,443] correlated with PD risk. A meta-analysis indicates a lack of association between cholesterol intake and PD [444]. Lower plasma cholesterol has been associated with PD [229,232,445–447], and confirmed by a meta-analysis [238], and higher plasma cholesterol levels have been linked to reduced PD risk [235,448–452] and slower clinical progression of PD [453]. However, others, including a meta-analysis [454], have found no association between plasma cholesterol levels and PD [230,455] or PD risk [233,456]. Even higher plasma cholesterol levels in PD patients compared to controls [231,457] have been reported. The differential outcome of these studies could be attributed to factors such as age and gender, among others, since lower plasma cholesterol levels have been reported in PD male patients of more than 55 years compared to controls [458], a high total cholesterol baseline has been associated with increased risk of PD in subjects of 25–54 years (but not in those above 55) [459], and female PD patients seem to have higher cholesterol levels compared to male PD patients [460]. Thus, proper patient stratification is necessary to determine whether plasma cholesterol is associated with PD, which would point towards defects in cholesterol metabolism.

In PD patients, no significant changes in cholesterol levels have been observed in the putamen [383], SN [270] or frontal cortex lipid rafts [33], while elevated levels of cholesterol have been found in the visual cortex [225]. Finally, decreased cholesterol biosynthesis has been described in fibroblasts from PD patients [461]. The differences in these observations could be related to tissue or brain-region specificities, technique sensitivity, and/or choice of patients. Thus, validation studies and larger cohorts are needed to determine the relevance of cholesterol changes in PD patients and their pathology. Additionally, some studies [13,462,463] have found an association between PD and a SNP near the gene *SREBF1*, which encodes a transcription factor that regulates cholesterol biosynthesis, although other studies could not confirm the findings [464].

6.1.2. Animal and Cellular Studies on Cholesterol

In animal and cell model studies, the link between PD and cholesterol has been demonstrated multiple times. For example, the cholesterol biosynthetic pathway controls *PRKN* expression [465], which in turn regulates fat (and cholesterol) uptake in *PRKN* mutant mice and human cells [466]. Additionally, *DJ-1* KO mouse embryonic fibroblasts and astrocytes display lower cellular (but not plasma [467]) cholesterol levels and impaired endocytosis [468], which can be rescued by increased membrane cholesterol [469]. In contrast, *GBA* KO and *PRKN* KO cells have increased cholesterol levels [470,471], and the N370S *GBA* mutation leads to cholesterol accumulation in lysosomes [472], while *LRRK2* KO rats have higher serum cholesterol levels [473]. Thus, cholesterol biosynthesis seems to be impaired in PD, but the direction of the change differs among PD etiologies, which could explain part of the variation observed in different studies with PD patients.

Increased cholesterol reduces cell death [474] and modulates presynaptic DA phenotype by increasing TH and VMAT2 expression in SH-SY5Y cells [475] and enhancing ligand binding of DAT and VMAT2 in the brains from rats and monkeys [476]. However, hypercholesterolemia seems to cause DA neuronal loss and oxidative stress in the SN and the striatum, leading to motor impairment [477–479]. Together with the observation that cholesterol treatment of (MPP+-treated) SH-SY5Y cells reduces their viability [480], this finding suggests that the effect of cholesterol levels on PD is dose dependent.

6.1.3. Alpha-Synuclein and Cholesterol

Alpha-synuclein interacts with cholesterol [481] and cholesterol-containing vesicles [482], but it is unclear whether cholesterol facilitates the binding of α-synuclein to charge-neutral membranes [483,484]. Alpha-synuclein-cholesterol interaction seems to be associated with α-synuclein accumulation [474,485] and aggregation [486] and is a determining factor in α-synuclein's ability to form pores [487,488]. Accordingly, reducing cholesterol levels leads to decreased α-synuclein accumulation and damage in the synapse [489–491]. Hence, high levels of cholesterol aggravate α-synuclein-associated pathology. Furthermore, α-synuclein potentiates cholesterol efflux [492], antagonizes cholesterol in lipid rafts [493], and enhances production of oxidative cholesterol metabolites [494]. Finally, A53T-α-synuclein-overexpressing mice have increased levels of serum cholesterol [239], while WT-α-synuclein-overexpressing mice have upregulation of genes involved in cholesterol biosynthesis in DA neurons from the SN [495], indicating a tight reciprocal relationship between α-synuclein and cholesterol metabolism.

6.1.4. Statins

Statins are cholesterol-lowering drugs that have been described to decrease [496–501] or not affect [502–504] PD risk. Interestingly, lipophilic, but not hydrophilic, statins increase PD risk [505]. In the current discussion on the contradictory findings regarding the effects of statins not enough attention is paid to confounding factors such as statin indication, statin-type effects or immortal time bias (span of cohort follow-up during which the outcome under study cannot occur), and healthy user effects [506]. In animal and cellular models, atorvastatin pretreatment seems to prevent early effects of MPTP administration in rats [507], and lovastatin has neuroprotective effects against MPP+ and 6-OHDA [474,508] and ameliorates α-synuclein accumulation [509,510]. Similarly, simvastatin is neuroprotective against 6-OHDA and MPTP treatments [511–514] and increases dopamine content in the striatum [515]. However, negative effects of simvastatin and atorvastatin on MPP+-mediated toxicity have also been reported [516], which could be explained by the fact that statin lactones, one of the statin metabolites, are able to inhibit mitochondrial complex III [517], potentiating MPP+ toxicity.

6.2. Cholesterol Precursors

In PD patients, the cholesterol-synthesizing enzymes isopentenyl diphosphate isomerases 1 and 2 have been observed in LB from the SN of PD patients [518]. The natural cholesterol intermediate squalene seems to prevent toxicity in the striatum of 6-OHDA-treated mice [519], whereas α-synuclein accumulation enhances squalene production [242], which could be a cellular response to oxidative damage. A derivative of squalene, squalane, exacerbates 6-OHDA toxicity [519]. The naturally occurring cholesterol precursor lanosterol induces mitochondrial uncoupling and protects DA neurons from cell death in the nigrostriatal region of MPTP-treated mice [520]. Thus, cholesterol precursors seem to have a protective role in PD. Interestingly, inhibitors of both geranylgeranyl transferase (GGTI) and farnesyl transferase (FTI), enzymes that transfer the prenyl group geranylgeranyl or farnesyl to proteins, protect nigrostriatal neurons in MPTP-intoxicated mice [521].

6.3. CEs

The esters between cholesterol and fatty acids, CEs, are synthesized from excess cholesterol in the cytosol by the enzyme acetyl-coA acetyltransferase 1, a process that can be reversed by the enzyme cholesteryl ester hydrolase. In PD patients, reduced cholesterol esterifying activity has been detected in fibroblasts [461] and CE 20:5 is reduced in their visual cortex [222]. Interestingly, in *C. elegans* the ortholog of neutral cholesteryl ester hydrolase 1 attenuates α-synuclein neurotoxicity when sufficient CE is present, while knockdown leads to neurodegeneration [522]. However, GBA KO cells have increased levels of CE 15:1, 22:6, and 24:1 [470], which could reflect either a protective or a pathological mechanism.

6.4. Oxysterols

The products of cholesterol oxidation, 7beta- and 27-hydroxycholesterol, and 7-ketocholesterol, are elevated in plasma from PD patients [162]. Additionally, 27-hydroxycholesterol CSF levels are increased in a subgroup of PD patients [523] Moreover, increased cholesterol lipid hydroperoxides have been observed in the SN of PD patients [524]. The CSF levels of 24-hydroxycholesterol appear to be correlated with PD duration [523], but higher levels have also been observed in early stage PD [525]. Conversely, 24-hydroxycholesterol esters are reduced in plasma from PD patients [526]. Of note, TH levels are increased by 24-hydroxycholesterol [527], while 27-hydroxycholesterol seems to reduce TH expression and increases α-synuclein levels [527–530]. An unexpected finding was that both 24- and 27-hydroxycholesterol seem to protect against staurosporine-induced cell death [531]. Interestingly, oxysterols, and more specifically 24(S),25-epoxycholesterol, increase DA neuronal differentiation via liver X receptors in both mouse and human embryonic stem cells [532,533].

7. Lipoproteins

Lipoproteins transport triglycerides and cholesteryl esters. Together, these lipids form the core of the lipoprotein, which is further surrounded by glycerophospholipids and free cholesterol [534]. Lipoproteins are classified according to their density, and thus their composition, as high-density lipoproteins (HDL), intermediate-density lipoproteins (IDL), low-density lipoproteins (LDL) or VLDL. Here, we will specifically review the current findings concerning HDL, LDL and VLDL (Supplementary Table S1).

7.1. HDL

The assembly complex HDL is composed of proteins (around 40%, mainly apolipoprotein A1 (ApoA1), but also apolipoprotein C (ApoC), apolipoprotein E (ApoE), and apolipoprotein J (ApoJ)) and lipids (including around 30% of glycerophospholipids, 25% of cholesterol/CE, and 5% of TAG). The main biological role of HDL is in cargo transport, in particular of lipids and proteins, but it is now also known to bring miRNAs to recipient cells [535]. Lower plasma HDL and ApoA1 levels have been associated with earlier PD onset [536] and higher PD risk [237,537–539], and HDL levels are positively correlated with disease duration [540]. Plasma levels of HDL-cholesterol are lower [229,385,541] or not different [230,233,446,447,457] in PD patients compared to controls. This controversial relationship is complex, as both sex [460] and *APOE* polymorphisms [231] seem to affect HDL-cholesterol levels in PD patients.

7.2. LDL

About 20% of LDL consists of proteins (mainly apolipoprotein B (ApoB)) and the remainder consists of lipids (including about 22% of glycerophospholipids, 50% of cholesterol/CE, and 8% of TAG). High LDL-cholesterol levels in plasma are protective for PD and associated with preserved executive and fine motor functions in PD [452,455,457,542], while lower LDL-cholesterol levels are associated with higher PD risk [229,445–447,500,543,544]. One study reported that plasma LDL levels are not different between PD patients and controls [237]. A number of other studies have reported no difference in baseline LDL-cholesterol [230,497], and two meta-analyses have found no association [238,496]. Furthermore, in contrast to HDL, LDL-cholesterol levels do not differ between male and female PD patients [460,540]. However, one study reported increased LDL-cholesterol levels in PD patients compared to controls [231]. Interestingly, compared to controls PD patients seem to have higher levels of oxidized LDL [545], which is able to enter neuronal cells and elicit neurotoxicity [546]. Finally, male *DJ-1* KO mice have higher LDL-cholesterol levels in serum, which could be due to the fact that the LDLR is a transcriptional target of DJ-1 [467].

7.3. VLDL

Very low-density lipoproteins are mainly composed of lipids (including around 15% of glycerophospholipids, 20% of cholesterol/CE, and 60% of TAG) and only minor amounts of protein (around 5%, mainly ApoB and ApoC). Parkinson's disease patients appear to have lower levels of both VLDL [230] and VLDL-cholesterol [231] than controls, but the role of VLDL in PD remains unclear.

8. The Cellular Lipidome

Above we have given an overview of the changes in lipid composition that have been observed in multiple studies involving PD patients, and animal and cellular PD models. The question arises what the significance of such changes is from a biological point of view. In mammalian cells, about 5% of the genes are involved in the generation and transport of an estimate of 10,000 individual lipid species [547,548], which have structural [549], signaling [549,550], and energy storage [17,18] roles. More specifically, above-mentioned molecules such as glycerophospholipids, sphingolipids, and sterols represent the main components of the cell's plasma and mitochondrial membranes, endoplasmic reticulum, the Golgi complex, and endosomes. In a dynamic manner, lipid composition defines organelle identity [547], controls the recruitment of proteins, and lipid bilayer properties, such as thickness, elastic compression, and intrinsic curvature, can be an allosteric regulator of membrane protein function [551]. Alterations in membranes thus dynamically control important processes such as (synaptic) vesicle trafficking, endocytosis-exocytosis [552] or α-synuclein aggregation [553], processes that have already been associated with PD [554–556].

Lipids also play an important role in intracellular and intercellular signaling in the brain by direct interaction with receptors and other signal-transducing proteins [549,550] that regulate integral physiological processes linked to PD. For example, PUFA are involved in inflammation, neurogenesis, and neuroprotection [56]. Endocannabinoids are lipid-based retrograde neurotransmitters that modulate synaptic plasticity [557], and LPA modulates processes like proliferation, survival and migration [249]. Additionally, although most energy consumed by brain cells comes from glucose, lipids have been suggested to provide up to 20% of the total energy consumption of the adult brain [549,558,559]. Therefore, changes in lipid composition or content, such as the ones that have been described here for PD, can have vast consequences for key processes in the maintenance of normal neuronal and brain function. However, unlike what holds for genes and proteins, most lipid species cannot be associated with specific functions: their role is dictated by the concentration and location of individual lipid species, and, most importantly, by their interaction with other lipid species. Since most of the available information is a description of changes in lipid concentration, firm conclusions regarding the effects of these changes are hard to draw.

This lack of precise knowledge regarding the (patho)biological significance of lipidome abnormalities is predominantly caused by the fact that lipids form a vast and enormously complex group of biomolecules. This creates two major challenges. First, it is currently very difficult—if not impossible—to characterize all lipids present in the lipidome of a sample, due to limitations in the separation methods. This precludes simultaneous analysis of all lipid classes, which is especially hindered by the presence of isomeric (i.e., same mass) lipids. Second, no methodology is currently available to accurately determine the concentrations of the various lipid species [560]. This lack of information hampers the interpretation of lipidomic studies and the creation of reliable databases that, on its turn, impedes the identification of pathways in which a combination of lipid species plays a role [561].

As mentioned, not only their composition in the lipidome but also the tissue distribution and intracellular localization of individual lipids are crucial for their function, which makes it of great importance to develop techniques to identify and quantify lipids at the single-cell level and with the spatial organization of the cell still intact. These developments will help to elucidate the interplay of different lipid species in a time- and location-dependent manner both in health and disease. Indeed, it would allow us to obtain more information about (i) the lipidomes of various cell types, which are now identified in growing numbers within different tissues and organs by RNAseq [562], and (ii) the dynamic changes in lipidome composition that are associated with disease progression. Unfortunately, proper sample preparation, even more than the detection limits for lipids in mass-spectrometry, currently forms the biggest barrier to develop effective single-cell lipidomics [563]. Moreover, the interplay between lipids and other biomolecules necessitates the integration of lipidomics with other omics strategies [564].

It is also important to note that the lipidome composition is not only defined by the activity of genes involved in lipid metabolism, but also strongly depends on exogenous factors. These include (1) the direct dietary intake of lipids and lipid precursors from food, (2) life-style factors, i.e., exercise, sleep patterns, and intrinsic and extrinsic motivation factors that determine the choice of food composition, and (3) the effects of drugs that affect metabolism or cell behavior. For example, accumulating evidence suggests that tight bidirectional interactions exist between dietary lipids and composition and structure of the gastrointestinal tract microbiota [565–567]. This could be especially relevant for PD, since dysbiosis (i.e., the change in microbiota structure relative to that found in healthy individuals [568]), has been repeatedly observed in PD patients [569–574] from early stages of the disease onwards [575].

Filling in the current gaps in lipidomics technology and knowledge is crucial to exploit its potential to help us further understand the molecular mechanisms underlying PD, better define its stages and classification, and identify biomarkers, create dietary interventions, or perform compound screening, preclinical testing and monitoring of drug responses [576,577].

9. Conclusions

From this review, it is clear that a strong correlation exists between PD and abnormalities in lipid metabolism. More specifically, there is an association between PD and the levels of fatty acyls (SFA, MUFA, PUFA, a number of eicosanoids, and acylcarnitine), glycerolipids (MAG, DAG, and TAG), glycerophospholipids (PA, LPA, PE, PS, PC, LPC, PI, PIPx, PG, and CL), sphingolipids (sphingosine(-1P), ceramide, SM, cerebrosides, gangliosides, and sulfatides), sterols (cholesterol precursors, cholesterol, CE, and oxysterols) and lipoproteins (HDL, LDL, and VLDL). Furthermore, there is a conspicuous relationship between the folding, aggregation, and distribution of α-synuclein and the lipids that drive some of the neuropathological features of PD. Yet, it is presently unclear whether links exist between PD and some eicosanoids (eoxins, thromboxanes, oxoeicoanoids, hepoxilins, lipoxins, and epoxyeiconsatetraenoic acid), glycerophospholipids (lysoPE, lysoPS, lysoPI, lysoPG, lysoCL, and Bis(monoacylglycero)phosphate), sphingolipids (globosides), and lipoproteins (IDL).

One of the main concerns regarding the findings summarized in this review is that most lipid classes have not been consistently found to be associated with PD. Variables such as sex, age, PD etiology, specific DNA polymorphisms or the microbiome may have influenced the findings. Thus, proper stratification of PD patients is necessary to understand the biological implications of the lipid changes observed. Additionally, more accurate description of the lipid profiles of plasma, CSF and/or fibroblasts from PD patients will help to classify the patients more accurately.

A further concern is that most studies have focused on plasma levels of lipids, but these may not correlate with their brain levels, e.g., levels of ganglioside species are increased in plasma but not in brains of PD patients [270,385]. Thus, CSF (and brain) lipidomes of PD patients have to be determined to get insight into the actual pathological lipid composition and processes. Moreover, it is often unclear whether the changes in the levels of lipid species reflect a pathological or rather a compensatory mechanism. Finally, studies on cellular and animal PD models do not always show the same directionality of lipid level changes as found in studies on PD patients.

In conclusion, ample evidence for a central role of lipids in PD is available, but current data yield a picture that is still too fragmented. This hinders the unraveling of the specific pathological mechanisms in which lipids are involved. Technological advances to better characterize the lipidome and explore the functions of specific lipid species, together with additional studies on CSF and/or brain tissue from PD patients are now urgently needed to further our understanding of the pathobiology of the relationship between PD and lipids and will help us to identify biomarkers and druggable targets for the development of disease-modifying therapies for this devastating neurodegenerative disease.

References

1. Tysnes, O.B.; Storstein, A. Epidemiology of Parkinson's disease. *J. Neural Transm.* **2017**, *124*, 901–905. [CrossRef] [PubMed]
2. Xia, R.; Mao, Z.H. Progression of motor symptoms in Parkinson's disease. *Neurosci. Bull.* **2012**, *28*, 39–48. [CrossRef] [PubMed]
3. Chaudhuri, K.R.; Schapira, A.H. Non-motor symptoms of Parkinson's disease: Dopaminergic pathophysiology and treatment. *Lancet Neurol.* **2009**, *8*, 464–474. [CrossRef]
4. Dexter, D.T.; Jenner, P. Parkinson disease: From pathology to molecular disease mechanisms. *Free Radic. Biol. Med.* **2013**, *62*, 132–144. [CrossRef] [PubMed]
5. Deng, H.; Wang, P.; Jankovic, J. The genetics of Parkinson disease. *Ageing Res. Rev.* **2018**, *42*, 72–85. [CrossRef] [PubMed]
6. Vila, M.; Przedborski, S. Targeting programmed cell death in neurodegenerative diseases. *Nat. Rev. Neurosci.* **2003**, *4*, 365–375. [CrossRef] [PubMed]
7. Simola, N.; Morelli, M.; Carta, A.R. The 6-hydroxydopamine model of Parkinson's disease. *Neurotox. Res.* **2007**, *11*, 151–167. [CrossRef]
8. Cicchetti, F.; Drouin-Ouellet, J.; Gross, R.E. Environmental toxins and Parkinson's disease: What have we learned from pesticide-induced animal models? *Trends Pharmacol. Sci.* **2009**, *30*, 475–483. [CrossRef]
9. Klemann, C.J.H.M.; Martens, G.J.M.; Sharma, M.; Martens, M.B.; Isacson, O.; Gasser, T.; Visser, J.E.; Poelmans, G. Integrated molecular landscape of Parkinson's disease. *NPJ Park. Dis.* **2017**, *3*, 14. [CrossRef]
10. Houlden, H.; Singleton, A.B. The genetics and neuropathology of Parkinson's disease. *Acta Neuropathol.* **2012**, *124*, 325–338. [CrossRef]
11. Nichols, W.C.; Pankratz, N.; Marek, D.K.; Pauciulo, M.W.; Elsaesser, V.E.; Halter, C.A.; Rudolph, A.; Wojcieszek, J.; Pfeiffer, R.F.; Foround, T.; et al. Mutations in GBA are associated with familial Parkinson disease susceptibility and age at onset. *Neurology* **2009**, *72*, 310–316. [CrossRef] [PubMed]
12. Sidransky, E.; Lopez, G. The link between the GBA gene and parkinsonism. *Lancet Neurol.* **2012**, *11*, 986–998. [CrossRef]
13. Do, C.B.; Tung, J.Y.; Dorfman, E.; Kiefer, A.K.; Drabant, E.M.; Francke, U.; Mountain, J.L.; Goldman, S.M.; Tanner, C.M.; Landston, J.W.; et al. Web-based genome-wide association study identifies two novel loci and a substantial genetic component for Parkinson's disease. *PLoS Genet.* **2011**, *7*, e1002141. [CrossRef] [PubMed]
14. Pankratz, N.; Wilk, J.B.; Latourelle, J.C.; DeStefano, A.L.; Halter, C.; Pugh, E.W.; Doheny, K.F.; Gausella, J.F.; Nichols, W.C.; Foroud, T.; et al. Genomewide association study for susceptibility genes contributing to familial Parkinson disease. *Hum. Genet.* **2009**, *124*, 593–605. [CrossRef] [PubMed]
15. Robak, L.A.; Jansen, I.E.; van Rooij, J.; Uitterlinden, A.G.; Kraaij, R.; Jankovic, J.; Heutink, P.; Shulman, J.M.; Nalls, M.A.; Plagnol, V.; et al. Excessive burden of lysosomal storage disorder gene variants in Parkinson's disease. *Brain* **2017**, *140*, 3191–3203. [CrossRef] [PubMed]
16. Gan-Or, Z.; Ozelius, L.J.; Bar-Shira, A.; Saunders-Pullman, R.; Mirelman, A.; Kornreich, R.; Gana-Weisz, M.; Raymond, D.; Rozenkrantz, L.; Deik, A.; et al. The p.L302P mutation in the lysosomal enzyme gene SMPD1 is a risk factor for Parkinson disease. *Neurology* **2013**, *80*, 1606–1610. [CrossRef]
17. Horton, T.J.; Drougas, H.; Brachey, A.; Reed, G.W.; Peters, J.C.; Hill, J.O. Fat and carbohydrate overfeeding in humans: Different effects on energy storage. *Am. J. Clin. Nutr.* **1995**, *62*, 19–29. [CrossRef]
18. Lass, A.; Zimmermann, R.; Oberer, M.; Zechner, R. Lipolysis—A highly regulated multi-enzyme complex mediates the catabolism of cellular fat stores. *Prog. Lipid Res.* **2011**, *50*, 14–27. [CrossRef]
19. Holthuis, J.C.M.; Menon, A.K. Lipid landscapes and pipelines in membrane homeostasis. *Nature* **2014**, *510*, 48–57. [CrossRef]
20. Fernandis, A.Z.; Wenk, M.R. Membrane lipids as signaling molecules. *Curr. Opin. Lipidol.* **2007**, *18*, 121–128. [CrossRef]
21. Bieberich, E. It's a lipid's world: Bioactive lipid metabolism and signaling in neural stem cell differentiation. *Neurochem. Res.* **2012**, *37*, 1208–1229. [CrossRef] [PubMed]
22. Welte, M.A.; Gould, A.P. Lipid droplet functions beyond energy storage. *Biochim. Biophys. Acta* **2017**, *1862*, 1260–1272. [CrossRef] [PubMed]
23. Welte, M.A. Expanding roles for lipid droplets. *Curr. Biol.* **2015**, *25*, R470–R481. [CrossRef] [PubMed]

24. Fahy, E.; Subramaniam, S.; Murphy, R.C.; Nishijima, M.; Raetz, C.R.H.; Shimizu, T.; Spener, F.; van Meer, D.; Wakelam, M.J.; Dennis, E.A. Update of the LIPID MAPS comprehensive classification system for lipids. *J. Lipid Res.* **2009**, *50*, S9–S14. [CrossRef] [PubMed]
25. Sud, M.; Fahy, E.; Cotter, D.; Brown, A.; Dennis, E.A.; Glass, C.K.; Merrill, A.H.; Murphy, R.C.; Raetz, C.R.; Russell, D.W.; et al. LMSD: LIPID MAPS structure database. *Nucleic Acids Res.* **2007**, *35*, D527–D532. [CrossRef] [PubMed]
26. Das, U.N. Essential Fatty acids—A review. *Curr. Pharm. Biotechnol.* **2006**, *7*, 467–482. [CrossRef]
27. Calder, P.C. Functional Roles of Fatty Acids and Their Effects on Human Health. *JPEN J. Parenter Enteral Nutr.* **2015**, *39*, 18S–32S. [CrossRef] [PubMed]
28. Fritsche, K.L. The science of fatty acids and inflammation. *Adv. Nutr.* **2015**, *6*, 293S–301S. [CrossRef]
29. Tan, L.C.; Methawasin, K.; Tan, E.K.; Tan, J.H.; Au, W.L.; Yuan, J.M.; Koh, W.P. Dietary cholesterol, fats and risk of Parkinson's disease in the Singapore Chinese Health Study. *J. Neurol. Neurosurg. Psychiatry* **2016**, *87*, 86–92. [CrossRef]
30. Miyake, Y.; Sasaki, S.; Tanaka, K.; Fukushima, W.; Kiyohara, C.; Tsuboi, Y.; Yamada, T.; Oeda, T.; Miki, T.; Kawamura, N.; et al. Dietary fat intake and risk of Parkinson's disease: A case-control study in Japan. *J. Neurol. Sci.* **2010**, *288*, 117–122. [CrossRef]
31. De Lau, L.M.L.; Bornebroek, M.; Witteman, J.C.M.; Hofman, A.; Koudstaal, P.J.; Breteler, M.M.B. Dietary fatty acids and the risk of Parkinson disease: The Rotterdam study. *Neurology* **2005**, *64*, 2040–2045. [CrossRef]
32. Kamel, F.; Goldman, S.M.; Umbach, D.M.; Chen, H.; Richardson, G.; Barber, M.R.; Meng, C.; Marras, C.; Koerll, M.; Kasten, M.; et al. Dietary fat intake, pesticide use, and Parkinson's disease. *Park. Relat. Disord.* **2014**, *20*, 82–87. [CrossRef] [PubMed]
33. Fabelo, N.; Martín, V.; Santpere, G.; Marín, R.; Torrent, L.; Ferrer, I.; Díaz, M. Severe alterations in lipid composition of frontal cortex lipid rafts from Parkinson's disease and incidental Parkinson's disease. *Mol. Med.* **2011**, *17*, 1107–1118. [CrossRef] [PubMed]
34. Julien, C.; Berthiaume, L.; Hadj-Tahar, A.; Rajput, A.H.; Bédard, P.J.; Di Paolo, T.; Julien, P.; Calon, F. Postmortem brain fatty acid profile of levodopa-treated Parkinson disease patients and parkinsonian monkeys. *Neurochem. Int.* **2006**, *48*, 404–414. [CrossRef] [PubMed]
35. Lee, J.Y.; Sohn, K.H.; Rhee, S.H.; Hwang, D. Saturated fatty acids, but not unsaturated fatty acids, induce the expression of cyclooxygenase-2 mediated through Toll-like receptor 4. *J. Biol. Chem.* **2001**, *276*, 16683–16689. [CrossRef] [PubMed]
36. Senyilmaz, D.; Virtue, S.; Xu, X.; Tan, C.Y.; Griffin, J.L.; Miller, A.K.; Vidal-Puig, A.; Teleman, A.A. Regulation of mitochondrial morphology and function by stearoylation of TFR1. *Nature* **2015**, *525*, 124–128. [CrossRef] [PubMed]
37. Bajracharya, R.; Bustamante, S.; Ballard, J.W.O. Stearic acid supplementation in high protein to carbohydrate (P:C) ratio diet improves physiological and mitochondrial functions of *Drosophila melanogaster parkin* null mutants. *J. Gerontol. A Biol. Sci. Med. Sci.* **2017**. [CrossRef] [PubMed]
38. Bajracharya, R.; Ballard, J.W.O. Dietary management and physical exercise can improve climbing defects and mitochondrial activity in *Drosophila melanogaster parkin* null mutants. *Fly* **2018**, *12*, 95–104. [CrossRef] [PubMed]
39. Joniec-Maciejak, I.; Wawer, A.; Turzyńska, D.; Sobolewska, A.; Maciejak, P.; Szyndler, J.; Mirowska-Guzel, D.; Płaźnik, A. Octanoic acid prevents reduction of striatal dopamine in the MPTP mouse model of Parkinson's disease. *Pharmacol. Rep.* **2018**, *70*, 988–992. [CrossRef] [PubMed]
40. Ng, Y.W.; Say, Y.H. Palmitic acid induces neurotoxicity and gliatoxicity in SH-SY5Y human neuroblastoma and T98G human glioblastoma cells. *PeerJ* **2018**, *6*, e4696. [CrossRef] [PubMed]
41. Morselli, E.; Fuente-Martin, E.; Finan, B.; Kim, M.; Frank, A.; Garcia-Caceres, C.; Navas, C.; Gordillo, R.; Neinast, M.; Kalainayakan, S.P.; et al. Hypothalamic PGC-1α protects against high-fat diet exposure by regulating ERα. *Cell Rep.* **2014**, *9*, 633–645. [CrossRef] [PubMed]
42. Gupta, S.; Knight, A.G.; Gupta, S.; Keller, J.N.; Bruce-Keller, A.J. Saturated long-chain fatty acids activate inflammatory signaling in astrocytes. *J. Neurochem.* **2012**, *120*, 1060–1071. [CrossRef] [PubMed]
43. Su, X.; Chu, Y.; Kordower, J.H.; Li, B.; Cao, H.; Huang, L.; Nishida, M.; Song, L.; Wang, D.; Federoff, H.J. PGC-1α Promoter Methylation in Parkinson's Disease. *PLoS ONE* **2015**, *10*, e0134087. [CrossRef] [PubMed]
44. Wang, Q.; Liu, Y.; Zhou, J. Neuroinflammation in Parkinson's disease and its potential as therapeutic target. *Transl. Neurodegener.* **2015**, *4*, 19. [CrossRef] [PubMed]

45. Bartels, A.L.; Leenders, K.L. Cyclooxygenase and neuroinflammation in Parkinson's disease neurodegeneration. *Curr. Neuropharmacol.* **2010**, *8*, 62–68. [CrossRef] [PubMed]
46. Golovko, M.Y.; Faergeman, N.J.; Cole, N.B.; Castagnet, P.I.; Nussbaum, R.L.; Murphy, E.J. Alpha-synuclein gene deletion decreases brain palmitate uptake and alters the palmitate metabolism in the absence of alpha-synuclein palmitate binding. *Biochemistry* **2005**, *44*, 8251–8259. [CrossRef] [PubMed]
47. Schmid, S.P.; Schleicher, E.D.; Cegan, A.; Deuschle, C.; Baur, S.; Hauser, A.K.; Synofzik, M.; Srulijes, K.; Brockmann, K.; Berg, D.; et al. Cerebrospinal fluid fatty acids in glucocerebrosidase-associated Parkinson's disease. *Mov. Disord.* **2012**, *27*, 288–292. [CrossRef]
48. Heller, A.; Won, L.; Bubula, N.; Hessefort, S.; Kurutz, J.W.; Reddy, G.A.; Gross, M. Long-chain fatty acids increase cellular dopamine in an immortalized cell line (MN9D) derived from mouse mesencephalon. *Neurosci. Lett.* **2005**, *376*, 35–39. [CrossRef]
49. Sergeeva, O.A.; De Luca, R.; Mazur, K.; Chepkova, A.N.; Haas, H.L.; Bauer, A. N-oleoyldopamine modulates activity of midbrain dopaminergic neurons through multiple mechanisms. *Neuropharmacology* **2017**, *119*, 111–122. [CrossRef]
50. Sharon, R.; Goldberg, M.S.; Bar-Josef, I.; Betensky, R.A.; Shen, J.; Selkoe, D.J. alpha-Synuclein occurs in lipid-rich high molecular weight complexes, binds fatty acids, and shows homology to the fatty acid-binding proteins. *Proc. Natl. Acad. Sci. USA* **2001**, *98*, 9110–9115. [CrossRef]
51. Kubo, S.; Nemani, V.M.; Chalkley, R.J.; Anthony, M.D.; Hattori, N.; Mizuno, Y.; Edwards, R.H.; Fortin, D.L. A combinatorial code for the interaction of alpha-synuclein with membranes. *J. Biol. Chem.* **2005**, *280*, 31664–31672. [CrossRef] [PubMed]
52. Tvrzicka, E.; Kremmyda, L.S.; Stankova, B.; Zak, A. Fatty acids as biocompounds: Their role in human metabolism, health and disease—A review. Part 1: Classification, dietary sources and biological functions. *Biomed. Pap. Med. Fac. Univ. Palacky Olomouc Czech Repub.* **2011**, *155*, 117–130. [CrossRef] [PubMed]
53. Raphael, W.; Sordillo, L.M. Dietary polyunsaturated fatty acids and inflammation: The role of phospholipid biosynthesis. *Int. J. Mol. Sci.* **2013**, *14*, 21167–21188. [CrossRef] [PubMed]
54. Burdge, G.C.; Lillycrop, K.A. Fatty acids and epigenetics. *Curr. Opin. Clin. Nutr. Metab. Care* **2014**, *17*, 156–161. [CrossRef] [PubMed]
55. Janssen, C.I.F.F.; Kiliaan, A.J. Long-chain polyunsaturated fatty acids (LCPUFA) from genesis to senescence: the influence of LCPUFA on neural development, aging, and neurodegeneration. *Prog. Lipid Res.* **2014**, *53*, 1–17. [CrossRef] [PubMed]
56. Bazinet, R.P.; Layé, S. Polyunsaturated fatty acids and their metabolites in brain function and disease. *Nat. Rev. Neurosci.* **2014**, *15*, 771–785. [CrossRef] [PubMed]
57. Dong, J.; Beard, J.D.; Umbach, D.M.; Park, Y.; Huang, X.; Blair, A.; Kamel, F.; Chen, H. Dietary fat intake and risk for Parkinson's disease. *Mov. Disord.* **2014**, *29*, 1623–1630. [CrossRef]
58. Chen, H.; Zhang, S.M.; Hernán, M.A.; Willett, W.C.; Ascherio, A. Dietary intakes of fat and risk of Parkinson's disease. *Am. J. Epidemiol.* **2003**, *157*, 1007–1014. [CrossRef]
59. Schulte, E.C.; Altmaier, E.; Berger, H.S.; Do, K.T.; Kastenmüller, G.; Wahl, S.; Adamski, J.; Peters, A.; Krumsiek, J.; Suhre, K.; et al. Alterations in Lipid and Inositol Metabolisms in Two Dopaminergic Disorders. *PLoS ONE* **2016**, *11*, e0147129. [CrossRef]
60. Selley, M.L. (E)-4-hydroxy-2-nonenal may be involved in the pathogenesis of Parkinson's disease. *Free Radic. Biol. Med.* **1998**, *25*, 169–174. [CrossRef]
61. Abbott, S.K.; Jenner, A.M.; Spiro, A.S.; Batterham, M.; Halliday, G.M.; Garner, B. Fatty acid composition of the anterior cingulate cortex indicates a high susceptibility to lipid peroxidation in Parkinson's disease. *J. Park. Dis.* **2015**, *5*, 175–185. [CrossRef]
62. Sharon, R.; Bar-Joseph, I.; Mirick, G.E.; Serhan, C.N.; Selkoe, D.J. Altered fatty acid composition of dopaminergic neurons expressing alpha-synuclein and human brains with alpha-synucleinopathies. *J. Biol. Chem.* **2003**, *278*, 49874–49881. [CrossRef] [PubMed]
63. Bousquet, M.; Saint-Pierre, M.; Julien, C.; Salem, N.; Cicchetti, F.; Calon, F. Beneficial effects of dietary omega-3 polyunsaturated fatty acid on toxin-induced neuronal degeneration in an animal model of Parkinson's disease. *FASEB J.* **2008**, *22*, 1213–1225. [CrossRef] [PubMed]

64. Bousquet, M.; Gibrat, C.; Saint-Pierre, M.; Julien, C.; Calon, F.; Cicchetti, F. Modulation of brain-derived neurotrophic factor as a potential neuroprotective mechanism of action of omega-3 fatty acids in a parkinsonian animal model. *Prog. Neuropsychopharmacol. Biol. Psychiatry* **2009**, *33*, 1401–1408. [CrossRef] [PubMed]
65. Barros, A.S.; Crispim, R.Y.G.; Cavalcanti, J.U.; Souza, R.B.; Lemos, J.C.; Cristino Filho, G.; Bezerra, M.M.; Pinheiro, T.F.M.; de Vasconcelos, S.M.M.; Macêdo, D.S.; et al. Impact of the Chronic Omega-3 Fatty Acids Supplementation in Hemiparkinsonism Model Induced by 6-Hydroxydopamine in Rats. *Basic Clin. Pharmacol. Toxicol.* **2017**, *120*, 523–531. [CrossRef] [PubMed]
66. Lu, Z.; Wang, J.; Li, M.; Liu, Q.; Wei, D.; Yang, M.; Kong, L. (1)H NMR-based metabolomics study on a goldfish model of Parkinson's disease induced by 1-methyl-4-phenyl-1, 2, 3, 6-tetrahydropyridine (MPTP). *Chem. Biol. Interact.* **2014**, *223*, 18–26. [CrossRef]
67. Cardoso, H.D.; dos Santos Junior, E.F.; de Santana, D.F.; Gonçalves-Pimentel, C.; Angelim, M.K.; Isaac, A.R.; Lagranha, C.J.; Guedes, R.C.; Beltrão, E.I.; Morya, E.; et al. Omega-3 deficiency and neurodegeneration in the substantia nigra: Involvement of increased nitric oxide production and reduced BDNF expression. *Biochim. Biophys. Acta* **2014**, *1840*, 1902–1912. [CrossRef] [PubMed]
68. Delattre, A.M.; Carabelli, B.; Mori, M.A.; Kempe, P.G.; Rizzo de Souza, L.E.; Zanata, S.M.; Machado, R.B.; Suchecki, D.; Andrade da Costa, B.L.S.; Lima, M.M.S.; et al. Maternal Omega-3 Supplement Improves Dopaminergic System in Pre- and Postnatal Inflammation-Induced Neurotoxicity in Parkinson's Disease Model. *Mol. Neurobiol.* **2017**, *54*, 2090–2106. [CrossRef]
69. Tanriover, G.; Seval-Celik, Y.; Ozsoy, O.; Akkoyunlu, G.; Savcioglu, F.; Hacioglu, G.; Demir, N.; Agar, A. The effects of docosahexaenoic acid on glial derived neurotrophic factor and neurturin in bilateral rat model of Parkinson's disease. *Folia Histochem. Cytobiol.* **2010**, *48*, 434–441. [CrossRef]
70. Hacioglu, G.; Seval-Celik, Y.; Tanriover, G.; Ozsoy, O.; Saka-Topcuoglu, E.; Balkan, S.; Agar, A. Docosahexaenoic acid provides protective mechanism in bilaterally MPTP-lesioned rat model of Parkinson's disease. *Folia Histochem. Cytobiol.* **2012**, *50*, 228–238. [CrossRef]
71. Ozkan, A.; Parlak, H.; Tanriover, G.; Dilmac, S.; Ulker, S.N.; Birsen, I.; Agar, A. The protective mechanism of docosahexaenoic acid in mouse model of Parkinson: The role of hemeoxygenase. *Neurochem. Int.* **2016**, *101*, 110–119. [CrossRef] [PubMed]
72. Lee, H.J.; Han, J.; Jang, Y.; Kim, S.J.; Park, J.H.; Seo, K.S.; Jeong, S.; Shin, S.; Lim, K.; Heo, J.Y.; et al. Docosahexaenoic acid prevents paraquat-induced reactive oxygen species production in dopaminergic neurons via enhancement of glutathione homeostasis. *Biochem. Biophys. Res. Commun.* **2015**, *457*, 95–100. [CrossRef] [PubMed]
73. Serrano-García, N.; Fernández-Valverde, F.; Luis-Garcia, E.R.; Granados-Rojas, L.; Juárez-Zepeda, T.E.; Orozco-Suárez, S.A.; Pedraza-Chaverri, J.; Orozco-Ibarra, M.; Jiménez-Anguiano, A. Docosahexaenoic acid protection in a rotenone induced Parkinson's model: Prevention of tubulin and synaptophysin loss, but no association with mitochondrial function. *Neurochem. Int.* **2018**, *121*, 26–37. [CrossRef] [PubMed]
74. Shashikumar, S.; Pradeep, H.; Chinnu, S.; Rajini, P.S.; Rajanikant, G.K. Alpha-linolenic acid suppresses dopaminergic neurodegeneration induced by 6-OHDA in *C. elegans*. *Physiol. Behav.* **2015**, *151*, 563–569. [CrossRef] [PubMed]
75. Coulombe, K.; Saint-Pierre, M.; Cisbani, G.; St-Amour, I.; Gibrat, C.; Giguère-Rancourt, A.; Calon, F.; Cicchetti, F. Partial neurorescue effects of DHA following a 6-OHDA lesion of the mouse dopaminergic system. *J. Nutr. Biochem.* **2016**, *30*, 133–142. [CrossRef] [PubMed]
76. Hernando, S.; Requejo, C.; Herran, E.; Ruiz-Ortega, J.A.; Morera-Herreras, T.; Lafuente, J.V.; Ugedo, L.; Gainza, E.; Pedraz, J.L.; Igartua, M.; et al. Beneficial effects of n-3 polyunsaturated fatty acids administration in a partial lesion model of Parkinson's disease: The role of glia and NRf2 regulation. *Neurobiol. Dis.* **2019**, *121*, 252–262. [CrossRef] [PubMed]
77. Gómez-Soler, M.; Cordobilla, B.; Morató, X.; Fernández-Dueñas, V.; Domingo, J.C.; Ciruela, F. Triglyceride Form of Docosahexaenoic Acid Mediates Neuroprotection in Experimental Parkinsonism. *Front. Neurosci.* **2018**, *12*, 604. [CrossRef]
78. Chang, Y.L.; Chen, S.J.; Kao, C.L.; Hung, S.C.; Ding, D.C.; Yu, C.C.; Chen, Y.J.; Ku, H.H.; Lin, C.P.; Lee, K.H.; et al. Docosahexaenoic acid promotes dopaminergic differentiation in induced pluripotent stem cells and inhibits teratoma formation in rats with Parkinson-like pathology. *Cell Transplant.* **2012**, *21*, 313–332. [CrossRef]

79. Parlak, H.; Ozkan, A.; Dilmac, S.; Tanriover, G.; Ozsoy, O.; Agar, A. Neuronal nitric oxide synthase phosphorylation induced by docosahexaenoic acid protects dopaminergic neurons in an experimental model of Parkinson's disease. *Folia Histochem. Cytobiol.* **2015**, *56*, 27–37. [CrossRef]
80. Luchtman, D.W.; Meng, Q.; Wang, X.; Shao, D.; Song, C. ω-3 fatty acid eicosapentaenoic acid attenuates MPP+-induced neurodegeneration in fully differentiated human SH-SY5Y and primary mesencephalic cells. *J. Neurochem.* **2013**, *124*, 855–868. [CrossRef]
81. Meng, Q.; Luchtman, D.W.; El Bahh, B.; Zidichouski, J.A.; Yang, J.; Song, C. Ethyl-eicosapentaenoate modulates changes in neurochemistry and brain lipids induced by parkinsonian neurotoxin 1-methyl-4-phenylpyridinium in mouse brain slices. *Eur. J. Pharmacol.* **2010**, *649*, 127–134. [CrossRef] [PubMed]
82. Luchtman, D.W.; Meng, Q.; Song, C. Ethyl-eicosapentaenoate (E-EPA) attenuates motor impairments and inflammation in the MPTP-probenecid mouse model of Parkinson's disease. *Behav. Brain Res.* **2012**, *226*, 386–396. [CrossRef]
83. Mori, M.A.; Delattre, A.M.; Carabelli, B.; Pudell, C.; Bortolanza, M.; Staziaki, P.V.; Visentainer, J.V.; Montanher, P.F.; Del Bel, E.A.; Ferraz, A.C. Neuroprotective effect of omega-3 polyunsaturated fatty acids in the 6-OHDA model of Parkinson's disease is mediated by a reduction of inducible nitric oxide synthase. *Nutr. Neurosci.* **2018**, *21*, 341–351. [CrossRef] [PubMed]
84. Delattre, A.M.; Kiss, A.; Szawka, R.E.; Anselmo-Franci, J.A.; Bagatini, P.B.; Xavier, L.L.; Rigon, P.; Achaval, M.; Iagher, F.; de David, C.; et al. Evaluation of chronic omega-3 fatty acids supplementation on behavioral and neurochemical alterations in 6-hydroxydopamine-lesion model of Parkinson's disease. *Neurosci. Res.* **2010**, *66*, 256–264. [CrossRef] [PubMed]
85. Kabuto, H.; Amakawa, M.; Mankura, M.; Yamanushi, T.T.; Mori, A. Docosahexaenoic acid ethyl ester enhances 6-hydroxydopamine-induced neuronal damage by induction of lipid peroxidation in mouse striatum. *Neurochem. Res.* **2009**, *34*, 1299–1303. [CrossRef] [PubMed]
86. Anderson, E.J.; Katunga, L.A.; Willis, M.S. Mitochondria as a source and target of lipid peroxidation products in healthy and diseased heart. *Clin. Exp. Pharmacol. Physiol.* **2012**, *39*, 179–193. [CrossRef] [PubMed]
87. Shamoto-Nagai, M.; Hisaka, S.; Naoi, M.; Maruyama, W. Modification of α-synuclein by lipid peroxidation products derived from polyunsaturated fatty acids promotes toxic oligomerization: Its relevance to Parkinson disease. *J. Clin. Biochem. Nutr.* **2018**, *62*, 207–212. [CrossRef] [PubMed]
88. Angelova, P.R.; Horrocks, M.H.; Klenerman, D.; Gandhi, S.; Abramov, A.Y.; Shchepinov, M.S. Lipid peroxidation is essential for α-synuclein-induced cell death. *J. Neurochem.* **2015**, *133*, 582–589. [CrossRef]
89. Shchepinov, M.S.; Chou, V.P.; Pollock, E.; Langston, J.W.; Cantor, C.R.; Molinari, R.J.; Manning-Boğ, A.B. Isotopic reinforcement of essential polyunsaturated fatty acids diminishes nigrostriatal degeneration in a mouse model of Parkinson's disease. *Toxicol. Lett.* **2011**, *207*, 97–103. [CrossRef]
90. Kinghorn, K.J.; Castillo-Quan, J.I.; Bartolome, F.; Angelova, P.R.; Li, L.; Pope, S.; Cochemé, H.M.; Khan, S.; Asghari, S.; Bhatia, K.P.; et al. Loss of PLA2G6 leads to elevated mitochondrial lipid peroxidation and mitochondrial dysfunction. *Brain* **2015**, *138 Pt 7*, 1801–1816. [CrossRef]
91. Denny Joseph, K.M.; Muralidhara. Combined oral supplementation of fish oil and quercetin enhances neuroprotection in a chronic rotenone rat model: Relevance to Parkinson's disease. *Neurochem. Res.* **2015**, *40*, 894–905. [CrossRef] [PubMed]
92. Lee, H.J.; Bazinet, R.P.; Rapoport, S.I.; Bhattacharjee, A.K. Brain arachidonic acid cascade enzymes are upregulated in a rat model of unilateral Parkinson disease. *Neurochem. Res.* **2010**, *35*, 613–619. [CrossRef] [PubMed]
93. Chalimoniuk, M.; Stolecka, A.; Ziemińska, E.; Stepień, A.; Langfort, J.; Strosznajder, J.B. Involvement of multiple protein kinases in cPLA2 phosphorylation, arachidonic acid release, and cell death in in vivo and in vitro models of 1-methyl-4-phenylpyridinium-induced parkinsonism—The possible key role of PKG. *J. Neurochem.* **2009**, *110*, 307–317. [CrossRef]
94. Tang, K.S. Protective effect of arachidonic acid and linoleic acid on 1-methyl-4-phenylpyridinium-induced toxicity in PC12 cells. *Lipids Health Dis.* **2014**, *13*, 197. [CrossRef] [PubMed]
95. Shioda, N.; Yabuki, Y.; Kobayashi, Y.; Onozato, M.; Owada, Y.; Fukunaga, K. FABP3 protein promotes α-synuclein oligomerization associated with 1-methyl-1, 2, 3, 6-tetrahydropiridine-induced neurotoxicity. *J. Biol. Chem.* **2014**, *289*, 18957–18965. [CrossRef]

96. Wang, Y.; Plastina, P.; Vincken, J.P.; Jansen, R.; Balvers, M.; Ten Klooster, J.P.; Gruppen, H.; Witkamp, R.; Meijerink, J. N-Docosahexaenoyl Dopamine, an Endocannabinoid-like Conjugate of Dopamine and the n-3 Fatty Acid Docosahexaenoic Acid, Attenuates Lipopolysaccharide-Induced Activation of Microglia and Macrophages via COX-2. *ACS Chem. Neurosci.* **2017**, *8*, 548–557. [CrossRef] [PubMed]
97. Ben Gedalya, T.; Loeb, V.; Israeli, E.; Altschuler, Y.; Selkoe, D.J.; Sharon, R. Alpha-synuclein and polyunsaturated fatty acids promote clathrin-mediated endocytosis and synaptic vesicle recycling. *Traffic* **2009**, *10*, 218–234. [CrossRef]
98. Darios, F.; Ruipérez, V.; López, I.; Villanueva, J.; Gutierrez, L.M.; Davletov, B. Alpha-synuclein sequesters arachidonic acid to modulate SNARE-mediated exocytosis. *EMBO Rep.* **2010**, *11*, 528–533. [CrossRef] [PubMed]
99. Karube, H.; Sakamoto, M.; Arawaka, S.; Hara, S.; Sato, H.; Ren, C.H.; Goto, S.; Koyama, S.; Wada, M.; Kawanami, T.; et al. N-terminal region of alpha-synuclein is essential for the fatty acid-induced oligomerization of the molecules. *FEBS Lett.* **2008**, *582*, 3693–3700. [CrossRef]
100. Israeli, E.; Sharon, R. Beta-synuclein occurs in vivo in lipid-associated oligomers and forms hetero-oligomers with alpha-synuclein. *J. Neurochem.* **2009**, *108*, 465–474. [CrossRef]
101. Sharon, R.; Bar-Joseph, I.; Frosch, M.P.; Walsh, D.M.; Hamilton, J.A.; Selkoe, D.J. The formation of highly soluble oligomers of alpha-synuclein is regulated by fatty acids and enhanced in Parkinson's disease. *Neuron* **2003**, *37*, 583–595. [CrossRef]
102. Perrin, R.J.; Woods, W.S.; Clayton, D.F.; George, J.M. Exposure to long chain polyunsaturated fatty acids triggers rapid multimerization of synucleins. *J. Biol. Chem.* **2001**, *276*, 41958–41962. [CrossRef] [PubMed]
103. Assayag, K.; Yakunin, E.; Loeb, V.; Selkoe, D.J.; Sharon, R. Polyunsaturated fatty acids induce alpha-synuclein-related pathogenic changes in neuronal cells. *Am. J. Pathol.* **2007**, *171*, 2000–2011. [CrossRef] [PubMed]
104. Broersen, K.; van den Brink, D.; Fraser, G.; Goedert, M.; Davletov, B. Alpha-synuclein adopts an alpha-helical conformation in the presence of polyunsaturated fatty acids to hinder micelle formation. *Biochemistry* **2006**, *45*, 15610–15616. [CrossRef] [PubMed]
105. Yakunin, E.; Loeb, V.; Kisos, H.; Biala, Y.; Yehuda, S.; Yaari, Y.; Selkoe, D.J.; Sharon, R. A-synuclein neuropathology is controlled by nuclear hormone receptors and enhanced by docosahexaenoic acid in a mouse model for Parkinson's disease. *Brain Pathol.* **2012**, *22*, 280–294. [CrossRef] [PubMed]
106. Fu, Y.; Zhen, J.; Lu, Z. Synergetic Neuroprotective Effect of Docosahexaenoic Acid and Aspirin in SH-Y5Y by Inhibiting miR-21 and Activating RXRα and PPARα. *DNA Cell Biol.* **2017**, *36*, 482–489. [CrossRef] [PubMed]
107. De Franceschi, G.; Frare, E.; Pivato, M.; Relini, A.; Penco, A.; Greggio, E.; Bubacco, L.; Fontana, A.; de Laureto, P.P. Structural and morphological characterization of aggregated species of α-synuclein induced by docosahexaenoic acid. *J. Biol. Chem.* **2011**, *286*, 22262–22274. [CrossRef]
108. Fecchio, C.; De Franceschi, G.; Relini, A.; Greggio, E.; Dalla Serra, M.; Bubacco, L.; de Laureto, P.P. α-Synuclein oligomers induced by docosahexaenoic acid affect membrane integrity. *PLoS ONE* **2013**, *8*, e82732. [CrossRef]
109. De Franceschi, G.; Frare, E.; Bubacco, L.; Mammi, S.; Fontana, A.; de Laureto, P.P. Molecular insights into the interaction between alpha-synuclein and docosahexaenoic acid. *J. Mol. Biol.* **2009**, *394*, 94–107. [CrossRef]
110. Broersen, K.; Ruiperez, V.; Davletov, B. Structural and Aggregation Properties of Alpha-Synuclein Linked to Phospholipase A2 Action. *Protein Pept. Lett.* **2018**, *25*, 368–378. [CrossRef]
111. Iljina, M.; Tosatto, L.; Choi, M.L.; Sang, J.C.; Ye, Y.; Hughes, C.D.; Bryant, C.E.; Gandhi, S.; Klenerman, D. Arachidonic acid mediates the formation of abundant alpha-helical multimers of alpha-synuclein. *Sci. Rep.* **2016**, *6*, 33928. [CrossRef]
112. Jiang, P.; Gan, M.; Yen, S.H.C. Dopamine prevents lipid peroxidation-induced accumulation of toxic α-synuclein oligomers by preserving autophagy-lysosomal function. *Front. Cell. Neurosci.* **2013**, *7*, 81. [CrossRef] [PubMed]
113. Liu, X.; Yamada, N.; Maruyama, W.; Osawa, T. Formation of dopamine adducts derived from brain polyunsaturated fatty acids: Mechanism for Parkinson disease. *J. Biol. Chem.* **2008**, *283*, 34887–34895. [CrossRef] [PubMed]
114. Muntané, G.; Janué, A.; Fernandez, N.; Odena, M.A.; Oliveira, E.; Boluda, S.; Porero-Otin, M.; Naudí, A.; Boada, J.; Pamplona, R.; et al. Modification of brain lipids but not phenotype in alpha-synucleinopathy transgenic mice by long-term dietary n-3 fatty acids. *Neurochem. Int.* **2010**, *56*, 318–328. [CrossRef] [PubMed]

115. Coulombe, K.; Kerdiles, O.; Tremblay, C.; Emond, V.; Lebel, M.; Boulianne, A.S.; Plourde, M.; Cicchetti, F.; Calon, F. Impact of DHA intake in a mouse model of synucleinopathy. *Exp. Neurol.* **2018**, *301*, 39–49. [CrossRef] [PubMed]
116. De Franceschi, G.; Fecchio, C.; Sharon, R.; Schapira, A.H.V.; Proukakis, C.; Bellotti, V.; de Laureto, P.P. α-Synuclein structural features inhibit harmful polyunsaturated fatty acid oxidation, suggesting roles in neuroprotection. *J. Biol. Chem.* **2017**, *292*, 6927–6937. [CrossRef] [PubMed]
117. Fecchio, C.; Palazzi, L.; de Laureto, P.P. α-Synuclein and Polyunsaturated Fatty Acids: Molecular Basis of the Interaction and Implication in Neurodegeneration. *Molecules* **2018**, *23*, 1531. [CrossRef]
118. Dennis, E.A.; Norris, P.C. Eicosanoid storm in infection and inflammation. *Nat. Rev. Immunol.* **2015**, *15*, 511–523. [CrossRef] [PubMed]
119. Pretorius, E.; Swanepoel, A.C.; Buys, A.V.; Vermeulen, N.; Duim, W.; Kell, D.B. Eryptosis as a marker of Parkinson's disease. *Aging* **2014**, *6*, 788–819. [CrossRef]
120. Teismann, P.; Tieu, K.; Choi, D.K.; Wu, D.C.; Naini, A.; Hunot, S.; Vila, M.; Jackson-Lewis, V.; Przedborski, S. Cyclooxygenase-2 is instrumental in Parkinson's disease neurodegeneration. *Proc. Natl. Acad. Sci. USA* **2003**, *100*, 5473–5478. [CrossRef]
121. Yu, S.Y.; Zuo, L.J.; Wang, F.; Chen, Z.J.; Hu, Y.; Wang, Y.J.; Wang, X.M.; Zhang, W. Potential biomarkers relating pathological proteins, neuroinflammatory factors and free radicals in PD patients with cognitive impairment: A cross-sectional study. *BMC Neurol.* **2014**, *14*, 113. [CrossRef] [PubMed]
122. Mattammal, M.B.; Strong, R.; Lakshmi, V.M.; Chung, H.D.; Stephenson, A.H. Prostaglandin H synthetase-mediated metabolism of dopamine: Implication for Parkinson's disease. *J. Neurochem.* **1995**, *64*, 1645–1654. [CrossRef] [PubMed]
123. Geng, Y.; Fang, M.; Wang, J.; Yu, H.; Hu, Z.; Yew, D.T.; Chen, W. Triptolide down-regulates COX-2 expression and PGE2 release by suppressing the activity of NF-κB and MAP kinases in lipopolysaccharide-treated PC12 cells. *Phytother. Res.* **2012**, *26*, 337–343. [CrossRef] [PubMed]
124. Zeng, K.W.; Zhang, T.; Fu, H.; Liu, G.X.; Wang, X.M. Schisandrin B exerts anti-neuroinflammatory activity by inhibiting the Toll-like receptor 4-dependent MyD88/IKK/NF-κB signaling pathway in lipopolysaccharide-induced microglia. *Eur. J. Pharmacol.* **2012**, *692*, 29–37. [CrossRef] [PubMed]
125. Bai, L.; Zhang, X.; Li, X.; Liu, N.; Lou, F.; Ma, H.; Luo, X.; Ren, Y. Somatostatin prevents lipopolysaccharide-induced neurodegeneration in the rat substantia nigra by inhibiting the activation of microglia. *Mol. Med. Rep.* **2015**, *12*, 1002–1008. [CrossRef] [PubMed]
126. Fu, Q.; Song, R.; Yang, Z.; Shan, Q.; Chen, W. 6-Hydroxydopamine induces brain vascular endothelial inflammation. *IUBMB Life* **2017**, *69*, 887–895. [CrossRef] [PubMed]
127. Wang, H.M.; Zhang, T.; Li, Q.; Huang, J.K.; Chen, R.F.; Sun, X.J. Inhibition of glycogen synthase kinase-3β by lithium chloride suppresses 6-hydroxydopamine-induced inflammatory response in primary cultured astrocytes. *Neurochem. Int.* **2013**, *63*, 345–353. [CrossRef] [PubMed]
128. Yildirim, F.B.; Ozsoy, O.; Tanriover, G.; Kaya, Y.; Ogut, E.; Gemici, B.; Dilmac, S.; Ozkan, A.; Agar, A.; Aslan, M. Mechanism of the beneficial effect of melatonin in experimental Parkinson's disease. *Neurochem. Int.* **2014**, *79*, 1–11. [CrossRef]
129. Zhou, F.; Yao, H.H.; Wu, J.Y.; Ding, J.H.; Sun, T.; Hu, G. Opening of microglial K(ATP) channels inhibits rotenone-induced neuroinflammation. *J. Cell. Mol. Med.* **2008**, *12*, 1559–1570. [CrossRef]
130. Hu, J.H.; Zhu, X.Z. Rotenone-induced neurotoxicity of THP-1 cells requires production of reactive oxygen species and activation of phosphatidylinositol 3-kinase. *Brain Res.* **2007**, *1153*, 12–19. [CrossRef]
131. Wang, T.; Pei, Z.; Zhang, W.; Liu, B.; Langenbach, R.; Lee, C.; Wilson, B.; Reece, J.M.; Miller, D.S.; Hong, J.S. MPP+-induced COX-2 activation and subsequent dopaminergic neurodegeneration. *FASEB J.* **2005**, *19*, 1134–1136. [CrossRef] [PubMed]
132. Ozsoy, O.; Tanriover, G.; Derin, N.; Uysal, N.; Demir, N.; Gemici, B.; Kencebay, C.; Yargicoglu, P.; Agar, A.; Aslan, M. The effect of docosahexaenoic Acid on visual evoked potentials in a mouse model of Parkinson's disease: The role of cyclooxygenase-2 and nuclear factor kappa-B. *Neurotox. Res.* **2011**, *20*, 250–262. [CrossRef] [PubMed]
133. Liu, J.; Zhou, Y.; Wang, Y.; Fong, H.; Murray, T.M.; Zhang, J. Identification of proteins involved in microglial endocytosis of alpha-synuclein. *J. Proteome Res.* **2007**, *6*, 3614–3627. [CrossRef]

134. Zhang, W.; Wang, T.; Pei, Z.; Miller, D.S.; Wu, X.; Block, M.L.; Wilson, B.; Zhang, W.; Zhou, Y.; Hong, J.S.; et al. Aggregated alpha-synuclein activates microglia: A process leading to disease progression in Parkinson's disease. *FASEB J.* **2005**, *19*, 533–542. [CrossRef]
135. Branchi, I.; D'Andrea, I.; Armida, M.; Carnevale, D.; Ajmone-Cat, M.A.; Pèzzola, A.; Potenza, R.L.; Morgese, M.G.; Cassano, T.; Minghetti, L.; et al. Striatal 6-OHDA lesion in mice: Investigating early neurochemical changes underlying Parkinson's disease. *Behav. Brain Res.* **2010**, *208*, 137–143. [CrossRef] [PubMed]
136. Dey, I.; Lejeune, M.; Chadee, K. Prostaglandin E2 receptor distribution and function in the gastrointestinal tract. *Br. J. Pharmacol.* **2006**, *149*, 611–623. [CrossRef]
137. Ahmad, A.S.; Maruyama, T.; Narumiya, S.; Doré, S. PGE2 EP1 receptor deletion attenuates 6-OHDA-induced Parkinsonism in mice: Old switch, new target. *Neurotox. Res.* **2013**, *23*, 260–266. [CrossRef] [PubMed]
138. Carrasco, E.; Casper, D.; Werner, P. PGE(2) receptor EP1 renders dopaminergic neurons selectively vulnerable to low-level oxidative stress and direct PGE(2) neurotoxicity. *J. Neurosci. Res.* **2007**, *85*, 3109–3117. [CrossRef]
139. Carrasco, E.; Werner, P.; Casper, D. Prostaglandin receptor EP2 protects dopaminergic neurons against 6-OHDA-mediated low oxidative stress. *Neurosci. Lett.* **2008**, *441*, 44–49. [CrossRef]
140. Pradhan, S.S.; Salinas, K.; Garduno, A.C.; Johansson, J.U.; Wang, Q.; Manning-Bog, A.; Andreasson, K.I. Anti-Inflammatory and Neuroprotective Effects of PGE2 EP4 Signaling in Models of Parkinson's Disease. *J. Neuroimmune Pharmacol.* **2017**, *12*, 292–304. [CrossRef] [PubMed]
141. Ashley, A.K.; Hinds, A.I.; Hanneman, W.H.; Tjalkens, R.B.; Legare, M.E. DJ-1 mutation decreases astroglial release of inflammatory mediators. *Neurotoxicology* **2016**, *52*, 198–203. [CrossRef] [PubMed]
142. Parga, J.A.; García-Garrote, M.; Martínez, S.; Raya, Á.; Labandeira-García, J.L.; Rodríguez-Pallares, J. Prostaglandin EP2 Receptors Mediate Mesenchymal Stromal Cell-Neuroprotective Effects on Dopaminergic Neurons. *Mol. Neurobiol.* **2018**, *55*, 4763–4776. [CrossRef]
143. Wang, X.; Qin, Z.H.; Leng, Y.; Wang, Y.; Jin, X.; Chase, T.N.; Bennett, M.C. Prostaglandin A1 inhibits rotenone-induced apoptosis in SH-SY5Y cells. *J. Neurochem.* **2002**, *83*, 1094–1102. [CrossRef] [PubMed]
144. Fujimori, K.; Fukuhara, A.; Inui, T.; Allhorn, M. Prevention of paraquat-induced apoptosis in human neuronal SH-SY5Y cells by lipocalin-type prostaglandin D synthase. *J. Neurochem.* **2012**, *120*, 279–291. [CrossRef] [PubMed]
145. Tsai, M.J.; Weng, C.F.; Yu, N.C.; Liou, D.Y.; Kuo, F.S.; Huang, M.C.; Tam, K.; Shyue, S.K.; Cheng, H. Enhanced prostacyclin synthesis by adenoviral gene transfer reduced glial activation and ameliorated dopaminergic dysfunction in hemiparkinsonian rats. *Oxid. Med. Cell. Longev.* **2013**, *2013*, 649809. [CrossRef] [PubMed]
146. Ogburn, K.D.; Figueiredo-Pereira, M.E. Cytoskeleton/endoplasmic reticulum collapse induced by prostaglandin J2 parallels centrosomal deposition of ubiquitinated protein aggregates. *J. Biol. Chem.* **2006**, *281*, 23274–23284. [CrossRef] [PubMed]
147. Shivers, K.Y.; Nikolopoulou, A.; Machlovi, S.I.; Vallabhajosula, S.; Figueiredo-Pereira, M.E. PACAP27 prevents Parkinson-like neuronal loss and motor deficits but not microglia activation induced by prostaglandin J2. *Biochim. Biophys. Acta* **2014**, *1842*, 1707–1719. [CrossRef]
148. Pierre, S.R.; Lemmens, M.A.M.; Figueiredo-Pereira, M.E. Subchronic infusion of the product of inflammation prostaglandin J2 models sporadic Parkinson's disease in mice. *J. Neuroinflamm.* **2009**, *6*, 18. [CrossRef]
149. Zhao, H.; Wang, C.; Zhao, N.; Li, W.; Yang, Z.; Liu, X.; Le, W.; Zhang, X. Potential biomarkers of Parkinson's disease revealed by plasma metabolic profiling. *J. Chromatogr. B Anal. Technol. Biomed. Life Sci.* **2018**, *1081–1082*, 101–108. [CrossRef]
150. Kang, K.H.; Liou, H.H.; Hour, M.J.; Liou, H.C.; Fu, W.M. Protection of dopaminergic neurons by 5-lipoxygenase inhibitor. *Neuropharmacology* **2013**, *73*, 380–387. [CrossRef]
151. Nagarajan, V.B.; Marathe, P.A. Effect of montelukast in experimental model of Parkinson's disease. *Neurosci. Lett.* **2018**, *682*, 100–105. [CrossRef] [PubMed]
152. Mansour, R.M.; Ahmed, M.A.E.; El-Sahar, A.E.; El Sayed, N.S. Montelukast attenuates rotenone-induced microglial activation/p38 MAPK expression in rats: Possible role of its antioxidant, anti-inflammatory and antiapoptotic effects. *Toxicol. Appl. Pharmacol.* **2018**, *358*, 76–85. [CrossRef] [PubMed]
153. Chou, V.P.; Ko, N.; Holman, T.R.; Manning-Boğ, A.B. Gene-environment interaction models to unmask susceptibility mechanisms in Parkinson's disease. *J. Vis. Exp.* **2014**, e50960. [CrossRef] [PubMed]

154. Searles Nielsen, S.; Bammler, T.K.; Gallagher, L.G.; Farin, F.M.; Longstreth, W.; Franklin, G.M.; Swanson, P.D.; Checkoway, H. Genotype and age at Parkinson disease diagnosis. *Int. J. Mol. Epidemiol. Genet.* **2013**, *4*, 61–69. [PubMed]
155. Lakkappa, N.; Krishnamurthy, P.T.; Hammock, B.D.; Velmurugan, D.; Bharath, M.M.S. Possible role of Epoxyeicosatrienoic acid in prevention of oxidative stress mediated neuroinflammation in Parkinson disorders. *Med. Hypotheses* **2016**, *93*, 161–165. [CrossRef]
156. Terashvili, M.; Sarkar, P.; Nostrand, M.V.; Falck, J.R.; Harder, D.R. The protective effect of astrocyte-derived 14, 15-epoxyeicosatrienoic acid on hydrogen peroxide-induced cell injury in astrocyte-dopaminergic neuronal cell line co-culture. *Neuroscience* **2012**, *223*, 68–76. [CrossRef]
157. Qin, X.; Wu, Q.; Lin, L.; Sun, A.; Liu, S.; Li, X.; Cao, X.; Gao, T.; Luo, P.; et al. Soluble Epoxide Hydrolase Deficiency or Inhibition Attenuates MPTP-Induced Parkinsonism. *Mol. Neurobiol.* **2015**, *52*, 187–195. [CrossRef] [PubMed]
158. Ren, Q.; Ma, M.; Yang, J.; Nonaka, R.; Yamaguchi, A.; Ishikawa, K.I.; Kobayashi, K.; Murayama, S.; Hwang, S.H.; Saiki, S.; et al. Soluble epoxide hydrolase plays a key role in the pathogenesis of Parkinson's disease. *Proc. Natl. Acad. Sci. USA* **2018**, *115*, E5815–E5823. [CrossRef] [PubMed]
159. Lakkappa, N.; Krishnamurthy, P.T.; Yamjala, K.; Hwang, S.H.; Hammock, B.D.; Babu, B. Evaluation of antiparkinson activity of PTUPB by measuring dopamine and its metabolites in Drosophila melanogaster: LC-MS/MS method development. *J. Pharm. Biomed. Anal.* **2018**, *149*, 457–464. [CrossRef]
160. Connolly, J.; Siderowf, A.; Clark, C.M.; Mu, D.; Pratico, D. F2 isoprostane levels in plasma and urine do not support increased lipid peroxidation in cognitively impaired Parkinson disease patients. *Cogn. Behav. Neurol.* **2008**, *21*, 83–86. [CrossRef]
161. Lee, C.Y.J.; Seet, R.C.S.; Huang, S.H.; Long, L.H.; Halliwell, B. Different patterns of oxidized lipid products in plasma and urine of dengue fever, stroke, and Parkinson's disease patients: Cautions in the use of biomarkers of oxidative stress. *Antioxid. Redox Signal.* **2009**, *11*, 407–420. [CrossRef] [PubMed]
162. Seet, R.C.S.; Lee, C.Y.J.; Lim, E.C.H.; Tan, J.J.H.; Quek, A.M.L.; Chong, W.L.; Looi, W.F.; Huang, S.H.; Wang, H.; Chang, Y.H.; et al. Oxidative damage in Parkinson disease: Measurement using accurate biomarkers. *Free Radic. Biol. Med.* **2010**, *48*, 560–566. [CrossRef] [PubMed]
163. Irizarry, M.C.; Yao, Y.; Hyman, B.T.; Growdon, J.H.; Praticò, D. Plasma F2A isoprostane levels in Alzheimer's and Parkinson's disease. *Neurodegener. Dis.* **2007**, *4*, 403–405. [CrossRef] [PubMed]
164. Fessel, J.P.; Hulette, C.; Powell, S.; Roberts, L.J.; Zhang, J. Isofurans, but not F2-isoprostanes, are increased in the substantia nigra of patients with Parkinson's disease and with dementia with Lewy body disease. *J. Neurochem.* **2003**, *85*, 645–650. [CrossRef] [PubMed]
165. Neely, M.D.; Davison, C.A.; Aschner, M.; Bowman, A.B. From the Cover: Manganese and Rotenone-Induced Oxidative Stress Signatures Differ in iPSC-Derived Human Dopamine Neurons. *Toxicol. Sci.* **2017**, *159*, 366–379. [CrossRef] [PubMed]
166. Xu, J.; Gao, X.; Yang, C.; Chen, L.; Chen, Z. Resolvin D1 Attenuates Mpp+-Induced Parkinson Disease via Inhibiting Inflammation in PC12 Cells. *Med. Sci. Monit.* **2017**, *23*, 2684–2691. [CrossRef] [PubMed]
167. Tian, Y.; Zhang, Y.; Zhang, R.; Qiao, S.; Fan, J. Resolvin D2 recovers neural injury by suppressing inflammatory mediators expression in lipopolysaccharide-induced Parkinson's disease rat model. *Biochem. Biophys. Res. Commun.* **2015**, *460*, 799–805. [CrossRef]
168. Traina, G. The neurobiology of acetyl-L-carnitine. *Front. Biosci.* **2016**, *21*, 1314–1329. [CrossRef]
169. Saiki, S.; Hatano, T.; Fujimaki, M.; Ishikawa, K.I.; Mori, A.; Oji, Y.; Okuzumi, A.; Fukuhara, T.; Koinuma, T.; Imamichi, Y.; et al. Decreased long-chain acylcarnitines from insufficient β-oxidation as potential early diagnostic markers for Parkinson's disease. *Sci. Rep.* **2017**, *7*, 7328. [CrossRef]
170. Crooks, S.A.; Bech, S.; Halling, J.; Christiansen, D.H.; Ritz, B.; Petersen, M.S. Carnitine levels and mutations in the SLC22A5 gene in Faroes patients with Parkinson's disease. *Neurosci. Lett.* **2018**, *675*, 116–119. [CrossRef]
171. Jiménez-Jiménez, F.J.; Rubio, J.C.; Molina, J.A.; Martín, M.A.; Campos, Y.; Benito-León, J.; Ortí-Pareja, M.; Gassalla, T.; Arenas, J. Cerebrospinal fluid carnitine levels in patients with Parkinson's disease. *J. Neurol. Sci.* **1997**, *145*, 183–185. [CrossRef]
172. Zhang, H.; Jia, H.; Liu, J.; Ao, N.; Yan, B.; Shen, W.; Wang, X.; Li, X.; Luo, C.; Liu, J. Combined R-alpha-lipoic acid and acetyl-L-carnitine exerts efficient preventative effects in a cellular model of Parkinson's disease. *J. Cell. Mol. Med.* **2010**, *14*, 215–225. [CrossRef] [PubMed]

173. Wang, C.; Sadovova, N.; Ali, H.K.; Duhart, H.M.; Fu, X.; Zou, X.; Patterson, T.A.; Ninienda, Z.K.; Virmani, A.; Paule, M.G.; et al. L-carnitine protects neurons from 1-methyl-4-phenylpyridinium-induced neuronal apoptosis in rat forebrain culture. *Neuroscience* **2007**, *144*, 46–55. [CrossRef] [PubMed]
174. Gill, E.L.; Raman, S.; Yost, R.A.; Garrett, T.J.; Vedam-Mai, V. l-Carnitine Inhibits Lipopolysaccharide-Induced Nitric Oxide Production of SIM-A9 Microglia Cells. *ACS Chem. Neurosci.* **2018**, *9*, 901–905. [CrossRef]
175. Singh, S.; Mishra, A.; Mishra, S.K.; Shukla, S. ALCAR promote adult hippocampal neurogenesis by regulating cell-survival and cell death-related signals in rat model of Parkinson's disease like-phenotypes. *Neurochem. Int.* **2017**, *108*, 388–396. [CrossRef]
176. Afshin-Majd, S.; Bashiri, K.; Kiasalari, Z.; Baluchnejadmojarad, T.; Sedaghat, R.; Roghani, M. Acetyl-L-carnitine protects dopaminergic nigrostriatal pathway in 6-hydroxydopamine-induced model of Parkinson's disease in the rat. *Biomed. Pharmacother.* **2017**, *89*, 1–9. [CrossRef] [PubMed]
177. Singh, S.; Mishra, A.; Srivastava, N.; Shukla, R.; Shukla, S. Acetyl-L-Carnitine via Upegulating Dopamine D1 Receptor and Attenuating Microglial Activation Prevents Neuronal Loss and Improves Memory Functions in Parkinsonian Rats. *Mol. Neurobiol.* **2018**, *55*, 583–602. [CrossRef]
178. Sarkar, S.; Gough, B.; Raymick, J.; Beaudoin, M.A.; Ali, S.F.; Virmani, A.; Binienda, Z.K. Histopathological and electrophysiological indices of rotenone-evoked dopaminergic toxicity: Neuroprotective effects of acetyl-L-carnitine. *Neurosci. Lett.* **2015**, *606*, 53–59. [CrossRef]
179. Zaitone, S.A.; Abo-Elmatty, D.M.; Shaalan, A.A. Acetyl-L-carnitine and α-lipoic acid affect rotenone-induced damage in nigral dopaminergic neurons of rat brain, implication for Parkinson's disease therapy. *Pharmacol. Biochem. Behav.* **2012**, *100*, 347–360. [CrossRef]
180. Bodis-Wollner, I.; Chung, E.; Ghilardi, M.F.; Glover, A.; Onofrj, M.; Pasik, P.; Samson, Y. Acetyl-levo-carnitine protects against MPTP-induced parkinsonism in primates. *J. Neural Transm. Park Dis. Dement. Sect.* **1991**, *3*, 63–72. [CrossRef]
181. Vetel, S.; Sérrière, S.; Vercouillie, J.; Vergote, J.; Chicheri, G.; Deloye, J.B.; Dollé, F.; Bodard, S.; Tronel, C.; Nadal-Desbarats, L.; et al. Extensive exploration of a novel rat model of Parkinson's disease using partial 6-hydroxydopamine lesion of dopaminergic neurons suggests new therapeutic approaches. *Synapse* **2018**. [CrossRef] [PubMed]
182. Li, X.; Zhang, S.; Lu, F.; Liu, C.; Wang, Y.; Bai, Y.; Wang, N.; Liu, S.M. Cerebral metabonomics study on Parkinson's disease mice treated with extract of Acanthopanax senticosus harms. *Phytomedicine* **2013**, *20*, 1219–1229. [CrossRef] [PubMed]
183. Tu-Sekine, B.; Goldschmidt, H.; Raben, D.M. Diacylglycerol, phosphatidic acid, and their metabolic enzymes in synaptic vesicle recycling. *Adv. Biol. Regul.* **2015**, *57*, 147–152. [CrossRef] [PubMed]
184. Almena, M.; Mérida, I. Shaping up the membrane: Diacylglycerol coordinates spatial orientation of signaling. *Trends Biochem. Sci.* **2011**, *36*, 593–603. [CrossRef]
185. Ahmadian, M.; Duncan, R.E.; Jaworski, K.; Sarkadi-Nagy, E.; Sul, H.S. Triacylglycerol metabolism in adipose tissue. *Future Lipidol.* **2007**, *2*, 229–237. [CrossRef] [PubMed]
186. Navarrete, F.; García-Gutiérrez, M.S.; Aracil-Fernández, A.; Lanciego, J.L.; Manzanares, J. Cannabinoid CB1 and CB2 Receptors, and Monoacylglycerol Lipase Gene Expression Alterations in the Basal Ganglia of Patients with Parkinson's Disease. *Neurotherapeutics* **2018**, *15*, 459–469. [CrossRef] [PubMed]
187. Fearnley, J.M.; Lees, A.J. Ageing and Parkinson's disease: Substantia nigra regional selectivity. *Brain* **1991**, *114 Pt 5*, 2283–2301. [CrossRef]
188. Kish, S.J.; Shannak, K.; Hornykiewicz, O. Uneven pattern of dopamine loss in the striatum of patients with idiopathic Parkinson's disease. Pathophysiologic and clinical implications. *N. Engl. J. Med.* **1988**, *318*, 876–880. [CrossRef]
189. Aymerich, M.S.; Rojo-Bustamante, E.; Molina, C.; Celorrio, M.; Sánchez-Arias, J.A.; Franco, R. Neuroprotective Effect of JZL184 in MPP(+)-Treated SH-SY5Y Cells Through CB2 Receptors. *Mol. Neurobiol.* **2016**, *53*, 2312–2319. [CrossRef]
190. Pasquarelli, N.; Porazik, C.; Bayer, H.; Buck, E.; Schildknecht, S.; Weydt, P.; Witting, A.; Ferger, B. Contrasting effects of selective MAGL and FAAH inhibition on dopamine depletion and GDNF expression in a chronic MPTP mouse model of Parkinson's disease. *Neurochem. Int.* **2017**, *110*, 14–24. [CrossRef]

191. Fernández-Suárez, D.; Celorrio, M.; Riezu-Boj, J.I.; Ugarte, A.; Pacheco, R.; González, H.; Oyarzabal, J.; Hillard, C.J.; Franco, R.; Aymerich, M.S. Monoacylglycerol lipase inhibitor JZL184 is neuroprotective and alters glial cell phenotype in the chronic MPTP mouse model. *Neurobiol. Aging* **2014**, *35*, 2603–2616. [CrossRef] [PubMed]
192. Stampanoni Bassi, M.; Sancesario, A.; Morace, R.; Centonze, D.; Iezzi, E. Cannabinoids in Parkinson's Disease. *Cannabis Cannabinoid Res.* **2017**, *2*, 21–29. [CrossRef] [PubMed]
193. Pisani, A.; Fezza, F.; Galati, S.; Battista, N.; Napolitano, S.; Finazzi-Agrò, A.; Bernardi, G.; Bursa, L.; Pierantozzi, M.; Stanzione, P.; et al. High endogenous cannabinoid levels in the cerebrospinal fluid of untreated Parkinson's disease patients. *Ann. Neurol.* **2005**, *57*, 777–779. [CrossRef] [PubMed]
194. Pisani, V.; Moschella, V.; Bari, M.; Fezza, F.; Galati, S.; Bernardi, G.; Stacione, P.; Pisani, A.; Maccarrone, M. Dynamic changes of anandamide in the cerebrospinal fluid of Parkinson's disease patients. *Mov. Disord.* **2010**, *25*, 920–924. [CrossRef] [PubMed]
195. Chagas, M.H.N.; Zuardi, A.W.; Tumas, V.; Pena-Pereira, M.A.; Sobreira, E.T.; Bergamaschi, M.M.; dos Santos, A.C.; Teixeira, A.L.; Hallack, J.E.; Crippa, J.A. Effects of cannabidiol in the treatment of patients with Parkinson's disease: An exploratory double-blind trial. *J. Psychopharmacol.* **2014**, *28*, 1088–1098. [CrossRef] [PubMed]
196. Oki, M.; Kaneko, S.; Morise, S.; Takenouchi, N.; Hashizume, T.; Tsuge, A.; Nakamura, M.; Wate, R.; Kausaka, H. Zonisamide ameliorates levodopa-induced dyskinesia and reduces expression of striatal genes in Parkinson model rats. *Neurosci. Res.* **2017**, *122*, 45–50. [CrossRef] [PubMed]
197. Mackovski, N.; Liao, J.; Weng, R.; Wei, X.; Wang, R.; Chen, Z.; Liu, X.; Yu, Y.; Meyer, B.J.; Xia, Y.; et al. Reversal effect of simvastatin on the decrease in cannabinoid receptor 1 density in 6-hydroxydopamine lesioned rat brains. *Life Sci.* **2016**, *155*, 123–132. [CrossRef]
198. Casteels, C.; Lauwers, E.; Baitar, A.; Bormans, G.; Baekelandt, V.; Van Laere, K. In vivo type 1 cannabinoid receptor mapping in the 6-hydroxydopamine lesion rat model of Parkinson's disease. *Brain Res.* **2010**, *1316*, 153–162. [CrossRef]
199. Walsh, S.; Mnich, K.; Mackie, K.; Gorman, A.M.; Finn, D.P.; Dowd, E. Loss of cannabinoid CB1 receptor expression in the 6-hydroxydopamine-induced nigrostriatal terminal lesion model of Parkinson's disease in the rat. *Brain Res. Bull.* **2010**, *81*, 543–548. [CrossRef]
200. Chaves-Kirsten, G.P.; Mazucanti, C.H.Y.; Real, C.C.; Souza, B.M.; Britto, L.R.G.; Torrão, A.S. Temporal changes of CB1 cannabinoid receptor in the basal ganglia as a possible structure-specific plasticity process in 6-OHDA lesioned rats. *PLoS ONE* **2013**, *8*, e76874. [CrossRef]
201. Silverdale, M.A.; McGuire, S.; McInnes, A.; Crossman, A.R.; Brotchie, J.M. Striatal cannabinoid CB1 receptor mRNA expression is decreased in the reserpine-treated rat model of Parkinson's disease. *Exp. Neurol.* **2001**, *169*, 400–406. [CrossRef] [PubMed]
202. González, S.; Mena, M.A.; Lastres-Becker, I.; Serrano, A.; de Yébenes, J.G.; Ramos, J.A.; Fernández-Ruiz, J. Cannabinoid CB(1) receptors in the basal ganglia and motor response to activation or blockade of these receptors in parkin-null mice. *Brain Res.* **2005**, *1046*, 195–206. [CrossRef] [PubMed]
203. Madeo, G.; Schirinzi, T.; Maltese, M.; Martella, G.; Rapino, C.; Fezza, F.; Mastrengelo, N.; Bonsi, P.; Maccarrone, M.; Pisani, A. Dopamine-dependent CB1 receptor dysfunction at corticostriatal synapses in homozygous PINK1 knockout mice. *Neuropharmacology* **2016**, *101*, 460–470. [CrossRef] [PubMed]
204. Concannon, R.M.; Okine, B.N.; Finn, D.P.; Dowd, E. Upregulation of the cannabinoid CB2 receptor in environmental and viral inflammation-driven rat models of Parkinson's disease. *Exp. Neurol.* **2016**, *283*, 204–212. [CrossRef] [PubMed]
205. Concannon, R.M.; Okine, B.N.; Finn, D.P.; Dowd, E. Differential upregulation of the cannabinoid CB_2 receptor in neurotoxic and inflammation-driven rat models of Parkinson's disease. *Exp. Neurol.* **2015**, *269*, 133–141. [CrossRef]
206. Palomo-Garo, C.; Gómez-Gálvez, Y.; García, C.; Fernández-Ruiz, J. Targeting the cannabinoid CB2 receptor to attenuate the progression of motor deficits in LRRK2-transgenic mice. *Pharmacol. Res.* **2016**, *110*, 181–192. [CrossRef]
207. Javed, H.; Azimullah, S.; Haque, M.E.; Ojha, S.K. Cannabinoid Type 2 (CB2) Receptors Activation Protects against Oxidative Stress and Neuroinflammation Associated Dopaminergic Neurodegeneration in Rotenone Model of Parkinson's Disease. *Front. Neurosci.* **2016**, *10*, 321. [CrossRef]

208. García-Arencibia, M.; González, S.; de Lago, E.; Ramos, J.A.; Mechoulam, R.; Fernández-Ruiz, J. Evaluation of the neuroprotective effect of cannabinoids in a rat model of Parkinson's disease: Importance of antioxidant and cannabinoid receptor-independent properties. *Brain Res.* **2007**, *1134*, 162–170. [CrossRef]
209. Fernandez-Espejo, E.; Caraballo, I.; Rodriguez de Fonseca, F.; Ferrer, B.; El Banoua, F.; Flores, J.A.; Galan-Rodriguez, B. Experimental parkinsonism alters anandamide precursor synthesis, and functional deficits are improved by AM404: A modulator of endocannabinoid function. *Neuropsychopharmacology* **2004**, *29*, 1134–1142. [CrossRef]
210. Maccarrone, M.; Gubellini, P.; Bari, M.; Picconi, B.; Battista, N.; Centonze, D.; Bernardi, G.; Finazzi-Agrò, A.; Calabresi, P. Levodopa treatment reverses endocannabinoid system abnormalities in experimental parkinsonism. *J. Neurochem.* **2003**, *85*, 1018–1025. [CrossRef]
211. Gubellini, P.; Picconi, B.; Bari, M.; Battista, N.; Calabresi, P.; Centonze, D.; Bernardi, G.; Finazzi-Agrò, A.; Maccarrone, M. Experimental parkinsonism alters endocannabinoid degradation: Implications for striatal glutamatergic transmission. *J. Neurosci.* **2002**, *22*, 6900–6907. [CrossRef] [PubMed]
212. Ferrer, B.; Asbrock, N.; Kathuria, S.; Piomelli, D.; Giuffrida, A. Effects of levodopa on endocannabinoid levels in rat basal ganglia: Implications for the treatment of levodopa-induced dyskinesias. *Eur. J. Neurosci.* **2003**, *18*, 1607–1614. [CrossRef] [PubMed]
213. Van der Stelt, M.; Fox, S.H.; Hill, M.; Crossman, A.R.; Petrosino, S.; Di Marzo, V.; Brotchie, J.M. A role for endocannabinoids in the generation of parkinsonism and levodopa-induced dyskinesia in MPTP-lesioned non-human primate models of Parkinson's disease. *FASEB J.* **2005**, *19*, 1140–1142. [CrossRef] [PubMed]
214. Viveros-Paredes, J.M.; Gonzalez-Castañeda, R.E.; Escalante-Castañeda, A.; Tejeda-Martínez, A.R.; Castañeda-Achutiguí, F.; Flores-Soto, M.E. Effect of inhibition of fatty acid amide hydrolase on MPTP-induced dopaminergic neuronal damage. *Neurologia* **2017**. [CrossRef]
215. Celorrio, M.; Fernández-Suárez, D.; Rojo-Bustamante, E.; Echeverry-Alzate, V.; Ramírez, M.J.; Hillard, C.J.; López-Morenoa, J.A.; Maldonado, R.; Oyarzábal, J.; Franco, R.; et al. Fatty acid amide hydrolase inhibition for the symptomatic relief of Parkinson's disease. *Brain Behav. Immun.* **2016**, *57*, 94–105. [CrossRef] [PubMed]
216. Mnich, K.; Finn, D.P.; Dowd, E.; Gorman, A.M. Inhibition by anandamide of 6-hydroxydopamine-induced cell death in PC12 cells. *Int. J. Cell Biol.* **2010**, *2010*, 818497. [CrossRef]
217. Mounsey, R.B.; Mustafa, S.; Robinson, L.; Ross, R.A.; Riedel, G.; Pertwee, R.G.; Tesimann, P. Increasing levels of the endocannabinoid 2-AG is neuroprotective in the 1-methyl-4-phenyl-1, 2, 3, 6-tetrahydropyridine mouse model of Parkinson's disease. *Exp. Neurol.* **2015**, *273*, 36–44. [CrossRef]
218. Hracskó, Z.; Baranyi, M.; Csölle, C.; Gölöncsér, F.; Madarász, E.; Kittel, A.; Sperlágh, B. Lack of neuroprotection in the absence of P2X7 receptors in toxin-induced animal models of Parkinson's disease. *Mol. Neurodegener.* **2011**, *6*, 28. [CrossRef]
219. Di Marzo, V.; Hill, M.P.; Bisogno, T.; Crossman, A.R.; Brotchie, J.M. Enhanced levels of endogenous cannabinoids in the globus pallidus are associated with a reduction in movement in an animal model of Parkinson's disease. *FASEB J.* **2000**, *14*, 1432–1438.
220. Freestone, P.S.; Guatteo, E.; Piscitelli, F.; di Marzo, V.; Lipski, J.; Mercuri, N.B. Glutamate spillover drives endocannabinoid production and inhibits GABAergic transmission in the Substantia Nigra pars compacta. *Neuropharmacology* **2014**, *79*, 467–475. [CrossRef]
221. Kreitzer, A.C.; Malenka, R.C. Endocannabinoid-mediated rescue of striatal LTD and motor deficits in Parkinson's disease models. *Nature* **2007**, *445*, 643–647. [CrossRef] [PubMed]
222. Zhang, J.; Zhang, X.; Wang, L.; Yang, C. High Performance Liquid Chromatography-Mass Spectrometry (LC-MS) Based Quantitative Lipidomics Study of Ganglioside-NANA-3 Plasma to Establish Its Association with Parkinson's Disease Patients. *Med. Sci. Monit.* **2017**, *23*, 5345–5353. [CrossRef] [PubMed]
223. Chan, R.B.; Perotte, A.J.; Zhou, B.; Liong, C.; Shorr, E.J.; Marder, K.S.; et al. Elevated GM3 plasma concentration in idiopathic Parkinson's disease: A lipidomic analysis. *PLoS ONE* **2017**, *12*, e0172348. [CrossRef] [PubMed]
224. Wood, P.L.; Tippireddy, S.; Feriante, J.; Woltjer, R.L. Augmented frontal cortex diacylglycerol levels in Parkinson's disease and Lewy Body Disease. *PLoS ONE* **2018**, *13*, e0191815. [CrossRef] [PubMed]
225. Cheng, D.; Jenner, A.M.; Shui, G.; Cheong, W.F.; Mitchell, T.W.; Nealon, J.R.; Kim, W.S.; McCann, H.; Wenk, M.R.; Halliday, G.M.; et al. Lipid pathway alterations in Parkinson's disease primary visual cortex. *PLoS ONE* **2011**, *6*, e17299. [CrossRef] [PubMed]

226. Simón-Sánchez, J.; van Hilten, J.J.; van de Warrenburg, B.; Post, B.; Berendse, H.W.; Arepalli, S.; Hernandez, D.G.; de Bie, R.M.; Velseboar, D.; Scheffer, H.; et al. Genome-wide association study confirms extant PD risk loci among the Dutch. *Eur. J. Hum. Genet.* **2011**, *19*, 655–661. [CrossRef] [PubMed]
227. Chen, Y.P.; Song, W.; Huang, R.; Chen, K.; Zhao, B.; Li, J.; Yang, Y.; Shang, H.F. GAK rs1564282 and DGKQ rs11248060 increase the risk for Parkinson's disease in a Chinese population. *J. Clin. Neurosci.* **2013**, *20*, 880–883. [CrossRef]
228. Sakane, F.; Mizuno, S.; Takahashi, D.; Sakai, H. Where do substrates of diacylglycerol kinases come from? Diacylglycerol kinases utilize diacylglycerol species supplied from phosphatidylinositol turnover-independent pathways. *Adv. Biol. Regul.* **2018**, *67*, 101–108. [CrossRef]
229. Guo, X.; Song, W.; Chen, K.; Chen, X.; Zheng, Z.; Cao, B.; Huang, R.; Zhao, B.; Wu, Y.; Shang, H.F. The serum lipid profile of Parkinson's disease patients: A study from China. *Int. J. Neurosci.* **2015**, *125*, 838–844. [CrossRef]
230. Wei, Q.; Wang, H.; Tian, Y.; Xu, F.; Chen, X.; Wang, K. Reduced serum levels of triglyceride, very low density lipoprotein cholesterol and apolipoprotein B in Parkinson's disease patients. *PLoS ONE* **2013**, *8*, e75743. [CrossRef] [PubMed]
231. Gregório, M.L.; Pinhel, M.A.S.; Sado, C.L.; Longo, G.S.; Oliveira, F.N.; Amorim, G.S.; Nakazone, M.A.; Florim, G.M.; Mazeti, C.M.; Martins, D.P.; et al. Impact of genetic variants of apolipoprotein E on lipid profile in patients with Parkinson's disease. *Biomed. Res. Int.* **2013**, *2013*, 641515. [CrossRef] [PubMed]
232. Cereda, E.; Cassani, E.; Barichella, M.; Spadafranca, A.; Caccialanza, R.; Bertoli, S.; Battezzati, A.; Pezzoli, G. Low cardiometabolic risk in Parkinson's disease is independent of nutritional status, body composition and fat distribution. *Clin. Nutr.* **2012**, *31*, 699–704. [CrossRef] [PubMed]
233. Sääksjärvi, K.; Knekt, P.; Männistö, S.; Lyytinen, J.; Heliövaara, M. Prospective study on the components of metabolic syndrome and the incidence of Parkinson's disease. *Park. Relat. Disord.* **2015**, *21*, 1148–1155. [CrossRef] [PubMed]
234. Vikdahl, M.; Bäckman, L.; Johansson, I.; Forsgren, L.; Håglin, L. Cardiovascular risk factors and the risk of Parkinson's disease. *Eur. J. Clin. Nutr.* **2015**, *69*, 729–733. [CrossRef] [PubMed]
235. Scigliano, G.; Musicco, M.; Soliveri, P.; Piccolo, I.; Ronchetti, G.; Girotti, F. Reduced risk factors for vascular disorders in Parkinson disease patients: A case-control study. *Stroke* **2006**, *37*, 1184–1188. [CrossRef] [PubMed]
236. Fukui, Y.; Hishikawa, N.; Shang, J.; Sato, K.; Nakano, Y.; Morihara, R.; Ohta, Y.; Yamashita, T.; Abe, K. Peripheral arterial endothelial dysfunction of neurodegenerative diseases. *J. Neurol. Sci.* **2016**, *366*, 94–99. [CrossRef]
237. Ya, L.; Lu, Z. Differences in ABCA1 R219K Polymorphisms and Serum Indexes in Alzheimer and Parkinson Diseases in Northern China. *Med. Sci. Monit.* **2017**, *23*, 4591–4600. [CrossRef]
238. Wei, Z.; Li, X.; Li, X.; Liu, Q.; Cheng, Y. Oxidative Stress in Parkinson's Disease: A Systematic Review and Meta-Analysis. *Front. Mol. Neurosci.* **2018**, *11*, 236. [CrossRef]
239. Guerreiro, P.S.; Coelho, J.E.; Sousa-Lima, I.; Macedo, P.; Lopes, L.V.; Outeiro, T.F.; Pais, T.F. Mutant A53T α-Synuclein Improves Rotarod Performance Before Motor Deficits and Affects Metabolic Pathways. *Neuromol. Med.* **2017**, *19*, 113–121. [CrossRef]
240. Meng, X.; Zheng, R.; Zhang, Y.; Qiao, M.; Liu, L.; Jing, P.; Wang, L.; Liu, J.; Gao, Y. An activated sympathetic nervous system affects white adipocyte differentiation and lipolysis in a rat model of Parkinson's disease. *J. Neurosci. Res.* **2015**, *93*, 350–360. [CrossRef]
241. He, Q.; Wang, M.; Petucci, C.; Gardell, S.J.; Han, X. Rotenone induces reductive stress and triacylglycerol deposition in C2C12 cells. *Int. J. Biochem. Cell Biol.* **2013**, *45*, 2749–2755. [CrossRef]
242. Sere, Y.Y.; Regnacq, M.; Colas, J.; Berges, T. A Saccharomyces cerevisiae strain unable to store neutral lipids is tolerant to oxidative stress induced by α-synuclein. *Free Radic. Biol. Med.* **2010**, *49*, 1755–1764. [CrossRef] [PubMed]
243. Cole, N.B.; Murphy, D.D.; Grider, T.; Rueter, S.; Brasaemle, D.; Nussbaum, R.L. Lipid droplet binding and oligomerization properties of the Parkinson's disease protein alpha-synuclein. *J. Biol. Chem.* **2002**, *277*, 6344–6352. [CrossRef]
244. Sánchez Campos, S.; Alza, N.P.; Salvador, G.A. Lipid metabolism alterations in the neuronal response to A53T α-synuclein and Fe-induced injury. *Arch. Biochem. Biophys.* **2018**, *655*, 43–54. [CrossRef] [PubMed]

245. Antonny, B.; Vanni, S.; Shindou, H.; Ferreira, T. From zero to six double bonds: Phospholipid unsaturation and organelle function. *Trends Cell Biol.* **2015**, *25*, 427–436. [CrossRef] [PubMed]
246. Bohdanowicz, M.; Grinstein, S. Role of phospholipids in endocytosis, phagocytosis, and macropinocytosis. *Physiol. Rev.* **2013**, *93*, 69–106. [CrossRef]
247. Zhang, Q.; Tamura, Y.; Roy, M.; Adachi, Y.; Iijima, M.; Sesaki, H. Biosynthesis and roles of phospholipids in mitochondrial fusion, division and mitophagy. *Cell. Mol. Life Sci.* **2014**, *71*, 3767–3778. [CrossRef]
248. Boss, W.F.; Im, Y.J. Phosphoinositide signaling. *Annu. Rev. Plant Biol.* **2012**, *63*, 409–429. [CrossRef]
249. Yung, Y.C.; Stoddard, N.C.; Mirendil, H.; Chun, J. Lysophosphatidic Acid signaling in the nervous system. *Neuron* **2015**, *85*, 669–682. [CrossRef]
250. Kay, J.G.; Grinstein, S. Phosphatidylserine-mediated cellular signaling. *Adv. Exp. Med. Biol.* **2013**, *991*, 177–193. [CrossRef]
251. Musille, P.M.; Kohn, J.A.; Ortlund, E.A. Phospholipid—Driven gene regulation. *FEBS Lett.* **2013**, *587*, 1238–1246. [CrossRef] [PubMed]
252. Liu, Y.; Su, Y.; Wang, X. Phosphatidic acid-mediated signaling. *Adv. Exp. Med. Biol.* **2013**, *991*, 159–176. [CrossRef] [PubMed]
253. Ammar, M.R.; Kassas, N.; Bader, M.F.; Vitale, N. Phosphatidic acid in neuronal development: A node for membrane and cytoskeleton rearrangements. *Biochimie* **2014**, *107*, 51–57. [CrossRef] [PubMed]
254. Yang, C.Y.; Frohman, M.A. Mitochondria: Signaling with phosphatidic acid. *Int. J. Biochem. Cell Biol.* **2012**, *44*, 1346–1350. [CrossRef] [PubMed]
255. Jiang, Z.; Hess, S.K.; Heinrich, F.; Lee, J.C. Molecular details of α-synuclein membrane association revealed by neutrons and photons. *J. Phys. Chem. B* **2015**, *119*, 4812–4823. [CrossRef] [PubMed]
256. Perrin, R.J.; Woods, W.S.; Clayton, D.F.; George, J.M. Interaction of human alpha-Synuclein and Parkinson's disease variants with phospholipids. Structural analysis using site-directed mutagenesis. *J. Biol. Chem.* **2000**, *275*, 34393–34398. [CrossRef] [PubMed]
257. Mizuno, S.; Sasai, H.; Kume, A.; Takahashi, D.; Satoh, M.; Kado, S.; Sakane, F. Dioleoyl-phosphatidic acid selectively binds to α-synuclein and strongly induces its aggregation. *FEBS Lett.* **2017**, *591*, 784–791. [CrossRef]
258. Soper, J.H.; Kehm, V.; Burd, C.G.; Bankaitis, V.A.; Lee, V.M.Y. Aggregation of α-synuclein in S. cerevisiae is associated with defects in endosomal trafficking and phospholipid biosynthesis. *J. Mol. Neurosci.* **2011**, *43*, 391–405. [CrossRef]
259. Holemans, T.; Sørensen, D.M.; van Veen, S.; Martin, S.; Hermans, D.; Kemmer, G.C.; Van den Haute, C.; Baekelandt, V.; Günther Pomorski, T.; Agostinis, P.; et al. A lipid switch unlocks Parkinson's disease-associated ATP13A2. *Proc. Natl. Acad. Sci. USA* **2015**, *112*, 9040–9045. [CrossRef]
260. Martin, S.; van Veen, S.; Holemans, T.; Demirsoy, S.; van den Haute, C.; Baekelandt, V.; Agostinis, P.; Eggermont, J.; Vangheluwe, P. Protection against Mitochondrial and Metal Toxicity Depends on Functional Lipid Binding Sites in ATP13A2. *Park. Dis.* **2016**, *2016*, 9531917. [CrossRef] [PubMed]
261. Mendez-Gomez, H.R.; Singh, J.; Meyers, C.; Chen, W.; Gorbatyuk, O.S.; Muzyczka, N. The Lipase Activity of Phospholipase D2 is Responsible for Nigral Neurodegeneration in a Rat Model of Parkinson's Disease. *Neuroscience* **2018**, *377*, 174–183. [CrossRef]
262. Binder, B.Y.K.; Williams, P.A.; Silva, E.A.; Leach, J.K. Lysophosphatidic Acid and Sphingosine-1-Phosphate: A Concise Review of Biological Function and Applications for Tissue Engineering. *Tissue Eng. Part B Rev.* **2015**, *21*, 531–542. [CrossRef]
263. Sheng, X.; Yung, Y.C.; Chen, A.; Chun, J. Lysophosphatidic acid signalling in development. *Development* **2015**, *142*, 1390–1395. [CrossRef]
264. Aikawa, S.; Hashimoto, T.; Kano, K.; Aoki, J. Lysophosphatidic acid as a lipid mediator with multiple biological actions. *J. Biochem.* **2015**, *157*, 81–89. [CrossRef]
265. Yang, X.Y.; Zhao, E.Y.; Zhuang, W.X.; Sun, F.X.; Han, H.L.; Han, H.R.; Lin, Z.J.; Pan, Z.F.; Qu, M.H.; Zeng, X.W.; et al. LPA signaling is required for dopaminergic neuron development and is reduced through low expression of the LPA1 receptor in a 6-OHDA lesion model of Parkinson's disease. *Neurol. Sci.* **2015**, *36*, 2027–2033. [CrossRef]
266. Choi, J.H.; Jang, M.; Oh, S.; Nah, S.Y.; Cho, I.H. Multi-Target Protective Effects of Gintonin in 1-Methyl-4-phenyl-1, 2, 3, 6-tetrahydropyridine-Mediated Model of Parkinson's Disease via Lysophosphatidic Acid Receptors. *Front. Pharmacol.* **2018**, *9*, 515. [CrossRef]

267. Vance, J.E. Phosphatidylserine and phosphatidylethanolamine in mammalian cells: Two metabolically related aminophospholipids. *J. Lipid Res.* **2008**, *49*, 1377–1387. [CrossRef]
268. Guedes, L.C.; Chan, R.B.; Gomes, M.A.; Conceição, V.A.; Machado, R.B.; Soares, T.; Xu, Y.; Gaspar, P.; Carriço, J.A.; Alacaly, R.N.; et al. Serum lipid alterations in GBA-associated Parkinson's disease. *Park. Relat. Disord.* **2017**, *44*, 58–65. [CrossRef]
269. Riekkinen, P.; Rinne, U.K.; Pelliniemi, T.T.; Sonninen, V. Interaction between dopamine and phospholipids. Studies of the substantia nigra in Parkinson disease patients. *Arch. Neurol.* **1975**, *32*, 25–27. [CrossRef]
270. Seyfried, T.N.; Choi, H.; Chevalier, A.; Hogan, D.; Akgoc, Z.; Schneider, J.S. Sex-Related Abnormalities in Substantia Nigra Lipids in Parkinson's Disease. *ASN Neuro* **2018**, *10*. [CrossRef]
271. Ross, B.M.; Mamalias, N.; Moszczynska, A.; Rajput, A.H.; Kish, S.J. Elevated activity of phospholipid biosynthetic enzymes in substantia nigra of patients with Parkinson's disease. *Neuroscience* **2001**, *102*, 899–904. [CrossRef]
272. Jo, E.; McLaurin, J.; Yip, C.M.; St George-Hyslop, P.; Fraser, P.E. alpha-Synuclein membrane interactions and lipid specificity. *J. Biol. Chem.* **2000**, *275*, 34328–34334. [CrossRef] [PubMed]
273. Zakharov, S.D.; Hulleman, J.D.; Dutseva, E.A.; Antonenko, Y.N.; Rochet, J.C.; Cramer, W.A. Helical alpha-synuclein forms highly conductive ion channels. *Biochemistry* **2007**, *46*, 14369–14379. [CrossRef] [PubMed]
274. Wang, S.; Zhang, S.; Liou, L.C.; Ren, Q.; Zhang, Z.; Caldwell, G.A.; Witt, S.N. Phosphatidylethanolamine deficiency disrupts α-synuclein homeostasis in yeast and worm models of Parkinson disease. *Proc. Natl. Acad. Sci. USA* **2014**, *111*, E3976–E3985. [CrossRef] [PubMed]
275. Wang, S.; Zhang, S.; Xu, C.; Barron, A.; Galiano, F.; Patel, D.; Lee, Y.J.; Caldwell, G.A.; Caldwell, K.A.; Witt, S.N. Chemical Compensation of Mitochondrial Phospholipid Depletion in Yeast and Animal Models of Parkinson's Disease. *PLoS ONE* **2016**, *11*, e0164465. [CrossRef]
276. Lee, E.S.; Charlton, C.G. 1-Methyl-4-phenyl-pyridinium increases S-adenosyl-L-methionine dependent phospholipid methylation. *Pharmacol. Biochem. Behav.* **2001**, *70*, 105–114. [CrossRef]
277. Lee, E.S.Y.; Chen, H.; Charlton, C.G.; Soliman, K.F.A. The role of phospholipid methylation in 1-methyl-4-phenyl-pyridinium ion (MPP+)-induced neurotoxicity in PC12 cells. *Neurotoxicology* **2005**, *26*, 945–957. [CrossRef]
278. Leventis, P.A.; Grinstein, S. The distribution and function of phosphatidylserine in cellular membranes. *Annu. Rev. Biophys.* **2010**, *39*, 407–427. [CrossRef]
279. Kim, H.Y.; Huang, B.X.; Spector, A.A. Phosphatidylserine in the brain: Metabolism and function. *Prog. Lipid Res.* **2014**, *56*, 1–18. [CrossRef]
280. Lobasso, S.; Tanzarella, P.; Vergara, D.; Maffia, M.; Cocco, T.; Corcelli, A. Lipid profiling of parkin-mutant human skin fibroblasts. *J. Cell. Physiol.* **2017**, *232*, 3540–3551. [CrossRef]
281. Valadas, J.S.; Esposito, G.; Vandekerkhove, D.; Miskiewicz, K.; Deaulmerie, L.; Raitano, S.; Seibler, P.; Lein, C.; Verstreken, P. ER Lipid Defects in Neuropeptidergic Neurons Impair Sleep Patterns in Parkinson's Disease. *Neuron* **2018**, *98*, 1155–1169.e6. [CrossRef]
282. Wei, L.; Sun, C.; Lei, M.; Li, G.; Yi, L.; Luo, F.; Li, Y.; Ding, L.; Liu, Z.; Li, S.; et al. Activation of Wnt/β-catenin pathway by exogenous Wnt1 protects SH-SY5Y cells against 6-hydroxydopamine toxicity. *J. Mol. Neurosci.* **2013**, *49*, 105–115. [CrossRef]
283. Lupescu, A.; Jilani, K.; Zbidah, M.; Lang, F. Induction of apoptotic erythrocyte death by rotenone. *Toxicology* **2012**, *300*, 132–137. [CrossRef]
284. González-Polo, R.A.; Niso-Santano, M.; Ortíz-Ortíz, M.A.; Gómez-Martín, A.; Morán, J.M.; García-Rubio, L.; Francisco-Mocillo, J.; Zaragoza, C.; Soler, G.; Fuentes, J.M. Inhibition of paraquat-induced autophagy accelerates the apoptotic cell death in neuroblastoma SH-SY5Y cells. *Toxicol. Sci.* **2007**, *97*, 448–458. [CrossRef]
285. Ye, S.; Koon, H.K.; Fan, W.; Xu, Y.; Wei, W.; Xu, C.; Cai, J. Effect of a Traditional Chinese Herbal Medicine Formulation on Cell Survival and Apoptosis of MPP+-Treated MES 23.5 Dopaminergic Cells. *Park. Dis.* **2017**, *2017*, 4764212. [CrossRef]
286. Flower, T.R.; Chesnokova, L.S.; Froelich, C.A.; Dixon, C.; Witt, S.N. Heat shock prevents alpha-synuclein-induced apoptosis in a yeast model of Parkinson's disease. *J. Mol. Biol.* **2005**, *351*, 1081–1100. [CrossRef]
287. Emmrich, J.V.; Hornik, T.C.; Neher, J.J.; Brown, G.C. Rotenone induces neuronal death by microglial phagocytosis of neurons. *FEBS J.* **2013**, *280*, 5030–5038. [CrossRef]

288. Almandoz-Gil, L.; Lindström, V.; Sigvardson, J.; Kahle, P.J.; Lannfelt, L.; Ingelsson, M.; Bergström, J. Mapping of Surface-Exposed Epitopes of In Vitro and In Vivo Aggregated Species of Alpha-Synuclein. *Cell. Mol. Neurobiol.* **2017**, *37*, 1217–1226. [CrossRef]
289. Bartels, T.; Kim, N.C.; Luth, E.S.; Selkoe, D.J. N-alpha-acetylation of α-synuclein increases its helical folding propensity, GM1 binding specificity and resistance to aggregation. *PLoS ONE* **2014**, *9*, e103727. [CrossRef]
290. Araki, K.; Sugawara, K.; Hayakawa, E.H.; Ubukawa, K.; Kobayashi, I.; Wakui, H.; Takahasi, N.; Sawada, K.; Mochizuki, H.; Nunomura, W. The localization of α-synuclein in the process of differentiation of human erythroid cells. *Int. J. Hematol.* **2018**, *108*, 130–138. [CrossRef]
291. Hu, R.; Diao, J.; Li, J.; Tang, Z.; Li, X.; Leitz, J.; Long, J.; Liu, J.; Yu, D.; Zhao, Q. Intrinsic and membrane-facilitated α-synuclein oligomerization revealed by label-free detection through solid-state nanopores. *Sci. Rep.* **2016**, *6*, 20776. [CrossRef]
292. Lou, X.; Kim, J.; Hawk, B.J.; Shin, Y.K. α-Synuclein may cross-bridge v-SNARE and acidic phospholipids to facilitate SNARE-dependent vesicle docking. *Biochem. J.* **2017**, *474*, 2039–2049. [CrossRef]
293. Mejia, E.M.; Hatch, G.M. Mitochondrial phospholipids: Role in mitochondrial function. *J. Bioenerg. Biomembr.* **2016**, *48*, 99–112. [CrossRef]
294. Treede, I.; Braun, A.; Sparla, R.; Kühnel, M.; Giese, T.; Turner, J.R.; Anes, E.; Kulaksiz, H.; Füllekrug, J.; Stremmel, W.; et al. Anti-inflammatory effects of phosphatidylcholine. *J. Biol. Chem.* **2007**, *282*, 27155–27164. [CrossRef]
295. Lagace, T.A. Phosphatidylcholine: Greasing the Cholesterol Transport Machinery. *Lipid Insights* **2015**, *8* (Suppl. 1), 65–73. [CrossRef]
296. Marcucci, H.; Paoletti, L.; Jackowski, S.; Banchio, C. Phosphatidylcholine biosynthesis during neuronal differentiation and its role in cell fate determination. *J. Biol. Chem.* **2010**, *285*, 25382–25393. [CrossRef]
297. Li, T.; Tang, W.; Zhang, L. Monte Carlo cross-validation analysis screens pathway cross-talk associated with Parkinson's disease. *Neurol. Sci.* **2016**, *37*, 1327–1333. [CrossRef]
298. Farmer, K.; Smith, C.A.; Hayley, S.; Smith, J. Major Alterations of Phosphatidylcholine and Lysophosphotidylcholine Lipids in the Substantia Nigra Using an Early Stage Model of Parkinson's Disease. *Int. J. Mol. Sci.* **2015**, *16*, 18865–18877. [CrossRef]
299. Stöckl, M.; Fischer, P.; Wanker, E.; Herrmann, A. Alpha-synuclein selectively binds to anionic phospholipids embedded in liquid-disordered domains. *J. Mol. Biol.* **2008**, *375*, 1394–1404. [CrossRef]
300. Jiang, Z.; de Messieres, M.; Lee, J.C. Membrane remodeling by α-synuclein and effects on amyloid formation. *J. Am. Chem. Soc.* **2013**, *135*, 15970–15973. [CrossRef]
301. Di Pasquale, E.; Fantini, J.; Chahinian, H.; Maresca, M.; Taïeb, N.; Yahi, N. Altered ion channel formation by the Parkinson's-disease-linked E46K mutant of alpha-synuclein is corrected by GM3 but not by GM1 gangliosides. *J. Mol. Biol.* **2010**, *397*, 202–218. [CrossRef]
302. O'Leary, E.I.; Jiang, Z.; Strub, M.P.; Lee, J.C. Effects of phosphatidylcholine membrane fluidity on the conformation and aggregation of N-terminally acetylated α-synuclein. *J. Biol. Chem.* **2018**, *293*, 11195–11205. [CrossRef]
303. Hollie, N.I.; Cash, J.G.; Matlib, M.A.; Wortman, M.; Basford, J.E.; Abplanalp, W.; Hui, D.Y. Micromolar changes in lysophosphatidylcholine concentration cause minor effects on mitochondrial permeability but major alterations in function. *Biochim. Biophys. Acta* **2014**, *1841*, 888–895. [CrossRef]
304. Li, X.; Fang, P.; Li, Y.; Kuo, Y.M.; Andrews, A.J.; Nanayakkara, G.; Johnson, C.; Fu, H.; Shan, H.; Du, F.; et al. Mitochondrial Reactive Oxygen Species Mediate Lysophosphatidylcholine-Induced Endothelial Cell Activation. *Arterioscler. Thromb. Vasc. Biol.* **2016**, *36*, 1090–1100. [CrossRef]
305. Hung, N.D.; Sok, D.E.; Kim, M.R. Prevention of 1-palmitoyl lysophosphatidylcholine-induced inflammation by polyunsaturated acyl lysophosphatidylcholine. *Inflamm. Res.* **2012**, *61*, 473–483. [CrossRef]
306. Lee, E.S.Y.; Chen, H.; Shepherd, K.R.; Lamango, N.S.; Soliman, K.F.A.; Charlton, C.G. Inhibitory effects of lysophosphatidylcholine on the dopaminergic system. *Neurochem. Res.* **2004**, *29*, 1333–1342. [CrossRef]
307. Lee, E.S.Y.; Soliman, K.F.A.; Charlton, C.G. Lysophosphatidylcholine decreases locomotor activities and dopamine turnover rate in rats. *Neurotoxicology* **2005**, *26*, 27–38. [CrossRef]
308. Pacheco, M.A.; Jope, R.S. Phosphoinositide signaling in human brain. *Prog. Neurobiol.* **1996**, *50*, 255–273. [CrossRef]

309. Chalimoniuk, M.; Snoek, G.T.; Adamczyk, A.; Małecki, A.; Strosznajder, J.B. Phosphatidylinositol transfer protein expression altered by aging and Parkinson disease. *Cell. Mol. Neurobiol.* **2006**, *26*, 1153–1166. [CrossRef]
310. Cockcroft, S. The diverse functions of phosphatidylinositol transfer proteins. *Curr. Top. Microbiol. Immunol.* **2012**, *362*, 185–208. [CrossRef]
311. Lee, E.N.; Lee, S.Y.; Lee, D.; Kim, J.; Paik, S.R. Lipid interaction of alpha-synuclein during the metal-catalyzed oxidation in the presence of Cu^{2+} and H_2O_2. *J. Neurochem.* **2003**, *84*, 1128–1142. [CrossRef]
312. Narayanan, V.; Guo, Y.; Scarlata, S. Fluorescence studies suggest a role for alpha-synuclein in the phosphatidylinositol lipid signaling pathway. *Biochemistry* **2005**, *44*, 462–470. [CrossRef]
313. Sekar, S.; Taghibiglou, C. Elevated nuclear phosphatase and tensin homolog (PTEN) and altered insulin signaling in substantia nigral region of patients with Parkinson's disease. *Neurosci. Lett.* **2018**, *666*, 139–143. [CrossRef]
314. Demirsoy, S.; Martin, S.; Motamedi, S.; van Veen, S.; Holemans, T.; Van den Haute, C.; Jordanova, A.; Baekelandt, V.; Vangheluwe, P.; Agostinis, P. ATP13A2/PARK9 regulates endo-/lysosomal cargo sorting and proteostasis through a novel PI(3, 5)P2-mediated scaffolding function. *Hum. Mol. Genet.* **2017**, *26*, 1656–1669. [CrossRef]
315. Horvath, S.E.; Daum, G. Lipids of mitochondria. *Prog. Lipid Res.* **2013**, *52*, 590–614. [CrossRef]
316. Ramakrishnan, M.; Jensen, P.H.; Marsh, D. Association of alpha-synuclein and mutants with lipid membranes: Spin-label ESR and polarized IR. *Biochemistry* **2006**, *45*, 3386–3395. [CrossRef]
317. Jiang, Z.; Flynn, J.D.; Teague, W.E.; Gawrisch, K.; Lee, J.C. Stimulation of α-synuclein amyloid formation by phosphatidylglycerol micellar tubules. *Biochim. Biophys. Acta* **2018**. [CrossRef]
318. Bédard, L.; Lefèvre, T.; Morin-Michaud, É.; Auger, M. Besides fibrillization: Putative role of the peptide fragment 71-82 on the structural and assembly behavior of α-synuclein. *Biochemistry* **2014**, *53*, 6463–6472. [CrossRef]
319. Pandey, A.P.; Haque, F.; Rochet, J.C.; Hovis, J.S. Clustering of alpha-synuclein on supported lipid bilayers: Role of anionic lipid, protein, and divalent ion concentration. *Biophys. J.* **2009**, *96*, 540–551. [CrossRef]
320. Stefanovic, A.N.D.; Claessens, M.M.A.E.; Blum, C.; Subramaniam, V. Alpha-synuclein amyloid oligomers act as multivalent nanoparticles to cause hemifusion in negatively charged vesicles. *Small* **2015**, *11*, 2257–2262. [CrossRef]
321. Van Rooijen, B.D.; Claessens, M.M.A.E.; Subramaniam, V. Lipid bilayer disruption by oligomeric alpha-synuclein depends on bilayer charge and accessibility of the hydrophobic core. *Biochim. Biophys. Acta* **2009**, *1788*, 1271–1278. [CrossRef] [PubMed]
322. Van Rooijen, B.D.; Claessens, M.M.A.E.; Subramaniam, V. Membrane Permeabilization by Oligomeric α-Synuclein: In Search of the Mechanism. *PLoS ONE* **2010**, *5*, e14292. [CrossRef] [PubMed]
323. Zhu, M.; Qin, Z.J.; Hu, D.; Munishkina, L.A.; Fink, A.L. Alpha-synuclein can function as an antioxidant preventing oxidation of unsaturated lipid in vesicles. *Biochemistry* **2006**, *45*, 8135–8142. [CrossRef] [PubMed]
324. Ren, M.; Phoon, C.K.L.; Schlame, M. Metabolism and function of mitochondrial cardiolipin. *Prog. Lipid Res.* **2014**, *55*, 1–16. [CrossRef] [PubMed]
325. Paradies, G.; Paradies, V.; Ruggiero, F.M.; Petrosillo, G. Cardiolipin and mitochondrial function in health and disease. *Antioxid. Redox. Signal.* **2014**, *20*, 1925–1953. [CrossRef] [PubMed]
326. Paradies, G.; Paradies, V.; De Benedictis, V.; Ruggiero, F.M.; Petrosillo, G. Functional role of cardiolipin in mitochondrial bioenergetics. *Biochim. Biophys. Acta* **2014**, *1837*, 408–417. [CrossRef]
327. Vos, M.; Geens, A.; Böhm, C.; Deaulmerie, L.; Swerts, J.; Rossi, M.; Craessaerts, K.; Leites, E.P.; Seibler, P.; Rakovic, A.; et al. Cardiolipin promotes electron transport between ubiquinone and complex I to rescue PINK1 deficiency. *J. Cell Biol.* **2017**, *216*, 695–708. [CrossRef]
328. Tyurina, Y.Y.; Winnica, D.E.; Kapralova, V.I.; Kapralov, A.A.; Tyurin, V.A.; Kagan, V.E. LC/MS characterization of rotenone induced cardiolipin oxidation in human lymphocytes: Implications for mitochondrial dysfunction associated with Parkinson's disease. *Mol. Nutr. Food Res.* **2013**, *57*, 1410–1422. [CrossRef]
329. Tyurina, Y.Y.; Polimova, A.M.; Maciel, E.; Tyurin, V.A.; Kapralova, V.I.; Winnica, D.E.; Vikulina, A.S.; Domigues, M.R.; McCoy, J.; Samders, L.H.; et al. LC/MS analysis of cardiolipins in substantia nigra and plasma of rotenone-treated rats: Implication for mitochondrial dysfunction in Parkinson's disease. *Free Radic. Res.* **2015**, *49*, 681–691. [CrossRef]

330. Zigoneanu, I.G.; Yang, Y.J.; Krois, A.S.; Haque, E.; Pielak, G.J. Interaction of α-synuclein with vesicles that mimic mitochondrial membranes. *Biochim. Biophys. Acta* **2012**, *1818*, 512–519. [CrossRef]
331. Robotta, M.; Gerding, H.R.; Vogel, A.; Hauser, K.; Schildknecht, S.; Karreman, C.; Leist, M.; Subramaniam, V.; Drescher, M. Alpha-synuclein binds to the inner membrane of mitochondria in an α-helical conformation. *Chembiochem* **2014**, *15*, 2499–2502. [CrossRef] [PubMed]
332. Ryan, T.; Bamm, V.V.; Stykel, M.G.; Coackley, C.L.; Humphries, K.M.; Jamieson-Williams, R.; Ambasudhan, R.; Mosser, D.D.; Lipton, S.A.; Harauz, G.; et al. Cardiolipin exposure on the outer mitochondrial membrane modulates α-synuclein. *Nat. Commun.* **2018**, *9*, 817. [CrossRef] [PubMed]
333. Nakamura, K.; Nemani, V.M.; Azarbal, F.; Skibinski, G.; Levy, J.M.; Egami, K.; Munushkina, L.; Zhang, J.; Gardner, B.; Wakabayashi, J.; et al. Direct membrane association drives mitochondrial fission by the Parkinson disease-associated protein alpha-synuclein. *J. Biol. Chem.* **2011**, *286*, 20710–20726. [CrossRef] [PubMed]
334. Shen, J.; Du, T.; Wang, X.; Duan, C.; Gao, G.; Zhang, J.; Lu, L.; Yang, H. α-Synuclein amino terminus regulates mitochondrial membrane permeability. *Brain Res.* **2014**, *1591*, 14–26. [CrossRef] [PubMed]
335. Gao, G.; Wang, Z.; Lu, L.; Duan, C.; Wang, X.; Yang, H. Morphological analysis of mitochondria for evaluating the toxicity of α-synuclein in transgenic mice and isolated preparations by atomic force microscopy. *Biomed. Pharmacother.* **2017**, *96*, 1380–1388. [CrossRef]
336. Bayir, H.; Kapralov, A.A.; Jiang, J.; Huang, Z.; Tyurina, Y.Y.; Tyurin, V.A.; Zhao, Q.; Belikova, N.A.; Vlasova, I.I.; Maeda, A.; et al. Peroxidase mechanism of lipid-dependent cross-linking of synuclein with cytochrome C: Protection against apoptosis versus delayed oxidative stress in Parkinson disease. *J. Biol. Chem.* **2009**, *284*, 15951–15969. [CrossRef]
337. Ghio, S.; Kamp, F.; Cauchi, R.; Giese, A.; Vassallo, N. Interaction of α-synuclein with biomembranes in Parkinson's disease—Role of cardiolipin. *Prog. Lipid Res.* **2016**, *61*, 73–82. [CrossRef]
338. Chu, C.T.; Bayır, H.; Kagan, V.E. LC3 binds externalized cardiolipin on injured mitochondria to signal mitophagy in neurons: Implications for Parkinson disease. *Autophagy* **2014**, *10*, 376–378. [CrossRef]
339. Bartke, N.; Hannun, Y.A. Bioactive sphingolipids: Metabolism and function. *J. Lipid Res.* **2009**, *50*, S91–S96. [CrossRef]
340. Xu, Y.H.; Barnes, S.; Sun, Y.; Grabowski, G.A. Multi-system disorders of glycosphingolipid and ganglioside metabolism. *J. Lipid Res.* **2010**, *51*, 1643–1675. [CrossRef]
341. Proia, R.L.; Hla, T. Emerging biology of sphingosine-1-phosphate: Its role in pathogenesis and therapy. *J. Clin. Investig.* **2015**, *125*, 1379–1387. [CrossRef]
342. Martin, R.; Sospedra, M. Sphingosine-1 phosphate and central nervous system. *Curr. Top. Microbiol. Immunol.* **2014**, *378*, 149–170. [CrossRef]
343. Taguchi, Y.V.; Liu, J.; Ruan, J.; Pacheco, J.; Zhang, X.; Abbasi, J.; Keutzer, J.; Mistry, P.K.; Chandra, S.S. Glucosylsphingosine Promotes α-Synuclein Pathology in Mutant GBA-Associated Parkinson's Disease. *J. Neurosci.* **2017**, *37*, 9617–9631. [CrossRef]
344. Motyl, J.; Wencel, P.L.; Cieślik, M.; Strosznajder, R.P.; Strosznajder, J.B. Alpha-synuclein alters differently gene expression of Sirts, PARPs and other stress response proteins: Implications for neurodegenerative disorders. *Mol. Neurobiol.* **2018**, *55*, 727–740. [CrossRef]
345. Zhang, L.; Okada, T.; Badawy, S.M.M.; Hirai, C.; Kajimoto, T.; Nakamura, S.I. Extracellular α-synuclein induces sphingosine 1-phosphate receptor subtype 1 uncoupled from inhibitory G-protein leaving β-arrestin signal intact. *Sci. Rep.* **2017**, *7*, 44248. [CrossRef] [PubMed]
346. Badawy, S.M.M.; Okada, T.; Kajimoto, T.; Hirase, M.; Matovelo, S.A.; Nakamura, S.; Yoshida, D.; Ijuin, T.; Nakamura, S.I. Extracellular α-synuclein drives sphingosine 1-phosphate receptor subtype 1 out of lipid rafts, leading to impaired inhibitory G-protein signaling. *J. Biol. Chem.* **2018**, *293*, 8208–8216. [CrossRef]
347. Sivasubramanian, M.; Kanagaraj, N.; Dheen, S.T.; Tay, S.S.W. Sphingosine kinase 2 and sphingosine-1-phosphate promotes mitochondrial function in dopaminergic neurons of mouse model of Parkinson's disease and in MPP+-treated MN9D cells in vitro. *Neuroscience* **2015**, *290*, 636–648. [CrossRef]
348. Pyszko, J.A.; Strosznajder, J.B. The key role of sphingosine kinases in the molecular mechanism of neuronal cell survival and death in an experimental model of Parkinson's disease. *Folia Neuropathol.* **2014**, *52*, 260–269. [CrossRef] [PubMed]

349. Pyszko, J.; Strosznajder, J.B. Sphingosine kinase 1 and sphingosine-1-phosphate in oxidative stress evoked by 1-methyl-4-phenylpyridinium (MPP+) in human dopaminergic neuronal cells. *Mol. Neurobiol.* **2014**, *50*, 38–48. [CrossRef] [PubMed]
350. Zhao, P.; Yang, X.; Yang, L.; Li, M.; Wood, K.; Liu, Q.; Zhu, X. Neuroprotective effects of fingolimod in mouse models of Parkinson's disease. *FASEB J.* **2017**, *31*, 172–179. [CrossRef]
351. Mencarelli, C.; Martinez-Martinez, P. Ceramide function in the brain: When a slight tilt is enough. *Cell. Mol. Life Sci.* **2013**, *70*, 181–203. [CrossRef]
352. Castro, B.M.; Prieto, M.; Silva, L.C. Ceramide: A simple sphingolipid with unique biophysical properties. *Prog. Lipid Res.* **2014**, *54*, 53–67. [CrossRef] [PubMed]
353. Kogot-Levin, A.; Saada, A. Ceramide and the mitochondrial respiratory chain. *Biochimie* **2014**, *100*, 88–94. [CrossRef] [PubMed]
354. Mielke, M.M.; Maetzler, W.; Haughey, N.J.; Bandaru, V.V.R.; Savica, R.; Deuschle, C.; Gasser, T.; Hauser, A.K.; Gräber-Sultan, S.; Schleicher, E.; et al. Plasma ceramide and glucosylceramide metabolism is altered in sporadic Parkinson's disease and associated with cognitive impairment: A pilot study. *PLoS ONE* **2013**, *8*, e73094. [CrossRef] [PubMed]
355. Atashrazm, F.; Hammond, D.; Perera, G.; Dobson-Stone, C.; Mueller, N.; Pickford, R.; Kim, W.S.; Kwok, J.B.; Lewis, S.J.G.; Halliday, G.M.; et al. Reduced glucocerebrosidase activity in monocytes from patients with Parkinson's disease. *Sci. Rep.* **2018**, *8*, 15446. [CrossRef]
356. Murphy, K.E.; Gysbers, A.M.; Abbott, S.K.; Tayebi, N.; Kim, W.S.; Sidransky, E.; Cooper, A.; Garner, B.; Halliday, G.M. Reduced glucocerebrosidase is associated with increased α-synuclein in sporadic Parkinson's disease. *Brain* **2014**, *137 Pt 3*, 834–848. [CrossRef]
357. Kim, M.J.; Jeon, S.; Burbulla, L.F.; Krainc, D. Acid ceramidase inhibition ameliorates α-synuclein accumulation upon loss of GBA1 function. *Hum. Mol. Genet.* **2018**, *27*, 1972–1988. [CrossRef]
358. Abbott, S.K.; Li, H.; Muñoz, S.S.; Knoch, B.; Batterham, M.; Murphy, K.E.; Halliday, G.M.; Garner, B. Altered ceramide acyl chain length and ceramide synthase gene expression in Parkinson's disease. *Mov. Disord.* **2014**, *29*, 518–526. [CrossRef]
359. Halliday, G.M.; McCann, H. The progression of pathology in Parkinson's disease. *Ann. N. Y. Acad. Sci.* **2010**, *1184*, 188–195. [CrossRef]
360. Lin, G.; Lee, P.T.; Chen, K.; Mao, D.; Tan, K.L.; Zuo, Z.; Lin, W.W.; Wang, L.; Bellen, H.J. Phospholipase PLA2G6, a Parkinsonism-Associated Gene, Affects Vps26 and Vps35, Retromer Function, and Ceramide Levels, Similar to α-Synuclein Gain. *Cell Metab.* **2018**. [CrossRef]
361. Ferrazza, R.; Cogo, S.; Melrose, H.; Bubacco, L.; Greggio, E.; Guella, G.; Civiero, L.; Plotegher, N. LRRK2 deficiency impacts ceramide metabolism in brain. *Biochem. Biophys. Res. Commun.* **2016**, *478*, 1141–1146. [CrossRef] [PubMed]
362. Torres-Odio, S.; Key, J.; Hoepken, H.H.; Canet-Pons, J.; Valek, L.; Roller, B.; Walter, M.; Morales-Gordo, B.; Meierhofer, D.; Harter, P.N.; et al. Progression of pathology in PINK1-deficient mouse brain from splicing via ubiquitination, ER stress, and mitophagy changes to neuroinflammation. *J. Neuroinflamm.* **2017**, *14*, 154. [CrossRef]
363. Arboleda, G.; Cárdenas, Y.; Rodríguez, Y.; Morales, L.C.; Matheus, L.; Arboleda, H. Differential regulation of AKT, MAPK and GSK3β during C2-ceramide-induced neuronal death. *Neurotoxicology* **2010**, *31*, 687–693. [CrossRef] [PubMed]
364. Martinez, T.N.; Chen, X.; Bandyopadhyay, S.; Merrill, A.H.; Tansey, M.G. Ceramide sphingolipid signaling mediates Tumor Necrosis Factor (TNF)-dependent toxicity via caspase signaling in dopaminergic neurons. *Mol. Neurodegener.* **2012**, *7*, 45. [CrossRef] [PubMed]
365. France-Lanord, V.; Brugg, B.; Michel, P.P.; Agid, Y.; Ruberg, M. Mitochondrial free radical signal in ceramide-dependent apoptosis: A putative mechanism for neuronal death in Parkinson's disease. *J. Neurochem.* **1997**, *69*, 1612–1621. [CrossRef] [PubMed]
366. Arboleda, G.; Waters, C.; Gibson, R. Inhibition of caspases but not of calpains temporarily protect against C2-ceramide-induced death of CAD cells. *Neurosci. Lett.* **2007**, *421*, 245–249. [CrossRef] [PubMed]
367. Da Costa, C.A.; Ancolio, K.; Checler, F. Wild-type but not Parkinson's disease-related ala-53→Thr mutant alpha -synuclein protects neuronal cells from apoptotic stimuli. *J. Biol. Chem.* **2000**, *275*, 24065–24069. [CrossRef] [PubMed]

368. Sánchez-Mora, R.M.; Arboleda, H.; Arboleda, G. PINK1 overexpression protects against C2-ceramide-induced CAD cell death through the PI3K/AKT pathway. *J. Mol. Neurosci.* **2012**, *47*, 582–594. [CrossRef]
369. Rojas-Charry, L.; Cookson, M.R.; Niño, A.; Arboleda, H.; Arboleda, G. Downregulation of Pink1 influences mitochondrial fusion-fission machinery and sensitizes to neurotoxins in dopaminergic cells. *Neurotoxicology* **2014**, *44*, 140–148. [CrossRef]
370. Jaramillo-Gómez, J.; Niño, A.; Arboleda, H.; Arboleda, G. Overexpression of DJ-1 protects against C2-ceramide-induced neuronal death through activation of the PI3K/AKT pathway and inhibition of autophagy. *Neurosci. Lett.* **2015**, *603*, 71–76. [CrossRef]
371. Jung, J.S.; Shin, K.O.; Lee, Y.M.; Shin, J.A.; Park, E.M.; Jeong, J.; Kim, D.H.; Choi, J.W.; Kim, H.S. Anti-inflammatory mechanism of exogenous C2 ceramide in lipopolysaccharide-stimulated microglia. *Biochim. Biophys. Acta* **2013**, *1831*, 1016–1026. [CrossRef] [PubMed]
372. Yang, W.; Wang, X.; Duan, C.; Lu, L.; Yang, H. Alpha-synuclein overexpression increases phospho-protein phosphatase 2A levels via formation of calmodulin/Src complex. *Neurochem. Int.* **2013**, *63*, 180–194. [CrossRef]
373. Wang, Y.; Liu, J.; Chen, M.; Du, T.; Duan, C.; Gao, G.; Yang, H. The novel mechanism of rotenone-induced α-synuclein phosphorylation via reduced protein phosphatase 2A activity. *Int. J. Biochem. Cell Biol.* **2016**, *75*, 34–44. [CrossRef] [PubMed]
374. Nixon, G.F. Sphingolipids in inflammation: Pathological implications and potential therapeutic targets. *Br. J. Pharmacol.* **2009**, *158*, 982–993. [CrossRef]
375. Norris, G.H.; Blesso, C.N. Dietary and Endogenous Sphingolipid Metabolism in Chronic Inflammation. *Nutrients* **2017**, *9*, 1180. [CrossRef]
376. Kiraz, Y.; Adan, A.; Kartal Yandim, M.; Baran, Y. Major apoptotic mechanisms and genes involved in apoptosis. *Tumour Biol.* **2016**, *37*, 8471–8486. [CrossRef] [PubMed]
377. Jazvinšćak Jembrek, M.; Hof, P.R.; Šimić, G. Ceramides in Alzheimer's Disease: Key Mediators of Neuronal Apoptosis Induced by Oxidative Stress and Aβ Accumulation. *Oxid. Med. Cell. Longev.* **2015**, *2015*, 346783. [CrossRef]
378. Tommasino, C.; Marconi, M.; Ciarlo, L.; Matarrese, P.; Malorni, W. Autophagic flux and autophagosome morphogenesis require the participation of sphingolipids. *Apoptosis* **2015**, *20*, 645–657. [CrossRef]
379. Foo, J.N.; Liany, H.; Bei, J.X.; Yu, X.-Q.; Liu, J.; Au, W.L.; Prakash, K.M.; Tan, L.C.; Tan, E.K. Rare lysosomal enzyme gene SMPD1 variant (p.R591C) associates with Parkinson's disease. *Neurobiol. Aging* **2013**, *34*, 2890.e13-5. [CrossRef]
380. Mao, C.Y.; Yang, J.; Wang, H.; Zhang, S.Y.; Yang, Z.H.; Luo, H.Y.; Li, F.; Shi, M.; Liu, Y.T.; Zhuang, Z.P.; et al. SMPD1 variants in Chinese Han patients with sporadic Parkinson's disease. *Park. Relat. Disord.* **2017**, *34*, 59–61. [CrossRef]
381. Kim, W.S.; Halliday, G.M. Changes in sphingomyelin level affect alpha-synuclein and ABCA5 expression. *J. Park. Dis.* **2012**, *2*, 41–46. [CrossRef]
382. Den Jager, W.A. Sphingomyelin in Lewy inclusion bodies in Parkinson's disease. *Arch. Neurol.* **1969**, *21*, 615–619. [CrossRef] [PubMed]
383. Gegg, M.E.; Sweet, L.; Wang, B.H.; Shihabuddin, L.S.; Sardi, S.P.; Schapira, A.H.V. No evidence for substrate accumulation in Parkinson brains with GBA mutations. *Mov. Disord.* **2015**, *30*, 1085–1089. [CrossRef] [PubMed]
384. Merrill, A.H. Sphingolipid and glycosphingolipid metabolic pathways in the era of sphingolipidomics. *Chem. Rev.* **2011**, *111*, 6387–6422. [CrossRef] [PubMed]
385. Kurup, R.K.; Kurup, P.A. Hypothalamic digoxin-mediated model for Parkinson's disease. *Int. J. Neurosci.* **2003**, *113*, 515–536. [CrossRef] [PubMed]
386. Boutin, M.; Sun, Y.; Shacka, J.J.; Auray-Blais, C. Tandem Mass Spectrometry Multiplex Analysis of Glucosylceramide and Galactosylceramide Isoforms in Brain Tissues at Different Stages of Parkinson Disease. *Anal. Chem.* **2016**, *88*, 1856–1863. [CrossRef] [PubMed]
387. Gegg, M.E.; Schapira, A.H.V. The role of glucocerebrosidase in Parkinson disease pathogenesis. *FEBS J.* **2018**. [CrossRef]

388. Marshall, M.S.; Bongarzone, E.R. Beyond Krabbe's disease: The potential contribution of galactosylceramidase deficiency to neuronal vulnerability in late-onset synucleinopathies. *J. Neurosci. Res.* **2016**, *94*, 1328–1332. [CrossRef]
389. Kim, S.; Yun, S.P.; Lee, S.; Umanah, G.E.; Bandaru, V.V.R.; Yin, X.; Rhee, P.; Karuppagounder, S.S.; Kwon, S.H.; Lee, H.; et al. GBA1 deficiency negatively affects physiological α-synuclein tetramers and related multimers. *Proc. Natl. Acad. Sci. USA* **2018**, *115*, 798–803. [CrossRef]
390. Zunke, F.; Moise, A.C.; Belur, N.R.; Gelyana, E.; Stojkovska, I.; Dzaferbegovic, H.; Toker, N.J.; Jeon, S.; Fredriksen, K.; Mazzulli, J.R. Reversible Conformational Conversion of α-Synuclein into Toxic Assemblies by Glucosylceramide. *Neuron* **2017**, *97*, 92–107.e10. [CrossRef]
391. Xu, Y.H.; Sun, Y.; Ran, H.; Quinn, B.; Witte, D.; Grabowski, G.A. Accumulation and distribution of α-synuclein and ubiquitin in the CNS of Gaucher disease mouse models. *Mol. Genet. Metab.* **2011**, *102*, 436–447. [CrossRef] [PubMed]
392. Suzuki, M.; Fujikake, N.; Takeuchi, T.; Kohyama-Koganeya, A.; Nakajima, K.; Hirabayashi, Y.; Wada, K.; Nagai, Y. Glucocerebrosidase deficiency accelerates the accumulation of proteinase K-resistant α-synuclein and aggravates neurodegeneration in a Drosophila model of Parkinson's disease. *Hum. Mol. Genet.* **2015**, *24*, 6675–6686. [CrossRef]
393. Mazzulli, J.R.; Xu, Y.H.; Sun, Y.; Knight, A.L.; McLean, P.J.; Caldwell, G.A.; Sidransky, E.; Grabowski, G.A.; Krainc, D. Gaucher disease glucocerebrosidase and α-synuclein form a bidirectional pathogenic loop in synucleinopathies. *Cell* **2011**, *146*, 37–52. [CrossRef] [PubMed]
394. Sardi, S.P.; Viel, C.; Clarke, J.; Treleaven, C.M.; Richards, A.M.; Park, H.; Olszewski, M.A.; Dodge, J.C.; Marshall, J.; Makino, E.; et al. Glucosylceramide synthase inhibition alleviates aberrations in synucleinopathy models. *Proc. Natl. Acad. Sci. USA* **2017**, *114*, 2699–2704. [CrossRef] [PubMed]
395. Noelker, C.; Lu, L.; Höllerhage, M.; Vulinovic, F.; Sturn, A.; Roscher, R.; Höglinger, G.U.; Hirsch, E.C.; Oertel, W.H.; Alvarez-Fisher, D.; et al. Glucocerebrosidase deficiency and mitochondrial impairment in experimental Parkinson disease. *J. Neurol. Sci.* **2015**, *356*, 129–136. [CrossRef] [PubMed]
396. Hallett, P.J.; Huebecker, M.; Brekk, O.R.; Moloney, E.B.; Rocha, E.M.; Priestman, D.A.; Platt, F.M.; Isacson, O. Glycosphingolipid levels and glucocerebrosidase activity are altered in normal aging of the mouse brain. *Neurobiol. Aging* **2018**, *67*, 189–200. [CrossRef] [PubMed]
397. Yu, R.K.; Tsai, Y.T.; Ariga, T.; Yanagisawa, M. Structures, biosynthesis, and functions of gangliosides—An overview. *J. Oleo Sci.* **2011**, *60*, 537–544. [CrossRef]
398. Palmano, K.; Rowan, A.; Guillermo, R.; Guan, J.; McJarrow, P. The role of gangliosides in neurodevelopment. *Nutrients* **2015**, *7*, 3891–3913. [CrossRef]
399. Schnaar, R.L. Gangliosides of the Vertebrate Nervous System. *J. Mol. Biol.* **2016**, *428*, 3325–3336. [CrossRef]
400. Wu, G.; Lu, Z.H.; Kulkarni, N.; Ledeen, R.W. Deficiency of ganglioside GM1 correlates with Parkinson's disease in mice and humans. *J. Neurosci. Res.* **2012**, *90*, 1997–2008. [CrossRef]
401. Schneider, J.S. Altered expression of genes involved in ganglioside biosynthesis in substantia nigra neurons in Parkinson's disease. *PLoS ONE* **2018**, *13*, e0199189. [CrossRef] [PubMed]
402. Schneider, J.S.; Cambi, F.; Gollomp, S.M.; Kuwabara, H.; Brašić, J.R.; Leiby, B.; Sendek, S.; Wong, D.F. GM1 ganglioside in Parkinson's disease: Pilot study of effects on dopamine transporter binding. *J. Neurol. Sci.* **2015**, *356*, 118–123. [CrossRef] [PubMed]
403. Schneider, J.S.; Gollomp, S.M.; Sendek, S.; Colcher, A.; Cambi, F.; Du, W. A randomized, controlled, delayed start trial of GM1 ganglioside in treated Parkinson's disease patients. *J. Neurol. Sci.* **2013**, *324*, 140–148. [CrossRef] [PubMed]
404. Schneider, J.S.; Sendek, S.; Daskalakis, C.; Cambi, F. GM1 ganglioside in Parkinson's disease: Results of a five year open study. *J. Neurol. Sci.* **2010**, *292*, 45–51. [CrossRef] [PubMed]
405. Schneider, J.S. GM1 ganglioside in the treatment of Parkinson's disease. *Ann. N. Y. Acad. Sci.* **1998**, *845*, 363–373. [CrossRef] [PubMed]
406. Schneider, J.S.; Roeltgen, D.P.; Mancall, E.L.; Chapas-Crilly, J.; Rothblat, D.S.; Tatarian, G.T. Parkinson's disease: Improved function with GM1 ganglioside treatment in a randomized placebo-controlled study. *Neurology* **1998**, *50*, 1630–1636. [CrossRef] [PubMed]
407. Schneider, J.S.; Roeltgen, D.P.; Rothblat, D.S.; Chapas-Crilly, J.; Seraydarian, L.; Rao, J. GM1 ganglioside treatment of Parkinson's disease: An open pilot study of safety and efficacy. *Neurology* **1995**, *45*, 1149–1154. [CrossRef]

408. Schneider, J.S.; Seyfried, T.N.; Choi, H.S.; Kidd, S.K. Intraventricular Sialidase Administration Enhances GM1 Ganglioside Expression and Is Partially Neuroprotective in a Mouse Model of Parkinson's Disease. *PLoS ONE* **2015**, *10*, e0143351. [CrossRef]

409. Xu, R.; Zhou, Y.; Fang, X.; Lu, Y.; Li, J.; Zhang, J.; Deng, X.; Li, S. The possible mechanism of Parkinson's disease progressive damage and the preventive effect of GM1 in the rat model induced by 6-hydroxydopamine. *Brain Res.* **2014**, *1592*, 73–81. [CrossRef]

410. Goettl, V.M.; Wemlinger, T.A.; Duchemin, A.M.; Neff, N.H.; Hadjiconstantinou, M. GM1 ganglioside restores dopaminergic neurochemical and morphological markers in aged rats. *Neuroscience* **1999**, *92*, 991–1000. [CrossRef]

411. Emborg, M.E.; Colombo, J.A. Long-term MPTP-treated monkeys are resistant to GM1 systemic therapy. *Mol. Chem. Neuropathol.* **1994**, *21*, 75–82. [CrossRef] [PubMed]

412. Pope-Coleman, A.; Schneider, J.S. Effects of Chronic GM1 Ganglioside Treatment on Cognitieve and Motor Deficits in a Slowly Progressing Model of Parkinsonism in Non-Human Primates. *Restor. Neurol. Neurosci.* **1998**, *12*, 255–266. [PubMed]

413. Herrero, M.T.; Perez-Otaño, I.; Oset, C.; Kastner, A.; Hirsch, E.C.; Agid, Y.; Luguin, M.R.; Obeso, J.A.; Del Río, J. GM-1 ganglioside promotes the recovery of surviving midbrain dopaminergic neurons in MPTP-treated monkeys. *Neuroscience* **1993**, *56*, 965–972. [CrossRef]

414. Rothblat, D.S.; Schneider, J.S. Effects of GM1 ganglioside treatment on dopamine innervation of the striatum of MPTP-treated mice. *Ann. N. Y. Acad. Sci.* **1998**, *845*, 274–277. [CrossRef] [PubMed]

415. Schneider, J.S.; Kean, A.; DiStefano, L. GM1 ganglioside rescues substantia nigra pars compacta neurons and increases dopamine synthesis in residual nigrostriatal dopaminergic neurons in MPTP-treated mice. *J. Neurosci. Res.* **1995**, *42*, 117–123. [CrossRef] [PubMed]

416. Kastner, A.; Herrero, M.T.; Hirsch, E.C.; Guillen, J.; Luquin, M.R.; Javoy-Agid, F.; Obeso, J.A.; Agid, Y. Decreased tyrosine hydroxylase content in the dopaminergic neurons of MPTP-intoxicated monkeys: Effect of levodopa and GM1 ganglioside therapy. *Ann. Neurol.* **1994**, *36*, 206–214. [CrossRef] [PubMed]

417. Herrero, M.T.; Kastner, A.; Perez-Otaño, I.; Hirsch, E.C.; Luquin, M.R.; Javoy-Agid, F.; Del Río, J.; Obeso, J.A.; Agid, Y. Gangliosides and parkinsonism. *Neurology* **1993**, *43*, 2132–2134. [CrossRef] [PubMed]

418. Schneider, J.S.; Pope, A.; Simpson, K.; Taggart, J.; Smith, M.G.; DiStefano, L. Recovery from experimental parkinsonism in primates with GM1 ganglioside treatment. *Science* **1992**, *256*, 843–846. [CrossRef] [PubMed]

419. Schneider, J.S. MPTP-induced parkinsonism: Acceleration of biochemical and behavioral recovery by GM1 ganglioside treatment. *J. Neurosci. Res.* **1992**, *31*, 112–119. [CrossRef]

420. Fazzini, E.; Durso, R.; Davoudi, H.; Szabo, G.K.; Albert, M.L. GM1 gangliosides alter acute MPTP-induced behavioral and neurochemical toxicity in mice. *J. Neurol. Sci.* **1990**, *99*, 59–68. [CrossRef]

421. Gupta, M.; Schwarz, J.; Chen, X.L.; Roisen, F.J. Gangliosides prevent MPTP toxicity in mice—An immunocytochemical study. *Brain Res.* **1990**, *527*, 330–334. [CrossRef]

422. Schneider, J.S.; Yuwiler, A. GM1 ganglioside treatment promotes recovery of striatal dopamine concentrations in the mouse model of MPTP-induced parkinsonism. *Exp. Neurol.* **1989**, *105*, 177–183. [CrossRef]

423. Hadjiconstantinou, M.; Mariani, A.P.; Neff, N.H. GM1 ganglioside-induced recovery of nigrostriatal dopaminergic neurons after MPTP: An immunohistochemical study. *Brain Res.* **1989**, *484*, 297–303. [CrossRef]

424. Ba, X. Therapeutic effects of GM1 on Parkinson's disease in rats and its mechanism. *Int. J. Neurosci.* **2016**, *126*, 163–167. [CrossRef] [PubMed]

425. Lundius, E.G.; Vukojevic, V.; Hertz, E.; Stroth, N.; Cederlund, A.; Hiraiwa, M.; Terenius, L.; Svennigsson, P. GPR37 protein trafficking to the plasma membrane regulated by prosaposin and GM1 gangliosides promotes cell viability. *J. Biol. Chem.* **2014**, *289*, 4660–4673. [CrossRef] [PubMed]

426. Park, J.Y.; Kim, K.S.; Lee, S.B.; Ryu, J.S.; Chung, K.C.; Choo, Y.K.; Jou, I.; Kim, J.; Park, S.M. On the mechanism of internalization of alpha-synuclein into microglia: Roles of ganglioside GM1 and lipid raft. *J. Neurochem.* **2009**, *110*, 400–411. [CrossRef]

427. Grey, M.; Dunning, C.J.; Gaspar, R.; Grey, C.; Brundin, P.; Sparr, E.; Linse, S. Acceleration of α-synuclein aggregation by exosomes. *J. Biol. Chem.* **2015**, *290*, 2969–2982. [CrossRef]

428. Martinez, Z.; Zhu, M.; Han, S.; Fink, A.L. GM1 specifically interacts with alpha-synuclein and inhibits fibrillation. *Biochemistry* **2007**, *46*, 1868–1877. [CrossRef]

429. Garten, M.; Prévost, C.; Cadart, C.; Gautier, R.; Bousset, L.; Melki, R.; Bassereau, P.; Vanni, S. Methyl-branched lipids promote the membrane adsorption of α-synuclein by enhancing shallow lipid-packing defects. *Phys. Chem. Chem. Phys.* **2015**, *17*, 15589–15597. [CrossRef]
430. Wu, G.; Lu, Z.H.; Kulkarni, N.; Amin, R.; Ledeen, R.W. Mice lacking major brain gangliosides develop parkinsonism. *Neurochem. Res.* **2011**, *36*, 1706–1714. [CrossRef]
431. Suzuki, K.; Iseki, E.; Togo, T.; Yamaguchi, A.; Katsuse, O.; Katsuyama, K.; Kanzaki, S.; Shiozaki, K.; Kawanishi, C.; Yamashita, S.; et al. Neuronal and glial accumulation of alpha- and beta-synucleins in human lipidoses. *Acta Neuropathol.* **2007**, *114*, 481–489. [CrossRef] [PubMed]
432. Akkhawattanangkul, Y.; Maiti, P.; Xue, Y.; Aryal, D.; Wetsel, W.C.; Hamilton, D.; Fowler, S.C.; McDonald, M.P. Targeted deletion of GD3 synthase protects against MPTP-induced neurodegeneration. *Genes Brain Behav.* **2017**, *16*, 522–536. [CrossRef] [PubMed]
433. Ryu, J.K.; Shin, W.H.; Kim, J.; Joe, E.H.; Lee, Y.B.; Cho, K.G.; Oh, Y.J.; Kim, S.U.; Jun, B.K. Trisialoganglioside GT1b induces in vivo degeneration of nigral dopaminergic neurons: Role of microglia. *Glia* **2002**, *38*, 15–23. [CrossRef] [PubMed]
434. Wei, J.; Fujita, M.; Nakai, M.; Waragai, M.; Sekigawa, A.; Sugama, S.; Takenouchi, T.; Masliah, E.; Hashimoto, M. Protective role of endogenous gangliosides for lysosomal pathology in a cellular model of synucleinopathies. *Am. J. Pathol.* **2009**, *174*, 1891–1909. [CrossRef] [PubMed]
435. Xiao, S.; Finkielstein, C.V.; Capelluto, D.G.S. The enigmatic role of sulfatides: New insights into cellular functions and mechanisms of protein recognition. *Adv. Exp. Med. Biol.* **2013**, *991*, 27–40. [CrossRef]
436. Antelmi, E.; Rizzo, G.; Fabbri, M.; Capellari, S.; Scaglione, C.; Martinelli, P. Arylsulphatase A activity in familial parkinsonism: A pathogenetic role? *J. Neurol.* **2014**, *261*, 1803–1809. [CrossRef] [PubMed]
437. Martinelli, P.; Ippoliti, M.; Montanari, M.; Martinelli, A.; Mochi, M.; Giuliani, S.; Sangiorgi, S. Arylsulphatase A (ASA) activity in parkinsonism and symptomatic essential tremor. *Acta Neurol. Scand.* **1994**, *89*, 171–174. [CrossRef]
438. Cheng, H.; Xu, J.; McKeel, D.W.; Han, X. Specificity and potential mechanism of sulfatide deficiency in Alzheimer's disease: An electrospray ionization mass spectrometric study. *Cell. Mol. Biol. (Noisy-le-grand)* **2003**, *49*, 809–818.
439. Spann, N.J.; Glass, C.K. Sterols and oxysterols in immune cell function. *Nat. Immunol.* **2013**, *14*, 893–900. [CrossRef] [PubMed]
440. Hannich, J.T.; Umebayashi, K.; Riezman, H. Distribution and Functions of Sterols and Sphingolipids. *Cold Spring Harb. Perspect. Biol.* **2011**, *3*, a004762. [CrossRef] [PubMed]
441. Powers, K.M.; Smith-Weller, T.; Franklin, G.M.; Longstreth, W.T.; Swanson, P.D.; Checkoway, H. Dietary fats, cholesterol and iron as risk factors for Parkinson's disease. *Park. Relat. Disord.* **2009**, *15*, 47–52. [CrossRef]
442. Johnson, C.C.; Gorell, J.M.; Rybicki, B.A.; Sanders, K.; Peterson, E.L. Adult nutrient intake as a risk factor for Parkinson's disease. *Int. J. Epidemiol.* **1999**, *28*, 1102–1109. [CrossRef] [PubMed]
443. Abbott, R.D.; Ross, G.W.; White, L.R.; Sanderson, W.T.; Burchfiel, C.M.; Kashon, M.; Sharp, D.S.; Masaki, K.H.; Curb, J.D.; Petrovitch, H. Environmental, life-style, and physical precursors of clinical Parkinson's disease: Recent findings from the Honolulu-Asia Aging Study. *J. Neurol.* **2003**, *250* (Suppl. 3), III30–III39. [CrossRef]
444. Wang, A.; Lin, Y.; Wu, Y.; Zhang, D. Macronutrients intake and risk of Parkinson's disease: A meta-analysis. *Geriatr. Gerontol. Int.* **2015**, *15*, 606–616. [CrossRef]
445. Zhang, L.; Wang, X.; Wang, M.; Sterling, N.W.; Du, G.; Lewis, M.M.; Yao, T.; Mailman, R.B.; Li, R.; Huang, X. Circulating Cholesterol Levels May Link to the Factors Influencing Parkinson's Risk. *Front. Neurol.* **2017**, *8*, 501. [CrossRef]
446. Kirbas, A.; Kirbas, S.; Cure, M.C.; Tufekci, A. Paraoxonase and arylesterase activity and total oxidative/anti-oxidative status in patients with idiopathic Parkinson's disease. *J. Clin. Neurosci.* **2014**, *21*, 451–455. [CrossRef] [PubMed]
447. Ikeda, K.; Nakamura, Y.; Kiyozuka, T.; Aoyagi, J.; Hirayama, T.; Nagata, R.; Ito, H.; Iwamoto, K.; Murata, K.; Yoshii, Y.; et al. Serological profiles of urate, paraoxonase-1, ferritin and lipid in Parkinson's disease: Changes linked to disease progression. *Neurodegener. Dis.* **2011**, *8*, 252–258. [CrossRef]
448. Huang, X.; Alonso, A.; Guo, X.; Umbach, D.M.; Lichtenstein, M.L.; Ballantyne, C.M.; Mailman, R.B.; Mosley, T.H.; Chen, H. Statins, plasma cholesterol, and risk of Parkinson's disease: A prospective study. *Mov. Disord.* **2015**, *30*, 552–559. [CrossRef]

449. Miyake, Y.; Tanaka, K.; Fukushima, W.; Sasaki, S.; Kiyohara, C.; Tsuboi, Y.; Yamada, T.; Oeda, T.; Miki, T.; Kawamura, N.; et al. Case-control study of risk of Parkinson's disease in relation to hypertension, hypercholesterolemia, and diabetes in Japan. *J. Neurol. Sci.* **2010**, *293*, 82–86. [CrossRef] [PubMed]
450. Simon, K.C.; Chen, H.; Schwarzschild, M.; Ascherio, A. Hypertension, hypercholesterolemia, diabetes, and risk of Parkinson disease. *Neurology* **2007**, *69*, 1688–1695. [CrossRef]
451. De Lau, L.M.L.; Koudstaal, P.J.; Hofman, A.; Breteler, M.M.B. Serum cholesterol levels and the risk of Parkinson's disease. *Am. J. Epidemiol.* **2006**, *164*, 998–1002. [CrossRef] [PubMed]
452. Rozani, V.; Gurevich, T.; Giladi, N.; El-Ad, B.; Tsamir, J.; Hemo, B.; Peretz, C. Higher serum cholesterol and decreased Parkinson's disease risk: A statin-free cohort study. *Mov. Disord.* **2018**, *33*, 1298–1305. [CrossRef] [PubMed]
453. Huang, X.; Auinger, P.; Eberly, S.; Oakes, D.; Schwarzschild, M.; Ascherio, A.; Mailman, R.; Chen, H.; Parkinson Study Group DATATOP Investigators. Serum cholesterol the progression of Parkinson's disease: Results from DATATOP. *PLoS ONE* **2011**, *6*, e22854. [CrossRef] [PubMed]
454. Gudala, K.; Bansal, D.; Muthyala, H. Role of serum cholesterol in Parkinson's disease: A meta-analysis of evidence. *J. Park. Dis.* **2013**, *3*, 363–370. [CrossRef]
455. Sterling, N.W.; Lichtenstein, M.; Lee, E.Y.; Lewis, M.M.; Evans, A.; Eslinger, P.J.; Du, G.; Gao, X.; Chen, H.; Kong, L.; et al. Higher Plasma LDL-Cholesterol is Associated with Preserved Executive and Fine Motor Functions in Parkinson's Disease. *Aging Dis.* **2016**, *7*, 237–245. [CrossRef] [PubMed]
456. Savica, R.; Grossardt, B.R.; Ahlskog, J.E.; Rocca, W.A. Metabolic markers or conditions preceding Parkinson's disease: A case-control study. *Mov. Disord.* **2012**, *27*, 974–979. [CrossRef] [PubMed]
457. Singh, N.K.; Banerjee, B.D.; Bala, K.; Mitrabasu Dung Dung, A.A.; Chhillar, N. APOE and LRPAP1 gene polymorphism and risk of Parkinson's disease. *Neurol. Sci.* **2014**, *35*, 1075–1081. [CrossRef] [PubMed]
458. Mollenhauer, B.; Trautmann, E.; Sixel-Döring, F.; Wicke, T.; Ebentheuer, J.; Schaumburg, M.; Lang, E.; Focke, N.K.; Kumar, K.R.; Lohmann, K.; et al. Nonmotor and diagnostic findings in subjects with de novo Parkinson disease of the DeNoPa cohort. *Neurology* **2013**, *81*, 1226–1234. [CrossRef]
459. Hu, G.; Antikainen, R.; Jousilahti, P.; Kivipelto, M.; Tuomilehto, J. Total cholesterol and the risk of Parkinson disease. *Neurology* **2008**, *70*, 1972–1979. [CrossRef] [PubMed]
460. Cereda, E.; Cassani, E.; Barichella, M.; Caccialanza, R.; Pezzoli, G. Anthropometric indices of fat distribution and cardiometabolic risk in Parkinson's disease. *Nutr. Metab. Cardiovasc. Dis.* **2013**, *23*, 264–271. [CrossRef] [PubMed]
461. Musanti, R.; Parati, E.; Lamperti, E.; Ghiselli, G. Decreased cholesterol biosynthesis in fibroblasts from patients with Parkinson disease. *Biochem. Med. Metab. Biol.* **1993**, *49*, 133–142. [CrossRef]
462. Shulman, J.M.; Yu, L.; Buchman, A.S.; Evans, D.A.; Schneider, J.A.; Bennett, D.A.; De Jager, P.L. Association of Parkinson Disease Risk Loci With Mild Parkinsonian Signs in Older Persons. *JAMA Neurol.* **2014**, *71*, 429. [CrossRef] [PubMed]
463. Lou, F.; Li, M.; Liu, N.; Li, X.; Ren, Y.; Luo, X. The Polymorphism of SREBF1 Gene rs11868035 G/A Is Associated with susceptibility to Parkinson's disease in a Chinese Population. *Int. J. Neurosci.* **2018**, 1–18. [CrossRef] [PubMed]
464. Yuan, X.; Cao, B.; Wu, Y.; Chen, Y.; Wei, Q.; Ou, R.; Yang, J.; Chen, X.; Zhao, B.; Song, W.; et al. Association analysis of SNP rs11868035 in SREBF1 with sporadic Parkinson's disease, sporadic amyotrophic lateral sclerosis and multiple system atrophy in a Chinese population. *Neurosci. Lett.* **2018**, *664*, 128–132. [CrossRef] [PubMed]
465. Hasson, S.A.; Fogel, A.I.; Wang, C.; MacArthur, R.; Guha, R.; Heman-Ackah, S.; Martin, S.; Youle, R.J.; Inglese, J. Chemogenomic profiling of endogenous PARK2 expression using a genome-edited coincidence reporter. *ACS Chem. Biol.* **2015**, *10*, 1188–1197. [CrossRef] [PubMed]
466. Kim, K.Y.; Stevens, M.V.; Akter, M.H.; Rusk, S.E.; Huang, R.J.; Cohen, A.; Noguchi, A.; Springer, D.; Bocharov, A.V.; Eggerman, T.L.; et al. Parkin is a lipid-responsive regulator of fat uptake in mice and mutant human cells. *J. Clin. Investig.* **2011**, *121*, 3701–3712. [CrossRef]
467. Yamaguchi, S.; Yamane, T.; Takahashi-Niki, K.; Kato, I.; Niki, T.; Goldberg, M.S.; Shen, J.; Ishimoto, K.; Doi, T.; Iguchi-Ariga, S.M. Transcriptional activation of low-density lipoprotein receptor gene by DJ-1 and effect of DJ-1 on cholesterol homeostasis. *PLoS ONE* **2012**, *7*, e38144. [CrossRef]
468. Kim, J.M.; Cha, S.H.; Choi, Y.R.; Jou, I.; Joe, E.H.; Park, S.M. DJ-1 deficiency impairs glutamate uptake into astrocytes via the regulation of flotillin-1 and caveolin-1 expression. *Sci. Rep.* **2016**, *6*, 28823. [CrossRef]

469. Kyung, J.W.; Kim, J.M.; Lee, W.; Ha, T.Y.; Cha, S.H.; Chung, K.H.; Choi, D.J.; Joi, I.; Song, W.K.; Joe, E.H.; et al. DJ-1 deficiency impairs synaptic vesicle endocytosis and reavailability at nerve terminals. *Proc. Natl. Acad. Sci. USA* **2018**, *115*, 1629–1634. [CrossRef]
470. Magalhaes, J.; Gegg, M.E.; Migdalska-Richards, A.; Doherty, M.K.; Whitfield, P.D.; Schapira, A.H.V. Autophagic lysosome reformation dysfunction in glucocerebrosidase deficient cells: Relevance to Parkinson disease. *Hum. Mol. Genet.* **2016**, *25*, 3432–3445. [CrossRef]
471. Cha, S.H.; Choi, Y.R.; Heo, C.H.; Kang, S.J.; Joe, E.H.; Jou, I.; Kim, H.M.; Park, S.M. Loss of parkin promotes lipid rafts-dependent endocytosis through accumulating caveolin-1: Implications for Parkinson's disease. *Mol. Neurodegener.* **2015**, *10*, 63. [CrossRef] [PubMed]
472. García-Sanz, P.; Orgaz, L.; Bueno-Gil, G.; Espadas, I.; Rodríguez-Traver, E.; Kulisevsky, J.; Gutierrez, A.; Dávila, J.C.; González-Polo, R.A.; Fuentes, J.M.; et al. N370S-GBA1 mutation causes lysosomal cholesterol accumulation in Parkinson's disease. *Mov. Disord.* **2017**, *32*, 1409–1422. [CrossRef] [PubMed]
473. Baptista, M.A.S.; Dave, K.D.; Frasier, M.A.; Sherer, T.B.; Greeley, M.; Beck, M.J.; Varsho, J.S.; Parker, G.A.; Moore, C.; Churchill, M.J.; et al. Loss of leucine-rich repeat kinase 2 (LRRK2) in rats leads to progressive abnormal phenotypes in peripheral organs. *PLoS ONE* **2013**, *8*, e80705. [CrossRef] [PubMed]
474. Eriksson, I.; Nath, S.; Bornefall, P.; Giraldo, A.M.V.; Öllinger, K. Impact of high cholesterol in a Parkinson's disease model: Prevention of lysosomal leakage versus stimulation of α-synuclein aggregation. *Eur. J. Cell Biol.* **2017**, *96*, 99–109. [CrossRef] [PubMed]
475. Schmitt, M.; Dehay, B.; Bezard, E.; Garcia-Ladona, F.J. U18666A, an activator of sterol regulatory element binding protein pathway, modulates presynaptic dopaminergic phenotype of SH-SY5Y neuroblastoma cells. *Synapse* **2017**, *71*, e21980. [CrossRef] [PubMed]
476. Morissette, M.; Morin, N.; Rouillard, C.; Di Paolo, T. Membrane cholesterol removal and replenishment affect rat and monkey brain monoamine transporters. *Neuropharmacology* **2018**, *133*, 289–306. [CrossRef] [PubMed]
477. Paul, R.; Choudhury, A.; Kumar, S.; Giri, A.; Sandhir, R.; Borah, A. Cholesterol contributes to dopamine-neuronal loss in MPTP mouse model of Parkinson's disease: Involvement of mitochondrial dysfunctions and oxidative stress. *PLoS ONE* **2017**, *12*, e0171285. [CrossRef] [PubMed]
478. Paul, R.; Choudhury, A.; Chandra Boruah, D.; Devi, R.; Bhattacharya, P.; Choudhury, M.D.; Borach, A. Hypercholesterolemia causes psychomotor abnormalities in mice and alterations in cortico-striatal biogenic amine neurotransmitters: Relevance to Parkinson's disease. *Neurochem. Int.* **2017**, *108*, 15–26. [CrossRef] [PubMed]
479. Paul, R.; Dutta, A.; Phukan, B.C.; Mazumder, M.K.; Justin-Thenmozhi, A.; Manivasagam, T.; Bhattacharya, P.; Borah, A. Accumulation of Cholesterol and Homocysteine in the Nigrostriatal Pathway of Brain Contributes to the Dopaminergic Neurodegeneration in Mice. *Neuroscience* **2018**, *388*, 347–356. [CrossRef] [PubMed]
480. Raju, A.; Jaisankar, P.; Borah, A.; Mohanakumar, K.P. 1-Methyl-4-Phenylpyridinium-Induced Death of Differentiated SH-SY5Y Neurons Is Potentiated by Cholesterol. *Ann. Neurosci.* **2018**, *24*, 243–251. [CrossRef]
481. Fantini, J.; Carlus, D.; Yahi, N. The fusogenic tilted peptide (67-78) of α-synuclein is a cholesterol binding domain. *Biochim. Biophys. Acta* **2011**, *1808*, 2343–2351. [CrossRef] [PubMed]
482. Kamp, F.; Beyer, K. Binding of alpha-synuclein affects the lipid packing in bilayers of small vesicles. *J. Biol. Chem.* **2006**, *281*, 9251–9259. [CrossRef] [PubMed]
483. Van Maarschalkerweerd, A.; Vetri, V.; Vestergaard, B. Cholesterol facilitates interactions between α-synuclein oligomers and charge-neutral membranes. *FEBS Lett.* **2015**, *589*, 2661–2667. [CrossRef] [PubMed]
484. Shvadchak, V.V.; Falomir-Lockhart, L.J.; Yushchenko, D.A.; Jovin, T.M. Specificity and kinetics of alpha-synuclein binding to model membranes determined with fluorescent excited state intramolecular proton transfer (ESIPT) probe. *J. Biol. Chem.* **2011**, *286*, 13023–13032. [CrossRef] [PubMed]
485. Murphy, K.E.; Gysbers, A.M.; Abbott, S.K.; Spiro, A.S.; Furuta, A.; Cooper, A.; Garner, B.; Kabuta, T.; Halliday, G.M. Lysosomal-associated membrane protein 2 isoforms are differentially affected in early Parkinson's disease. *Mov. Disord.* **2015**, *30*, 1639–1647. [CrossRef]
486. Bar-On, P.; Crews, L.; Koob, A.O.; Mizuno, H.; Adame, A.; Spencer, B.; Masliah, E. Statins reduce neuronal alpha-synuclein aggregation in in vitro models of Parkinson's disease. *J. Neurochem.* **2008**, *105*, 1656–1667. [CrossRef] [PubMed]
487. Di Scala, C.; Yahi, N.; Boutemeur, S.; Flores, A.; Rodriguez, L.; Chahinian, H.; Fantini, J. Common molecular mechanism of amyloid pore formation by Alzheimer's β-amyloid peptide and α-synuclein. *Sci. Rep.* **2016**, *6*, 28781. [CrossRef] [PubMed]

488. Fantini, J.; Yahi, N. The driving force of alpha-synuclein insertion and amyloid channel formation in the plasma membrane of neural cells: Key role of ganglioside- and cholesterol-binding domains. *Adv. Exp. Med. Biol.* **2013**, *991*, 15–26. [CrossRef] [PubMed]
489. Bate, C.; Williams, A. α-Synuclein-induced synapse damage in cultured neurons is mediated by cholesterol-sensitive activation of cytoplasmic phospholipase A2. *Biomolecules* **2015**, *5*, 178–193. [CrossRef] [PubMed]
490. Bar-On, P.; Rockenstein, E.; Adame, A.; Ho, G.; Hashimoto, M.; Masliah, E. Effects of the cholesterol-lowering compound methyl-beta-cyclodextrin in models of alpha-synucleinopathy. *J. Neurochem.* **2006**, *98*, 1032–1045. [CrossRef] [PubMed]
491. Fortin, D.L.; Troyer, M.D.; Nakamura, K.; Kubo, S.; Anthony, M.D.; Edwards, R.H. Lipid rafts mediate the synaptic localization of alpha-synuclein. *J. Neurosci.* **2004**, *24*, 6715–6723. [CrossRef] [PubMed]
492. Hsiao, J.H.T.; Halliday, G.M.; Kim, W.S. α-Synuclein Regulates Neuronal Cholesterol Efflux. *Molecules* **2017**, *22*, 1769. [CrossRef] [PubMed]
493. Leftin, A.; Job, C.; Beyer, K.; Brown, M.F. Solid-state ^{13}C NMR reveals annealing of raft-like membranes containing cholesterol by the intrinsically disordered protein α-Synuclein. *J. Mol. Biol.* **2013**, *425*, 2973–2987. [CrossRef]
494. Bosco, D.A.; Fowler, D.M.; Zhang, Q.; Nieva, J.; Powers, E.T.; Wentworth, P.; Lerner, R.A.; Kelly, J.W. Elevated levels of oxidized cholesterol metabolites in Lewy body disease brains accelerate alpha-synuclein fibrilization. *Nat. Chem. Biol.* **2006**, *2*, 249–253. [CrossRef] [PubMed]
495. Chesselet, M.F.; Fleming, S.; Mortazavi, F.; Meurers, B. Strengths and limitations of genetic mouse models of Parkinson's disease. *Park. Relat. Disord.* **2008**, *14* (Suppl. 2), S84–S87. [CrossRef]
496. Sheng, Z.; Jia, X.; Kang, M. Statin use and risk of Parkinson's disease: A meta-analysis. *Behav. Brain Res.* **2016**, *309*, 29–34. [CrossRef]
497. Friedman, B.; Lahad, A.; Dresner, Y.; Vinker, S. Long-term statin use and the risk of Parkinson's disease. *Am. J. Manag. Care* **2013**, *19*, 626–632.
498. Gao, X.; Simon, K.C.; Schwarzschild, M.A.; Ascherio, A. Prospective study of statin use and risk of Parkinson disease. *Arch. Neurol.* **2012**, *69*, 380–384. [CrossRef]
499. Wolozin, B.; Wang, S.W.; Li, N.C.; Lee, A.; Lee, T.A.; Kazis, L.E. Simvastatin is associated with a reduced incidence of dementia and Parkinson's disease. *BMC Med.* **2007**, *5*, 20. [CrossRef]
500. Huang, X.; Chen, H.; Miller, W.C.; Mailman, R.B.; Woodard, J.L.; Chen, P.C.; Xiang, D.; Murrow, R.M.; Wand, Y.Z.; Poole, C. Lower low-density lipoprotein cholesterol levels are associated with Parkinson's disease. *Mov. Disord.* **2007**, *22*, 22. [CrossRef]
501. Wahner, A.D.; Bronstein, J.M.; Bordelon, Y.M.; Ritz, B. Statin use and the risk of Parkinson disease. *Neurology* **2008**, *70*, 1418–1422. [CrossRef] [PubMed]
502. Rozani, V.; Giladi, N.; El-Ad, B.; Gurevich, T.; Tsamir, J.; Hemo, B.; Peretz, C. Statin adherence and the risk of Parkinson's disease: A population-based cohort study. *PLoS ONE* **2017**, *12*, e0175054. [CrossRef] [PubMed]
503. Ritz, B.; Manthripragada, A.D.; Qian, L.; Schernhammer, E.; Wermuth, L.; Olsen, J.; Friss, S. Statin use and Parkinson's disease in Denmark. *Mov. Disord.* **2010**, *25*, 1210–1216. [CrossRef] [PubMed]
504. Becker, C.; Jick, S.S.; Meier, C.R. Use of statins and the risk of Parkinson's disease: A retrospective case-control study in the UK. *Drug Saf.* **2008**, *31*, 399–407. [CrossRef] [PubMed]
505. Liu, G.; Sterling, N.W.; Kong, L.; Lewis, N.M.; Mailman, R.B.; Chen, H.; Leslie, D.; Huang, X. Statins may facilitate Parkinson's disease: Insight gained from a large, national claims database. *Mov Disord* **2017**, *32*, 913–917. [CrossRef]
506. Bykov, K.; Yoshida, K.; Weisskopf, M.G.; Gagne, J.J. Confounding of the association between statins and Parkinson disease: Systematic review and meta-analysis. *Pharmacoepidemiol. Drug Saf.* **2017**, *26*, 294–300. [CrossRef]
507. Marques, N.F.; Castro, A.A.; Mancini, G.; Rocha, F.L.; Santos, A.R.S.; Prediger, R.D.; De Bem, A.F.; Tasca, C.I. Atorvastatin Prevents Early Oxidative Events and Modulates Inflammatory Mediators in the Striatum Following Intranasal 1-Methyl-4-phenyl-1, 2, 3, 6-tetrahydropyridine (MPTP) Administration in Rats. *Neurotox. Res.* **2018**, *33*, 549–559. [CrossRef]
508. Yan, J.Q.; Ma, Y.J.; Sun, J.C.; Bai, S.F.; Huang, L.N. Neuroprotective effect of lovastatin by inhibiting NMDA receptor1 in 6-hydroxydopamine treated PC12 cells. *Int. J. Clin. Exp. Med.* **2014**, *7*, 3313–3319.

509. Jiang, P.; Gan, M.; Lin, W.L.; Yen, S.H.C. Nutrient deprivation induces α-synuclein aggregation through endoplasmic reticulum stress response and SREBP2 pathway. *Front. Aging Neurosci.* **2014**, *6*, 268. [CrossRef]
510. Koob, A.O.; Ubhi, K.; Paulsson, J.F.; Kelly, J.; Rockenstein, E.; Mante, M.; Adame, A.; Masliah, E. Lovastatin ameliorates alpha-synuclein accumulation and oxidation in transgenic mouse models of alpha-synucleinopathies. *Exp. Neurol.* **2010**, *221*, 267–274. [CrossRef]
511. Xu, Y.Q.; Long, L.; Yan, J.Q.; Wei, L.; Pan, M.Q.; Gao, H.M.; Zhou, P.; Liu, M.; Zhu, C.S.; Tang, B.S.; et al. Simvastatin induces neuroprotection in 6-OHDA-lesioned PC12 via the PI3K/AKT/caspase 3 pathway and anti-inflammatory responses. *CNS Neurosci. Ther.* **2013**, *19*, 170–177. [CrossRef] [PubMed]
512. Kumar, A.; Sharma, N.; Gupta, A.; Kalonia, H.; Mishra, J. Neuroprotective potential of atorvastatin and simvastatin (HMG-CoA reductase inhibitors) against 6-hydroxydopamine (6-OHDA) induced Parkinson-like symptoms. *Brain Res.* **2012**, *1471*, 13–22. [CrossRef] [PubMed]
513. Ghosh, A.; Roy, A.; Matras, J.; Brahmachari, S.; Gendelman, H.E.; Pahan, K. Simvastatin inhibits the activation of p21ras and prevents the loss of dopaminergic neurons in a mouse model of Parkinson's disease. *J. Neurosci.* **2009**, *29*, 13543–13556. [CrossRef] [PubMed]
514. Selley, M.L. Simvastatin prevents 1-methyl-4-phenyl-1, 2, 3, 6-tetrahydropyridine-induced striatal dopamine depletion and protein tyrosine nitration in mice. *Brain. Res.* **2005**, *1037*, 1–6. [CrossRef] [PubMed]
515. Wang, Q.; Tang, X.N.; Wang, L.; Yenari, M.A.; Ying, W.; Goh, B.C.; Lee, H.S.; Wilder-Smith, E.P.; Wong, P.T. Effects of high dose of simvastatin on levels of dopamine and its reuptake in prefrontal cortex and striatum among SD rats. *Neurosci. Lett.* **2006**, *408*, 189–193. [CrossRef] [PubMed]
516. Kreisler, A.; Gelé, P.; Wiart, J.F.; Lhermitte, M.; Destée, A.; Bordet, R. Lipid-lowering drugs in the MPTP mouse model of Parkinson's disease: Fenofibrate has a neuroprotective effect, whereas bezafibrate and HMG-CoA reductase inhibitors do not. *Brain Res.* **2007**, *1135*, 77–84. [CrossRef] [PubMed]
517. Schirris, T.J.J.; Renkema, G.H.; Ritschel, T.; Voermans, N.C.; Bilos, A.; van Engelen, B.G.M.; Brandt, U.; Koopman, W.J.; Beyrath, J.D.; Rodenburg, R.J.; et al. Statin-Induced Myopathy Is Associated with Mitochondrial Complex III Inhibition. *Cell Metab.* **2015**, *22*, 399–407. [CrossRef] [PubMed]
518. Nakamura, K.; Mori, F.; Tanji, K.; Miki, Y.; Yamada, M.; Kakita, A.; Takahasi, H.; Utsumi, J.; Sasaki, H.; Wakabayashi, K. Isopentenyl diphosphate isomerase, a cholesterol synthesizing enzyme, is localized in Lewy bodies. *Neuropathology* **2015**, *35*, 432–440. [CrossRef]
519. Kabuto, H.; Yamanushi, T.T.; Janjua, N.; Takayama, F.; Mankura, M. Effects of squalene/squalane on dopamine levels, antioxidant enzyme activity, and fatty acid composition in the striatum of Parkinson's disease mouse model. *J. Oleo Sci.* **2013**, *62*, 21–28. [CrossRef]
520. Lim, L.; Jackson-Lewis, V.; Wong, L.C.; Shui, G.H.; Goh, A.X.H.; Kesavapany, S.; Jenner, A.M.; Fivaz, M.; Przedborski, S.; Wenk, M.R. Lanosterol induces mitochondrial uncoupling and protects dopaminergic neurons from cell death in a model for Parkinson's disease. *Cell Death Differ.* **2012**, *19*, 416–427. [CrossRef]
521. Roy, A.; Ghosh, A.; Jana, A.; Liu, X.; Brahmachari, S.; Gendelman, H.E.; Pahan, K. Sodium phenylbutyrate controls neuroinflammatory and antioxidant activities and protects dopaminergic neurons in mouse models of Parkinson's disease. *PLoS ONE* **2012**, *7*, e38113. [CrossRef] [PubMed]
522. Zhang, S.; Glukhova, S.A.; Caldwell, K.A.; Caldwell, G.A. NCEH-1 modulates cholesterol metabolism and protects against α-synuclein toxicity in a *C. elegans* model of Parkinson's disease. *Hum. Mol. Genet.* **2017**, *26*, 3823–3836. [CrossRef] [PubMed]
523. Björkhem, I.; Lövgren-Sandblom, A.; Leoni, V.; Meaney, S.; Brodin, L.; Salveson, L.; Winge, K.; Pålhagen, S.; Svenningsson, P. Oxysterols and Parkinson's disease: Evidence that levels of 24S-hydroxycholesterol in cerebrospinal fluid correlates with the duration of the disease. *Neurosci. Lett.* **2013**, *555*, 102–105. [CrossRef] [PubMed]
524. Dexter, D.T.; Holley, A.E.; Flitter, W.D.; Slater, T.F.; Wells, F.R.; Daniel, S.E.; Lee, A.J.; Jenner, P.; Marsden, C.D. Increased levels of lipid hydroperoxides in the parkinsonian substantia nigra: An HPLC and ESR study. *Mov. Disord.* **1994**, *9*, 92–97. [CrossRef] [PubMed]
525. Björkhem, I.; Patra, K.; Boxer, A.L.; Svenningsson, P. 24S-Hydroxycholesterol Correlates with Tau and Is Increased in Cerebrospinal Fluid in Parkinson's Disease and Corticobasal Syndrome. *Front. Neurol.* **2018**, *9*, 756. [CrossRef] [PubMed]
526. Di Natale, C.; Monaco, A.; Pedone, C.; Tessitore, A.; De Mase, A.; Tedeschi, G.; Netti, P.A.; Abrescia, P. The level of 24-hydroxycholesteryl esters decreases in plasma of patients with Parkinson's disease. *Neurosci. Lett.* **2018**, *672*, 108–112. [CrossRef] [PubMed]

527. Rantham Prabhakara, J.P.; Feist, G.; Thomasson, S.; Thompson, A.; Schommer, E.; Ghribi, O. Differential effects of 24-hydroxycholesterol and 27-hydroxycholesterol on tyrosine hydroxylase and alpha-synuclein in human neuroblastoma SH-SY5Y cells. *J. Neurochem.* **2008**, *107*, 1722–1729. [CrossRef]
528. Marwarha, G.; Rhen, T.; Schommer, T.; Ghribi, O. The oxysterol 27-hydroxycholesterol regulates α-synuclein and tyrosine hydroxylase expression levels in human neuroblastoma cells through modulation of liver X receptors and estrogen receptors—Relevance to Parkinson's disease. *J. Neurochem.* **2011**, *119*, 1119–1136. [CrossRef]
529. Cheng, D.; Kim, W.S.; Garner, B. Regulation of alpha-synuclein expression by liver X receptor ligands in vitro. *Neuroreport* **2008**, *19*, 1685–1689. [CrossRef]
530. Schommer, J.; Marwarha, G.; Schommer, T.; Flick, T.; Lund, J.; Ghribi, O. 27-Hydroxycholesterol increases α-synuclein protein levels through proteasomal inhibition in human dopaminergic neurons. *BMC Neurosci.* **2018**, *19*, 17. [CrossRef]
531. Emanuelsson, I.; Norlin, M. Protective effects of 27- and 24-hydroxycholesterol against staurosporine-induced cell death in undifferentiated neuroblastoma SH-SY5Y cells. *Neurosci. Lett.* **2012**, *525*, 44–48. [CrossRef]
532. Theofilopoulos, S.; Wang, Y.; Kitambi, S.S.; Sacchetti, P.; Sousa, K.M.; Bodin, K.; Kirk, J.; Saltó, C.; Gustafsson, M.; Toledo, E.M.; et al. Brain endogenous liver X receptor ligands selectively promote midbrain neurogenesis. *Nat. Chem. Biol.* **2013**, *9*, 126–133. [CrossRef] [PubMed]
533. Sacchetti, P.; Sousa, K.M.; Hall, A.C.; Liste, I.; Steffensen, K.R.; Theofilopoulos, S.; Parish, C.L.; Hazenber, C.; Richter, L.A.; Hovatta, O.; et al. Liver X receptors and oxysterols promote ventral midbrain neurogenesis in vivo and in human embryonic stem cells. *Cell Stem Cell* **2009**, *5*, 409–419. [CrossRef] [PubMed]
534. Ramasamy, I. Recent advances in physiological lipoprotein metabolism. *Clin. Chem. Lab. Med.* **2014**, *52*, 1695–1727. [CrossRef] [PubMed]
535. Kuai, R.; Li, D.; Chen, Y.E.; Moon, J.J.; Schwendeman, A. High-Density Lipoproteins: Nature's Multifunctional Nanoparticles. *ACS Nano* **2016**, *10*, 3015–3041. [CrossRef] [PubMed]
536. Swanson, C.R.; Berlyand, Y.; Xie, S.X.; Alcalay, R.N.; Chahine, L.M.; Chen-Plotkin, A.S. Plasma apolipoprotein A1 associates with age at onset and motor severity in early Parkinson's disease patients. *Mov. Disord.* **2015**, *30*, 1648–1656. [CrossRef] [PubMed]
537. Swanson, C.R.; Li, K.; Unger, T.L.; Gallagher, M.D.; Van Deerlin, V.M.; Agarwal, P.; Leverenz, J.; Roberts, J.; Samii, A.; Gross, R.G.; et al. Lower plasma apolipoprotein A1 levels are found in Parkinson's disease and associate with apolipoprotein A1 genotype. *Mov. Disord.* **2015**, *30*, 805–812. [CrossRef]
538. Lu, W.; Wan, X.; Liu, B.; Rong, X.; Zhu, L.; Li, P.; Li, J.; Wang, L.; Cui, L.; Wang, X. Specific changes of serum proteins in Parkinson's disease patients. *PLoS ONE* **2014**, *9*, e95684. [CrossRef]
539. Qiang, J.K.; Wong, Y.C.; Siderowf, A.; Hurtig, H.I.; Xie, S.X.; Lee, V.M.Y.; Trojanowski, J.Q.; Yearout, D.; Leverenz, J.; Montine, T.J.; et al. Plasma apolipoprotein A1 as a biomarker for Parkinson disease. *Ann. Neurol.* **2013**, *74*, 119–127. [CrossRef]
540. Cassani, E.; Cereda, E.; Barichella, M.; Madio, C.; Cancello, R.; Caccialanza, R.; Zini, M.; Cilia, R.; Pezzoli, G. Cardiometabolic factors and disease duration in patients with Parkinson's disease. *Nutrition* **2013**, *29*, 1331–1335. [CrossRef]
541. Kawata, M.; Nemoto, Y.; Asahina, M.; Moroo, I.; Shinomiya, M.; Yamada, T. Risk factors for cerebral arteriosclerosis in Parkinson's disease. *Park. Relat. Disord.* **1996**, *2*, 75–79. [CrossRef]
542. Du, G.; Lewis, M.M.; Shaffer, M.L.; Chen, H.; Yang, Q.X.; Mailman, R.B.; Huang, X. Serum cholesterol and nigrostriatal R2* values in Parkinson's disease. *PLoS ONE* **2012**, *7*, e35397. [CrossRef] [PubMed]
543. Huang, X.; Abbott, R.D.; Petrovitch, H.; Mailman, R.B.; Ross, G.W. Low LDL cholesterol and increased risk of Parkinson's disease: Prospective results from Honolulu-Asia Aging Study. *Mov. Disord.* **2008**, *23*, 1013–1018. [CrossRef] [PubMed]
544. Benn, M.; Nordestgaard, B.G.; Frikke-Schmidt, R.; Tybjærg-Hansen, A. Low LDL cholesterol, PCSK9 and HMGCR genetic variation, and risk of Alzheimer's disease and Parkinson's disease: Mendelian randomisation study. *BMJ* **2017**, *357*, j1648. [CrossRef] [PubMed]
545. Andican, G.; Konukoglu, D.; Bozluolcay, M.; Bayülkem, K.; Firtiına, S.; Burcak, G. Plasma oxidative and inflammatory markers in patients with idiopathic Parkinson's disease. *Acta Neurol. Belg.* **2012**, *112*, 155–159. [CrossRef] [PubMed]

546. Schroeter, H.; Williams, R.J.; Matin, R.; Iversen, L.; Rice-Evans, C.A. Phenolic antioxidants attenuate neuronal cell death following uptake of oxidized low-density lipoprotein. *Free Radic. Biol. Med.* **2000**, *29*, 1222–1233. [CrossRef]
547. Van Meer, G.; Voelker, D.R.; Feigenson, G.W. Membrane lipids: Where they are and how they behave. *Nat. Rev. Mol. Cell Biol.* **2008**, *9*, 112–124. [CrossRef]
548. Tumanov, S.; Kamphorst, J.J. Recent advances in expanding the coverage of the lipidome. *Curr. Opin. Biotechnol.* **2017**, *43*, 127–133. [CrossRef]
549. Tracey, T.J.; Steyn, F.J.; Wolvetang, E.J.; Ngo, S.T. Neuronal Lipid Metabolism: Multiple Pathways Driving Functional Outcomes in Health and Disease. *Front. Mol. Neurosci.* **2018**, *11*, 10. [CrossRef]
550. Piomelli, D.; Astarita, G.; Rapaka, R. A neuroscientist's guide to lipidomics. *Nat. Rev. Neurosci.* **2007**, *8*, 743–754. [CrossRef]
551. Andersen, O.S.; Koeppe, R.E. Bilayer thickness and membrane protein function: An energetic perspective. *Annu. Rev. Biophys. Biomol. Struct.* **2007**, *36*, 107–130. [CrossRef] [PubMed]
552. Rohrbough, J.; Broadie, K. Lipid regulation of the synaptic vesicle cycle. *Nat. Rev. Neurosci.* **2005**, *6*, 139–150. [CrossRef] [PubMed]
553. Galvagnion, C.; Brown, J.W.P.; Ouberai, M.M.; Flagmeier, P.; Vendruscolo, M.; Buell, A.K.; Sparr, E.; Dobson, C.M. Chemical properties of lipids strongly affect the kinetics of the membrane-induced aggregation of α-synuclein. *Proc. Natl. Acad. Sci. USA* **2016**, *113*, 7065–7070. [CrossRef]
554. Esposito, G.; Ana Clara, F.; Verstreken, P. Synaptic vesicle trafficking and Parkinson's disease. *Dev. Neurobiol.* **2012**, *72*, 134–144. [CrossRef] [PubMed]
555. Dijkstra, A.A.; Ingrassia, A.; de Menezes, R.X.; van Kesteren, R.E.; Rozemuller, A.J.M.; Heutink, P.; van de Berg, W.D. Evidence for Immune Response, Axonal Dysfunction and Reduced Endocytosis in the Substantia Nigra in Early Stage Parkinson's Disease. *PLoS ONE* **2015**, *10*, e0128651. [CrossRef] [PubMed]
556. Spillantini, M.G.; Crowther, R.A.; Jakes, R.; Hasegawa, M.; Goedert, M. alpha-Synuclein in filamentous inclusions of Lewy bodies from Parkinson's disease and dementia with lewy bodies. *Proc. Natl. Acad. Sci. USA* **1998**, *95*, 6469–6473. [CrossRef] [PubMed]
557. Castillo, P.E.; Younts, T.J.; Chávez, A.E.; Hashimotodani, Y. Endocannabinoid signaling and synaptic function. *Neuron* **2012**, *76*, 70–81. [CrossRef]
558. Panov, A.; Orynbayeva, Z.; Vavilin, V.; Lyakhovich, V. Fatty acids in energy metabolism of the central nervous system. *Biomed. Res. Int.* **2014**, *2014*, 472459. [CrossRef]
559. Ebert, D.; Haller, R.G.; Walton, M.E. Energy contribution of octanoate to intact rat brain metabolism measured by ^{13}C nuclear magnetic resonance spectroscopy. *J. Neurosci.* **2003**, *23*, 5928–5935. [CrossRef]
560. Rustam, Y.H.; Reid, G.E. Analytical Challenges and Recent Advances in Mass Spectrometry Based Lipidomics. *Anal. Chem.* **2018**, *90*, 374–397. [CrossRef]
561. Bou Khalil, M.; Hou, W.; Zhou, H.; Elisma, F.; Swayne, L.A.; Blanchard, A.P.; Yao, Z.; Bennett, S.A.; Figeys, D. Lipidomics era: Accomplishments and challenges. *Mass Spectrom. Rev.* **2010**, *29*, 877–929. [CrossRef]
562. Hu, P.; Fabyanic, E.; Kwon, D.Y.; Tang, S.; Zhou, Z.; Wu, H. Dissecting Cell-Type Composition and Activity-Dependent Transcriptional State in Mammalian Brains by Massively Parallel Single-Nucleus RNA-Seq. *Mol. Cell* **2017**, *68*, 1006–1015.e7. [CrossRef] [PubMed]
563. Rubakhin, S.S.; Romanova, E.V.; Nemes, P.; Sweedler, J.V. Profiling metabolites and peptides in single cells. *Nat. Methods* **2011**, *8*, S20–S29. [CrossRef] [PubMed]
564. Yang, K.; Han, X. Lipidomics: Techniques, Applications, and Outcomes Related to Biomedical Sciences. *Trends Biochem. Sci.* **2016**, *41*, 954–969. [CrossRef] [PubMed]
565. Caesar, R.; Tremaroli, V.; Kovatcheva-Datchary, P.; Cani, P.D.; Bäckhed, F. Crosstalk between Gut Microbiota and Dietary Lipids Aggravates WAT Inflammation through TLR Signaling. *Cell Metab.* **2015**, *22*, 658–668. [CrossRef]
566. Den Besten, G.; Lange, K.; Havinga, R.; van Dijk, T.H.; Gerding, A.; van Eunen, K.; Müller, M.; Groen, A.K.; Hooiveld, G.J.; Bakker, B.M.; et al. Gut-derived short-chain fatty acids are vividly assimilated into host carbohydrates and lipids. *Am. J. Physiol. Gastrointest. Liver Physiol.* **2013**, *305*, G900–G910. [CrossRef]
567. Caesar, R.; Nygren, H.; Orešič, M.; Bäckhed, F. Interaction between dietary lipids and gut microbiota regulates hepatic cholesterol metabolism. *J. Lipid Res.* **2016**, *57*, 474–481. [CrossRef]
568. Petersen, C.; Round, J.L. Defining dysbiosis and its influence on host immunity and disease. *Cell. Microbiol.* **2014**, *16*, 1024–1033. [CrossRef]

569. Hill-Burns, E.M.; Debelius, J.W.; Morton, J.T.; Wissemann, W.T.; Lewis, M.R.; Wallen, Z.D.; Peddada, S.D.; Factor, S.A.; Molho, E.; Zabetian, C.P.; et al. Parkinson's disease and Parkinson's disease medications have distinct signatures of the gut microbiome. *Mov. Disord.* **2017**, *32*, 739–749. [CrossRef]
570. Heintz-Buschart, A.; Pandey, U.; Wicke, T.; Sixel-Döring, F.; Janzen, A.; Sittig-Wiegand, E.; Trenkwalder, C.; Oertel, W.H.; Mollenhauer, B.; Wilmes, P. The nasal and gut microbiome in Parkinson's disease and idiopathic rapid eye movement sleep behavior disorder. *Mov. Disord.* **2018**, *33*, 88–98. [CrossRef]
571. Petrov, V.A.; Saltykova, I.V.; Zhukova, I.A.; Alifirova, V.M.; Zhukova, N.G.; Dorofeeva, Y.B.; Tyakht, A.V.; Kovarsky, B.A.; Alekseev, D.G.; Kostryukova, E.S.; et al. Analysis of Gut Microbiota in Patients with Parkinson's Disease. *Bull. Exp. Biol. Med.* **2017**, *162*, 734–737. [CrossRef] [PubMed]
572. Unger, M.M.; Spiegel, J.; Dillmann, K.U.; Grundmann, D.; Philippeit, H.; Bürmann, J.; Faßbender, K.; Schwiertz, A.; Schäfer, K.H. Short chain fatty acids and gut microbiota differ between patients with Parkinson's disease and age-matched controls. *Park. Relat. Disord.* **2016**, *32*, 66–72. [CrossRef] [PubMed]
573. Keshavarzian, A.; Green, S.J.; Engen, P.A.; Voigt, R.M.; Naqib, A.; Forsyth, C.B.; Mutlu, E.; Shannon, K.M. Colonic bacterial composition in Parkinson's disease. *Mov. Disord.* **2015**, *30*, 1351–1360. [CrossRef]
574. Minato, T.; Maeda, T.; Fujisawa, Y.; Tsuji, H.; Nomoto, K.; Ohno, K.; Hirayama, M. Progression of Parkinson's disease is associated with gut dysbiosis: Two-year follow-up study. *PLoS ONE* **2017**, *12*, e0187307. [CrossRef] [PubMed]
575. Cassani, E.; Barichella, M.; Cancello, R.; Cavanna, F.; Iorio, L.; Cereda, E.; Bolliri, C.; Zampella Maria, P.; Bianchi, F.; Cestaro, B.; et al. Increased urinary indoxyl sulfate (indican): New insights into gut dysbiosis in Parkinson's disease. *Park. Relat. Disord.* **2015**, *21*, 389–393. [CrossRef] [PubMed]
576. Lista, S.; Khachaturian, Z.S.; Rujescu, D.; Garaci, F.; Dubois, B.; Hampel, H. Application of Systems Theory in Longitudinal Studies on the Origin and Progression of Alzheimer's Disease. *Methods Mol. Biol.* **2016**, *1303*, 49–67. [CrossRef] [PubMed]
577. Dehairs, J.; Derua, R.; Rueda-Rincon, N.; Swinnen, J.V. Lipidomics in drug development. *Drug Discov. Today Technol.* **2015**, *13*, 33–38. [CrossRef]

8

Molecular Imaging of the Dopamine Transporter

Giovanni Palermo and Roberto Ceravolo *

Unit of Neurology, Department of Clinical and Experimental Medicine, University of Pisa, 56,126 Pisa, Italy
* Correspondence: r.ceravolo@med.unipi.it

Abstract: Dopamine transporter (DAT) single-photon emission tomography (SPECT) with (123)Ioflupane is a widely used diagnostic tool for patients with suspected parkinsonian syndromes, as it assists with differentiating between Parkinson's disease (PD) or atypical parkinsonisms and conditions without a presynaptic dopaminergic deficit such as essential tremor, vascular and drug-induced parkinsonisms. Recent evidence supports its utility as in vivo proof of degenerative parkinsonisms, and DAT imaging has been proposed as a potential surrogate marker for dopaminergic nigrostriatal neurons. However, the interpretation of DAT-SPECT imaging may be challenged by several factors including the loss of DAT receptor density with age and the effect of certain drugs on dopamine uptake. Furthermore, a clear, direct relationship between nigral loss and DAT decrease has been controversial so far. Striatal DAT uptake could reflect nigral neuronal loss once the loss exceeds 50%. Indeed, reduction of DAT binding seems to be already present in the prodromal stage of PD, suggesting both an early synaptic dysfunction and the activation of compensatory changes to delay the onset of symptoms. Despite a weak correlation with PD severity and progression, quantitative measurements of DAT binding at baseline could be used to predict the emergence of late-disease motor fluctuations and dyskinesias. This review addresses the possibilities and limitations of DAT-SPECT in PD and, focusing specifically on regulatory changes of DAT in surviving DA neurons, we investigate its role in diagnosis and its prognostic value for motor complications as disease progresses.

Keywords: [123I]FP-CIT-SPECT; DAT; nigral cells; Parkinson's disease; parkinsonisms

---

1. Dopamine Transporter Imaging: Role in Diagnosis

Parkinson's disease (PD) is a progressive neurodegenerative disease characterized primarily by the selective degeneration of dopaminergic neurons in the pars compacta of the substantia nigra (SN). Intraneuronal inclusions composed of aggregates of a-synuclein (α-syn), called Lewy bodies (LBs), are the other neuropathological hallmark [1]. Loss of the nigral neurons, first the lateral tier followed by the medial region, is extensive and characteristic for PD, and it leads to substantial reduction of the presynaptic dopamine transporter (DAT) [2]. DAT is a transmembrane sodium chloride dependent protein expressed only in presynaptic dopaminergic cells. DAT is responsible for reuptake of dopamine (DA) from the synaptic cleft, and it is critical in the spatial and temporal buffering of released DA levels [3]. DAT also has a role in the regulation of quantal DA release at endplates by influencing the DA storage in the synaptic vesicles and the mobility of vesicle pools for release [4,5]. In PD, a lower membrane DAT expression on presynaptic terminals may possibly reflect striatal dopamine terminal loss and is in direct proportion to the magnitude of the depletion of nigral cells. The DAT ligands for single-photon emission tomography (SPECT) have all shown significantly reduced striatal uptake in PD. Abnormal uptake progresses from putamen to caudate and matches, contralaterally, the clinically more affected side, which correlates well with disease severity and duration [6], as well as with both rigidity and bradykinesia, but not with tremor severity [7,8]. [123I]FP-CIT (123I-ioflupane) is one of the most used radiotracers for single-photon emission computed tomography (SPECT) imaging of DAT. Indeed, this [123I]-labelled cocaine analogue shows high specificity for the DAT and has fast kinetics, which allows it to be adequately imaged in clinical practice [9]. Other dopamine transporter ligands that examine dopaminergic systems in vivo include [99mTc]TRODAT [10], [123I] β-CIT [11], and [123I]IPT [12], and the differences among these tracers are mainly in their kinetic properties [13]. Dopamine transporter single-photon emission computed tomography (DAT-SPECT) allows us to evaluate the integrity of the nigrostriatal pathway in vivo, providing the promise of an objective and quantitative biomarker of neuronal degeneration in PD [14]. It is a valuable diagnostic tool to help differentiate essential tremor (ET) from tremor due to parkinsonian syndromes, and it is valuable for a differential diagnosis of degenerative parkinsonisms compared to vascular and drug-induced parkinsonisms [15] (Figure 1). However, the diagnosis of vascular as well as drug-induced parkinsonism might be misleading since evidence of vascular, even strategic, lesions or a positive history of long-term exposure to antidopaminergic drugs does not automatically rule out a concomitant disturbance of the nigrostriatal dopaminergic pathway. Thus, if DAT imaging in pure vascular parkinsonism is typically normal [16], other reports have described an involvement of the nigrostriatal system in patients with parkinsonism and brain vascular lesions [17–19]. Similarly, DAT imaging is likely to be normal in pure drug-induced parkinsonism caused by the D2-receptor blockade [20], but involvement of the nigrostriatal system is possible in up to 50% of patients with parkinsonism and with a long-term exposure to antidopaminergic drugs [21,22]. Furthermore, DAT imaging can reveal nigrostriatal impairment, even in isolated/atypical tremors, in which an abnormal SPECT can even predict clinical conversion to a fully blown parkinsonism [23–26]. All this evidence supports the conclusion that DAT imaging is the most reliable proof of degenerative parkinsonism, and, accordingly, a normal DaTSCAN has been incorporated in the new Movement Disorders Society (MDS) criteria as exclusion criteria for PD [27]. In this respect, cases of parkinsonism without evidence of dopaminergic deficit (so-called scans without evidence for dopaminergic deficit, SWEDD) are not to be considered PD [28], and the term SWEDD should be abandoned, although some data support the notion that an initial, normal DAT-SPECT cannot always exclude early degenerative parkinsonism, suggesting that such patients should be monitored over time [29–31]. To complicate matters further, in some degenerative parkinsonisms, such as corticobasal degeneration (CBD) [32] and dementia with Lewy body (DLB) [33,34], in which nigrostriatal dopaminergic dysfunction is usually present, initial DAT-SCANs can be normal, and patients may develop later alterations of DAT imaging in the course of the disease [34–38] (Figure 1).

Some attempts have been performed to use DAT-SCAN for a differential diagnosis between PD and atypical parkinsonisms, such as an evaluation of the asymmetric index and the caudate/putamen ratio, with no significant success in clinical practice [39–41]. In visual assessments, a burst striatum pattern was reportedly more common in patients with a clinical diagnosis of atypical parkinsonism, whereas an egg-shaped pattern, which is typical of putaminal degeneration, was reportedly more indicative for PD [42,43]. However, no usefulness on an individual basis has so far been reported in distinguishing PD from atypical parkinsonisms because in pathologically proven atypical parkinsonisms and PD, the DAT-SPECT patterns greatly overlap each other [44].

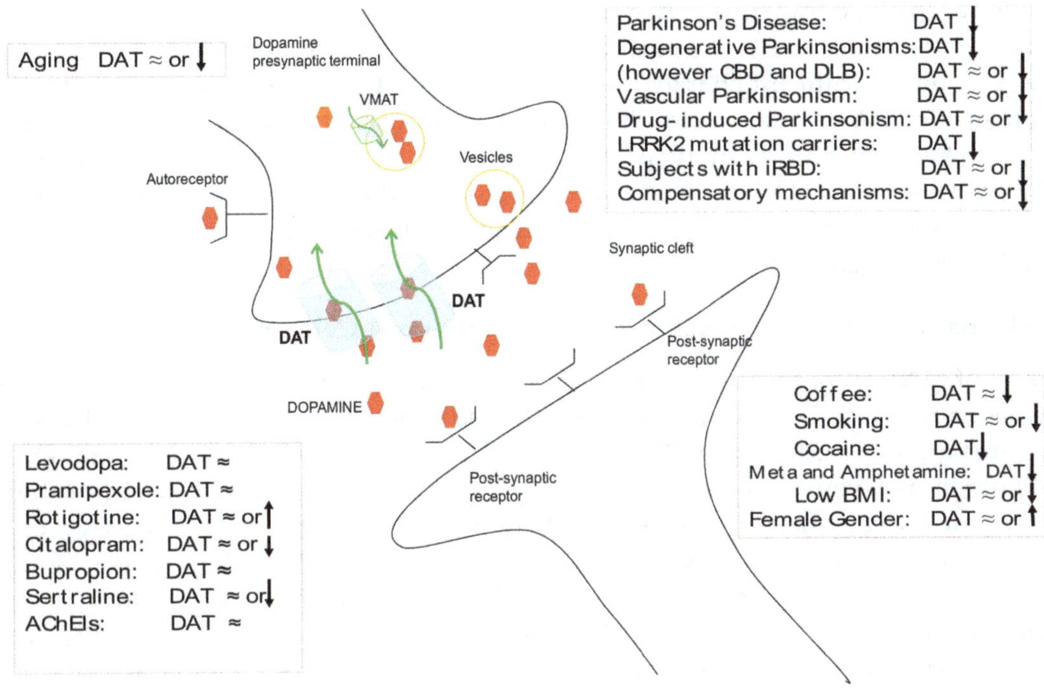

Figure 1. Schematic representation of a dopamine synapse with the dopamine transporter (DAT) and its interference by both external and disease-related factors. (AChEIs: acetylcholinesterase inhibitors; BMI: body mass index; iRBD: idiopathic REM sleep behavior disorder; LRRK2: leucine-rich repeat kinase 2; and VMAT: vesicular monoamine transporter).

2. Dopamine Transporter Imaging: Drug and Habit Interferences

DAT activity is under the control of different presynaptic proteins, including DA autoreceptors, that can promote both trafficking and the availability of the transporter at the plasma membrane [45] (Figure 1). In animal studies, DAT expression can be interfered by dopaminergic drugs. In rats and monkeys in particular, levodopa can induce an increase of DAT [46–48]. However, a decrease of DAT in rats has also been reported with levodopa [49]. Previous studies using DAT radioligands have demonstrated that levodopa appears to have no significant effect on its striatal uptake [50,51]. Similarly, dopamine agonists in standard dosages do not markedly affect the DAT binding capacity independently on the pharmacological profile [52,53]. Current data suggest that D2 autoreceptors induce recruitment of DAT to the membrane, facilitate DA uptake, and DA D2 agonists up-regulate DAT functions in vitro [54]. Recently, the effects of prolonged treatment with DA D3 agonists on DAT have been reported, and they indicated significant underexpression of DAT and the reduction of DA reuptake by dopaminergic neurons [55]. Because of this potential dual effect on D2 or D3 autoreceptors, a trial with pramipexole, a DA agonist, found its use to be ineffective on DAT at the doses usually employed in clinical practice [51]. Instead, the effect of DA D1 agonists on DAT is not entirely known; however, the inhibition of DAT by SKF-83566, a D1 receptor antagonist, has been well documented [56]. After short-term exposure to rotigotine, which has great affinity for D1 and D2 receptors, a mild DAT striatal up-regulation has been recently reported [57]. Based on these findings, PD patients are currently submitted to SPECT studies without the need to withdraw dopaminergic drugs [58].

Selective serotonin reuptake inhibitors (SSRIs) are the mainstay treatment for major depressive disorder, and their effect on DAT is not fully understood yet (Figure 1). Citalopram has been reported to significantly decrease DAT availability, whereas bupropion has no effect [59]. No changes in striatal DAT with different SSRIs have been recently reported [60]. However, some concerns still persist regarding sertraline and citalopram, which in clinical use are preferentially withdrawn before DAT imaging studies. Cholinesterase inhibitors, drugs used for dementia in Lewy body disease and PD with dementia, did not significantly influence DAT expression in a clinical study [61]. Coffee may interfere with dopaminergic transmission, and this action would possibly enhance motor activity and exert an antidyskinetic effect in PD. Nevertheless, in a recent study, chronic coffee consumption was not associated with any significant change in DAT expression [62]. Conversely, cigarette smoking was associated with striatal DAT reduction in PD smokers, as already reported in non-PD smokers [63,64]. However, results of a recent meta-analysis on nicotine did not show any significant effects between smokers and nonsmokers on DAT availability, whereas significant downregulation was reported for most studies investigating the effect of stimulant drugs on DAT such as cocaine, methamphetamine, and amphetamine [65] (Figure 1).

3. Dopamine Transporter Imaging: Relationship with Nigral Cell Counts

The density of DAT on presynaptic terminals is considered a surrogate marker for dopaminergic nigral cell counts and vitality, but a clear, direct relationship between nigral loss and DAT decrease has, so far, been controversial. Reduced DAT striatal availability parallels loss of striatal DA and loss of nigrostriatal fibers in monkeys chronically treated with 1-methyl-4-phenyl-1,2,3,6-tetrahydropyridine (MPTP) [66]. Recently, in a cohort of autopsy-confirmed neurodegenerative disease cases, researchers investigated whether changes in DAT images reflected loss of DAT, because of cell death or neurodegenerative pathology, by examining the influence of nigral neuronal loss as well as nigral (α-syn, tau) and striatal (α-syn, tau, amyloid-β) pathology on striatal uptake in 4 cases of Alzheimer's disease (AD), 7 cases of dementia with Lewy bodies (DLB), and 12 Parkinson's disease dementia (PDD) cases. Subjects underwent antemortem dopaminergic scanning and postmortem assessments (mean interval 3.7 years). In all striatal regions, tracer uptake was associated with nigral dopaminergic neuronal density but not α-syn, tau, or amyloid-β burden [67]. In another paper, postmortem SN cell counts in patients (PD, DLB, AD, multiple system atrophy, and CBD) who had previously undergone DAT-SPECT were correlated with striatal DAT values. They found a high correlation between striatal uptake binding and postmortem SN cell counts, confirming the validity of DAT imaging as an excellent in vivo marker of nigrostriatal dopaminergic degeneration [68]. In contrast, in nonhuman primates with experimental parkinsonism, striatal uptake of the DAT ligand did not faithfully reflect nigral cell counts throughout the full range of neuronal loss, with a flooring effect once nigral loss exceeded 50% [69]. To support this finding, in a recent study in which nigral neuron numbers were calculated for 18 patients (11 patients had neuropathologically confirmed PD) who had been examined with DAT-SPECT before death, postmortem SN–pars compacta (SNc) neuron counts were not associated with striatal DAT binding in PD. These results fit with the theory that there is no correlation between the number of SN neurons and striatal DA after a certain level of damage has occurred. Striatal DAT binding in PD may reflect axonal dysfunction or DAT expression rather than the number of viable neurons [70]. It should be considered that in these studies there is a possible bias represented by the time interval between the in vivo scan and the postmortem examination (3.5 years in the group with the shorter interval). A possible approach to overcome this limitation might be to correlate in vivo dopaminergic, neuromelanin-rich neurons in SN measured by magnetic resonance imaging (MRI) with DAT ligand striatal uptake. MRI measures of neuromelanin (both area and signal intensity) show decreased levels in the region corresponding to SNc in PD patients, with respect to healthy controls [71–78], and patients with ET [79] that paralleled disease severity and duration [80,81]. Moreover, a good correlation exists between signal intensity in neuromelanin-sensitive MRIs and the density of neuromelanin-containing neurons in the SNc [82,83]. Interestingly, a good correlation between neuromelanin-sensitive MRI SN measures and striatal DAT-SPECT values has been recently demonstrated [84,85].

Direct visualization of the SN is one of the unmet needs in neuroimaging of parkinsonisms, with controversial results obtained from the various attempted approaches [86]. By using susceptibility-weighted imaging (SWI) at 7T and even at 3T MRI, the loss of the hyperintense laminar or ovoid-shaped areas present in SN of healthy controls [87,88], identified as nigrosome-1, has been proposed as a morphological marker of nigral depletion in both PD and atypical parkinsonisms, with no significant difference on an individual basis [89].

These observations could also have implications in the potential detection of the premotor phase of PD in order to evaluate at-risk subjects. In this respect, carriers of dominant PD genes, such as leucine-rich repeat kinase 2 (LRRK2) mutations, or subjects with REM sleep behavior disorder (RBD) are ideal candidates to evaluate the detection of preclinical/premotor PD by means of nuclear medicine or MRI. The potential role of DAT imaging has been demonstrated [90–93], whereas data on SWI–MRI are only sparse. In a study of a family with three LRRK2 mutation carriers, the DAT-SPECT was abnormal for all family members, whereas the MRI signal in the SN was abnormal only in the unaffected mother and in the affected daughter, and it was normal in the unaffected son [94]. Moreover, in a series of 15 RBD subjects studied by 3T MRI, an alteration of nigrosome-1 was reported in two-thirds of them [95].

Emerging interest has pointed towards the mutual relationship between SN MRI anatomical changes and the assessment of dopaminergic nigrostriatal functions, as assessed by DAT imaging. A good agreement between SN-MRI abnormalities and [123I]FP-CIT-SPECT has been reported [96]. However, preservation of nigrosome-1 was reported in few patients with PD and degeneration of the nigrostriatal pathway. The disagreement between MRI and SPECT was also observed in patients with CBD [97], in a carrier of the LRRK2 mutation [94], and in one patient with RBD [98]. By assuming DAT-SPECT is a surrogate marker of degenerative parkinsonisms, such data could be interpreted as the effect of some inaccuracies of SWI–MRI, but some considerations could also be raised. If iron is assumed to be responsible for SN abnormality in SWI–MRI, it should be noted that iron is not invariably associated with degeneration. The two methods explore nigral neurons at different levels, and DAT-SPECT could overestimate degeneration because of early DAT down-regulation [99], or, alternatively, it might reflect synaptic degeneration prior to neuronal death [100].

At the molecular level, loss of dopaminergic axonal terminals seems to precede SN cell body loss [101]. Specific deregulated pathways linked to axonal degeneration have been found to occur prior to the onset of α-syn pathology in the SN of early PD subjects [102]. Neurodegeneration in PD might begin at the nerve terminal, and neuronal death may result from a "dying back" process [103]. Indeed, α-syn is a natively unfolded presynaptic protein involved in synaptic transmission and synaptic vesicle retrieval, and accumulation of misfolded fibrillar α-syn has been coupled with the severity of neurodegeneration [104]. Thus, the role of axonal destabilization as the putative mechanism that primarily causes neurotoxicity in PD has advanced. Notably, a decrease of DAT not only occurs early in the disease process, but it is greatest in the early stages of PD than in subsequent years, disappearing by year four of diagnosis [105]. In consonance with this, nigral cell loss appears to be profound and substantial soon after diagnosis, and it becomes modest to negligible for decades thereafter, suggesting a nonlinear reduction in tracer binding as well as in the rate of dopaminergic cell loss over the course of the illness [106]. Neuropathological studies in human brain samples, and both in vivo and in vitro models, support the hypothesis that nigrostriatal synapses may be affected at the earliest stages of the neurodegenerative process. Mechanisms leading to either structural or functional synaptic dysfunction are starting to be elucidated, and they include dysregulation of axonal transport, impairment of the exocytosis and endocytosis machinery, altered intracellular trafficking, and loss of corticostriatal synaptic plasticity. Recent data support the view that, in PD, early synaptic dysfunction is directly caused by α-syn oligomers through different and multiple mechanisms [107]. Although α-syn is expressed throughout the brain, DA neurons are the most vulnerable in PD, likely because α-syn directly regulates DA levels as a negative modulator by inhibiting enzymes responsible for its synthesis and by interacting with and reducing the activity of vesicular monoamine transporter 2 (VMAT2) and DAT [108].

4. Dopamine Transporter Imaging: A Window of Compensatory Mechanisms

Neuronal loss in the SN, with or without LBs, occurs with normal aging in a spectrum that spans from age-associated brain changes to structural alterations of neurodegenerative diseases like PD [109]. In general, there is abundant evidence in support of the association between aging and PD: the incidence and prevalence of PD strongly increase with age, and aging is the predominant risk factor [110]. PD affects 3–4% of individuals over the age of 65 years and reaches a prevalence of 4% in the highest age groups [111]. Furthermore, age at onset of PD significantly affects the phenotype and accelerated progression of disease, especially in the early–middle phase [112]. However, although there is a marked age-related decline in DA levels (about 7% per decade), aging alone cannot lead to the critical level of cell loss necessary for parkinsonian signs to emerge (estimated as approximately 80% striatal dopamine depletion). Moreover, the patterns of striatal dopamine loss in PD and normal aging are different, affecting mostly the ventrolateral segment of the SN in patients with PD [113]. Similarly, age-related striatal DAT decline does not seem to be exaggerated or accelerated in PD [114], even though activation of regulatory and compensatory mechanisms directed at maintaining DA uptake, superimposed upon disease-related DA terminal loss, may have masked significant age correlations [115,116] (Figure 1). As a consequence of this, currently PD is not merely considered the result of an accelerated aging process, even though the effects of aging on dopaminergic neurotransmission in the striatum might be consistent with a pre-PD state [117].

Similarly, Lewy pathology has been found at autopsy in the brains of elderly individuals without clinical PD in association with intermediate SN neuron loss between PD cases and controls, supporting the concept that subjects with incidental LB pathology represent preclinical PD [118]. Indeed, LB pathology and a nigrostriatal dopaminergic deficit are thought to antedate the clinical diagnosis by several years [119]. Furthermore, nigrostriatal dysfunction can also be detected in clinically unaffected members of kindred with familial PD [120]. In addition, reduction of DAT binding has been reported in subjects with a high genetic risk of PD, suggesting that nigrostriatal dopaminergic dysfunction may be considered a subclinical manifestation of the disease [121,122]. The SN is neither the earliest nor the most severely affected region, and PD-related lesions are seen in the lower brainstem and in olfactory bulb/anterior olfactory nucleus prior to involvement of the SN [123]. Nevertheless, once the involvement of the SN occurs, it still takes time before full parkinsonism can be diagnosed. Indeed, the cardinal motor signs of PD (bradykinesia, rigidity, and resting tremor), which are attributed mostly to dopaminergic cell loss, emerge when there is already a loss of approximately 30% of SN neurons and 50% to 70% of dopaminergic markers in the striatum [124].

Redundancy in basal ganglia functions and mechanisms of functional compensation could serve to delay the clinical onset of PD motor signs during early stages. From this point of view, a late appearance of parkinsonian symptoms would be due to failure of compensatory mechanisms to maintain dopaminergic control over striatal cell function [125]. It is possible that compensatory processes within and outside the basal ganglia, which may be particularly relevant in the early presymptomatic phase of PD, could interfere with the detection of functional impairments for a given period of time [126]. Indeed, regulatory changes that may occur to partially compensate for the loss of dopaminergic terminals include downregulation of the DAT to increase dopaminergic activity and maintain the basal ganglia output within normal limits. Thus, although DAT-SPECT imaging is able to detect reduction of DAT density in patients with PD, we cannot exclude that possible compensatory DAT downregulation at baseline may cause its underestimation and give more severe estimates of DA dysfunction [99]. However, the relevance of DAT downregulation to functional compensation in the preclinical phase is not clear, and it does not appear to reduce sensitivity of the SPECT technique [127].

More recently, adaptive changes in other neural structures than the nigrostriatal pathway, such as the subthalamic nucleus and the globus pallidus pars internalis, have been invoked to play a central role in the compensatory mechanisms that sustain dopaminergic functions in the early phases of PD [128]. It is probable that different mechanisms are likely to come into play at different stages of disease progression [129], and a better recognition of the key mechanisms underlying long-term compensation could lead to the development of therapeutic strategies that aim to further delay the clinical onset of PD [130]. Additionally, these compensatory early events could have long-term, deleterious effects on basal ganglia function and determine the response to drug treatment and the development of motor complications in PD. In particular, it is conceivable that the same mechanisms, which seem to be involved in delaying the onset of clinical symptoms in early disease, may have a different effect in more advanced disease when DA storage capacity and DA availability are further impaired.

Indeed, presynaptic and postsynaptic mechanisms have also been observed to persist in more advanced stages of the disease and have been predicted to promote the development of motor complications as the disease progresses [131]. Specifically, evidence suggests that increased dopamine turnover plays a major role in the pathogenesis of motor fluctuations and dyskinesias [132–134]. Functional downregulation of DAT expression has been related to an increased DA turnover and DA release, even accounting for the degree of denervation, contributing to oscillations in synaptic DA levels. Additionally, DA turnover increases after the initiation of levodopa treatment in early PD [135]. This is in keeping with motor complications arising from chronic pulsatile stimulation of DA receptors with levodopa treatment [136]. Such models could explain why patients with dyskinesias tend to express relatively lower levels of DAT per surviving nigrostriatal dopaminergic nerve terminal [137].

Several longitudinal observations indicate that the magnitude of DAT binding in early PD, especially in the posterior putamen, could be predictive for the development of dyskinesias [138] and motor complications in later disease [139]. Of note is that reinstating even a small amount of DAT expression in rats, by grafting cells that express DAT into the striatum of dyskinetic rats, might significantly ameliorate dyskinesias [140]. Furthermore, a recent study demonstrated that the initially low posterior putaminal DAT activity is closely coupled not only with dyskinesia development but also with the timing of levodopa-induced dyskinesia onset in patients with de novo PD [141]. Interestingly, an antero-posterior gradient of dopamine denervation in the putamen, most pronounced in the postero-dorsal area, is maintained throughout the course of Parkinson's disease, but it is more prominent in early disease [142]. Consistently, baseline DA turnover showed a similar regional pattern with elevated DA turnover in the putamen and further accentuation in its posterior part [139]. In contrast, the lower levels of DAT activity in the anterior putamen of de novo PD patients has been reported to be a significant predictor for future development of motor fluctuations [143]. It remains difficult to explain the underlying pathomechanisms for this regional difference between fluctuations and dyskinesias.

If the loss of DA storage due to presynaptic dopamine neuronal degeneration is preferentially associated with the development of motor fluctuations, dopamine neurons in the anterior putamen could have a greater role in storing DA [143]. However, overall, other functional neuroimaging results have been inconsistent, and the relationship between DAT binding and the emergence of late motor complications of PD remains unclear [144,145]. In particular, postsynaptic changes and the influence of neurotransmitters other than dopamine seem to be involved in their development [146].

Similarly, the relationship between DAT binding and PD progression remains unclear. Bearing in mind that reduced DAT expression might indicate neuronal and axonal dysfunction without alterations in dopaminergic neuronal counts, several authors investigated whether the baseline dopaminergic striatal uptake could predict the development of clinically important long-term motor and nonmotor outcomes. Ravina and colleagues reported that lower [123I]b-CIT SPECT at baseline is independently associated with motor-related disability as well as with nonmotor features such as cognitive impairment, psychosis, and depression [147]. Accordingly, in other individual studies, patients with a more severe reduction of striatal DAT uptake had higher incidence of anxiety [148], visual hallucinations [149], apathy [150], freezing of gait [151], and urinary symptoms [152]. However, imaging markers were not incorporated in the useful predictive model of PD motor progression constructed by Latourelle and colleagues because of their demonstrated scant prognostic utility [153].

The emergence of new DAT tracers promises to result in the development of powerful markers for PD. Furthermore, it is noteworthy to mention that SPECT imaging is limited by poor spatial and temporal resolution and has a worse quantitative capacity than PET. Recently, Li et al. identified ^{11}C-PE2, a highly specific DAT radioligand suitable for PET imaging, as a potential alternative surrogate biomarker for studies of PD progression for its ability to track changes in motor performance over time in 33 PD subjects [154]. Overall, findings about the prognostic implications of abnormal DAT imaging in early PD are discordant. Moreover, following PD progression with DAT imaging can be confounded by dopaminergic and other medications. In addition to aging, in terms of clinical confounding factors that determine the initial striatal DAT activity, gender [116] and body mass index [155] have also been suggested to have differential effects in the striatal binding of [123I]FP-CIT.

5. Conclusions

DAT Imaging is a reliable tool for investigating the presynaptic dopaminergic nigrostriatal pathway, and it is useful in clinical practice as proof of degenerative parkinsonisms. However, there are several findings that DAT expression is not merely related to the vitality of the nigral neurons, since synapsis involvement could antedate nigral cellular degeneration and DAT density could be modulated as a compensatory mechanism in preclinical/early PD. All of these points, along with several potential external interferences, make DAT imaging a less reliable marker of disease progression.

Author Contributions: Conceptualization, R.C. and G.P.; writing—original draft preparation, R.C. and G.P.; writing—review and editing, R.C. and G.P.

References

1. Dickson, D.W. Neuropathology of Parkinson disease. *Parkinsonism Relat. Disord.* **2018**, *46*, S30–S33.
2. Sulzer, D. Multiple hit hypothesis for dopamine neuron loss in Parkinson's disease. *Trends Neurosci.* **2007**, *30*, 244–250. [CrossRef] [PubMed]
3. Uhl, G.R. Dopamine transporter: Basic science and human variation of a key molecule for dopaminergic function, locomotion, and parkinsonism. *Mov. Disord.* **2003**, *18* (Suppl. 7), S71–S80. [CrossRef]
4. Sulzer, D.; Cragg, S.J.; Rice, M.E. Striatal dopamine neurotransmission: Regulation of release and uptake. *Basal Ganglia* **2016**, *6*, 123–148. [CrossRef] [PubMed]
5. Mulvihill, K.G. Presynaptic regulation of dopamine release: Role of the DAT and VMAT2 transporters. *Neurochem. Int.* **2019**, *122*, 94–105. [CrossRef] [PubMed]
6. Benamer, H.T.S.; Patterson, J.; Wyper, D.J.; Hadley, D.L.; Macphee, G.J.A.; Grosset, D.G. Correlation of Parkinsons disease severity and duration with [123I]FP-CIT SPECT striatal uptake. *Mov. Disord.* **2000**, *15*, 692–698. [CrossRef]
7. Spiegel, J.; Hellwig, D.; Samnick, S.; Jost, W.; Möllers, M.O.; Fassbender, K.; Kirsch, C.M.; Dillmann, U. Striatal FP-CIT uptake differs in the subtypes of early Parkinson's disease. *J. Neural Transm.* **2007**, *114*, 331–335. [CrossRef] [PubMed]
8. Rossi, C.; Frosini, D.; Volterrani, D.; De Feo, P.; Unti, E.; Nicoletti, V.; Kiferle, L.; Bonuccelli, U.; Ceravolo, R. Differences in nigro-striatal impairment in clinical variants of early Parkinson's disease: Evidence from a FP-CIT SPECT study. *Eur. J. Neurol.* **2010**, *17*, 626–630. [CrossRef]
9. Booij, T.; Tissingh, G.; Boer, G.; Speelman, J.D.; Stoof, J.C.; Janssen, A.G.; Wolters, E.C.; van Royen, E.A. [123I]FP-SPECT shows a pronounced decline of striatal dopamine transporter labelling in early and advanced Parkinson's disease. *J. Neurol. Neurosurg. Psychiatry* **1997**, *62*, 133–140. [CrossRef]
10. Mozley, P.D.; Schneider, J.S.; Acton, P.D.; Plössl, K.; Stern, M.B.; Siderowf, A.; Leopold, N.A.; Li, P.Y.; Alavi, A.; Kung, H.F. Binding of [99mTc-]TRODAT-1 to dopamine transporters in patients with Parkinson's disease and in healthy volunteers. *J. Nucl. Med.* **2000**, *41*, 584–589.
11. Marek, K.; Innis, R.B.; Van Dyck, C.H.; Fussell, B.; Early, M.; Eberly, S.; Oakes, D.; Seibyl, J. [123I]beta-CIT SPECT imaging assessment of the rate of Parkinson's disease progression. *Neurology* **2001**, *57*, 2089–2094. [CrossRef] [PubMed]
12. Kim, H.J.; Im, J.H.; Yang, S.O.; Moon, D.H.; Ryu, J.S.; Bong, J.K.; Nam, K.P.; Cheon, J.H.; Lee, M.C.; Lee, H.K. Imaging and quantitation of dopamine transporters with iodine-123-IPT in normal and Parkinson's disease subjects. *J. Nucl. Med.* **1997**, *38*, 1703–1711. [PubMed]
13. Tatsch, K.; Poepperl, G. Nigrostriatal dopamine terminal imaging with dopamine transporter SPECT: An update. *J. Nucl. Med.* **2013**, *54*, 1331–1338. [CrossRef] [PubMed]

14. Saeed, U.; Compagnone, J.; Aviv, R.I.; Strafella, A.P.; Black, S.E.; Lang, A.E.; Masellis, M. Imaging biomarkers in Parkinson's disease and Parkinsonian syndromes: Current and emerging concepts. *Transl. Neurodegener.* **2017**, *6*, 8. [CrossRef] [PubMed]
15. Bajaj, N.; Hauser, R.A.; Grachev, I.D. Clinical utility of dopamine transporter single photon emission CT (DaT-SPECT) with (123I) ioflupane in diagnosis of parkinsonian syndromes. *J. Neurol. Neurosurg. Psychiatry* **2013**, *84*, 1288–1295. [CrossRef] [PubMed]
16. Gerschlager, W.; Bencsits, G.; Pirker, W.; Bloem, B.R.; Asenbaum, S.; Prayer, D.; Zijlmans, J.C.; Hoffmann, M.; Brücke, T. [123I]beta-CIT SPECT distinguishes vascular parkinsonism from Parkinson's disease. *Mov. Disord.* **2002**, *17*, 518–523. [CrossRef]
17. Lorberboym, M.; Djaldetti, R.; Melamed, E.; Sadeh, M.; Lampl, Y. 123I-FP-CIT SPECT imaging of dopamine transporters in patients with cerebrovascular disease and clinical diagnosis of vascular parkinsonism. *J. Nucl. Med.* **2004**, *45*, 1688–1693. [PubMed]
18. Antonini, A.; Vitale, C.; Barone, P.; Cilia, R.; Righini, A.; Bonuccelli, U.; Abbruzzese, G.; Ramat, S.; Petrone, A.; Quatrale, R.; et al. The relationship between cerebral vascular disease and parkinsonism: The VADO study. *Parkinsonism Relat. Disord.* **2012**, *18*, 775–780. [CrossRef]
19. Benítez-Rivero, S.; Marín-Oyaga, V.A.; García-Solís, D.; Huertas-Fernández, I.; García-Gómez, F.J.; Jesús, S.; Cáceres, M.T.; Carrillo, F.; Ortiz, A.M.; Carballo, M.; et al. Clinical features and 123I-FP-CIT SPECT imaging in vascular parkinsonism and Parkinson's disease. *J. Neurol. Neurosurg. Psychiatry* **2013**, *84*, 122–129. [CrossRef]
20. Tolosa, E.; Coelho, M.; Gallardo, M. DAT imaging in drug-induced and psychogenic parkinsonism. *Mov. Disord.* **2003**, *18* (Suppl. 7), 28–33. [CrossRef]
21. Lorberboym, M.; Treves, T.A.; Melamed, E.; Lampl, Y.; Hellmann, M.; Djaldetti, E. [123I]-FP/CIT SPECT imaging for distinguishing drug-induced parkinsonism from Parkinson's disease. *Mov. Disord.* **2006**, *21*, 510–514. [CrossRef] [PubMed]
22. Tinazzi, M.; Cipriani, A.; Matinella, A.; Cannas, A.; Solla, P.; Nicoletti, A.; Zappia, M.; Morgante, L.; Morgante, F.; Pacchetti, C.; et al. [123I]FP-CIT single photon emission computed tomography findings in drug-induced Parkinsonism. *Schizophr. Res.* **2012**, *139*, 40–45. [CrossRef] [PubMed]
23. Ceravolo, R.; Antonini, A.; Volterrani, D.; Rossi, C.; Kiferle, L.; Frosini, D.; Lucetti, C.; Isaias, I.U.; Benti, R.; Murri, L.; et al. Predictive value of nigrostriatal dysfunction in isolated tremor: A clinical and SPECT study. *Mov. Disord.* **2008**, *23*, 2049–2054. [CrossRef] [PubMed]
24. Novellino, F.; Arabia, G.; Bagnato, A.; Cascini, G.L.; Salsone, M.; Nicoletti, G.; Messina, D.; Morelli, M.; Paglionico, S.; Giofrè, L.; et al. Combined use of DAT-SPECT and cardiac MIBG scintigraphy in mixed tremors. *Mov. Disord.* **2009**, *24*, 2242–2248. [CrossRef] [PubMed]
25. Arabia, G.; Novellino, F.; Morelli, M.; Paglionico, S.; Salsone, M.; Giofrè, L.; Pucci, F.; Bagnato, A.; Cascini, G.L.; Nicoletti, G.; et al. Mixed tremors with integrity of nigrostriatal system: A clinical and DAT-SPECT follow-up study. *Mov. Disord.* **2010**, *25*, 662–664. [CrossRef] [PubMed]
26. De Verdal, M.; Renard, D.; Collombier, L.; Boudousq, V.; Kotzki, P.O.; Labauge, P.; Castelnovo, G. I123-FP-CIT single-photon emission tomography in patients with long-standing mixed tremor. *Eur. J. Neurol.* **2013**, *20*, 382–388. [CrossRef] [PubMed]
27. Postuma, R.B.; Berg, D.; Stern, M.; Poewe, W.; Olanow, C.W.; Oertel, W.; Obeso, J.; Marek, K.; Litvan, I.; Lang, A.E.; et al. MDS clinical diagnostic criteria for Parkinson's disease. *Mov. Disord.* **2015**, *30*, 1591–1601. [CrossRef]
28. Erro, R.; Schneider, S.A.; Stamelou, M.; Quinn, N.P.; Bhatia, K.P. What do patients with scans without evidence of dopaminergic deficit (SWEDD) have? New evidence and continuing controversies. *J. Neurol. Neurosurg. Psychiatry* **2016**, *87*, 319–323. [CrossRef]
29. Sixel-Döring, F.; Liepe, K.; Mollenhauer, B.; Trautmann, E.; Trenkwalder, C. The role of 123I-FP-CIT-SPECT in the differential diagnosis of Parkinson and tremor syndromes: A critical assessment of 125 cases. *J. Neurol.* **2011**, *258*, 2147–2154. [CrossRef]
30. Marshall, V.L.; Patterson, J.; Hadley, D.M.; Grosset, K.A.; Grosset, D.G. Two-year follow-up in 150 consecutive cases with normal dopamine transporter imaging. *Nucl. Med. Commun.* **2006**, *27*, 933–937. [CrossRef]
31. Wyman-Chick, K.A.; Martin, P.K.; Minar, M.; Schroeder, R.W. Cognition in patients with a clinical diagnosis of Parkinson disease and scans without evidence of dopaminergic deficit (SWEDD): 2-year follow-up. *Cogn. Behav. Neurol.* **2016**, *29*, 190–196. [CrossRef] [PubMed]

32. Mille, E.; Levin, J.; Brendel, M.; Zach, C.; Barthel, H.; Sabri, O.; Bötzel, K.; Bartenstein, P.; Danek, A.; Rominger, A. Cerebral glucose metabolism and dopaminergic function in patients with corticobasal syndrome. *J. Neuroimaging* **2017**, *27*, 255–261. [CrossRef] [PubMed]
33. McKeith, I.; O'Brien, J.; Walker, Z.; Tatsch, K.; Booij, J.; Darcourt, J.; Padovani, A.; Giubbini, R.; Bonuccelli, U.; Volterrani, D.; et al. Sensitivity and specificity of dopamine transporter imaging with 123I-FP-CIT SPECT in dementia with Lewy bodies: A phase III, multicentre study. *Lancet Neurol.* **2007**, *6*, 305–313. [CrossRef]
34. Thomas, A.J.; Attems, J.; Colloby, S.J.; O'Brien, J.T.; McKeith, I.; Walker, R.; Lee, L.; Burn, D.; Lett, D.J.; Walker, Z. Autopsy validation of 123I-FP-CIT dopaminergic neuroimaging for the diagnosis of DLB. *Neurology* **2017**, *88*, 276–283. [CrossRef] [PubMed]
35. Cilia, R.; Rossi, C.; Frosini, D.; Volterrani, D.; Siri, C.; Pagni, C.; Benti, R.; Pezzoli, G.; Bonuccelli, U.; Antonini, A.; et al. Dopamine transporter SPECT imaging in corticobasal syndrome. *PLos ONE* **2011**, *6*, e18301. [CrossRef] [PubMed]
36. Ceravolo, R.; Rossi, C.; Cilia, R.; Tognoni, G.; Antonini, A.; Volterrani, D.; Bonuccelli, U. Evidence of delayed nigrostriatal dysfunction in corticobasal syndrome: A SPECT follow-up study. *Parkinsonism Relat. Disord.* **2013**, *19*, 557–559. [CrossRef] [PubMed]
37. Pirker, S.; Perju-Dumbrava, L.; Kovacs, G.G.; Traub-Weidinger, T.; Pirker, W. Progressive dopamine transporter binding loss in autopsy-confirmed corticobasal degeneration. *J. Parkinsons Dis.* **2015**, *5*, 907–912. [CrossRef] [PubMed]
38. Van der Zande, J.J.; Booij, J.; Scheltens, P.; Raijmakers, P.G.; Lemstra, A.W. [(123)]FP-CIT SPECT scans initially rated as normal became abnormal over time in patients with probable dementia with Lewy bodies. *Eur. J. Nucl. Med. Mol. Imaging* **2016**, *43*, 1060–1066. [CrossRef]
39. Filippi, L.; Manni, C.; Pierantozzi, M.; Brusa, L.; Danieli, R.; Stanzione, P.; Schillaci, O. 123I-FP-CIT in progressive supranuclear palsy and in Parkinson's disease: A SPECTsemiquantitative study. *Nucl. Med. Commun.* **2006**, *27*, 381–386. [CrossRef]
40. Perju-Dumbrava, L.D.; Kovacs, G.G.; Pirker, S.; Jellinger, K.; Hoffmann, M.; Asenbaum, S.; Pirker, W. Dopamine transporter imaging in autopsy-confirmed Parkinson's disease and multiple system atrophy. *Mov. Disord.* **2012**, *27*, 65–71. [CrossRef]
41. Matesan, M.; Gaddikeri, S.; Longfellow, K.; Miyaoka, R.; Elojeimy, S.; Elman, S.; Hu, S.C.; Minoshima, S.; Lewis, D. I-123 DaTscan SPECT Brain Imaging in Parkinsonian Syndromes: Utility of the Putamen-to-Caudate Ratio. *J. Neuroimaging* **2018**, *28*, 629–634. [CrossRef]
42. Kahraman, D.; Eggers, C.; Schicha, H.; Timmermann, L.; Schmidt, M. Visual assessment of dopaminergic degeneration pattern in 123I-FP-CIT SPECT differentiates patients with atypical parkinsonian syndromes and idiopathic Parkinson's disease. *J. Neurol.* **2012**, *259*, 251–260. [CrossRef] [PubMed]
43. Davidsson, A.; Georgiopoulos, C.; Dizdar, N.; Granerus, G.; Zachrisson, H. Comparison between visual assessment of dopaminergic degeneration pattern and semi-quantitative ratio calculations in patients with Parkinson's disease and atypical parkinsonian syndromes using DaTSCAN (R) SPECT. *Ann. Nucl. Med.* **2014**, *28*, 851–859. [CrossRef] [PubMed]
44. Brooks, D.J. Imaging of genetic and degenerative disorders primarily causing Parkinsonism. *Handb. Clin. Neurol.* **2016**, *135*, 493–505. [PubMed]
45. Torres, G.E. The dopamine transporter proteome. *J. Neurochem.* **2006**, *97* (Suppl 1), 3–10. [CrossRef]
46. Ikawa, K.; Watanabe, A.; Kaneno, S.; Toru, M. Modulation of [3H]mazindol binding sites in rat striatum by dopaminergic agents. *Eur. J. Pharmacol.* **1993**, *250*, 261–266. [CrossRef]
47. Rioux, L.; Frohna, P.A.; Joyce, J.N.; Schneider, J.S. The effects of chronic levodopa treatment on pre- and postsynaptic markers of dopaminergic function in striatum of parkinsonian monkeys. *Mov. Disord.* **1997**, *12*, 148–158. [CrossRef]
48. Murer, M.G.; Dziewczapolski, G.; Menalled, L.B.; García, M.C.; Agid, Y.; Gershanik, O.; Raisman-Vozari, R. Chronic levodopa is not toxic for remaining dopamine neurons, but instead promotes their recovery, in rats with moderate nigrostriatal lesions. *Ann. Neurol.* **1998**, *43*, 561–575. [CrossRef]
49. Gnanalingham, K.K.; Robertson, R.G. The effects of chronic continuous versus intermittent levodopa treatments on striatal and extrastriatal D1 and D2 dopamine receptors and dopamine uptake sites in the 6-hydroxydopamine lesioned rat–an autoradiographic study. *Brain Res.* **1994**, *640*, 185–194. [CrossRef]

50. Innis, R.B.; Marek, K.L.; Sheff, K.; Zoghbi, S.; Castronuovo, J.; Feigin, A.; Seibyl, J.P. Effect of treatment with Levodopa-Carbidopa or L-Selegiline on striatal dopamine transporter SPECT imaging with I [123]Beta-CIT. *Mov. Disord.* **1999**, *14*, 436–442. [CrossRef]
51. Guttman, M.; Stewart, D.; Hussey, D.; Wilson, A.; Houle, S.; Kish, S. Influence of L-Dopa and pramipexole on striatal dopamine transporter in early PD. *Neurology* **2001**, *56*, 1559–1564. [CrossRef] [PubMed]
52. Ahlskog, J.E.; Uitti, R.J.; O'Connor, M.K.; Maraganore, D.M.; Matsumoto, J.Y.; Stark, K.F.; Turk, M.F.; Burnett, O.L. The effect of dopamine agonist therapy on dopamine transporter imaging in Parkinson's disease. *Mov. Disord.* **1999**, *14*, 940–946. [CrossRef]
53. Parkinson Study Group. Dopamine transporter brain imaging to assess the effects of pramipexole vs. levodopa on Parkinson disease progression. *JAMA* **2002**, *287*, 1653–1661. [CrossRef] [PubMed]
54. Bolan, E.A.; Kivell, B.; Jaligam, V.; Oz, M.; Jayanthi, L.D.; Han, Y.; Sen, N.; Urizar, E.; Gomes, I.; Devi, L.A.; et al. D2 receptors regulate dopamine transporter function via an extracellular signal-regulated kinases 1 and 2-dependent and phosphoinositide 3 kinase-independent mechanism. *Mol. Pharmacol.* **2007**, *7*, 1222–1232. [CrossRef] [PubMed]
55. Castro-Hernandez, J.; Afonso-Oramas, D.; Cruz-Munos, I.; Salas-Hernández, J.; Barroso-Chinea, P.; Moratalla, R.; Millan, M.J.; González-Hernández, T. Prolonged treatment with pramipexole promotes physical interaction of striatal dopamine D3 autoreceptors with dopamine transporters to reduce dopamine uptake. *Neurobiol. Dis.* **2015**, *74*, 325–335. [CrossRef] [PubMed]
56. Stouffer, M.A.; Ali, S.; Reith, M.E.A.; Patel, J.C.; Sarti, F.; Carr, K.D.; Rice, M.E. SKF-83566, a D1 dopamine receptor antagonist, inhibits the dopamine transporter. *J. Neurochem.* **2011**, *118*, 714–720. [CrossRef] [PubMed]
57. Rossi, C.; Genovesi, D.; Marzullo, P.; Giorgetti, A.; Filidei, E.; Corsini, G.U.; Bonuccelli, U.; Ceravolo, R. Striatal dopamine transporter modulation after rotigotine: Results from a pilot single-photon emission computed tomography study in a group of early stage Parkinson disease patients. *Clin. Neuropharmacol.* **2017**, *40*, 34–36. [CrossRef]
58. Ikeda, K.; Ebina, J.; Kawabe, K.; Iwasaki, Y. Dopamine Transporter Imaging in Parkinson Disease: Progressive Changes and Therapeutic Modification after Anti-parkinsonian Medications. *Intern. Med.* **2019**, Epub ahead of print. [CrossRef]
59. Kugaya, A.; Seneca, N.M.; Snyder, P.J.; Williams, S.A.; Malison, R.T.; Baldwin, R.M.; Seibyl, J.P.; Innis, R.B. Changes in human in vivo serotonin and dopamine transporter availabilities during chronic antidepressant administration. *Neuropsychopharmacology* **2003**, *28*, 413–420. [CrossRef]
60. Wu, C.K.; Chin Chen, K.; See Chen, P.; Chiu, N.T.; Yeh, T.L.; Lee, I.H.; Yang, Y.K. No changes in striatal dopamine transporter in antidepressant-treated patients with major depression. *Int. Clin. Psychopharmacol.* **2013**, *28*, 141–144. [CrossRef]
61. Taylor, J.P.; Colloby, S.J.; McKeith, I.G.; Burn, D.J.; Williams, D.; Patterson, J.; O'Brien, J.T. Cholinesterase inhibitor use does not significantly influence the ability of 123I-FP-CIT imaging to distinguish Alzheimer's disease from dementia with Lewy bodies. *J. Neurol. Neurosurg. Psychiatry* **2007**, *78*, 1069–1071. [CrossRef] [PubMed]
62. Gigante, A.F.; Asabella, A.N.; Iliceto, G.; Martino, T.; Ferrari, C.; Defazio, G.; Rubini, G. Chronic coffee consumption and striatal DAT-SPECT findings in Parkinson's disease. *Neurol. Sci.* **2018**, *39*, 551–555. [CrossRef] [PubMed]
63. Yang, Y.K.; Yao, W.J.; Yeh, T.L.; Lee, I.H.; Chen, P.S.; Lu, R.B.; Chiu, N.T. Decreased dopamine transporter availability in male smokers—A dual isotope SPECT study. *Prog. Neuropsychopharmacol. Biol. Psychiatry* **2008**, *32*, 274–279. [CrossRef] [PubMed]
64. Ashok, A.H.; Mizuno, Y.; Howes, O.D. Tobacco smoking and dopaminergic function in humans: A meta-analysis of molecular imaging studies. *Psychopharmacology* **2019**, *236*, 1119–1129. [CrossRef] [PubMed]
65. Proebstl, L.; Kamp, F.; Manz, K.; Krause, D.; Adorjan, K.; Pogarell, O.; Koller, G.; Soyka, M.; Falkai, P.; Kambeitz, J. Effects of stimulant drug use on the dopaminergic system: A systematic review and meta-analysis of in vivo neuroimaging studies. *Eur. Psychiatry* **2019**, *59*, 15–24. [CrossRef] [PubMed]

66. Bezard, E.; Dovero, S.; Prunier, C.; Ravenscroft, P.; Chalon, S.; Guilloteau, D.; Crossman, A.R.; Bioulac, B.; Brotchie, J.M.; Gross, C.E. Relationship between the appearance of symptoms and the level of nigrostriatal degeneration in a progressive 1-methyl-4-phenyl-1,2,3,6 tetrahydropyridinelesioned macaque model of Parkinson's disease. *J. Neurosci.* **2001**, *21*, 6853–6861. [CrossRef] [PubMed]
67. Colloby, S.; McParland, S.; O'Brien, J.T.; Attems, J. Neuropathological correlates of dopaminergic imaging in Alzheimer's disease and Lewy body dementias. *Brain* **2012**, *135*, 2798–2808. [CrossRef] [PubMed]
68. Kraemmer, J.; Kovacs, G.G.; Perju-Dumbrava, L.; Pirker, S.; Traub-Weidinger, T.; Pirker, W. Correlation of striatal dopamine transporter imaging with post mortem substantia nigra cell counts. *Mov. Disord.* **2014**, *29*, 1767–1773. [CrossRef]
69. Karimi, M.; Tian, L.; Brown, C.A.; Flores, H.P.; Loftin, S.K.; Videen, T.O.; Moerlein, S.M.; Perlmutter, J.S. Validation of nigrostriatal positron emission tomography measures: Critical limits. *Ann. Neurol.* **2013**, *73*, 390–396. [CrossRef]
70. Saari, L.; Kivinen, K.; Gardberg, M.; Joutsa, J.; Noponen, T.; Kaasinen, V. Dopamine transporter imaging does not predict the number of nigral neurons in Parkinson disease. *Neurology* **2017**, *88*, 1461–1467. [CrossRef]
71. Sasaki, M.; Shibata, E.; Tohyama, K.; Takahashi, J.; Otsuka, K.; Tsuchiya, K.; Takahashi, S.; Ehara, S.; Terayama, Y.; Sakai, A. Neuromelanin magnetic resonance imaging of locus ceruleus and substantia nigra in Parkinson's disease. *Neuroreport* **2006**, *17*, 1215–1218. [CrossRef] [PubMed]
72. Ohtsuka, C.; Sasaki, M.; Konno, K.; Koide, M.; Kato, K.; Takahashi, J.; Takahashi, S.; Kudo, K.; Yamashita, F.; Terayama, Y. Changes in substantia nigra and locus coeruleus in patients with early-stage Parkinson's disease using neuromelanin-sensitive MR imaging. *Neurosci. Lett.* **2013**, *29*, 93–98. [CrossRef] [PubMed]
73. Ogisu, K.; Kudo, K.; Sasaki, M.; Sakushima, K.; Yabe, I.; Sasaki, H.; Terae, S.; Nakanishi, M.; Shirato, H. 3D neuromelanin-sensitive magnetic resonance imaging with semi-automated volume measurement of the sub stantia nigra pars compacta for diagnosis of Parkinson's disease. *Neuroradiology* **2013**, *55*, 719–724. [CrossRef] [PubMed]
74. Ohtsuka, C.; Sasaki, M.; Konno, K.; Kato, K.; Takahashi, J.; Yamashita, F.; Terayama, Y. Differentiation of early-stage parkinsonisms using neuromelanin-sensitive magnetic resonance imaging. *Parkinsonism Relat. Disord.* **2014**, *20*, 755–760. [CrossRef] [PubMed]
75. Reimao, S.; Pita Lobo, P.; Neutel, D.; Correia Guedes, L.; Coelho, M.; Rosa, M.M.; Ferreira, J.; Abreu, D.; Gonçalves, N.; Morgado, C.; et al. Substantia nigra neuromelanin magnetic resonance imaging in de novo Parkinson's disease patients. *Eur. J. Neurol.* **2015**, *22*, 540–546. [CrossRef] [PubMed]
76. Schwarz, S.T.; Xing, Y.; Tomar, P.; Bajaj, N.; Auer, D.P. In vivo assessment of brainstem depigmentation in Parkinson disease: Potential as a severity marker for multicenter studies. *Radiology* **2017**, *283*, 789–798. [CrossRef]
77. Prasad, S.; Stezin, A.; Lenka, A.; George, L.; Saini, J.; Yadav, R.; Pal, P.K. Three-dimensional neuromelanin-sensitive magnetic resonance imaging of the substantia nigra in Parkinson's disease. *Eur. J. Neurol.* **2018**, *25*, 680–686. [CrossRef]
78. Pavese, N.; Tai, Y.F. Nigrosome Imaging and Neuromelanin Sensitive MRI in Diagnostic Evaluation of Parkinsonism. *Mov. Disord. Clin. Pract.* **2018**, *22*, 131–140. [CrossRef]
79. Reimao, S.; Pita Lobo, P.; Neutel, D.; Guedes, L.C.; Coelho, M.; Rosa, M.M.; Azevedo, P.; Ferreira, J.; Abreu, D.; Gonçalves, N.; et al. Substantia nigra neuromelanin-MR imaging differentiates essential tremor from Parkinson's disease. *Mov. Disord.* **2015**, *30*, 953–959. [CrossRef]
80. Kashihara, K.; Shinya, T.; Higaki, F. Neuromelanin magnetic resonance imaging of nigral volume loss in patients with Parkinson's disease. *J. Clin. Neurosci.* **2011**, *18*, 1093–1096. [CrossRef]
81. Miyoshi, F.; Ogawa, T.; Kitao, S.I.; Kitayama, M.; Shinohara, Y.; Takasugi, M.; Fujii, S.; Kaminou, T. Evaluation of Parkinson disease and Alzheimer disease with the use of neuromelanin MR imaging and (123)Imetaiodobenzylguanidine scintigraphy. *AJNR Am. J. Neuroradiol.* **2013**, *34*, 2113–2118. [CrossRef]
82. Kitao, S.; Matsusue, E.; Fujii, S.; Miyoshi, F.; Kaminou, T.; Kato, S.; Ito, H.; Ogawa, T. Correlation between pathology and neuromelanin MR imaging in Parkinson's disease and dementia with Lewy bodies. *Neuroradiology* **2013**, *55*, 947–953. [CrossRef] [PubMed]
83. Martín-Bastida, A.; Lao-Kaim, N.P.; Roussakis, A.A.; Searle, G.E.; Xing, Y.; Gunn, R.N.; Schwarz, S.T.; Barker, R.A.; Auer, D.P.; Piccini, P. Relationship between neuromelanin and dopamine terminals within the Parkinson's nigrostriatal system. *Brain* **2019**, Epub ahead of print.

84. Kuya, K.; Shinohara, Y.; Miyoshi, F.; Fujii, S.; Tanabe, Y.; Ogawa, T. Correlation between neuromelanin-sensitive MR imaging and (123)I-FP-CIT SPECT in patients with parkinsonism. *Neuroradiology* **2016**, *58*, 351–356. [CrossRef] [PubMed]
85. Isaias, I.U.; Trujillo, P.; Summers, P.; Marotta, G.; Mainardi, L.; Pezzoli, G.; Zecca, L.; Costa, A. Neuromelanin imaging and dopaminergic loss in Parkinson's disease. *Front. Aging Neurosci.* **2016**, *8*, 196. [CrossRef] [PubMed]
86. Lehericy, S.; Sharman, M.A.; Dos Santos, C.L.; Paquin, R.; Gallea, C. Magnetic resonance imaging of the substantia nigra in Parkinson's disease. *Mov. Disord.* **2012**, *27*, 822–830. [CrossRef] [PubMed]
87. Cosottini, M.; Frosini, D.; Pesaresi, I.; Costagli, M.; Biagi, L.; Ceravolo, R.; Bonuccelli, U.; Tosetti, M. MR imaging of the substantia nigra at 7 T enables diagnosis of Parkinson disease. *Radiology* **2014**, *271*, 831–838. [CrossRef]
88. Blazejewska, A.I.; Schwarz, S.T.; Pitiot, A.; Stephenson, M.C.; Lowe, J.; Bajaj, N.; Bowtell, R.W.; Auer, D.P.; Gowland, P.A. Visualization of nigrosome 1 and its loss in PD: Pathoanatomical correlation and in vivo 7 T MRI. *Neurology* **2013**, *81*, 534–540. [CrossRef]
89. Lehericy, S.; Vaillancourt, D.E.; Seppi, K.; Monchi, O.; Rektorova, I.; Antonini, A.; McKeown, M.J.; Masellis, M.; Berg, D.; Rowe, J.B.; et al. The role of high-field magnetic resonance imaging in parkinsonian disorders: Pushing the boundaries forward. *Mov. Disord.* **2017**, *32*, 510–525. [CrossRef]
90. Postuma, R.B.; Berg, D. Advances in markers of prodromal Parkinson disease. *Nat. Rev. Neurol.* **2016**, *12*, 622–634. [CrossRef]
91. Iranzo, A.; Lomeña, F.; Stockner, H.; Valldeoriola, F.; Vilaseca, I.; Salamero, M.; Molinuevo, J.L.; Serradell, M.; Duch, J.; Pavía, J.; et al. Decreased striatal dopamine transporter uptake and substantia nigra hyperechogenicity as risk markers of synucleinopathy in patients with idiopathic rapid-eye-movement sleep behaviour disorder: A prospective study. *Lancet Neurol.* **2010**, *9*, 1070–1077. [CrossRef]
92. Li, Y.; Kang, W.; Yang, Q.; Zhang, L.; Zhang, L.; Dong, F.; Chen, S.; Liu, J. Predictive markers for early conversion of iRBD to neurodegenerative synucleinopathy diseases. *Neurology* **2017**, *88*, 1493–1500. [CrossRef] [PubMed]
93. Iranzo, A.; Santamaria, J.; Valldeoriola, F.; Serradell, M.; Salamero, M.; Gaig, C.; Niñerola-Baizán, A.; Sánchez-Valle, R.; Lladó, A.; De Marzi, R.; et al. Dopamine transporter imaging deficit predicts early transition to synucleinopathy in idiopathic REM sleep behavior disorder. *Ann. Neurol.* **2017**, *82*, 419–428. [CrossRef] [PubMed]
94. Ceravolo, R.; Antonini, A.; Frosini, D.; De Iuliis, A.; Weis, L.; Cecchin, D.; Tosetti, M.; Bonuccelli, U.; Cosottini, M. Nigral anatomy and striatal denervation in genetic parkinsonism: A family report. *Mov. Disord.* **2015**, *30*, 1148–1149. [CrossRef] [PubMed]
95. De Marzi, R.; Seppi, K.; Hogl, B.; Müller, C.; Scherfler, C.; Stefani, A.; Iranzo, A.; Tolosa, E.; Santamaria, J.; Gizewski, E.; et al. Loss of dorsolateral nigral hyperintensity on 3.0 tesla susceptibility-weighted imaging in idiopathic rapid eye movement sleep behavior disorder. *Ann. Neurol.* **2016**, *79*, 1026–1030. [CrossRef] [PubMed]
96. Bae, Y.J.; Kim, J.M.; Kim, E.; Lee, K.M.; Kang, S.Y.; Park, H.S.; Kim, K.J.; Kim, Y.E.; Oh, E.S.; Yun, J.Y.; et al. Loss of nigral hyperintensity on 3 Tesla MRI of parkinsonism: Comparison with (123) I-FP-CIT SPECT. *Mov. Disord.* **2016**, *31*, 684–692. [CrossRef] [PubMed]
97. Frosini, D.; Ceravolo, R.; Tosetti, M.; Bonuccelli, U.; Cosottini, M. Nigral involvement in atypical parkinsonisms: Evidence from a pilot study with ultra-high field MRI. *J. Neural Transm.* **2016**, *123*, 509–513. [CrossRef] [PubMed]
98. Frosini, D.; Cosottini, M.; Donatelli, G.; Costagli, M.; Biagi, L.; Pacchetti, C.; Terzaghi, M.; Cortelli, P.; Arnaldi, D.; Bonanni, E.; et al. Seven tesla MRI of the substantia nigra in patients with rapid eye movement sleep behavior disorder. *Parkinsonism Relat. Disord.* **2017**, *43*, 105–109. [CrossRef]
99. Lee, C.S.; Samii, A.; Sossi, V.; Ruth, T.J.; Schulzer, M.; Holden, J.E.; Wudel, J.; Pal, P.K.; de la Fuente-Fernandez, R.; Calne, D.B.; et al. In vivo positron emission tomographic evidence for compensatory changes in presynaptic dopaminergic nerve terminals in Parkinson's disease. *Ann. Neurol.* **2000**, *47*, 493–503. [CrossRef]
100. Schirinzi, T.; Madeo, G.; Martella, G.; Maltese, M.; Picconi, B.; Calabresi, P.; Pisani, A. Early synaptic dysfunction in Parkinson's disease: Insights from animal models. *Mov. Disord.* **2016**, *31*, 802–813. [CrossRef]
101. Cheng, H.C.; Ulane, C.M.; Burke, R.E. Clinical progression in Parkinson disease and the neurobiology of axons. *Ann. Neurol.* **2010**, *67*, 715–725. [CrossRef] [PubMed]

102. Dijkstra, A.A.; Ingrassia, A.; de Menezes, R.X.; van Kesteren, R.E.; Rozemuller, A.J.; Heutink, P.; van de Berg, W.D. Evidence for Immune Response, Axonal Dysfunction and Reduced Endocytosis in the Substantia Nigra in Early Stage Parkinson's Disease. *PLos ONE* **2015**, *10*, e0128651. [CrossRef] [PubMed]
103. Hornykiewicz, O. Biochemical aspects of Parkinson's disease. *Neurology* **1998**, *51*, S2–S9. [CrossRef] [PubMed]
104. Sian-Hulsmann, J.; Monoranu, C.; Strobel, S.; Riederer, P. Lewy bodies: A spectator or salient killer? *Cns Neurol. Disord. Drug Targets* **2015**, *14*, 947–955. [CrossRef] [PubMed]
105. Simuni, T.; Siderowf, A.; Lasch, S.; Coffey, C.S.; Caspell-Garcia, C.; Jennings, D.; Tanner, C.M.; Trojanowski, J.Q.; Shaw, L.M.; Seibyl, J.; et al. Parkinson's Progression Marker Initiative. Longitudinal change of clinical and biological measures in early Parkinson's disease: Parkinson's Progression Markers Initiative cohort. *Mov. Disord.* **2018**, *33*, 771–782. [CrossRef] [PubMed]
106. Kordower, J.H.; Olanow, C.W.; Dodiya, H.B.; Chu, Y.; Beach, T.G.; Adler, C.H.; Halliday, G.M.; Bartus, R.T. Disease duration and the integrity of the nigrostriatal system in Parkinson's disease. *Brain* **2013**, *136*, 2419–2431. [CrossRef] [PubMed]
107. Bridi, J.C.; Hirth, F. Mechanisms of α-Synuclein Induced Synaptopathy in Parkinson's Disease. *Front. Neurosci.* **2018**, *12*, 80. [CrossRef]
108. Phan, J.A.; Stokholm, K.; Zareba-Paslawska, J.; Jakobsen, S.; Vang, K.; Gjedde, A.; Landau, A.M.; Romero-Ramos, M. Early synaptic dysfunction induced by α-synuclein in a rat model of Parkinson's disease. *Sci. Rep.* **2017**, *7*, 6363. [CrossRef]
109. Ross, G.W.; Petrovitch, H.; Abbott, R.D.; Nelson, J.; Markesbery, W.; Davis, D.; Hardman, J.; Launer, L.; Masaki, K.; Tanner, C.M.; et al. Parkinsonian signs and substantia nigra neuron density in decendents elders without PD. *Ann. Neurol.* **2004**, *56*, 532–539. [CrossRef]
110. Dorsey, E.R.; Sherer, T.; Okun, M.S.; Bloem, B.R. The Emerging Evidence of the Parkinson Pandemic. *J. Parkinsons Dis.* **2018**, *8*, S3–S8. [CrossRef]
111. Tysnes, O.B.; Storstein, A. Epidemiology of Parkinson's disease. *J. Neural Transm.* **2017**, *124*, 901–905. [CrossRef] [PubMed]
112. Kempster, P.A.; O'Sullivan, S.S.; Holton, J.L.; Revesz, T.; Lees, A.J. Relationships between age and late progression of Parkinson's disease: A clinico-pathological study. *Brain* **2010**, *133*, 1755–1762. [CrossRef] [PubMed]
113. Kish, S.J.; Shannak, K.; Rajput, A.; Deck, J.H.; Hornykiewicz, O. Aging produces a specific pattern of striatal dopamine loss: Implications for the etiology of idiopathic Parkinson's disease. *J. Neurochem.* **1992**, *58*, 642–648. [CrossRef] [PubMed]
114. Cruz-Muros, I.; Afonso-Oramas, D.; Abreu, P.; Pérez-Delgado, M.M.; Rodríguez, M.; González-Hernández, T. Aging effects on the dopamine transporter expression and compensatory mechanisms. *Neurobiol. Aging* **2009**, *30*, 973–986. [CrossRef] [PubMed]
115. Lee, C.S.; Kim, S.J.; Oh, S.J.; Kim, H.O.; Yun, S.C.; Doudet, D.; Kim, J.S. Uneven age effects of [(18)F]FP-CIT binding in the striatum of Parkinson's disease. *Ann. Nucl. Med.* **2014**, *28*, 874–879. [CrossRef]
116. Kaasinen, V.; Joutsa, J.; Noponen, T.; Johansson, J.; Seppänen, M. Effects of aging and gender on striatal and extrastriatal [123I]FP-CIT binding in Parkinson's disease. *Neurobiol. Aging* **2015**, *36*, 1757–1763. [CrossRef] [PubMed]
117. Darbin, O. The aging striatal dopamine function. *Parkinsonism Relat. Disord.* **2012**, *18*, 426–432. [CrossRef]
118. Iacono, D.; Geraci-Erck, M.; Rabin, M.L.; Adler, C.H.; Serrano, G.; Beach, T.G.; Kurlan, R. Parkinson disease and incidental Lewy body disease: Just a question of time? *Neurology* **2015**, *85*, 1670–1679. [CrossRef]
119. Gaig, C.; Tolosa, E. When does Parkinson's disease begin? *Mov. Disord.* **2009**, *24* (Suppl. 2), S656–S664. [CrossRef]
120. Piccini, P.; Morrish, P.K.; Turjanski, N.; Sawle, G.V.; Burn, D.J.; Weeks, R.A.; Mark, M.H.; Maraganore, D.M.; Lees, A.J.; Brooks, D.J. Dopaminergic function in familial Parkinson's disease: A clinical and [18F]dopa positron emission tomography study. *Ann. Neurol.* **1997**, *41*, 222–229. [CrossRef]
121. Piccini, P.; Burn, D.J.; Ceravolo, R.; Maraganore, D.; Brooks, D.J. The role of inheritance in sporadic Parkinson's disease: Evidence from a longitudinal study of dopaminergic function in twins. *Ann. Neurol.* **1999**, *45*, 577–582. [CrossRef]
122. Adams, J.R.; van Netten, H.; Schulzer, M.; Mak, E.; Mckenzie, J.; Strongosky, A.; Sossi, V.; Ruth, T.J.; Lee, C.S.; Farrer, M.; et al. PET inLRRK2 mutations: Comparison to sporadic Parkinson's disease and evidence for presymptomatic compensation. *Brain* **2005**, *128*, 2777–2785. [CrossRef] [PubMed]

123. Beach, T.G.; Adler, C.H.; Lue, L.; Sue, L.I.; Bachalakuri, J.; Henry-Watson, J.; Sasse, J.; Boyer, S.; Shirohi, S.; Brooks, R.; et al. Unified staging system for Lewy body disorders: Correlation with nigrostriatal degeneration, cognitive impairment and motor dysfunction. *Acta Neuropathol.* **2009**, *117*, 613–634. [CrossRef] [PubMed]
124. Greffard, S.; Verny, M.; Bonnet, A.M.; Beinis, J.Y.; Gallinari, C.; Meaume, S.; Piette, F.; Hauw, J.J.; Duyckaerts, C. Motor score of the Unified Parkinson Disease Rating Scale as a good predictor of Lewy body-associated neuronal loss in the substantia nigra. *Arch. Neurol.* **2006**, *63*, 584–588. [CrossRef] [PubMed]
125. Zigmond, M.J.; Abercrombie, E.D.; Berger, T.W.; Grace, A.A.; Stricker, E.M. Compensations after lesions of central dopaminergic neurons: Some clinical and basic implications. *Trends Neurosci.* **1990**, *13*, 290–296. [CrossRef]
126. Obeso, J.A.; Rodriguez-Oroz, M.C.; Lanciego, J.L.; Diaz, M.R.; Rodriguez Diaz, M. How does Parkinson's disease begin? The role of compensatory mechanisms. *Trends Neurosci.* **2004**, *27*, 125–127. [CrossRef] [PubMed]
127. Kagi, G.; Bhatia, K.P.; Tolosa, E. The role of DAT-SPECT in movement disorders. *J. Neurol. Neurosurg. Psychiatry* **2010**, *81*, 5–12. [CrossRef]
128. Bezard, E.; Gross, C.E.; Brotchie, J.M. Presymptomatic compensation in Parkinson's disease is not dopamine-mediated. *Trends Neurosci.* **2003**, *26*, 215–221. [CrossRef]
129. Brotchie, J.; Fitzer-Attas, C. Mechanisms compensating for dopamine loss in early Parkinson disease. *Neurology* **2009**, *72* (Suppl. 7), S32–S38. [CrossRef]
130. Blesa, J.; Trigo-Damas, I.; Dileone, M.; Del Rey, N.L.; Hernandez, L.F.; Obeso, J.A. Compensatory mechanisms in Parkinson's disease: Circuits adaptations and role in disease modification. *Exp. Neurol.* **2017**, *298*, 148–161. [CrossRef]
131. De la Fuente-Fernández, R.; Pal, P.K.; Vingerhoets, F.J.; Kishore, A.; Schulzer, M.; Mak, E.K.; Ruth, T.J.; Snow, B.J.; Calne, D.B.; Stoessl, A.J. Evidence for impaired presynaptic dopamine function in parkinsonian patients with motor fluctuations. *J. Neural Transm.* **2000**, *107*, 49–57. [CrossRef] [PubMed]
132. De la Fuente-Fernàndez, R.; Lu, J.Q.; Sossi, V.; Jivan, S.; Schulzer, M.; Holden, J.E.; Lee, C.S.; Ruth, T.J.; Calne, D.B.; Stoessl, A.J. Biochemical variations in the synaptic level of dopamine precede motor fluctuations in Parkinson's disease: PET evidence of increased dopamine turnover. *Ann. Neurol.* **2001**, *49*, 298–303. [CrossRef] [PubMed]
133. De la Fuente-Fernàndez, R.; Sossi, V.; Huang, Z.; Furtado, S.; Lu, J.Q.; Calne, D.B.; Ruth, T.J.; Stoessl, A.J. Levodopa induced changes in synaptic dopamine levels increase with progression of Parkinson's disease: Implications for dyskinesias. *Brain* **2004**, *127*, 2747–2754. [CrossRef] [PubMed]
134. De la Fuente-Fernàndez, R.; Schulzer, M.; Mak, E.; Calne, D.B.; Stoessl, A.J. Presynaptic mechanisms of motor fluctuations in Parkinson's disease: A probabilistic model. *Brain* **2004**, *127*, 888–899. [CrossRef] [PubMed]
135. Sossi, V.; de la Fuente-Fernández, R.; Schulzer, M.; Troiano, A.R.; Ruth, T.J.; Stoessl, A.J. Dopamine transporter relation to dopamine turnover in Parkinson's disease: A positron emission tomography study. *Ann. Neurol.* **2007**, *62*, 468–474. [CrossRef] [PubMed]
136. Stoessl, A.J. Central pharmacokinetics of levodopa: Lessons from imaging studies. *Mov. Disord.* **2015**, *30*, 73–79. [CrossRef] [PubMed]
137. Troiano, A.R.; de la Fuente-Fernàndez, R.; Sossi, V.; Schulzer, M.; Mak, E.; Ruth, T.J.; Stoessl, A.J. PET demonstrates reduced dopamine transporter expression in PD with dyskinesias. *Neurology* **2009**, *72*, 1211–1216. [CrossRef] [PubMed]
138. Hong, J.Y.; Oh, J.S.; Lee, I.; Sunwoo, M.K.; Ham, J.H.; Lee, J.E.; Sohn, Y.H.; Kim, J.S.; Lee, P.H. Presynaptic dopamine depletion predicts levodopa-induced dyskinesia in de novo Parkinson disease. *Neurology* **2014**, *82*, 1597–1604. [CrossRef]
139. Löhle, M.; Mende, J.; Wolz, M.; Beuthien-Baumann, B.; Oehme, L.; van den Hoff, J.; Kotzerke, J.; Reichmann, H.; Storch, A. Putaminal dopamine turnover in de novo Parkinson disease predicts later motor complications. *Neurology* **2016**, *86*, 231–240. [CrossRef]
140. Tomas, D.; Stanic, D.; Chua, H.K.; White, K.; Boon, W.C.; Horne, M. Restoration of the Dopamine Transporter through Cell Therapy Improves Dyskinesia in a Rat Model of Parkinson's Disease. *PLos ONE* **2016**, *11*, e0153424. [CrossRef]
141. Yoo, H.S.; Chung, S.J.; Chung, S.J.; Moon, H.; Oh, J.S.; Kim, J.S.; Hong, J.Y.; Ye, B.S.; Sohn, Y.H.; Lee, P.H. Presynaptic dopamine depletion determines the timing of levodopa-induced dyskinesia onset in Parkinson's disease. *Eur. J. Nucl. Med. Mol. Imaging* **2018**, *45*, 423–443. [CrossRef] [PubMed]

142. Nandhagopal, R.; Kuramoto, L.; Schulzer, M.; Mak, E.; Cragg, J.; Lee, C.S.; McKenzie, J.; McCormick, S.; Samii, A.; Troiano, A.; et al. Longitudinal progression of sporadic Parkinson's disease: A multitracer positron emission tomography study. *Brain* **2009**, *132*, 2970–2979. [CrossRef]
143. Chung, S.J.; Lee, Y.; Oh, J.S.; Kim, J.S.; Lee, P.H.; Sohn, Y.H. Putaminal dopamine depletion in de novo Parkinson's disease predicts future development of wearing-off. *Parkinsonism Relatdisord.* **2018**, *53*, 96–100. [CrossRef] [PubMed]
144. Linazasoro, G.; Van Blercom, N.; Bergaretxe, A.; Iñaki, F.M.; Laborda, E.; Ruiz Ortega, J.A. Levodopa-induced dyskinesias in Parkinson disease are independent of the extent of striatal dopaminergic denervation: A pharmacological and SPECT study. *Clin. Neuropharmacol.* **2009**, *32*, 326–329. [CrossRef]
145. Djaldetti, R.; Rigbi, A.; Greenbaum, L.; Reiner, J.; Lorberboym, M. Can early dopamine transporter imaging serve as a predictor of Parkinson's disease progression and late motor complications? *J. Neurol. Sci.* **2018**, *390*, 255–260. [CrossRef] [PubMed]
146. Calabresi, P.; Di Filippo, M.; Ghiglieri, V.; Tambasco, N.; Picconi, B. Levodopa-induced dyskinesias in patients with Parkinson's disease: Filling the bench-to-bedside gap. *Lancet Neurol.* **2010**, *9*, 1106–1117. [CrossRef]
147. Ravina, B.; Marek, K.; Eberly, S.; Oakes, D.; Kurlan, R.; Ascherio, A.; Beal, F.; Beck, J.; Flagg, E.; Galpern, W.R.; et al. Dopamine transporter imaging is associated with long-term outcomes in Parkinson's disease. *Mov. Disord.* **2012**, *27*, 1392–1397. [CrossRef]
148. Erro, R.; Pappatà, S.; Amboni, M.; Vicidomini, C.; Longo, K.; Santangelo, G.; Picillo, M.; Vitale, C.; Moccia, M.; Giordano, F.; et al. Anxiety is associated with striatal dopamine transporter availability in newly diagnosed untreated Parkinson's disease patients. *Parkinsonism Relat. Disord.* **2012**, *18*, 1034–1038. [CrossRef]
149. Kiferle, L.; Ceravolo, R.; Giuntini, M.; Linsalata, G.; Puccini, G.; Volterrani, D.; Bonuccelli, U. Caudate dopaminergic denervation and visual hallucinations: Evidence from a 123I-FP-CIT SPECT study. *Parkinsonism Relat. Disord.* **2014**, *20*, 761–765. [CrossRef]
150. Santangelo, G.; Vitale, C.; Picillo, M.; Cuoco, S.; Moccia, M.; Pezzella, D.; Erro, R.; Longo, K.; Vicidomini, C.; Pellecchia, M.T.; et al. Apathy and striatal dopamine transporter levels in de-novo, untreated Parkinson's disease patients. *Parkinsonism Relat. Disord.* **2015**, *21*, 489–493. [CrossRef]
151. Kim, R.; Lee, J.; Kim, Y.; Kim, A.; Jang, M.; Kim, H.J.; Jeon, B.; Kang, U.J.; Fahn, S. Presynaptic striatal dopaminergic depletion predicts the later development of freezing of gait in de novo Parkinson's disease: An analysis of the PPMI cohort. *Parkinsonism Relat. Disord.* **2018**, *51*, 49–54. [CrossRef] [PubMed]
152. Kim, R.; Jun, J.S. Association of autonomic symptoms with presynaptic striatal dopamine depletion in drug-naive Parkinson's disease: An analysis of the PPMI data. *Auton. Neurosci.* **2019**, *216*, 59–62. [CrossRef] [PubMed]
153. Latourelle, J.C.; Beste, M.T.; Hadzi, T.C.; Miller, R.E.; Oppenheim, J.N.; Valko, M.P.; Wuest, D.M.; Church, B.W.; Khalil, I.G.; Hayete, B.; et al. Large-scale identification of clinical and genetic predictors of motor progression in patients with newly diagnosed Parkinson's disease: A longitudinal cohort study and validation. *Lancet Neurol.* **2017**, *16*, 908–916. [CrossRef]
154. Li, W.; Lao-Kaim, N.P.; Roussakis, A.A.; Martín-Bastida, A.; Valle-Guzman, N.; Paul, G.; Loane, C.; Widner, H.; Politis, M.; Foltynie, T. 11 C-PE2I and 18 F-Dopa PET for assessing progression rate in Parkinson's: A longitudinal study. *Mov. Disord.* **2018**, *33*, 117–127. [CrossRef] [PubMed]
155. Lee, J.J.; Oh, J.S.; Ham, J.H.; Lee, D.H.; Lee, I.; Sohn, Y.H.; Kim, J.S.; Lee, P.H. Association of body mass index and the depletion of nigrostriatal dopamine in Parkinson's disease. *Neurobiol. Aging* **2016**, *38*, 197–204. [CrossRef]

9

Are Proteinopathy and Oxidative Stress Two Sides of the Same Coin?

Nihar J. Mehta [1], Praneet Kaur Marwah [2] and David Njus [2,*]

1 Radiation Oncology Branch, Center for Cancer Research, National Cancer Institute, National Institutes of Health, Bethesda, MD 20892, USA; nihar.mehta@nih.gov

2 Department of Biological Sciences, Wayne State University, Detroit, MI 48202, USA; praneet.marwah@wayne.edu

* Correspondence: dnjus@wayne.edu

Abstract: Parkinson's disease, like other neurodegenerative diseases, exhibits two common features: Proteinopathy and oxidative stress, leading to protein aggregation and mitochondrial damage respectively. Because both protein aggregates and dysfunctional mitochondria are eliminated by autophagy, we suggest that inadequate clearance may couple the two phenomena. If a neuron's autophagy machinery is overwhelmed, whether by excessive oxidative stress or by excessive protein aggregation, protein aggregates and dysfunctional mitochondria will both accumulate. Parkinson's disease may provide a unique window into this because there is evidence that both sides contribute. Mutations amplifying the aggregation of α-synuclein are associated with Parkinson's disease. Likewise, mutations in Parkin and PINK1, proteins involved in mitophagy, suggest that impaired mitochondrial clearance is also a contributing factor. Many have suggested that dopamine oxidation products lead to oxidative stress accounting for the dopaminergic selectivity of the disease. We have presented evidence for the specific involvement of hypochlorite-oxidized cysteinyl-dopamine (HOCD), a redox-cycling benzothiazine derivative. While toxins like 6-hydroxydopamine and 1-methyl-4-phenyl pyridinium (MPP+) have been used to study mitochondrial involvement in Parkinson's disease, HOCD may provide a more physiologically relevant approach. Understanding the role of mitochondrial dysfunction and oxidative stress in Parkinson's disease and their relation to α-synuclein proteinopathy is important to gain a full picture of the cause, especially for the great majority of cases which are idiopathic.

Keywords: autophagy; cysteinyl-dopamine; hypochlorite; oxidative stress; Parkinson's disease; redox cycling

---

1. Introduction

Neurodegenerative diseases, such as Alzheimer's, Parkinson's, Huntington's, and amyotrophic lateral sclerosis (ALS), are commonly associated with both protein aggregation and oxidative stress. The protein deposits are clearly visible and have attracted considerable attention. Oxidative stress has been much more elusive, and cellular processes contributing to it have been vaguely defined. The dopaminergic neurons that die in Parkinson's disease are unusually prone to mutations in mitochondrial quality-control factors, such as Parkin and PINK1 [1,2], to mitochondrial toxins like rotenone [3], and to dopamine oxidation products [4–11]. As a consequence, Parkinson's disease may offer a unique window into the role of oxidative stress in neurodegenerative diseases generally.

While oxidative stress and proteinopathy are usually studied separately, an integrated perspective may help synthesize what we know about neurodegenerative diseases. Oxidative stress and proteinopathy are linked by autophagy, which is a normal cellular mechanism for clearing both dysfunctional mitochondria and aggregated proteins (Figure 1). This implies that a problem with autophagy will result in accumulation of both protein aggregates and dysfunctional mitochondria, consistent with the coincident occurrence of proteinopathy and oxidative stress. A further implication is that autophagy overload may be caused by either excessive protein aggregation or extreme oxidative stress. In neurodegenerative diseases, what might cause the normal cycling of material by autophagy to spiral out of control leading to the death of neurons?

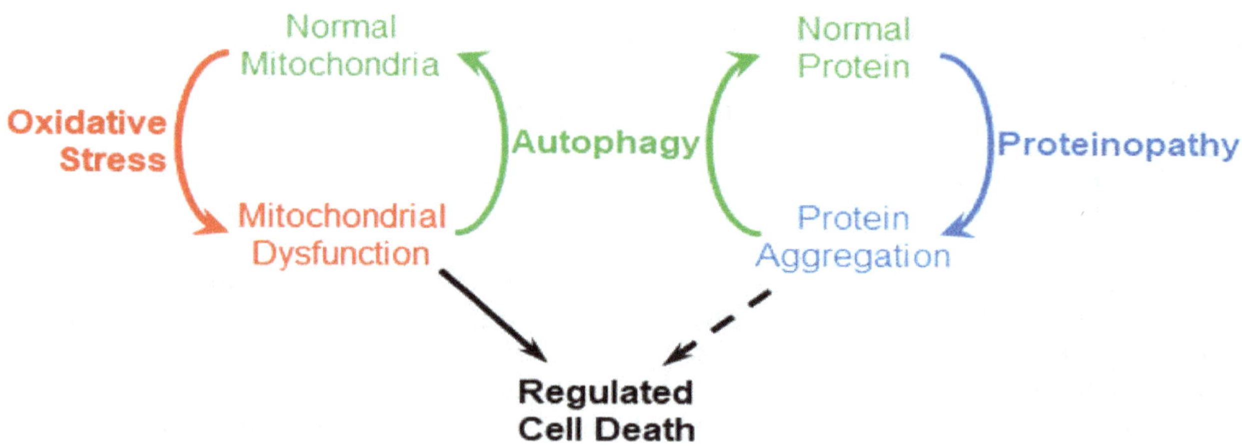

Figure 1. Autophagy (specifically macroautophagy) couples proteinopathy and oxidative stress.

In the context of Parkinson's disease (PD), the unique vulnerability of dopamine neurons to oxidative stress has received considerable attention. Much of that has focused on dopamine, the defining component of dopaminergic neurons, and on potentially toxic products formed by its oxidation [6–11]. Although candidate toxins have been elusive, we recently described hypochlorite-oxidized cysteinyl-dopamine (HOCD), a cytotoxin formed by exposing cysteinyl-dopamine to hypochlorite [4]. Hypochlorite is itself a contributor to oxidative stress, and the enzyme that produces it, myeloperoxidase, is induced by low concentrations of rotenone and by HOCD itself. HOCD is a potent redox cycler and may increase oxidative stress and accelerate the formation of dysfunctional mitochondria. Therefore, HOCD along with α-synuclein aggregation may contribute to an excessive and unsustainable demand for autophagy, ultimately triggering regulated cell death. To put this in a broader context, we will begin with brief discussions of autophagy/proteinopathy and mitophagy/oxidative stress and then consider how HOCD may contribute to the selective death of dopaminergic neurons.

2. Proteinopathy and Autophagy

It is well known that proteinopathies or formation of protein deposits are a common feature of neurodegenerative diseases. The plaques and tangles formed by deposits of amyloid β and tau proteins are hallmarks of Alzheimer's disease. The cytoplasmic inclusion bodies called Lewy bodies containing aggregates of α-synuclein are characteristic of Parkinson's disease. Aggregation of polyglutamate variants of huntingtin is a cause of Huntington's disease. And aggregation of TDP-43 and/or superoxide dismutase 1 is observed in ALS. Mutations in these proteins are associated with familial forms of these diseases arguing that the protein aggregates can contribute to neurodegeneration. Accordingly, much work has focused on the toxicity of specific protein aggregates.

In all proteinopathies, it can be debated whether lethality is attributable to intrinsic toxicity of the protein or to the aggregated state itself. In either case, however, the problem lies in the accumulation of the protein. Evidence supports the view that protein aggregates are removed by autophagy [12–20]. Moreover, there is good evidence that enhancing autophagy is beneficial for treating neurodegenerative diseases [13–15,17,21]. Therefore, we can imagine that protein aggregation leads to autophagy, and problems arise when the capacity of the neuron to clear these deposits by autophagy is exceeded.

Following common practice, we will use the term autophagy to refer specifically to macroautophagy. Autophagy is a highly selective process in which damaged/aggregated proteins or damaged organelles are marked by ubiquitin for elimination. When mitochondria are the target, the process is commonly called mitophagy. In either case, the process begins with the formation of a double-membrane structure called the isolation membrane or phagophore. It wraps around the material to be eliminated forming a closed structure called the autophagosome. Finally, lysosomes fuse with the autophagosome creating an autolysosome, and lysosomal hydrolases then degrade the material inside. During the process, a cytosolic protein, microtubule-associated protein 1A/1B light chain 3B (LC3-I) is recruited to the autophagosomal membrane where it is lipidated with phosphatidylethanolamine to form LC3-phosphatidylethanolamine conjugate (LC3-II). LC3-II, therefore, is useful as a marker for autophagosomes. For reviews of autophagy see References [22,23].

Autophagy/mitophagy is a receptor-mediated process; specific receptors including p62 and optineurin simultaneously interact with the cargo and LC3-II on the isolation membrane, thus recruiting the damaged protein/organelle to the autophagosome. These receptors have an LC3-interacting region (LIR) and a ubiquitin-binding domain (UBD) that binds to ubiquitin chains on the target protein/organelle. Ubiquitin and p62 seem to be more commonly associated with protein deposits. For example, p62 has been reported in α-synuclein deposits in neuron-specific autophagy-deficient mice [24]. On the other hand, in mitophagy, optineurin seems to be the favored receptor [25,26].

With specific relevance to Parkinson's disease, both glucocerebrosidase and leucine-rich repeat kinase 2 seem to be needed for proper functioning of the autophagy/lysosome pathway [27–29]. Mutations in genes for both proteins (*GBA1* and *LRRK2*) are associated with familial cases of Parkinson's disease.

What happens when a cell cannot keep pace with the accumulation of cellular debris? An emerging view is that autophagy is just the normal first response. If a cell is so dysfunctional that its autophagic machinery cannot keep up, that cell is eliminated by regulated cell death as a last resort [22,23]. Historically, regulated cell death was more or less synonymous with apoptosis. As new pathways of regulated cell death were discovered, however, new nomenclature was required. The Nomenclature Committee on Cell Death 2018 has defined twelve major cell death subroutines [30]. Two, intrinsic apoptosis and parthanatos, are of interest here.

Intrinsic apoptosis has been studied extensively. It involves mitochondrial outer membrane permeabilization (MOMP), followed by release of cytochrome c and other factors into the cytosol. This activates a caspase cascade ultimately resulting in cell death. A multitude of regulatory proteins are also involved. Given the detail with which intrinsic apoptosis is understood, it is apparent that it and autophagy are mutually inhibitory [22,23]. Autophagy inhibits apoptosis by eliminating dysfunctional mitochondria that might otherwise trigger the intrinsic apoptotic pathway. Autophagy also degrades components of the apoptosis cascade including caspases and the BH3-only protein NOXA. Apoptosis, on the other hand, inhibits autophagy because apoptotic caspases degrade Beclin-1, a protein involved in expansion of the isolation membrane in autophagy [31]. Thus, the two processes are coordinated so they do not occur concurrently. Autophagy is the normal response to clear unwanted material; if that fails, death by apoptosis may occur.

Parthanatos is a mode of regulated cell death involving hyperactivation of poly(ADP-ribose)polymerase 1 (PARP1) and occurs in response to DNA damage and oxidative stress [30]. Hyperactivation of PARP1 results in the production of poly(ADP-ribose) which binds to apoptosis inducing factor (AIF), causing its release from mitochondria and translocation to the nucleus where it promotes DNA fragmentation. Significantly, protein aggregates, including α-synuclein [32] and amyloid β [33] seem to trigger hyperactivation of PARP1. The dopaminergic neurotoxins MPTP and 6-hydroxydopamine also affect PARP1 [34,35], and we have found that HOCD causes cleavage of PARP1 in PC12 cells [36]. All of this marks parthanatos as a prime contender for the cause of neuronal death in neurodegenerative disease. The interplay between autophagy and parthanatos is not yet clear. One would presume, as with intrinsic apoptosis, that parthanatos is the last resort after autophagy and other protective mechanisms fail. How parthanatos interacts with autophagy and its role in neuronal death are clearly significant questions.

3. Mitochondrial Dysfunction and Mitophagy

Several lines of evidence show that mitochondrial dysfunction plays a pivotal role in the pathogenesis of Parkinson's disease and other neurodegenerative disorders [37–40]. Disruption of electron flow through the respiratory chain as well as other metabolic reactions in the mitochondrion can produce reactive oxygen (ROS) and reactive nitrogen species (RNS), thus contributing to oxidative stress [41]. This ROS/RNS generation can cause irreversible damage to DNA, lipids, and proteins. This is especially significant in mitochondria, which lack many of the repair mechanisms available in the cytosol and nucleus and which are prone to oxidation because of the relatively high pH in the matrix. Mitochondrial damage, especially inhibition of complex I and other enzymes involved in the respiratory chain, has been suggested as one of the fundamental causes of Parkinson's disease [42,43]. In this connection, Complex I inhibition by rotenone increases ROS production [44], and low concentrations of rotenone selectively kill dopaminergic neurons [3].

To prevent the accumulation of damaged, ROS-producing mitochondria, elimination of dysfunctional mitochondria is essential. Mitophagy uses the machinery of autophagy for this selective degradation of senescent and damaged mitochondria, in order to maintain a healthy mitochondrial pool. Several types of mitophagy have been described differing in the mechanism by which mitochondria are engulfed by the autophagosome prior to degradation in the lysosomes. The most well characterized is mitophagy mediated by two proteins—PTEN-induced kinase 1 or PINK1 (a serine/threonine kinase) and Parkin (an E3-ubiquitin ligase). In healthy and polarized mitochondria, PINK1 is imported to the inner membrane where it is cleaved by mitochondrial proteases such as mitochondrial processing peptidase (MPP) and presenilin-associated rhomboid-like protein (PARL) [45–47]. However, mitochondrial inner membrane depolarization, a sign of a damaged mitochondrion, stabilizes PINK1 [48], which then phosphorylates serine 65 of ubiquitin and the N-terminal ubiquitin-like domain of Parkin [49–51]. Phosphorylation of parkin activates its E3 ubiquitin ligase activity, resulting in ubiquitination of mitochondrial proteins, targeting them for degradation by autophagy.

Genetic studies have revealed that mutation in the genes *PRKN* and *PARK6*, which encode for Parkin and PINK1 respectively, are linked to autosomal recessive cases of early-onset or juvenile forms of PD [1,2]. In addition, the *PARK7* gene, also linked to autosomal recessive early-onset cases of PD, encodes for the protein deglycase DJ-1, which also promotes autophagy and maintenance of mitochondrial function [52]. The identification of these mutations in familial forms of PD clearly suggests that impaired mitochondrial turnover is a key feature in the pathogenesis of PD. Moreover, mitophagy is not only impaired in PD, but accumulating evidence suggests that dysfunctional autophagy/mitophagy is also manifested in other neurodegenerative disorders such as Alzheimer's disease [53,54], Huntington's disease [14,55], and ALS [25,56,57].

As Parkin/PINK1-mediated mitophagy depends on the loss of mitochondrial inner membrane potential, it is not surprising that mitophagy is initiated by a variety of mitochondrial toxins. These include the protonophore FCCP, the respiratory chain inhibitor antimycin, and the ATP synthase inhibitor oligomycin. Others include the dopaminergic toxins 6-hydroxydopamine and 1-methyl-4-phenylpyridinium (MPP+) and the pesticide rotenone [58].

Pioneering work by the Greenamyre group established that chronic, systemic exposure to rotenone can produce two major hallmarks of Parkinson's disease: Selective dopaminergic neuron degeneration and α-synuclein accumulation in cytoplasmic inclusions resembling Lewy bodies [3].

Because rotenone is an inhibitor of Complex I of the mitochondrial respiratory chain, this has been considered evidence for the involvement of mitochondrial dysfunction in PD. Rotenone treatment has other effects as well, however. Especially interesting is a link between rotenone and myeloperoxidase expression. Chang et al. [59] demonstrated that rotenone-induced neurotoxicity can be mitigated by modulating myeloperoxidase levels. Moreover, we have reported that rotenone increases the expression of myeloperoxidase in PC12 cells which, by forming hypochlorite, leads to the formation of a toxic redox cycler, HOCD [4]. HOCD formation is exclusive to dopaminergic neurons since it is formed by hypochlorite-mediated oxidation of cysteinyl-dopamine, a product of dopamine oxidation. Interestingly, myeloperoxidase is a lysosomal enzyme, and this may account for its upregulation by agents such as rotenone that promote autophagy/mitophagy.

4. Dopamine Oxidation and HOCD

Following the discovery that Parkinson's disease is associated with the extensive loss of dopamine neurons in the substantia nigra, there has been considerable speculation that dopamine oxidation leads to the formation of toxic products. Some of this has focused on normal products of dopamine metabolism, in particular 3,4-dihydroxyphenylacetaldehyde (DOPAL), which is the immediate product of the enzyme monoamine oxidase (Figure 2). The aldehyde is normally converted to 3,4-dihydroxyphenylacetic acid (DOPAC) by aldehyde dehydrogenase. The aldehyde, however, can conjugate with amines in proteins altering the activity of those proteins [60], and inhibition of aldehyde dehydrogenase does lead to increased toxicity of dopamine [61].

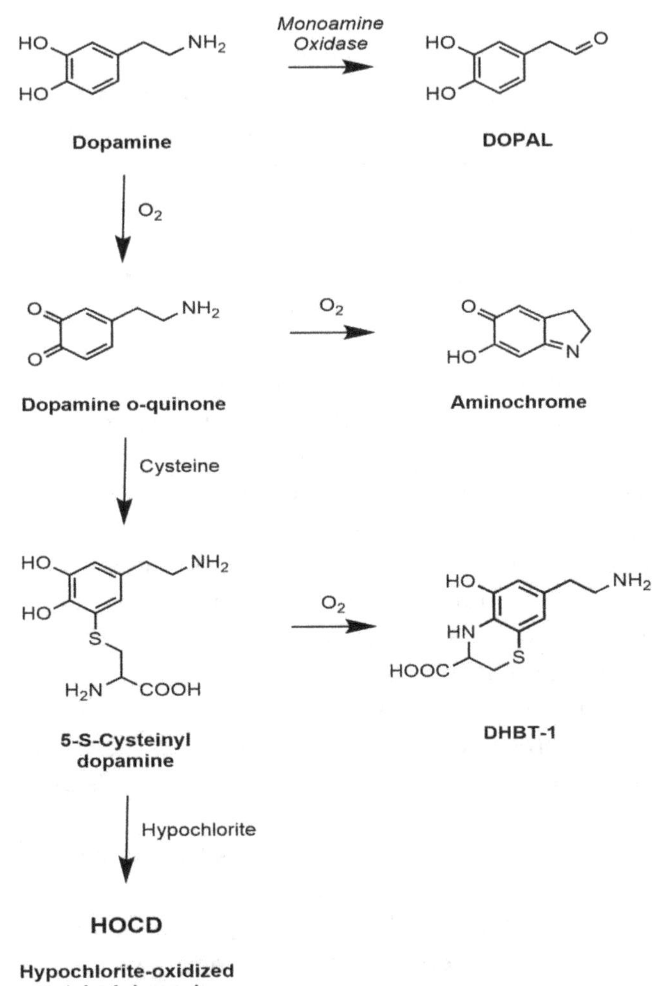

Figure 2. Products of dopamine oxidation.

Most attention, however, has focused on the non-enzymatic oxidation of dopamine. Using induced pluripotent stem cells from genetic and sporadic PD patients, Burbulla et al. [9] found that elevated mitochondrial oxidative stress levels can trigger accumulation of dopamine oxidation adducts which, together with mutation in DJ-1, initiates a toxic cascade resulting in α-synuclein accumulation. Dopamine undergoes spontaneous auto-oxidation to form the dopamine quinone. This is accelerated in the presence of metal ions such as iron or copper, so these would be expected to exacerbate effects of dopamine oxidation. The dopamine quinone itself has been cited as a toxin [10], but it is unstable and either cyclizes to form aminochrome or conjugates with thiols to form products such as 5-S-cysteinyl-dopamine (Figure 2). Aminochrome continues to receive attention [11], but it is neither a very potent neurotoxin nor the main product of dopamine oxidation in vivo. The predominant product in vivo, given the pervasive presence of cysteine, is cysteinyl-dopamine. Carlsson and his colleagues [62] detected cysteinyl-dopamine in the cerebrospinal fluid of PD patients, in dopamine-rich regions of the brain such as the caudate nucleus, putamen, globus pallidus, and substantia nigra, and in neuromelanin. Cysteinyl-dopamine has been reported to kill neuronal cells [6,8], but it is uncertain whether it is cytotoxic itself or metabolizes to toxic products. Dryhurst and colleagues [63] identified many products formed by the oxidation of dopamine in the presence of cysteine. They found that DHBT-1 (7-(2-aminoethyl)-3,4-dihydro-5-hydroxy-2H-1,4-benzothiazine-3-carboxylic acid) is the principal product formed by air oxidation of cysteinyl-dopamine. It inhibits mitochondrial Complex I but is only weakly cytotoxic requiring millimolar concentrations. Treatment of cells with cysteinyl-dopamine can result in oxidative damage, a rise in intracellular calcium, and ultimately apoptosis. Recently, Vauzour et al. [8] attributed its toxicity to combined effects of cysteinyl-dopamine itself and DHBT-1.

Oxidized dopamine and cysteinyl-dopamine also polymerize to form neuromelanin. Neuromelanin is a stable substance, and it may confer protection by acting as a sink for oxidized dopamine products and by chelating iron [64]. However, neuromelanin may also contribute to increased susceptibility of melanized neurons due to accumulation of increased loads of iron and toxic metabolites. Moreover, microglia activation by neuromelanin released from degenerating neurons can further contribute to neurodegeneration.

Rather than examine toxicity of specific products, we chose to approach the problem by looking for an activity: Redox cycling. The process of redox cycling involves alternating reduction and oxidation reactions continuing until either molecular oxygen or reducing equivalents are exhausted. This leads to the proliferation of a variety of reactive oxygen species including superoxide and hydrogen peroxide, so redox cycling agents can induce oxidative stress and mitochondrial dysfunction. Because cysteinyl-dopamine is the primary product of dopamine oxidation in vivo, we chose to seek redox cycling products formed from cysteinyl-dopamine. We [4] discovered that treatment of cysteinyl-dopamine with hypochlorite yields a product with very high redox cycling activity (Figure 3), and we refer to this product as HOCD (hypochlorite-oxidized cysteinyl-dopamine).

Using PC12 cells, we confirmed that cysteinyl-dopamine is toxic. However, HOCD is toxic at lower concentrations. Moreover, two lines of evidence suggest that the toxicity of cysteinyl-dopamine depends on its conversion to HOCD. First, including taurine in the medium protects PC12 cells against cysteinyl-dopamine but not against HOCD. Taurine scavenges hypochlorite and blocks the hypochlorite-dependent conversion of cysteinyl-dopamine into HOCD in vitro. Thus, it is likely that taurine also prevents this conversion in vivo, thereby protecting cells against cysteinyl-dopamine but not against HOCD.

The second line of evidence is that rotenone potentiates the toxicity of cysteinyl-dopamine but not of HOCD. Consistent with reports of others, we found that low concentrations of rotenone increase myeloperoxidase expression in PC12 cells. The resulting increase in hypochlorite due to increased myeloperoxidase activity would be expected to yield more effective conversion of cysteinyl-dopamine to HOCD, thereby increasing toxicity of cysteinyl-dopamine but not of HOCD. In this connection, Gellhaar et al. [65] found that brain regions affected in PD patients show significant increases in myeloperoxidase immunoreactivity, providing further evidence that myeloperoxidase may mediate the selective vulnerability of dopaminergic neurons to oxidative stress.

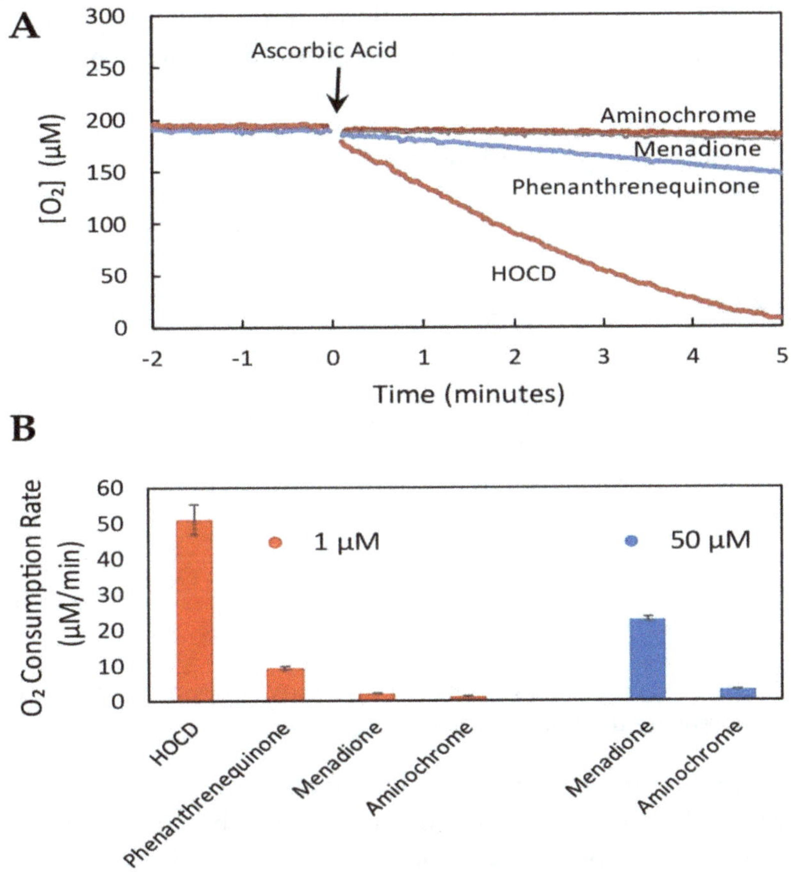

Figure 3. Hypochlorite-oxidized cysteinyl-dopamine (HOCD) undergoes extremely rapid redox cycling. (**A**) Oxygen consumption mediated by 1 μM concentrations of the indicated redox cyclers following addition of 2.5 mM ascorbic acid (arrow). Oxygen consumption was measured in aqueous solution (0.2 M potassium phosphate, 1 μM EDTA, pH 7.4) at 37 °C. (**B**) Comparison of redox cycling rates (initial slopes of plots shown in 3A) by 1 μM concentrations of redox cyclers (red bars) or 50 μM concentrations (blue bars). Averages (± standard deviation) of three replicate samples are shown (authors' unpublished data).

We have succeeded in scaling up the synthesis of HOCD and have purified the redox cycling product by chromatography through Dowex 50Wx8. It appears to be a 1,4-benzothiazine with the dopamine oxygens on the 7 and 8 carbons. This puts an O para to the N and allows the compound to undergo facile oxidation/reduction. In fact, HOCD redox cycles faster than any compound we have tested using our standard redox cycling assay (measured as rate of oxygen consumption in 0.2 M potassium phosphate, 1 μM EDTA, pH 7.4 in the presence of 2.5 mM ascorbic acid at 37 °C). HOCD is two orders of magnitude faster than menadione and three orders of magnitude faster than aminochrome (Figure 3).

Because of this fast redox-cycling, we suspect that HOCD contributes to oxidative stress and damage to mitochondria. This is suggested by the fact that other redox active compounds, such as 6-hydroxydopamine and aminochrome, do the same. Moreover, HOCD causes a rapid increase in superoxide levels in PC12 cells as observed using the fluorescent mitochondrial superoxide indicator MitoSOX Red [4]. Finally, HOCD causes an increase in expression and activity of the lysosomal enzyme myeloperoxidase, perhaps by increasing mitophagy. A consequence of the upregulation of myeloperoxidase is increased hypochlorite production, which should increase HOCD formation. This self-reinforcing feedback should cause oxidative stress to spiral out of control

(Figure 4). For neurodegenerative diseases, this self-reinforcing feedback is especially significant. Something must change a manageable situation into one that accelerates uncontrollably. Our results with PC12 cells indicate that neuronal myeloperoxidase may produce enough hypochlorite to convert cysteinyl-dopamine to HOCD, but it is also possible that microglial myeloperoxidase plays a significant or even dominant role in vivo. Thus, inflammation stimulated by α-synuclein deposits may also promote HOCD formation.

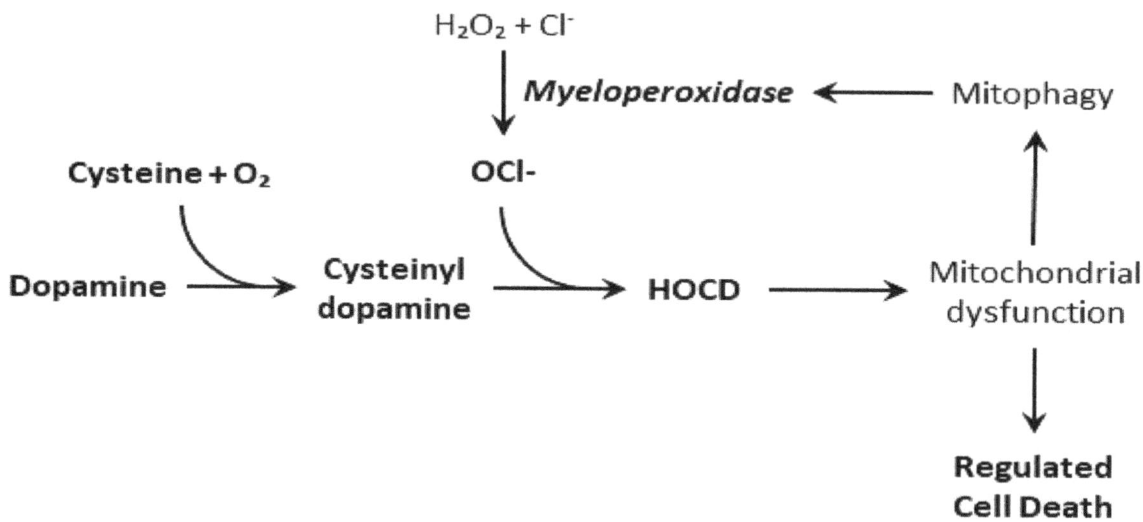

Figure 4. Formation of hypochlorite-oxidized cysteinyl-dopamine from dopamine and its self-enhancement by increasing myeloperoxidase expression.

A question that is important but difficult to answer now is whether HOCD reaches toxic concentrations under physiological conditions. In most cases, the answer is clearly no, because most people do not suffer from Parkinson's disease. But toxicity depends on context. How robust or how compromised are cellular protective mechanisms including autophagy/mitophagy? To what extent are neurons under assault by other factors such as protein aggregation or other sources of oxidative stress? And do these factors interact amplifying the total stress on the neuron? Then, HOCD may be the extra burden that makes dopaminergic neurons uniquely vulnerable in Parkinson's disease.

5. Conclusions

In summary, we view Parkinson's disease as a problem of autophagy overload caused by the combined accumulation of dysfunctional mitochondria and aggregated α-synuclein. Under normal circumstances, mitochondrial degeneration and protein aggregation occur at rates slow enough for autophagy to maintain homeostasis. Under pathological conditions, however, something causes these to accelerate out of control. The two factors may contribute differently in different patients. In familial cases involving mutations in α-synuclein or duplication or triplication of the *SNCA* gene, excessive protein aggregation is likely the dominant contributor. In cases involving mutations in Parkin or PINK1, inadequate mitophagy is the obvious culprit. In the great majority of idiopathic cases, environmental toxins or dopamine oxidation products such as HOCD may make a significant contribution. Protein aggregation and oxidative stress may also interact. Oxidative stress-induced damage of α-synuclein can enhance its oligomerization and aggregation [66,67]. Moreover, α-synuclein aggregation may exacerbate oxidative stress. Therefore, proteinopathy and oxidative stress may be synergistic, not simply additive, mutually escalating the rates at which they occur. This kind of positive feedback is essential to push the normal clearance of material by autophagy out of control. Then, if overwhelming autophagy is the signal for regulated cell death in Parkinson's and other neurodegenerative diseases, proteinopathy and oxidative stress must be considered as a whole; they are two sides of the same coin.

References

1. Kitada, T.; Asakawa, S.; Hattori, N.; Matsumine, H.; Yamamura, Y.; Minoshima, S.; Yokochi, M.; Mizuno, Y.; Shimizu, N. Mutations in the parkin gene cause autosomal recessive juvenile parkinsonism. *Nature* **1998**, *392*, 605–608. [CrossRef]
2. Valente, E.M.; Abou-Sleiman, P.M.; Caputo, V.; Muqit, M.M.; Harvey, K.; Gispert, S.; Ali, Z.; Del Turco, D.; Bentivoglio, A.R.; Healy, D.G.; et al. Hereditary early-onset Parkinson's disease caused by mutations in PINK1. *Science* **2004**, *304*, 1158–1160. [CrossRef]
3. Betarbet, R.; Sherer, T.B.; MacKenzie, G.; Garcia-Osuna, M.; Panov, A.V.; Greenamyre, J.T. Chronic systemic pesticide exposure reproduces features of Parkinson's disease. *Nat. Neurosci.* **2000**, *3*, 1301–1306. [CrossRef] [PubMed]
4. Mehta, N.J.; Asmaro, K.; Hermiz, D.J.; Njus, M.M.; Saleh, A.H.; Beningo, K.A.; Njus, D. Hypochlorite converts cysteinyl-dopamine into a cytotoxic product: A possible factor in Parkinson's Disease. *Free Radic. Biol. Med.* **2016**, *101*, 44–52. [CrossRef]
5. Genova, M.L.; Abd-Elsalam, N.M.; Mahdy, E.S.M.E.; Bernacchia, A.; Lucarini, M.; Pedulli, G.F.; Lenaz, G. Redox cycling of adrenaline and adrenochrome catalyzed by mitochondrial Complex I. *Arch. Biochem. Biophys.* **2006**, *447*, 167–173. [CrossRef] [PubMed]
6. Mosca, L.; Tempera, I.; Lendaro, E.; Di Francesco, L.; d'Erme, M. Characterization of catechol-thioether-induced apoptosis in human SH-SY5Y neuroblastoma cells. *J. Neurosci. Res.* **2008**, *86*, 954–960. [CrossRef]
7. Hauser, D.N.; Dukes, A.A.; Mortimer, A.D.; Hastings, T.G. Dopamine quinone modifies and decreases the abundance of the mitochondrial selenoprotein glutathione peroxidase 4. *Free Radic. Biol. Med.* **2013**, *65*, 419–427. [CrossRef] [PubMed]
8. Vauzour, D.; Pinto, J.T.; Cooper, A.J.L.; Spencer, J.P.E. The neurotoxicity of 5-S -cysteinyl-dopamine is mediated by the early activation of ERK1/2 followed by the subsequent activation of ASK1/JNK1/2 pro-apoptotic signaling. *Biochem. J.* **2014**, *463*, 41–52. [CrossRef] [PubMed]
9. Burbulla, L.F.; Song, P.; Mazzulli, J.R.; Zampese, E.; Wong, Y.C.; Jeon, S.; Santos, D.P.; Blanz, J.; Obermaier, C.D.; Strojny, C.; et al. Dopamine oxidation mediates mitochondrial and lysosomal dysfunction in Parkinson's disease. *Science* **2017**, *357*, 1255–1261. [CrossRef] [PubMed]
10. Biosa, A.; Arduini, I.; Soriano, M.E.; Giorgio, V.; Bernardi, P.; Bisaglia, M.; Bubacco, L. Dopamine oxidation products as mitochondrial endotoxins, a potential molecular mechanism for preferential neurodegeneration in Parkinson's Disease. *ACS Chem. Neurosci.* **2018**, *9*, 2849–2858. [CrossRef]
11. Herrera, A.; Muñoz, P.; Steinbusch, H.W.M.; Segura-Aguilar, J. Are dopamine oxidation metabolites involved in the loss of dopaminergic neurons in the Nigrostriatal system in Parkinson's Disease? *ACS Chem. Neurosci.* **2017**, *8*, 702–711. [CrossRef]
12. Nixon, R.A.; Yang, D.S. Autophagy failure in Alzheimer's disease—Locating the primary defect. *Neurobiol. Dis.* **2011**, *43*, 38–45. [CrossRef] [PubMed]
13. Orr, M.E.; Oddo, S. Autophagic/lysosomal dysfunction in Alzheimer's Disease. *Alzheimers Res. Ther.* **2013**, *5*, 53. [CrossRef]
14. Martin, D.D.O.; Ladha, S.; Ehrnhoefer, D.E.; Hayden, M.R. Autophagy in Huntington disease and huntingtin in autophagy. *Trends Neurosci.* **2015**, *38*, 26–35. [CrossRef] [PubMed]
15. Lee, J.K.; Shin, J.H.; Lee, J.E.; Choi, E.J. Role of autophagy in the pathogenesis of amyotrophic lateral sclerosis. *Biochim. Biophys. Acta* **2015**, *1852*, 2517–2524. [CrossRef] [PubMed]
16. Wang, B.; Abraham, N.; Gao, G.; Yang, Q. Dysregulation of autophagy and mitochondrial function in Parkinson's disease. *Transl Neurodegener.* **2016**, *5*, 19. [CrossRef]
17. Moors, T.E.; Hoozemans, J.J.M.; Ingrassia, A.; Beccari, T.; Parnetti, L.; Chartier-Harlin, M.C.; van de Berg, W.D.J. Therapeutic potential of autophagy enhancing agents in Parkinson's disease. *Mol. Neurodegener.* **2017**, *12*, 11. [CrossRef]
18. Uddin, M.S.; Stachowiak, A.; Al Mamun, A.; Tzvetkov, N.T.; Takeda, S.; Atanasov, A.G.; Bergantin, L.B.; Abdel-Daim, M.M.; Stankiewicz, A.M. Autophagy and Alzheimer's disease: From molecular mechanisms to therapeutic implications. *Front. Aging Neurosci.* **2018**, *10*, 4. [CrossRef]

19. Pircs, K.; Petri, R.; Madsen, S.; Brattas, P.L.; Vuono, R.; Ottosson, D.R.; St-Amour, I.; Hersbach, B.A.; Matusiak-Bruckner, M.; Lundh, S.H.; et al. Huntingtin Aggregation Impairs Autophagy, Leading to Argonaute-2 Accumulation and Global MicroRNA Dysregulation. *Cell Rep.* **2018**, *24*, 1397–1406. [CrossRef]
20. Nguyen, D.K.H.; Thombre, R.; Wang, J. Autophagy as a common pathway in amyotrophic lateral sclerosis. *Neurosci. Lett.* **2018**. [CrossRef]
21. Hetz, C.; Thielen, P.; Matus, S.; Nassif, M.; Court, F.; Kiffin, R.; Martinez, G.; Cuervo, A.M.; Brown, R.H.; Glimcher, L.H. XBP-1 deficiency in the nervous system protects against amyotrophic lateral sclerosis by increasing autophagy. *Genes Dev.* **2009**, *23*, 2294–2306. [CrossRef]
22. Mukhopadhyay, S.; Panda, P.K.; Sinha, N.; Das, D.N.; Bhutia, S.K. Autophagy and apoptosis: Where do they meet? *Apoptosis* **2014**, *19*, 555–566. [CrossRef] [PubMed]
23. Marino, G.; Niso-Santano, M.; Baehrecke, E.H.; Kroemer, G. Self-consumption: The interplay of autophagy and apoptosis. *Nat. Rev. Mol. Cell Biol.* **2014**, *15*, 81–94. [CrossRef] [PubMed]
24. Sato, S.; Uchihara, T.; Fukuda, T.; Noda, S.; Kondo, H.; Saiki, S.; Komatsu, M.; Uchiyama, Y.; Tanaka, K.; Hattori, N. Loss of autophagy in dopaminergic neurons causes Lewy pathology and motor dysfunction in aged mice. *Sci. Rep.* **2018**, *8*, 2813. [CrossRef]
25. Wong, Y.C.; Holzbaur, E.L. Optineurin is an autophagy receptor for damaged mitochondria in parkin-mediated mitophagy that is disrupted by an ALS-linked mutation. *Proc. Natl. Acad. Sci. USA* **2014**, *111*, E4439–E4448. [CrossRef] [PubMed]
26. Lazarou, M.; Sliter, D.A.; Kane, L.A.; Sarraf, S.A.; Wang, C.; Burman, J.L.; Sideris, D.P.; Fogel, A.I.; Youle, R.J. The ubiquitin kinase PINK1 recruits autophagy receptors to induce mitophagy. *Nature* **2015**, *524*, 309–314. [CrossRef]
27. Murphy, K.E.; Gysbers, A.M.; Abbott, S.K.; Tayebi, N.; Kim, W.S.; Sidransky, E.; Cooper, A.; Garner, B.; Halliday, G.M. Reduced glucocerebrosidase is associated with increased α-synuclein in sporadic Parkinson's disease. *Brain* **2014**, *137*, 834–848. [CrossRef] [PubMed]
28. Magalhaes, J.; Gegg, M.E.; Migdalska-Richards, A.; Doherty, M.K.; Whitfield, P.D.; Schapira, A.H.V. Autophagic lysosome reformation dysfunction in glucocerebrosidase deficient cells: Relevance to Parkinson disease. *Hum. Mol. Gen.* **2016**, *25*, 3432–3445. [CrossRef]
29. Manzoni, C. The LRRK2-macroautophagy axis and its relevance to Parkinson's disease. *Biochem. Soc. Trans.* **2017**, *45*, 155–162. [CrossRef]
30. Galluzzi, L.; Vitale, I.; Aaronson, S.A.; Abrams, J.M.; Adam, D.; Agostinis, P.; Alnemri, E.S.; Altucci, L.; Amelio, I.; Andrews, D.W.; et al. Molecular mechanisms of cell death: Recommendations of the Nomenclature Committee on Cell Death 2018. *Cell Death Differ.* **2018**, *25*, 486–541. [CrossRef]
31. Wirawan, E.; Vande Walle, L.; Kersse, K.; Cornelis, S.; Claerhout, S.; Vanoverberghe, I.; Roelandt, R.; De Rycke, R.; Verspurten, J.; Declercq, W.; et al. Caspase-mediated cleavage of Beclin-1 inactivates Beclin-1-induced autophagy and enhances apoptosis by promoting the release of proapoptotic factors from mitochondria. *Cell Death Dis.* **2010**, *1*, e18. [CrossRef] [PubMed]
32. Kam, T.I.; Mao, X.; Park, H.; Chou, S.C.; Karuppagounder, S.S.; Umanah, G.E.; Yun, S.P.; Brahmachari, S.; Panicker, N.; Chen, R.; et al. Poly(ADP-ribose) drives pathologic α-synuclein neurodegeneration in Parkinson's disease. *Science* **2018**, *362*, 557. [CrossRef] [PubMed]
33. Martire, S.; Mosca, L.; d'Erme, M. PARP-1 involvement in neurodegeneration: A focus on Alzheimer's and Parkinson's diseases. *Mech. Ageing Dev.* **2015**, *146*, 53–64. [CrossRef]
34. Mandir, A.S.; Przedborski, S.; Jackson-Lewis, V.; Wang, Z.Q.; Simbulan-Rosenthal, C.M.; Smulson, M.E.; Hoffman, B.E.; Guastella, D.B.; Dawson, V.L.; Dawson, T.M. Poly(ADP-ribose) polymerase activation mediates 1-methyl-4-phenyl-1, 2,3,6-tetrahydropyridine (MPTP)-induced parkinsonism. *Proc. Natl. Acad. Sci. USA* **1999**, *96*, 5774–5779. [CrossRef] [PubMed]
35. Kim, T.W.; Cho, H.M.; Choi, S.Y.; Suguira, Y.; Hayasaka, T.; Setou, M.; Koh, H.C.; Mi Hwang, E.; Park, J.Y.; Kang, S.J.; et al. (ADP-ribose) polymerase 1 and AMP-activated protein kinase mediate progressive dopaminergic neuronal degeneration in a mouse model of Parkinson's disease. *Cell Death Dis.* **2013**, *4*, e919. [CrossRef] [PubMed]
36. Mehta, N.J. Understanding the mechanism of oxidative stress generation by dopamine oxidized metabolites: Implications in Parkinson's disease. Ph.D. Thesis, Wayne State University, Detroit, MI, USA, 2017.
37. Keane, P.C.; Kurzawa, M.; Blain, P.G.; Morris, C.M. Mitochondrial dysfunction in Parkinson's disease. *Parkinsons Dis.* **2011**, *2011*, 716871. [CrossRef]

38. Johri, A.; Beal, M.F. Mitochondrial dysfunction in neurodegenerative diseases. *J. Pharmacol. Exp. Ther.* **2012**, *342*, 619–630. [CrossRef]
39. Hroudová, J.; Singh, N.; Fišar, Z. Mitochondrial dysfunctions in neurodegenerative diseases: Relevance to Alzheimer's Disease. *BioMed Res. Int.* **2014**, *2014*, 9. [CrossRef]
40. Golpich, M.; Amini, E.; Mohamed, Z.; Ali, R.A.; Ibrahim, N.M.; Ahmadiani, A. Mitochondrial dysfunction and biogenesis in neurodegenerative diseases: Pathogenesis and Treatment. *CNS Neurosci. Ther.* **2017**, *23*, 5–22. [CrossRef]
41. Murphy, M.P. How mitochondria produce reactive oxygen species. *Biochem. J.* **2009**, *417*, 1–13. [CrossRef]
42. Schapira, A.H.; Cooper, J.M.; Dexter, D.; Clark, J.B.; Jenner, P.; Marsden, C.D. Mitochondrial complex I deficiency in Parkinson's disease. *J. Neurochem.* **1990**, *54*, 823–827. [CrossRef]
43. Parker, W.D., Jr.; Parks, J.K.; Swerdlow, R.H. Complex I deficiency in Parkinson's disease frontal cortex. *Brain Res.* **2008**, *1189*, 215–218. [CrossRef] [PubMed]
44. Fato, R.; Bergamini, C.; Bortolus, M.; Maniero, A.L.; Leoni, S.; Ohnishi, T.; Lenaz, G. Differential effects of mitochondrial Complex I inhibitors on production of reactive oxygen species. *Biochim. Biophys. Acta* **2009**, *1787*, 384–392. [CrossRef] [PubMed]
45. Jin, S.M.; Lazarou, M.; Wang, C.; Kane, L.A.; Narendra, D.P.; Youle, R.J. Mitochondrial membrane potential regulates PINK1 import and proteolytic destabilization by PARL. *J. Cell Biol.* **2010**, *191*, 933–942. [CrossRef]
46. Greene, A.W.; Grenier, K.; Aguileta, M.A.; Muise, S.; Farazifard, R.; Haque, M.E.; McBride, H.M.; Park, D.S.; Fon, E.A. Mitochondrial processing peptidase regulates PINK1 processing, import and Parkin recruitment. *EMBO Rep.* **2012**, *13*, 378–385. [CrossRef] [PubMed]
47. Yamano, K.; Youle, R.J. PINK1 is degraded through the N-end rule pathway. *Autophagy* **2013**, *9*, 1758–1769. [CrossRef] [PubMed]
48. Narendra, D.P.; Jin, S.M.; Tanaka, A.; Suen, D.F.; Gautier, C.A.; Shen, J.; Cookson, M.R.; Youle, R.J. PINK1 is selectively stabilized on impaired mitochondria to activate Parkin. *PLoS Biol.* **2010**, *8*, e1000298. [CrossRef]
49. Kane, L.A.; Lazarou, M.; Fogel, A.I.; Li, Y.; Yamano, K.; Sarraf, S.A.; Banerjee, S.; Youle, R.J. PINK1 phosphorylates ubiquitin to activate Parkin E3 ubiquitin ligase activity. *J. Cell Biol.* **2014**, *205*, 143–153. [CrossRef]
50. Kondapalli, C.; Kazlauskaite, A.; Zhang, N.; Woodroof, H.I.; Campbell, D.G.; Gourlay, R.; Burchell, L.; Walden, H.; Macartney, T.J.; Deak, M.; et al. PINK1 is activated by mitochondrial membrane potential depolarization and stimulates Parkin E3 ligase activity by phosphorylating Serine 65. *Open Biol.* **2012**, *2*, 120080. [CrossRef]
51. Koyano, F.; Okatsu, K.; Kosako, H.; Tamura, Y.; Go, E.; Kimura, M.; Kimura, Y.; Tsuchiya, H.; Yoshihara, H.; Hirokawa, T.; et al. Ubiquitin is phosphorylated by PINK1 to activate parkin. *Nature* **2014**, *510*, 162. [CrossRef] [PubMed]
52. Thomas, K.J.; McCoy, M.K.; Blackinton, J.; Beilina, A.; van der Brug, M.; Sandebring, A.; Miller, D.; Maric, D.; Cedazo-Minguez, A.; Cookson, M.R. DJ-1 acts in parallel to the PINK1/parkin pathway to control mitochondrial function and autophagy. *Hum. Mol. Gen.* **2011**, *20*, 40–50. [CrossRef] [PubMed]
53. Nixon, R.A.; Wegiel, J.; Kumar, A.; Yu, W.H.; Peterhoff, C.; Cataldo, A.; Cuervo, A.M. Extensive involvement of autophagy in Alzheimer disease: An immuno-electron microscopy study. *J. Neuropathol. Exp. Neurol.* **2005**, *64*, 113–122. [CrossRef]
54. Lee, J.H.; Yu, W.H.; Kumar, A.; Lee, S.; Mohan, P.S.; Peterhoff, C.M.; Wolfe, D.M.; Martinez-Vicente, M.; Massey, A.C.; Sovak, G.; et al. Lysosomal proteolysis and autophagy require presenilin 1 and are disrupted by Alzheimer-related PS1 mutations. *Cell* **2010**, *141*, 1146–1158. [CrossRef]
55. Martinez-Vicente, M.; Talloczy, Z.; Wong, E.; Tang, G.; Koga, H.; Kaushik, S.; de Vries, R.; Arias, E.; Harris, S.; Sulzer, D.; et al. Cargo recognition failure is responsible for inefficient autophagy in Huntington's disease. *Nat. Neurosci.* **2010**, *13*, 567–576. [CrossRef] [PubMed]
56. Maruyama, H.; Morino, H.; Ito, H.; Izumi, Y.; Kato, H.; Watanabe, Y.; Kinoshita, Y.; Kamada, M.; Nodera, H.; Suzuki, H.; et al. Mutations of optineurin in amyotrophic lateral sclerosis. *Nature* **2010**, *465*, 223–226. [CrossRef] [PubMed]
57. Majcher, V.; Goode, A.; James, V.; Layfield, R. Autophagy receptor defects and ALS-FTLD. *Mol. Cell Neurosci.* **2015**, *66*, 43–52. [CrossRef] [PubMed]
58. Georgakopoulos, N.D.; Wells, G.; Campanella, M. The pharmacological regulation of cellular mitophagy. *Nat. Chem. Biol.* **2017**, *13*, 136–146. [CrossRef]
59. Chang, C.Y.; Choi, D.K.; Lee, D.K.; Hong, Y.J.; Park, E.J. Resveratrol confers protection against rotenone-induced neurotoxicity by modulating myeloperoxidase levels in glial cells. *PLoS ONE* **2013**, *8*, e60654. [CrossRef]

60. Jinsmaa, Y.; Sharabi, Y.; Sullivan, P.; Isonaka, R.; Goldstein, D.S. 3,4-Dihydroxyphenyl acetaldehyde-induced protein modifications and their mitigation by N-acetylcysteine. *J. Pharmacol. Exp. Ther.* **2018**, *366*, 113–124. [CrossRef]
61. Fitzmaurice, A.G.; Rhodes, S.L.; Lulla, A.; Murphy, N.P.; Lam, H.A.; O'Donnell, K.C.; Barnhill, L.; Casida, J.E.; Cockburn, M.; Sagasti, A.; et al. Aldehyde dehydrogenase inhibition as a pathogenic mechanism in Parkinson disease. *Proc. Natl. Acad. Sci. USA* **2013**, *110*, 636–641. [CrossRef]
62. Rosengren, E.; Linder-Eliasson, E.; Carlsson, A. Detection of 5-S-cysteinyldopamine in human brain. *J. Neural Transm.* **1985**, *63*, 247–253. [CrossRef] [PubMed]
63. Shen, X.M.; Dryhurst, G. Further insights into the influence of L-cysteine on the oxidation chemistry of dopamine: Reaction pathways of potential relevance to Parkinson's Disease. *Chem. Res. Toxicol.* **1996**, *9*, 751–763. [CrossRef]
64. Zucca, F.A.; Segura-Aguilar, J.; Ferrari, E.; Munoz, P.; Paris, I.; Sulzer, D.; Sarna, T.; Casella, L.; Zecca, L. Interactions of iron, dopamine, and neuromelanin pathways in brain aging and Parkinson's disease. *Prog. Neurobiol.* **2017**, *155*, 96–119. [CrossRef] [PubMed]
65. Gellhaar, S.; Sunnemark, D.; Eriksson, H.; Olson, L.; Galter, D. Myeloperoxidase-immunoreactive cells are significantly increased in brain areas affected by neurodegeneration in Parkinson's and Alzheimer's disease. *Cell Tissue Res.* **2017**, *369*, 445–454. [CrossRef]
66. Scudamore, O.; Ciossek, T. Increased oxidative stress exacerbates alpha-Synuclein aggregation in vivo. *J. Neuropath. Exp. Neurol.* **2018**, *77*, 443–453. [CrossRef]
67. Xiang, W.; Schlachetzki, J.C.M.; Helling, S.; Bussmann, J.C.; Berlinghof, M.; Schäffer, T.E.; Marcus, K.; Winkler, J.; Klucken, J.; Becker, C.-M. Oxidative stress-induced posttranslational modifications of alpha-synuclein: Specific modification of alpha-synuclein by 4-hydroxy-2-nonenal increases dopaminergic toxicity. *Mol. Cell. Neurosci.* **2013**, *54*, 71–83. [CrossRef] [PubMed]

10

GBA, Gaucher Disease and Parkinson's Disease: From Genetic to Clinic to New Therapeutic Approaches

Giulietta M. Riboldi [1,2,*] and Alessio B. Di Fonzo [2,3]

1 The Marlene and Paolo Fresco Institute for Parkinson's and Movement Disorders, NYU Langone Health, New York, NY 10017, USA
2 Dino Ferrari Center, Neuroscience Section, Department of Pathophysiology and Transplantation, University of Milan, 20122 Milan, Italy; alessio.difonzo@policlinico.mi.it
3 Foundation IRCCS Ca' Granda Ospedale Maggiore Policlinico, Neurology Unit, 20122 Milan, Italy
* Correspondence: giulietta.riboldi@nyulangone.org or giulietta.riboldi@unimi.it

Abstract: Parkinson's disease (PD) is the second most common degenerative disorder. Although the disease was described more than 200 years ago, its pathogenetic mechanisms have not yet been fully described. In recent years, the discovery of the association between mutations of the *GBA* gene (encoding for the lysosomal enzyme glucocerebrosidase) and PD facilitated a better understating of this disorder. *GBA* mutations are the most common genetic risk factor of the disease. However, mutations of this gene can be found in different phenotypes, such as Gaucher's disease (GD), PD, dementia with Lewy bodies (DLB) and rapid eye movements (REM) sleep behavior disorders (RBDs). Understanding the pathogenic role of this mutation and its different manifestations is crucial for geneticists and scientists to guide their research and to select proper cohorts of patients. Moreover, knowing the implications of the *GBA* mutation in the context of PD and the other associated phenotypes is also important for clinicians to properly counsel their patients and to implement their care. With the present review we aim to describe the genetic, clinical, and therapeutic features related to the mutation of the *GBA* gene.

Keywords: glucocerebrosidase; Parkinson's disease; Gaucher's disease; Lewy Body Dementia; REM sleep behavior disorders

1. Introduction

GBA is a gene located on chromosome 1 (1q21) encoding for the glucocerebrosidase (GCase), a lysosomal enzyme involved in the metabolism of glucosylceramide. The mutation of this gene has been classically associated with Gaucher's disease, a systemic disorder with a variable degree of involvement of the central nervous system. Surprisingly, about 14 years ago it was observed that mutations in this same gene were associated with an increased incidence of Parkinson's disease (PD), in both Gaucher's patients as well as asymptomatic carriers [1–4]. PD is the second most common neurodegenerative disorder, affecting 2–3% of the world population over the age of 65 [5]. It is caused by the progressive loss of dopaminergic neurons in the substantia nigra. Classically it presents with a combination of bradykinesia, rigidity, resting tremor, and postural instability. However, a list of non-motor features, such as hyposmia, constipation, urinary symptoms, orthostatic hypotension, anxiety, depression, impaired sleep, and cognitive impairment can present as well in various degrees [5]. Since the first observations of *GBA* and PD, their association has been extensively explored. Different hypotheses have been formulated to explain the causative role of this mutation in PD [6]. First of all, GCase is part of the endolysosomal pathway, which seems to be particularly crucial in the pathogenesis of PD. Indeed, many different monogenic familial forms of PD are caused by genes involved in this pathway [7]. Moreover, mutated GCase is not able to fold properly and thus can accumulate in different cellular compartments of the dopaminergic neurons, causing a cell stress response that can be deleterious of the cells. In addition, impaired GCase activity seems to cause an accumulation of alpha-synuclein (for a comprehensive review see [8]).

Today we know that *GBA* mutations are the major genetic risk factor for PD. Impaired GCase activity has been identified also in idiopathic cases of PD patients who did not carry a mutation in the gene, suggesting a central role of this enzyme in the pathogenesis of the disease [9,10].

In the present review, we aim to summarize the genetic changes and the characteristic features associated with the mutations of this gene, spanning from Gaucher's disease to PD and the other described phenotypes. This will aid in a better understanding of the pathogenic role of this mutation. The identification of these phenotypes will allow for clinicians to offer more appropriate counseling to the patients and their families.

2. Pathogenetic Mutations of the *GBA* Gene

2.1. *GBA Mutation and Gaucher's Disease (GD)*

Gaucher's disease (GD) is a systemic disorder that can present with a various degree of systemic and neurological manifestations. According to the severity of the disease and the neurological involvement, three different types of GD have been identified. GD type 1 has been classically considered only a systemic disorder, with no neurological involvement whatsoever. Anemia, leukopenia, thrombocytopenia with frequent bleeding, osteopenia with bone pain, easy fractures, Erlenmeyer flask deformity, as well as hepatosplenomegaly, failure to grow, and puberty delay can be presenting features of this disease [11–14]. Monoclonal gammopathy has been reported as well [15]. The disease can manifest early in childhood but it may remain undiagnosed until adulthood when the phenotype is mild. The pathological hallmarks of the disease are the so-called Gaucher cells, macrophages engorged with aberrant lysosomes as a consequence of the GCase-impaired activity. Symptoms are caused by the infiltration of these cells in the reticuloendothelial system of the affected organs [16]. In recent years, the natural history of GD type 1 has dramatically changed since the introduction of target treatments, such as enzyme replacement therapy (ERT) (human recombinant enzyme to be administered intravenously every other week) and oral substrate reduction therapy (SRT) [17]. Treatments with these two approaches are able to address the majority of the systemic symptoms associated with GD type 1 and those in GD type 3. So far, SRT has been approved only for subjects over the age of 18 years. However, in the adult population it represents an important alternative first line treatment. Unfortunately, these therapies are not able to cross the blood-brain barrier and therefore they are not suitable for the treatment of the neurological complications associated with GD type 2 and 3. The two latest forms are also referred to as the acute (type 2) and chronic (type 3) neuronopathic form. Patients affected with GD type 2 start manifesting severe symptoms very early, usually within the first six months of life. They usually present a combination of severe neurological manifestations, with brainstem involvement (i.e., eye movement abnormalities, spasticity, hypotonia) and seizure, as well as life-threatening systemic symptoms, such as respiratory distress and aspiration pneumonia [18,19]. Skin manifestations, like ichthyosis or collodion abnormalities, as well as hydrops fetalis, can be present. Prognosis is very poor and death usually occurs before the age of four [20]. GD type 3 (chronic neuronopathic form) has been further classified as GD type 3a,b,c. GD type 3a presents a milder visceral phenotype, but can be associated with severe and life-threatening myoclonic seizures. GD type 3b, instead, is characterized by a more prominent visceral involvement [21]. Interestingly, one of the features that have been used to try to discriminate between patients with GD type 1 and the milder neuropathic form GD type 3 is the assessment of the eye movements. Indeed, patients with GD type 3, especially type b, present with characteristic eye movement abnormalities. In particular they show loss of horizontal before vertical gaze palsy and slowing of the saccades, suggesting involvement of the brainstem. GD type 3c, instead, is the only subtype of the disease presenting with cardiac mitral and aortic calcification and poor prognosis [21]. A particular cluster of patients with GD type 3 has been identified among the Swedish population. This is also referred as Norrbottnian form, because of its geographical distribution. It is associated with the c.1448T > G mutation and it presents with an early and severe splenomegaly and a combination in the first or second decade of ataxia, spastic paresis, horizontal supranuclear gaze palsy, kyphoscoliosis and other orthopedic abnormalities, cognitive impairment, and seizures [22].

Those different phenotypes are associated with discrete genetic mutations, as detailed below.

Different Pathogenic Mutations of *GBA* Associated with Gaucher's Disease (GD) Subtypes

More than 300 variants of the *GBA* gene have been associated with Gaucher's disease [23]. GD is an autosomal recessive disorder. In order for the disease to manifest, patients need to carry a pathogenic mutation on both alleles of the GBA gene, either in a homozygous or compound heterozygous fashion. Point mutations, insertion, deletion, missense mutations, splice junctions, and concomitant multiple mutations have been reported [24]. The different variants can be more represented in particular ethnic groups as well as in particular phenotypes. The c.1226A < G (N370S; or N409S according to the new nomenclature) mutation is the most common one among Ashkenazi Jew (AJ) patients, followed by the c.84dupG (84GG) mutation, which is more rare. The c.115 + 1G > A (IVS2 + 1), c.1504C > T (R463C), and c.1604G > A (R496H) are commonly found in AJ patients with GD type 1 [24]. On the contrary, the N370S mutation is rarely found among Chinese and Japanese patients [24] (Hruska et al., 2008). Among Asian ethnic groups, the c.1448T > C (L444P, or L483P according to the new nomenclature) and the c.754T > A (F252I), usually associated with GD type 2 and 3, are more prevalent, also explaining why among these populations the neuropathic forms of GD are more frequent [20]. c.1448T > C (L444P) is also the most frequent mutation among Caucasians with a non-Ashkenazi Jew ancestry [25] (Figure 1).

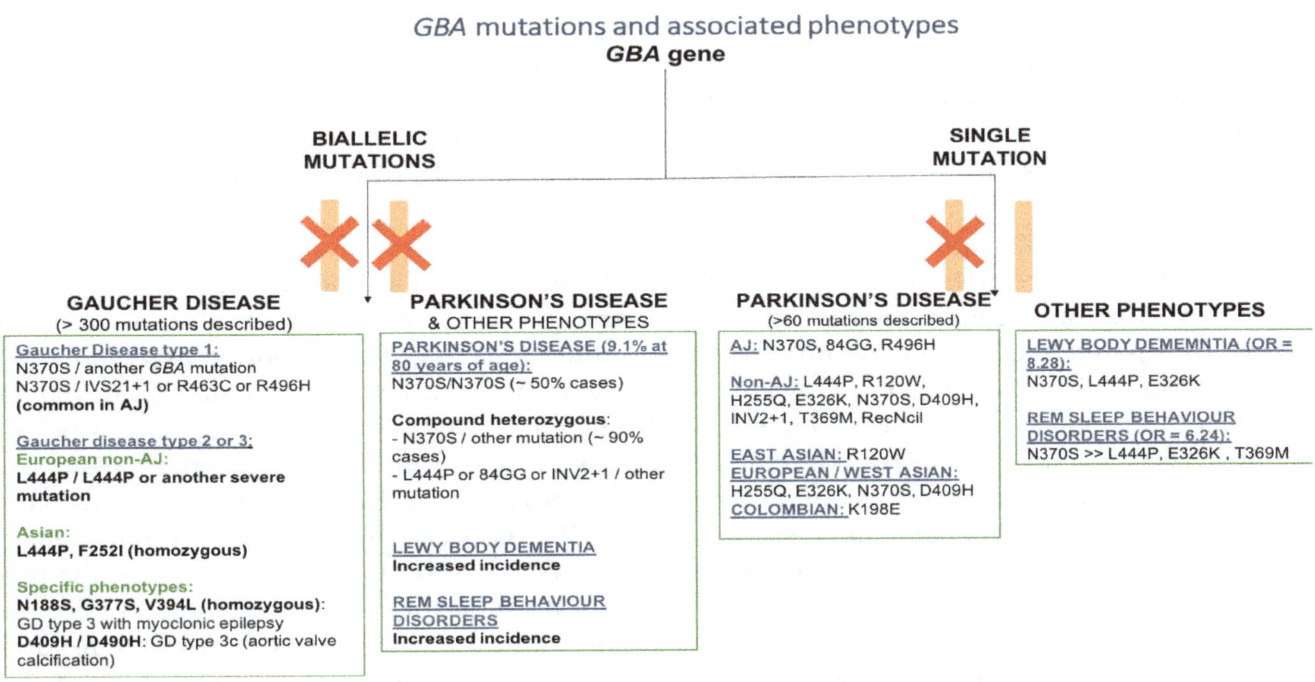

Figure 1. Schematic representation of the most common pathogenic mutations of the *GBA* genes and associated phenotypes. Phenotypes were grouped based on homozygous and heterozygous mutations, ancestry, and specific associated features.

Different mutations can lead to different phenotypes of GD. The c.1226A > G (N370S) mutation is associated only with Gaucher's disease type 1 and it seems to be protective for the development of the neurological involvement characteristic of GD type 2 and 3. Indeed, patients who present the c.1226A > G (N370S) mutation on at least one allele of the *GBA* gene will manifest only GD type 1 [24]. Interestingly, subjects who are homozygous for the N370S variant can also remain asymptomatic for the disease. On the other hand, the c.1448T > C (L444P) mutation is usually associated with GD type 2 or 3, even when presenting in a compound heterozygous state [19]. Homozygous c.1448T > C (L444P) mutation [c.1448T > C]1[c.1448T > C] (L444P/L444P) with no recombinant alleles can be associated with very severe but also milder phenotypes [26]. The c.1342G > C (D409H) variant is responsible for GD type 3c which presents with characteristic cardiac valve calcifications [27]. c.680A > G (N188S), c.1246G > A (G377S), and c.1297G > T (V394L) are more likely to be associated with myoclonic epilepsy [28–30]. Despite previously reported observations, it is commonly found that members of the same family report variability in the manifestation of symptoms even with an identical genotype, suggesting that a genotype/phenotype correlation is tentative still. Other reported mutations are uniquely rare and oftentimes private among specific families. [12]. Hence, it is difficult to make generalizations about these mutations specific phenotypical profiles.

Another interesting mutation is the c.1093G > A (E326K), which caused a lot of debate in the literature [31]. Indeed, it is not clear whether this mutation is really pathogenic for GD, since it was found also in a significant number of asymptomatic carriers in homozygosity [32,33]. However, when associated with other *GBA* mutations on the same allele, it can cause severe impairment of the GCase activity [34,35]. Interestingly enough, the same mutation seems to be significantly associated with an increased risk of PD [33].

2.2. GBA Mutation and Parkinson's Disease (PD)

2.2.1. Pathogenic Mutations of *GBA* Associated with PD

More than a decade ago, the association between an increased risk of developing PD and the presence of *GBA* mutations was initially noticed in large Gaucher's disease clinics. The incidence of PD among GD patients and their relatives, which were supposedly carriers for the mutation, seemed to be higher than the general population. Initially, only single case reports were suggesting this association. Interestingly, PD was noticed in patients with GD type 1, which has always been considered the non-neuropathic form of the disease [36–41]. It was only when larger populations of PD patients were screened for mutations of this gene that the important role of *GBA* in the pathogenesis of PD was assessed worldwide.

So far, more than 50 population studies have screened the *GBA* gene among PD patients, covering a large number of ancestries (reviewed in [42,43]). Overall, these studies demonstrated that the incidence of *GBA* mutations is significantly higher among PD patients, compared to non-affected subjects. Compared to GD, a smaller number of *GBA* mutations have been reported in patients with PD (about 130 *GBA* mutations) [42]. However, in many of these studies, only the mutations that are most commonly associated with PD were screened. Therefore, less frequent variants still associated with the disease could go undetected. Among all, the c.1226A > G (N370S) and the c.1448T > C (L444P) mutations are the two most common mutations worldwide. Indeed, in some populations they account for the 70–80% of the total number of variants of *GBA* associated with PD [44]. Among subjects from eastern Europe with an AJ ancestry, the c.1226A > G (N370S) mutation is definitely the most frequent one among PD patients, as already reported for GD (Figure 1). Among the non-AJ European descendants, the c.1448T > C (L444P) mutation is more common. Interestingly, it has been reported that some mutations are able to increase the risk of PD only in the context of specific ancestry [42]. This is the case of the c.84dupG (84GG) and c.1604G > A (R496H) for AJ subjects, the c. 475 C > T (R120W) for East Asian populations, and the c.882T > G (H255Q), c.1093G > A (E326K), c.1342G > C (D409H), and c.1226A < G (N370S), which are only found in subjects of European or West Asian ancestry [42] (Figure 1). A recent study identified an increased incidence of the K198E variant (previously described in GD1 and GD2 patients) in a population of PD patients from Columbia compared to controls [43].

It seems that severe *GBA* mutations (as classified according to the subtype of GD that they are associated with), such as c.84dupG (84GG), c.115 + 1G >A (IVS2 + 1), c.1297G > T (V394L), c.1342G > C (D409H), c.1448T > C (L444P), and c.1263del + RecTL, are associated with a higher risk of causing PD compared to milder mutations, such as the N370S and c.84dupG (84GG) [45]. Moreover, severe mutations are associated with an earlier age of onset, as well as a more rapid progression and increased involvement of cognitive functions [45–47]. In one study, the motor and some of the non-motor symptoms (such as depression, REM sleep behavior disorders, and hyposmia) were significantly worse in PD patients carrying severe *GBA* mutations compared to subjects carrying mild mutations or with idiopathic PD [48].

Interestingly, *GBA* represents only a risk factor for PD. This means that not every carrier will develop the disease. The reason for the reduced penetrance of these mutations has not yet been fully elucidated. Based on large population studies, today we know that, among *GBA* carriers, about 9.1% will develop PD. Some reports suggest that the penetrance of PD in GD patient is 30% at 80 years, but this data needs to be confirmed by further studies [49]. Patients with a homozygous mutation of *GBA*, thus affected with Gaucher's disease, have a higher risk of developing PD and usually with an earlier age of onset of symptoms [48]. Having said that, it is worth noticing that the majority of subjects with GD will never develop PD, even in the case of severe mutations. It is still controversial whether PD in patients with GD presents with a more severe phenotype compared to *GBA* carriers. Carriers of the *GBA* mutation harbor an increased risk of developing PD by five times in heterozygous carriers and 10–20 times in homozygous carriers [50–53]. *GBA* mutations are present in about 2–30% of PD patients [54]. Carrier frequency can be very different across different ancestry. Among AJs, it goes from 10 to 31%, while in Norwegian's it is only 2.3% [54]. In patients of European non-AJ ancestry, it ranges from 2.9 to 12% [54].

In the last few years, there has been a great effort to try to clarify the pathogenic role of the *GBA* mutations in PD and many different hypotheses have been formulated, as reported above (for review see [8]). It is important to note that a growing amount of data is suggesting a failure of the endolysosomal and of the autophagic pathways in PD [55]. These scavenger systems are crucial for the degradation of alpha-synuclein, whose accumulation in the dopaminergic neurons is one of the hallmarks of PD. In the lysosome, GCase plays an important contribution for these processes and, in particular, in the interplay with alpha-synuclein [56]. Therefore, it is not totally surprising that a dysfunction of this enzyme is related to PD. How the different mutations of *GBA* that have been described in PD patients are able to affect the activity of the GCase has not been fully understood. We know that the GCase has three active domains. PD-associated mutations are found in distinctive domains of the protein. The c.1342G > C (D409H) and c.1297G > T (V394L) variants are located in domain I. The c.84dupG (84GG) mutation causes a frameshift that can induce aberrantly shorter or longer proteins that are non-functioning [23]. Other mutations, instead, are not found in the functional domains but do interfere with the final structure of the enzyme, thus making it more unstable or affecting its interaction with other proteins. The c.1226A > G (N370S) and c.1448T > C (L444P) mutations are, for example, located in the proximity of the binding site of the Saposin C, an activator of GCase [57]. More importantly, SapC competes with the binding of alpha-synuclein to GCase, which would cause the inhibition of the enzyme [58,59]. Interestingly, the c.1226A > G (N370S) mutation also seems to affect the ability of the GCase to modify the conformation of one of its loops, loop 3, according to changes in pH [60,61]. Conformational changes in response to the changes of the cellular environment are critical for the proper function of the protein. Despite our knowledge about the structural effects of the different mutations, the exact correlation between the localizations of pathogenetic variants of the gene and the degree of expression of PD has not yet been fully described.

It is also worth noting that *GBA* presents a pseudogene (*GBAP1*) that shares a very high degree of homology—96% sequence identity–located in the proximity of the original gene [62,63]. Therefore, genetic analysis will have to take this into account and should be performed in a specialized laboratory in order to obtain reliable results. New technologies, such as the long-read sequencer, are on the horizon for even more in-depth identification of possible *GBA* mutations [64].

2.2.2. *GBA* Mutations and Parkinson's Disease Phenotype

PD patients carrying *GBA* mutations are not easily recognizable in most cases because they do not present exclusive features that would clearly distinguish them from patients with idiopathic PD (iPD). However, large population studies comparing carriers vs. non-carriers, mild vs. severe mutations, as well as heterozygous manifesting carriers vs. PD–GD patients, allowed the ability to define common traits in these subgroups of patients (for a comprehensive review see [8]). In particular, *GBA*–PD patients present an overall earlier age of onset compared to non-carriers. Disease manifests about 3–6 years earlier in heterozygous carriers, irrespectively of the severity of the mutation, and about 6–11 years earlier in subjects with homozygous mutations [45,46,48,54,65–70]. There are limited reports of *GBA* mutation carriers having an age of onset in the 20's. [31,54,71–73].

The progression of the disease has been characterized in many different studies by a more pronounced cognitive deficit in a significant percentage of these patients, with a risk of developing dementia up to three times higher compared to iPD, which is even more increased in patients with severe mutations [46,48,74]. Hallucinations and REM sleep behavior disorders (RBD) also are more common among *GBA* patients in a dose-dependent fashion, being more frequent in subjects with homozygous mutations and in patients carrying severe vs. milder mutations. However, other non-motor symptoms, such as depression and anxiety, constipation, urinary symptoms, orthostatic hypotension, and sexual dysfunctions are over-represented as well in *GBA* carriers compared to iPD, especially in the presence of severe mutations, but with no increased severity in GD patients [46,48,75,76]. An increased incidence of dysautonomic features has been suggested to be the main driver of the slightly reduced survival reported in these patients [77]. Motor complications, such as dysphagia, dysarthria, and freezing of gait, are more frequent as well in *GBA* carriers [46,67].

In patients with *GBA* mutations and PD, the rigid akinetic phenotype seems to be more common. Usually, these patients present a very good response to levodopa, although the progression of the motor symptoms can be slightly faster compared to iPD but without higher rates of motor fluctuations or dyskinesia. Therefore, no specific treatment approaches need to be considered for this subgroup of patients. Interestingly, a recent study evaluated the outcomes of treatment with deep brain stimulation (DBS) in a cohort of PD patients carrying *GBA* mutations [78]. After a follow up of 7.5 years on average, it was noticed that the het-*GBA* cohort presented similar outcomes compared to iPD in terms of motor symptoms, while cognitive impairment and non-motor symptoms were definitely more represented among carriers [78]. However, because of the beneficial effect on the motor symptoms, DBS should be considered as a suitable option for these patients.

2.2.3. *GBA* Mutations and Other Phenotypes

GBA mutations were identified also in cases of REM sleep behavior disorders (RBD) and in cases of dementia with Lewy bodies (DLB) [79].

GBA and Dementia with Lewy Bodies

A relatively low number of studies have been conducted to explore the incidence of the *GBA* mutation among patients affected with dementia with Lewy bodies (DLB), which was found to be even higher compared to the one in PD patients. In a cohort study of DLB patients, the frequency of *GBA* mutations was 7.49% with an odd ratio of 8.28 [79]. In another study in Spanish subjects, and in a number of autoptic brain tissues from pathologically proven DLB patients, a *GBA* mutation was identified in 12–13% of the cases [80]. Recent genome-wide association studies (GWAS) also confirmed the significant association between *GBA* mutations and DLB (particularly the rs35749011 variant) [81]. Among *GBA* carriers, the risk of developing DLB is about three times greater than developing PD [82].

As well as in PD patients, *GBA* mutations are associated with an earlier age of onset in DLB cases compared to non-carriers (of approximately five years) and a higher disease severity score [79,80]. The association between *GBA* mutations and DLB was found to be higher among male subjects compared to female [80]. These observations were confirmed also in a following study in a cohort of patients with DLB and AJ ancestry [83]. *GBA* mutation carriers (about 11% of the entire cohort) presented more severe symptoms, particularly in terms of increased hallucinations, worse RBD symptoms, and overall cognitive and motor features [83].

A number of different mutations of the *GBA* gene have been reported in DLB patients. Other than the two mutations most frequently associated with PD (c.1226A > G (N370S) and c.1448T > C (L444P)), the E326K variant is over-represented in this cohort of patients compared to controls [79,80]. Interestingly, the c.1093G > A (E326K) mutation also is frequently found in patients with PD dementia (PDD) [84].

Neuropathological data does not significantly differ between DLB patients with and without a *GBA* mutation [79]. However, *GBA* carriers present a reduced GCase activity as well as a more pronounced alteration of lipid profiles in the brain [85]. *GBA* expression profiles have been shown to be reduced in DLB and PDD cases in both specific brain regions (temporal cortex and caudate nucleus respectively) and in the peripheral blood [86]. *GBA* mutations are more significantly associated with Lewy bodies (LB) pathology (especially with a cortical localization) than with Alzheimer's disease (AD) pathology (i.e., beta-amyloid and neurofibrillary tangles inclusions) [87].

GBA and REM Sleep Behavior Disorders

REM sleep behavior disorders (RBDs) are considered one of the prodromal symptoms of PD and patients affected by this disorder may present with alpha-synuclein accumulation in the brain [88]. According to a recent metanalysis, patients affected with RBDs present an estimated risk of developing a neurodegenerative disorder up to 97% after more than 14 years of follow up [89]. The majority of the cases who present a phenoconversion will develop an alpha-synucleinopathy, represented by PD in the majority of the cases, but also Multiple System Atrophy (MSA), Dementia with Lewy Bodies (DLB), and PD with dementia [90]. In fact, subjects with RBD may present clinical symptoms fulfilling the criteria for prodromal PD in up to 74% of the cases, manifesting worse performances in both motor and non-motor assessments compared to non-affected subjects [91,92]. Notably, many of the studies in this field did not take into consideration the significance of a family history of a neurodegenerative disorder, therefore, it is probable that the percentage of patients that reported a neurodegenerative disease is misrepresented. It would be worth exploring this aspect in future studies.

RBD seems to be more frequent in PD patients with *GBA* mutations compared with patients without this mutation (OR 3.13) [48,65,67,76]. RBDs are also more frequent in PD patients with concomitant GD than in heterozygous carriers [48]. Based on these observations, a few studies explored the incidence of the *GBA* mutation among patients affected with RBD [65,91–93]. These studies reported that among patients with idiopathic RBDs there is an increased frequency of *GBA* mutations (2.6–11.6% of RBD patients vs. 0.4–1.8% of the controls) [65,91,93]. A number of different *GBA* mutations were identified in RBD patients [65,93]. Some of these mutations have already been reported in PD patients, while others still do not have a clear pathogenic role. Among all the reported mutations, the two more commonly found in PD (i.e., c.1226A > G (N370S) and c.1448T > C (L444P), with N370S >> L444P), together with the c.1093G > A (E326K) and the c.1223C > T (T369M), were the most frequently represented in subjects with RBD [65,91–93].

Subjects with homozygous *GBA* mutations, thus affected with GD, and heterozygous carriers with no PD, presented significant worsening of rapid eye movement sleep behavior disorder scores over a period of time of two years compared with non-carrier subjects [92]. Among *GBA* carriers, the odds ratio (OR) for RBD was 6.24 (95% CI 3.76–10.35, $P < 0.0001$) [65]. The presence of *GBA* mutations does not seem to increase the risk among RBD patients of phenoconverting into PD [93]. These observations all together suggest that *GBA* may play a role in the development of RBDs, but not necessarily in determining more severe phenotypes.

Interestingly, no mutations of the *LRRK2* gene, the other common genetic risk factor for PD, have been identified so far in patients with RBDs [91,94].

3. New Targeted Treatments for *GBA*–PD Patients

Despite the very successful treatments that are now available to address the systemic manifestations of Gaucher's disease, unfortunately these approaches (i.e., enzyme replacement therapy and substrate reduction therapy) are not able to reach the central nervous system and thus fail to address the neurological symptoms caused by the disease. Different companies have been working for years to try to address this issue, producing very promising results in cellular and animal models. We are now in a very exciting era where some of these experimental approaches are starting to reach the clinical scene. The treatments available so far in clinical trials try to address two main mechanisms that are thought to be detrimental in linking *GBA* mutations to PD. The first hypothesis is that mutated forms of *GBA* are not able to fold properly in the endoplasmic reticulum (ER) in the cells, causing the protein to accumulate in this cellular compartment [95]. This would trigger a stress response in the dopaminergic neurons leading to their damage and death [95]. Also, the entrapment of the beta GCase in the ER causes reduced levels of the enzyme in the cells, triggering alpha-synuclein accumulation [95]. In order to target this pathogenic mechanism, different chaperones, which are proteins able to facilitate the refolding of their substrates, were tested [95–99]. In 2016, a clinical study assessing the efficacy of ambroxol, one of these chaperones that showed very exciting preliminary results, was started (NCT02914366 study: https://www.clinicaltrials.gov/ct2/show/NCT02914366?cond=gba+parkinson&rank=7). This is a phase 2 clinical trial to assess the safety and the efficacy of this drug to improve motor and cognitive features of PD patients with a *GBA* mutation. The study is currently ongoing. Another similar approach has been tested in a phase 1 study by Allergan with LTI-291, a chaperone molecule able to increase the activity of GCase (https://lti-staging.squarespace.com/our-science/#lti-291). Isofagomine is another chaperone protein that has been tested in vitro and in vivo to assess its ability to modulate the phenotype induced by mutations of *GBA* [97]. This molecule is an inhibitory chaperone whose role would be the stabilization of the GCase. Clinical trials with this molecule are not available at the moment. It is also worth considering that small molecules, such as chaperones, can present different therapeutic profiles in carriers of the different mutations of *GBA* according to the effect of these variants on the protein [100].

The second mechanism that has been explored to treat *GBA*–PD patients is the accumulation in the dopaminergic neurons of glucosylceramide (the substrate normally degraded by the GCase) because of the mutation of *GBA* [101–103]. Genzyme recently started a multicenter, randomized, double-blind, placebo-controlled phase 2 study to assess the safety, pharmacokinetics, and pharmacodynamics of an oral compound, ibiglustat (GZ/SAR402671), which is able to reduce the levels of beta-glucocerebrosidase in *GBA* carriers with early-stage PD (MOVES-PD study: https://www.clinicaltrials.gov/ct2/show/NCT02906020?cond=gba+parkinson&rank=2). It is still a long road for the establishment of an effective treatment, but many paths have been established, giving hope for patients with PD.

Mutated GCase is more unstable compared to the wild-type form. Therefore, modulation of the degradation of GCase could be another suitable strategy to increase the activity of the enzyme and thus tackle alpha-synuclein accumulation and neurodegeneration. Hsp90β, together with other heat shock proteins (HSP), such as Hsp27, parkin, and the endoplasmic reticulum-associated pathway, are responsible for the degradation of misfolded GCase. In particular, histone deacetylase inhibitors (HDACis) and direct inhibitors of specific HSP are able to increase the GCase activity, reducing its degradation [104]. Indeed, HDACis prevent the interaction between Hdp90β and GCase through the hyperactivation of one of its domains [105].

GCase plays an important role in the autophagy-lysosomal pathway (ALP), where other genes that have been associated with PD, such as *ATP13A2*, scavenger receptor class B member 2 (*SCARB2*), sphingomyelin phosphodiesterase 1 (*SMPD1*), and others, are also involved (Moors et al., 2016). Failure of the ALP seems to be responsible for the accumulation of alpha-synuclein in neurons. Therefore, a number of pharmacological approaches directed to the ALP have been attempted in cellular and animal models of PD (for a comprehensive review see [106]). However, the autophagic pathway is broadly represented and active in different cell types and tissues in the organism. Therefore, the identification of approaches with a high selectivity for certain tissues (such as the dopaminergic neurons) or for specific mechanisms within ALP (such as GCase failure) is detrimental for the achievement of effective but also safe treatments for patients.

In order to restore GCase activity, whose failure seems to be responsible for its neuronal pathogenicity, gene therapy approaches are also in the pipeline. Preclinical studies showed that delivery of *GBA* using adeno-associated virus 1 (AAV1) in A53T–alpha-synuclein mice is able to reduce alpha-synuclein accumulation in the brain [107,108]. The field of gene therapy is now continuously growing in the context of the neurodegenerative disorders [109]. Clinical trials to assess the efficacy of this type of approach may soon be a reality in the context of PD and *GBA* mutations.

4. Conclusions

The discovery of the association between mutations of the *GBA* gene and PD allowed important considerations and discoveries that are contributing to a better understating of the pathogenesis of PD. Indeed, after this initial observation, the role of lysosomal impairment has been extensively explored in PD. A growing amount of emerging evidence supports the idea that the endolysosomal trafficking is involved in alpha-synuclein accumulation and dopaminergic neuron degeneration. A number of genes involved in monogenic forms of PD or genetic risk factors for the disease (such as *SNCA*, *ATP13A2*, *VPS35*, *DNAJC6*, *SYNJ1*, *LRRK2*, *RAB39B*) are part of this pathway (for review see [110]). Mutations of genes involved in the endolysosomal pathways are responsible for a group of disorders designated as Lysosomal Storage Disorders (LSD). These are typically rare autosomal recessive diseases which cause systemic involvements with variable degrees of severity and neurological involvement, usually presenting during childhood (reviewed in [111]). It is interesting to note that an increased burden of LSD-associated mutations has been identified in the screening of large PD populations compared to controls [112]. At the same time, among the 39 new gene loci associated with PD reported in the largest genome wide association study (GWAS) performed in PD patients so far, a number of these variants were found in LSD-associated genes (i.e., *NAGLU*, *GUSB*, *NEU1*, and *GRN*) [113].

The case of autosomal recessive conditions causing severe and rare disorders during childhood, which in turn present as genetic risk factors for common adult neurodegenerative disorders when in a heterozygous state, appears to be more and more frequent, usually presenting an incomplete penetrance. This is the case for a number of LSD in the context of PD or of a parkinsonian degeneration, such as *SMPD1* (sphingomyelin phosphodiesterase, Niemann–Pick disease), *ATP13A2* (P5-type ATPase, Kufor–Rakeb disease), *GALC* (galactosylceramidase, Krabbe disease), *NPC1* (Niemann–Pick type C), *NAGLU* (α-N-acetylglucosminidase, Sanfilippo syndrome B, or mucopoly-saccharidosis III disease B (MPS-IIIB)), *HEXB* (β-hexosaminidase B, Sandhoff disease (GM2 gangliosidosis)) (summarized in [114]). The association between *GBA* mutations, GD, and PD must be just the tip of the iceberg of a larger phenomenon, where the association between genes initially considered responsible only for autosomal recessive disorders turned out to be risk factors for common neurodegenerative conditions. This association may have been recognized first in GD patients because of the higher frequency of this disease compared with other LSD.

Interestingly, this is also the case for the *TREM2* gene (encoding for Triggering Receptor Expressed on Myeloid cells 2), which seems to be the most frequent genetic risk factor of another common neurodegenerative disorder, Alzheimer's disease (AD) [115]. Autosomal recessive mutations of *TREM2* are responsible for the rare, juvenile condition known as Polycystic lipomembranous osteodysplasia with sclerosing leukoencephalopathy. Of note, *TREM2* plays a crucial role in microglia cells as part of the phagocytic scavenger pathway [116].

The phenomena of one gene presenting with different phenotypes is becoming more common in the context of neurological disorders and in respect to common diseases, such as PD and AD. It is important for clinicians to be familiar with these concepts in order to be able to properly counsel their patients and their family members. Also, the identification of such patients will hopefully offer more effective treatments, once available.

These new insights into the understanding of neurodegenerative diseases and, in particular, PD open new scenarios that only a few years ago were still totally obscure. Hopefully, these discoveries will be important for a real discernment of these severe conditions and for the discovery of more effective therapeutic approaches.

Author Contributions: Conceptualization, G.M.R. and A.B.D.F.; writing—original draft preparation, G.M.R. and A.B.D.F.; writing—review and editing, G.M.R. and A.B.D.F.

Acknowledgments: We sincerely thank Brooklyn Henderson, Registered Nurse for her significant contribution in the editing of the manuscript.

References

1. Goker-Alpan, O.; Schiffmann, R.; LaMarca, M.E.; Nussbaum, R.L.; McInerney-Leo, A.; Sidransky, E. Parkinsonism among Gaucher disease carriers. *J. Med. Genet.* **2004**, *41*, 937–940. [CrossRef] [PubMed]
2. Lwin, A.; Orvisky, E.; Goker-Alpan, O.; LaMarca, M.E.; Sidransky, E. Glucocerebrosidase mutations in subjects with parkinsonism. *Mol. Genet. Metab.* **2004**, *81*, 70–73. [CrossRef]
3. Eblan, M.J.; Walker, J.M.; Sidransky, E. The glucocerebrosidase gene and Parkinson's disease in Ashkenazi Jews. *N. Engl. J. Med.* **2005**, *352*, 728–731. [CrossRef]
4. Sidransky, E. Heterozygosity for a Mendelian disorder as a risk factor for complex disease. *Clin. Genet.* **2006**, *70*, 275–282. [CrossRef]
5. Poewe, W.; Seppi, K.; Tanner, C.M.; Halliday, G.M.; Brundin, P.; Volkmann, J.; Schrag, A.E.; Lang, A.E. Parkinson disease. *Nat. Rev. Dis Primers* **2017**, *3*, 17013. [CrossRef] [PubMed]
6. Gegg, M.E.; Schapira, A.H.V. The role of glucocerebrosidase in Parkinson disease pathogenesis. *FEBS J.* **2018**, *285*, 3591–3603. [CrossRef]
7. Klein, A.D.; Mazzulli, J.R. Is Parkinson's disease a lysosomal disorder? *Brain* **2018**, *141*, 2255–2262. [CrossRef] [PubMed]
8. Blandini, F.; Cilia, R.; Cerri, S.; Pezzoli, G.; Schapira, A.H.V.; Mullin, S.; Lanciego, J.L. Glucocerebrosidase mutations and synucleinopathies: Toward a model of precision medicine. *Mov. Disord. Off. J. Mov. Disord. Soc.* **2019**, *34*, 9–21. [CrossRef] [PubMed]
9. Chiasserini, D.; Paciotti, S.; Eusebi, P.; Persichetti, E.; Tasegian, A.; Kurzawa-Akanbi, M.; Chinnery, P.F.; Morris, C.M.; Calabresi, P.; Parnetti, L.; et al. Selective loss of glucocerebrosidase activity in sporadic Parkinson's disease and dementia with Lewy bodies. *Mol. Neurodegener.* **2015**, *10*, 15. [CrossRef]
10. Parnetti, L.; Paciotti, S.; Eusebi, P.; Dardis, A.; Zampieri, S.; Chiasserini, D.; Tasegian, A.; Tambasco, N.; Bembi, B.; Calabresi, P.; et al. Cerebrospinal fluid beta-glucocerebrosidase activity is reduced in parkinson's disease patients. *Mov. Disord. Off. J. Mov. Disord. Soc.* **2017**, *32*, 1423–1431. [CrossRef]
11. Wenstrup, R.J.; Roca-Espiau, M.; Weinreb, N.J.; Bembi, B. Skeletal aspects of Gaucher disease: A review. *Br. J. Radiol.* **2002**, *75* (Suppl. 1), A2–A12. [CrossRef] [PubMed]
12. Elstein, D.; Abrahamov, A.; Dweck, A.; Hadas-Halpern, I.; Zimran, A. Gaucher disease: Pediatric concerns. *Paediatr. Drugs* **2002**, *4*, 417–426. [CrossRef]
13. Andersson, H.; Kaplan, P.; Kacena, K.; Yee, J. Eight-year clinical outcomes of long-term enzyme replacement therapy for 884 children with Gaucher disease type 1. *Pediatrics* **2008**, *122*, 1182–1190. [CrossRef]
14. Kauli, R.; Zaizov, R.; Lazar, L.; Pertzelan, A.; Laron, Z.; Galatzer, A.; Phillip, M.; Yaniv, Y.; Cohen, I.J. Delayed growth and puberty in patients with Gaucher disease type 1: Natural history and effect of splenectomy and/or enzyme replacement therapy. *ISR Med. Assoc. J.* **2000**, *2*, 158–163. [PubMed]

15. Arends, M.; van Dussen, L.; Biegstraaten, M.; Hollak, C.E. Malignancies and monoclonal gammopathy in Gaucher disease; a systematic review of the literature. *Br. J. Haematol.* **2013**, *161*, 832–842. [CrossRef] [PubMed]
16. Zimran, A.; Elstein, D. Gaucher disease and related Lysosomal Storage Diseases. In *Williams' Hematology*; Lichtman, M.A., Kaushansky, K., Prchal, J.T., Levi, M., Burns, L.J., Press, O.W., Caligiuri, M.A., Eds.; McGraw-Hill: New York, NY, USA, 2016; Volume 72, p. 1121.
17. Gupta, P.; Pastores, G. Pharmacological treatment of pediatric Gaucher disease. *Expert Rev. Clin. Pharm.* **2018**, *11*, 1183–1194. [CrossRef] [PubMed]
18. Beutler, E.; Grabowski, G. Gaucher disease. In *The Metabolic and Molecular Basis of Inherited Disease*, 8th ed.; Beaudet, A.L., Scriver, C.R., Sly, W.S., Valle, D., Childs, B., Kinzler, K.W., Vogelstein, B., Eds.; McGraw-Hill International Book Co.: New York, NY, USA, 2001; pp. 3635–3668.
19. Stirnemann, J.; Belmatoug, N.; Camou, F.; Serratrice, C.; Froissart, R.; Caillaud, C.; Levade, T.; Astudillo, L.; Serratrice, J.; Brassier, A.; et al. A Review of Gaucher Disease Pathophysiology, Clinical Presentation and Treatments. *Int. J. Mol. Sci.* **2017**, *18*, 441. [CrossRef] [PubMed]
20. Pastores, G.; Hughes, D. Gaucher Disease. In *GeneReviews® [Internet]*; Ardinger, H.H., Adam, M.P., Pagon, R.A., Wallace, S.E., Bean, L.J.H., Stephens, K., Amemiya, A., Eds.; University of Washington: Seattle, WA, USA, 2000; (updated 2018 June 21).
21. Bennett, L.L.; Mohan, D. Gaucher disease and its treatment options. *Ann. Pharm.* **2013**, *47*, 1182–1193. [CrossRef]
22. Dreborg, S.; Erikson, A.; Hagberg, B. Gaucher disease—Norrbottnian type. I. General clinical description. *Eur. J. Pediatrics* **1980**, *133*, 107–118. [CrossRef]
23. Smith, L.; Mullin, S.; Schapira, A.H.V. Insights into the structural biology of Gaucher disease. *Exp. Neurol.* **2017**, *298*, 180–190. [CrossRef]
24. Hruska, K.S.; LaMarca, M.E.; Scott, C.R.; Sidransky, E. Gaucher disease: Mutation and polymorphism spectrum in the glucocerebrosidase gene (GBA). *Hum. Mutat.* **2008**, *29*, 567–583. [CrossRef] [PubMed]
25. Grabowski, G.A.; Horowitz, M. Gaucher's disease: Molecular, genetic and enzymological aspects. *Baillieres Clin. Haematol.* **1997**, *10*, 635–656. [CrossRef]
26. Goker-Alpan, O.; Hruska, K.S.; Orvisky, E.; Kishnani, P.S.; Stubblefield, B.K.; Schiffmann, R.; Sidransky, E. Divergent phenotypes in Gaucher disease implicate the role of modifiers. *J. Med. Genet.* **2005**, *42*, e37. [CrossRef]
27. Cindik, N.; Ozcay, F.; Suren, D.; Akkoyun, I.; Gokdemir, M.; Varan, B.; Alehan, F.; Ozbek, N.; Tokel, K. Gaucher disease with communicating hydrocephalus and cardiac involvement. *Clin. Cardiol* **2010**, *33*, E26–E30. [CrossRef] [PubMed]
28. Koprivica, V.; Stone, D.L.; Park, J.K.; Callahan, M.; Frisch, A.; Cohen, I.J.; Tayebi, N.; Sidransky, E. Analysis and classification of 304 mutant alleles in patients with type 1 and type 3 Gaucher disease. *Am. J. Hum. Genet.* **2000**, *66*, 1777–1786. [CrossRef]
29. Kowarz, L.; Goker-Alpan, O.; Banerjee-Basu, S.; LaMarca, M.E.; Kinlaw, L.; Schiffmann, R.; Baxevanis, A.D.; Sidransky, E. Gaucher mutation N188S is associated with myoclonic epilepsy. *Hum. Mutat.* **2005**, *26*, 271–273. [CrossRef]
30. Park, J.K.; Orvisky, E.; Tayebi, N.; Kaneski, C.; Lamarca, M.E.; Stubblefield, B.K.; Martin, B.M.; Schiffmann, R.; Sidransky, E. Myoclonic epilepsy in Gaucher disease: Genotype-phenotype insights from a rare patient subgroup. *Pediatr. Res.* **2003**, *53*, 387–395. [CrossRef] [PubMed]
31. Duran, R.; Mencacci, N.E.; Angeli, A.V.; Shoai, M.; Deas, E.; Houlden, H.; Mehta, A.; Hughes, D.; Cox, T.M.; Deegan, P.; et al. The glucocerobrosidase E326K variant predisposes to Parkinson's disease, but does not cause Gaucher's disease. *Mov. Disord. Off. J. Mov. Disord. Soc.* **2013**, *28*, 232–236. [CrossRef] [PubMed]
32. Park, J.K.; Tayebi, N.; Stubblefield, B.K.; LaMarca, M.E.; MacKenzie, J.J.; Stone, D.L.; Sidransky, E. The E326K mutation and Gaucher disease: Mutation or polymorphism? *Clin. Genet.* **2002**, *61*, 32–34. [CrossRef]
33. Horowitz, M.; Pasmanik-Chor, M.; Ron, I.; Kolodny, E.H. The enigma of the E326K mutation in acid beta-glucocerebrosidase. *Mol. Genet. Metab.* **2011**, *104*, 35–38. [CrossRef]
34. Chabas, A.; Gort, L.; Diaz-Font, A.; Montfort, M.; Santamaria, R.; Cidras, M.; Grinberg, D.; Vilageliu, L. Perinatal lethal phenotype with generalized ichthyosis in a type 2 Gaucher disease patient with the [L444P;E326K]/P182L genotype: Effect of the E326K change in neonatal and classic forms of the disease. *Blood Cells Mol. Dis.* **2005**, *35*, 253–258. [CrossRef]

35. Liou, B.; Grabowski, G.A. Is E326K glucocerebrosidase a polymorphic or pathological variant? *Mol. Genet. Metab.* **2012**, *105*, 528–529. [CrossRef]
36. Neudorfer, O.; Giladi, N.; Elstein, D.; Abrahamov, A.; Turezkite, T.; Aghai, E.; Reches, A.; Bembi, B.; Zimran, A. Occurrence of Parkinson's syndrome in type I Gaucher disease. *QJM* **1996**, *89*, 691–694. [CrossRef]
37. Machaczka, M.; Rucinska, M.; Skotnicki, A.B.; Jurczak, W. Parkinson's syndrome preceding clinical manifestation of Gaucher's disease. *Am. J. Hematol.* **1999**, *61*, 216–217. [CrossRef]
38. Perez-Calvo, J.; Bernal, M.; Giraldo, P.; Torralba, M.A.; Civeira, F.; Giralt, M.; Pocovi, M. Co-morbidity in Gaucher's disease results of a nationwide enquiry in Spain. *Eur. J. Med. Res.* **2000**, *5*, 231–235.
39. Varkonyi, J.; Simon, Z.; Soos, K.; Poros, A. Gaucher disease type I complicated with Parkinson's syndrome. *Haematologia* **2002**, *32*, 271–275. [CrossRef]
40. Tayebi, N.; Walker, J.; Stubblefield, B.; Orvisky, E.; LaMarca, M.E.; Wong, K.; Rosenbaum, H.; Schiffmann, R.; Bembi, B.; Sidransky, E. Gaucher disease with parkinsonian manifestations: Does glucocerebrosidase deficiency contribute to a vulnerability to parkinsonism? *Mol. Genet. Metab.* **2003**, *79*, 104–109. [CrossRef]
41. Bembi, B.; Zambito Marsala, S.; Sidransky, E.; Ciana, G.; Carrozzi, M.; Zorzon, M.; Martini, C.; Gioulis, M.; Pittis, M.G.; Capus, L. Gaucher's disease with Parkinson's disease: Clinical and pathological aspects. *Neurology* **2003**, *61*, 99–101. [CrossRef]
42. Zhang, Y.; Shu, L.; Zhou, X.; Pan, H.; Xu, Q.; Guo, J.; Tang, B.; Sun, Q. A Meta-Analysis of GBA-Related Clinical Symptoms in Parkinson's Disease. *Parkinsons Dis.* **2018**, *2018*, 3136415. [CrossRef]
43. Velez-Pardo, C.; Lorenzo-Betancor, O.; Jimenez-Del-Rio, M.; Moreno, S.; Lopera, F.; Cornejo-Olivas, M.; Torres, L.; Inca-Martinez, M.; Mazzetti, P.; Cosentino, C.; et al. The distribution and risk effect of GBA variants in a large cohort of PD patients from Colombia and Peru. *Parkinsonism Relat. Disord.* **2019**. [CrossRef]
44. Lesage, S.; Anheim, M.; Condroyer, C.; Pollak, P.; Durif, F.; Dupuits, C.; Viallet, F.; Lohmann, E.; Corvol, J.C.; Honore, A.; et al. Large-scale screening of the Gaucher's disease-related glucocerebrosidase gene in Europeans with Parkinson's disease. *Hum. Mol. Genet.* **2011**, *20*, 202–210. [CrossRef] [PubMed]
45. Gan-Or, Z.; Giladi, N.; Rozovski, U.; Shifrin, C.; Rosner, S.; Gurevich, T.; Bar-Shira, A.; Orr-Urtreger, A. Genotype-phenotype correlations between GBA mutations and Parkinson disease risk and onset. *Neurology* **2008**, *70*, 2277–2283. [CrossRef]
46. Cilia, R.; Tunesi, S.; Marotta, G.; Cereda, E.; Siri, C.; Tesei, S.; Zecchinelli, A.L.; Canesi, M.; Mariani, C.B.; Meucci, N.; et al. Survival and dementia in GBA-associated Parkinson's disease: The mutation matters. *Ann. Neurol.* **2016**, *80*, 662–673. [CrossRef] [PubMed]
47. Liu, G.; Boot, B.; Locascio, J.J.; Jansen, I.E.; Winder-Rhodes, S.; Eberly, S.; Elbaz, A.; Brice, A.; Ravina, B.; van Hilten, J.J.; et al. Specifically neuropathic Gaucher's mutations accelerate cognitive decline in Parkinson's. *Ann. Neurol.* **2016**, *80*, 674–685. [CrossRef]
48. Thaler, A.; Gurevich, T.; Bar Shira, A.; Gana Weisz, M.; Ash, E.; Shiner, T.; Orr-Urtreger, A.; Giladi, N.; Mirelman, A. A "dose" effect of mutations in the GBA gene on Parkinson's disease phenotype. *Parkinsonism Relat. Disord.* **2017**, *36*, 47–51. [CrossRef]
49. Anheim, M.; Elbaz, A.; Lesage, S.; Durr, A.; Condroyer, C.; Viallet, F.; Pollak, P.; Bonaiti, B.; Bonaiti-Pellie, C.; Brice, A. Penetrance of Parkinson disease in glucocerebrosidase gene mutation carriers. *Neurology* **2012**, *78*, 417–420. [CrossRef] [PubMed]
50. Sidransky, E.; Nalls, M.A.; Aasly, J.O.; Aharon-Peretz, J.; Annesi, G.; Barbosa, E.R.; Bar-Shira, A.; Berg, D.; Bras, J.; Brice, A.; et al. Multicenter analysis of glucocerebrosidase mutations in Parkinson's disease. *N. Engl. J. Med.* **2009**, *361*, 1651–1661. [CrossRef] [PubMed]
51. Bultron, G.; Kacena, K.; Pearson, D.; Boxer, M.; Yang, R.; Sathe, S.; Pastores, G.; Mistry, P.K. The risk of Parkinson's disease in type 1 Gaucher disease. *J. Inherit. Metab. Dis.* **2010**, *33*, 167–173. [CrossRef] [PubMed]
52. McNeill, A.; Duran, R.; Hughes, D.A.; Mehta, A.; Schapira, A.H. A clinical and family history study of Parkinson's disease in heterozygous glucocerebrosidase mutation carriers. *J. Neurol. Neurosurg. Psychiatry* **2012**, *83*, 853–854. [CrossRef]
53. McNeill, A.; Duran, R.; Proukakis, C.; Bras, J.; Hughes, D.; Mehta, A.; Hardy, J.; Wood, N.W.; Schapira, A.H. Hyposmia and cognitive impairment in Gaucher disease patients and carriers. *Mov. Disord. Off. J. Mov. Disord. Soc.* **2012**, *27*, 526–532. [CrossRef]
54. Neumann, J.; Bras, J.; Deas, E.; O'Sullivan, S.S.; Parkkinen, L.; Lachmann, R.H.; Li, A.; Holton, J.; Guerreiro, R.; Paudel, R.; et al. Glucocerebrosidase mutations in clinical and pathologically proven Parkinson's disease. *Brain* **2009**, *132*, 1783–1794. [CrossRef]

55. Moors, T.; Paciotti, S.; Chiasserini, D.; Calabresi, P.; Parnetti, L.; Beccari, T.; van de Berg, W.D. Lysosomal Dysfunction and alpha-Synuclein Aggregation in Parkinson's Disease: Diagnostic Links. *Mov. Disord. Off. J. Mov. Disord. Soc.* **2016**, *31*, 791–801. [CrossRef] [PubMed]
56. Mazzulli, J.R.; Xu, Y.H.; Sun, Y.; Knight, A.L.; McLean, P.J.; Caldwell, G.A.; Sidransky, E.; Grabowski, G.A.; Krainc, D. Gaucher disease glucocerebrosidase and alpha-synuclein form a bidirectional pathogenic loop in synucleinopathies. *Cell* **2011**, *146*, 37–52. [CrossRef] [PubMed]
57. Lieberman, R.L.; Wustman, B.A.; Huertas, P.; Powe, A.C., Jr.; Pine, C.W.; Khanna, R.; Schlossmacher, M.G.; Ringe, D.; Petsko, G.A. Structure of acid beta-glucosidase with pharmacological chaperone provides insight into Gaucher disease. *Nat. Chem. Biol.* **2007**, *3*, 101–107. [CrossRef]
58. Yap, T.L.; Gruschus, J.M.; Velayati, A.; Sidransky, E.; Lee, J.C. Saposin C protects glucocerebrosidase against alpha-synuclein inhibition. *Biochemistry* **2013**, *52*, 7161–7163. [CrossRef] [PubMed]
59. Ouled Amar Bencheikh, B.; Leveille, E.; Ruskey, J.A.; Spiegelman, D.; Liong, C.; Fon, E.A.; Rouleau, G.A.; Dauvilliers, Y.; Dupre, N.; Alcalay, R.N.; et al. Sequencing of the GBA coactivator, Saposin C, in Parkinson disease. *Neurobiol. Aging* **2018**, *72*, 187.e1–187.e3. [CrossRef] [PubMed]
60. Offman, M.N.; Krol, M.; Silman, I.; Sussman, J.L.; Futerman, A.H. Molecular basis of reduced glucosylceramidase activity in the most common Gaucher disease mutant, N370S. *J. Biol. Chem.* **2010**, *285*, 42105–42114. [CrossRef]
61. Wei, R.R.; Hughes, H.; Boucher, S.; Bird, J.J.; Guziewicz, N.; Van Patten, S.M.; Qiu, H.; Pan, C.Q.; Edmunds, T. X-ray and biochemical analysis of N370S mutant human acid beta-glucosidase. *J. Biol. Chem.* **2011**, *286*, 299–308. [CrossRef]
62. Horowitz, M.; Wilder, S.; Horowitz, Z.; Reiner, O.; Gelbart, T.; Beutler, E. The human glucocerebrosidase gene and pseudogene: Structure and evolution. *Genomics* **1989**, *4*, 87–96. [CrossRef]
63. Imai, K.; Nakamura, M.; Yamada, M.; Asano, A.; Yokoyama, S.; Tsuji, S.; Ginns, E.I. A novel transcript from a pseudogene for human glucocerebrosidase in non-Gaucher disease cells. *Gene* **1993**, *136*, 365–368.
64. Leija-Salazar, M.; Sedlazeck, F.J.; Toffoli, M.; Mullin, S.; Mokretar, K.; Athanasopoulou, M.; Donald, A.; Sharma, R.; Hughes, D.; Schapira, A.H.V.; et al. Evaluation of the detection of GBA missense mutations and other variants using the Oxford Nanopore MinION. *Mol. Genet. Genom. Med.* **2019**, e564. [CrossRef]
65. Gan-Or, Z.; Amshalom, I.; Kilarski, L.L.; Bar-Shira, A.; Gana-Weisz, M.; Mirelman, A.; Marder, K.; Bressman, S.; Giladi, N.; Orr-Urtreger, A. Differential effects of severe vs. mild GBA mutations on Parkinson disease. *Neurology* **2015**, *84*, 880–887. [CrossRef] [PubMed]
66. Alcalay, R.N.; Dinur, T.; Quinn, T.; Sakanaka, K.; Levy, O.; Waters, C.; Fahn, S.; Dorovski, T.; Chung, W.K.; Pauciulo, M.; et al. Comparison of Parkinson risk in Ashkenazi Jewish patients with Gaucher disease and GBA heterozygotes. *JAMA Neurol.* **2014**, *71*, 752–757. [CrossRef]
67. Jesus, S.; Huertas, I.; Bernal-Bernal, I.; Bonilla-Toribio, M.; Caceres-Redondo, M.T.; Vargas-Gonzalez, L.; Gomez-Llamas, M.; Carrillo, F.; Calderon, E.; Carballo, M.; et al. GBA Variants Influence Motor and Non-Motor Features of Parkinson's Disease. *PLoS ONE* **2016**, *11*, e0167749. [CrossRef] [PubMed]
68. Malek, N.; Weil, R.S.; Bresner, C.; Lawton, M.A.; Grosset, K.A.; Tan, M.; Bajaj, N.; Barker, R.A.; Burn, D.J.; Foltynie, T.; et al. Features of GBA-associated Parkinson's disease at presentation in the UK Tracking Parkinson's study. *J. Neurol. Neurosurg. Psychiatry* **2018**, *89*, 702–709. [CrossRef] [PubMed]
69. Rosenbloom, B.; Balwani, M.; Bronstein, J.M.; Kolodny, E.; Sathe, S.; Gwosdow, A.R.; Taylor, J.S.; Cole, J.A.; Zimran, A.; Weinreb, N.J. The incidence of Parkinsonism in patients with type 1 Gaucher disease: Data from the ICGG Gaucher Registry. *Blood Cells Mol. Dis.* **2011**, *46*, 95–102. [CrossRef] [PubMed]
70. Lopez, G.; Kim, J.; Wiggs, E.; Cintron, D.; Groden, C.; Tayebi, N.; Mistry, P.K.; Pastores, G.M.; Zimran, A.; Goker-Alpan, O.; et al. Clinical course and prognosis in patients with Gaucher disease and parkinsonism. *Neurol Genet.* **2016**, *2*, e57. [CrossRef] [PubMed]
71. Tan, E.K.; Tong, J.; Fook-Chong, S.; Yih, Y.; Wong, M.C.; Pavanni, R.; Zhao, Y. Glucocerebrosidase mutations and risk of Parkinson disease in Chinese patients. *Arch. Neurol.* **2007**, *64*, 1056–1058. [CrossRef]
72. Eblan, M.J.; Nguyen, J.; Ziegler, S.G.; Lwin, A.; Hanson, M.; Gallardo, M.; Weiser, R.; De Lucca, M.; Singleton, A.; Sidransky, E. Glucocerebrosidase mutations are also found in subjects with early-onset parkinsonism from Venezuela. *Mov. Disord. Off. J. Mov. Disord. Soc.* **2006**, *21*, 282–283. [CrossRef] [PubMed]
73. Wu, Y.R.; Chen, C.M.; Chao, C.Y.; Ro, L.S.; Lyu, R.K.; Chang, K.H.; Lee-Chen, G.J. Glucocerebrosidase gene mutation is a risk factor for early onset of Parkinson disease among Taiwanese. *J. Neurol. Neurosurg. Psychiatry* **2007**, *78*, 977–979. [CrossRef]

74. Davis, M.Y.; Johnson, C.O.; Leverenz, J.B.; Weintraub, D.; Trojanowski, J.Q.; Chen-Plotkin, A.; Van Deerlin, V.M.; Quinn, J.F.; Chung, K.A.; Peterson-Hiller, A.L.; et al. Association of GBA Mutations and the E326K Polymorphism With Motor and Cognitive Progression in Parkinson Disease. *JAMA Neurol.* **2016**, *73*, 1217–1224. [CrossRef]
75. Li, Y.; Sekine, T.; Funayama, M.; Li, L.; Yoshino, H.; Nishioka, K.; Tomiyama, H.; Hattori, N. Clinicogenetic study of GBA mutations in patients with familial Parkinson's disease. *Neurobiol. Aging* **2014**, *35*. [CrossRef]
76. Brockmann, K.; Srulijes, K.; Hauser, A.K.; Schulte, C.; Csoti, I.; Gasser, T.; Berg, D. GBA-associated PD presents with nonmotor characteristics. *Neurology* **2011**, *77*, 276–280. [CrossRef]
77. Brockmann, K.; Srulijes, K.; Pflederer, S.; Hauser, A.K.; Schulte, C.; Maetzler, W.; Gasser, T.; Berg, D. GBA-associated Parkinson's disease: Reduced survival and more rapid progression in a prospective longitudinal study. *Mov. Disord. Off. J. Mov. Disord. Soc.* **2015**, *30*, 407–411. [CrossRef]
78. Lythe, V.; Athauda, D.; Foley, J.; Mencacci, N.E.; Jahanshahi, M.; Cipolotti, L.; Hyam, J.; Zrinzo, L.; Hariz, M.; Hardy, J.; et al. GBA-Associated Parkinson's Disease: Progression in a Deep Brain Stimulation Cohort. *J. Parkinsons Dis.* **2017**, *7*, 635–644. [CrossRef]
79. Nalls, M.A.; Duran, R.; Lopez, G.; Kurzawa-Akanbi, M.; McKeith, I.G.; Chinnery, P.F.; Morris, C.M.; Theuns, J.; Crosiers, D.; Cras, P.; et al. A multicenter study of glucocerebrosidase mutations in dementia with Lewy bodies. *JAMA Neurol.* **2013**, *70*, 727–735. [CrossRef]
80. Gamez-Valero, A.; Prada-Dacasa, P.; Santos, C.; Adame-Castillo, C.; Campdelacreu, J.; Rene, R.; Gascon-Bayarri, J.; Ispierto, L.; Alvarez, R.; Ariza, A.; et al. GBA Mutations Are Associated With Earlier Onset and Male Sex in Dementia With Lewy Bodies. *Mov. Disord. Off. J. Mov. Disord. Soc.* **2016**, *31*, 1066–1070. [CrossRef]
81. Guerreiro, R.; Ross, O.A.; Kun-Rodrigues, C.; Hernandez, D.G.; Orme, T.; Eicher, J.D.; Shepherd, C.E.; Parkkinen, L.; Darwent, L.; Heckman, M.G.; et al. Investigating the genetic architecture of dementia with Lewy bodies: A two-stage genome-wide association study. *Lancet Neurol.* **2018**, *17*, 64–74. [CrossRef]
82. Asselta, R.; Rimoldi, V.; Siri, C.; Cilia, R.; Guella, I.; Tesei, S.; Solda, G.; Pezzoli, G.; Duga, S.; Goldwurm, S. Glucocerebrosidase mutations in primary parkinsonism. *Parkinsonism Relat. Disord.* **2014**, *20*, 1215–1220. [CrossRef]
83. Shiner, T.; Mirelman, A.; Gana Weisz, M.; Bar-Shira, A.; Ash, E.; Cialic, R.; Nevler, N.; Gurevich, T.; Bregman, N.; Orr-Urtreger, A.; et al. High Frequency of GBA Gene Mutations in Dementia With Lewy Bodies Among Ashkenazi Jews. *JAMA Neurol.* **2016**, *73*, 1448–1453. [CrossRef]
84. Mata, I.F.; Leverenz, J.B.; Weintraub, D.; Trojanowski, J.Q.; Chen-Plotkin, A.; Van Deerlin, V.M.; Ritz, B.; Rausch, R.; Factor, S.A.; Wood-Siverio, C.; et al. GBA Variants are associated with a distinct pattern of cognitive deficits in Parkinson's disease. *Mov. Disord. Off. J. Mov. Disord. Soc.* **2016**, *31*, 95–102. [CrossRef]
85. Clark, L.N.; Chan, R.; Cheng, R.; Liu, X.; Park, N.; Parmalee, N.; Kisselev, S.; Cortes, E.; Torres, P.A.; Pastores, G.M.; et al. Gene-wise association of variants in four lysosomal storage disorder genes in neuropathologically confirmed Lewy body disease. *PLoS ONE* **2015**, *10*, e0125204. [CrossRef]
86. Perez-Roca, L.; Adame-Castillo, C.; Campdelacreu, J.; Ispierto, L.; Vilas, D.; Rene, R.; Alvarez, R.; Gascon-Bayarri, J.; Serrano-Munoz, M.A.; Ariza, A.; et al. Glucocerebrosidase mRNA is Diminished in Brain of Lewy Body Diseases and Changes with Disease Progression in Blood. *Aging Dis.* **2018**, *9*, 208–219. [CrossRef]
87. Clark, L.N.; Kartsaklis, L.A.; Wolf Gilbert, R.; Dorado, B.; Ross, B.M.; Kisselev, S.; Verbitsky, M.; Mejia-Santana, H.; Cote, L.J.; Andrews, H.; et al. Association of glucocerebrosidase mutations with dementia with lewy bodies. *Arch. Neurol.* **2009**, *66*, 578–583. [CrossRef]
88. Boeve, B.F.; Silber, M.H.; Ferman, T.J.; Lin, S.C.; Benarroch, E.E.; Schmeichel, A.M.; Ahlskog, J.E.; Caselli, R.J.; Jacobson, S.; Sabbagh, M.; et al. Clinicopathologic correlations in 172 cases of rapid eye movement sleep behavior disorder with or without a coexisting neurologic disorder. *Sleep Med.* **2013**, *14*, 754–762. [CrossRef]
89. Galbiati, A.; Verga, L.; Giora, E.; Zucconi, M.; Ferini-Strambi, L. The risk of neurodegeneration in REM sleep behavior disorder: A systematic review and meta-analysis of longitudinal studies. *Sleep Med. Rev.* **2019**, *43*, 37–46. [CrossRef] [PubMed]
90. Ferini-Strambi, L.; Marelli, S.; Galbiati, A.; Rinaldi, F.; Giora, E. REM Sleep Behavior Disorder (RBD) as a marker of neurodegenerative disorders. *Arch. Ital. Biol.* **2014**, *152*, 129–146. [CrossRef]
91. Barber, T.R.; Lawton, M.; Rolinski, M.; Evetts, S.; Baig, F.; Ruffmann, C.; Gornall, A.; Klein, J.C.; Lo, C.; Dennis, G.; et al. Prodromal Parkinsonism and Neurodegenerative Risk Stratification in REM Sleep Behavior Disorder. *Sleep* **2017**, *40*. [CrossRef]

92. Beavan, M.; McNeill, A.; Proukakis, C.; Hughes, D.A.; Mehta, A.; Schapira, A.H. Evolution of prodromal clinical markers of Parkinson disease in a GBA mutation-positive cohort. *JAMA Neurol.* **2015**, *72*, 201–208. [CrossRef] [PubMed]
93. Gamez-Valero, A.; Iranzo, A.; Serradell, M.; Vilas, D.; Santamaria, J.; Gaig, C.; Alvarez, R.; Ariza, A.; Tolosa, E.; Beyer, K. Glucocerebrosidase gene variants are accumulated in idiopathic REM sleep behavior disorder. *Parkinsonism Relat. Disord.* **2018**, *50*, 94–98. [CrossRef]
94. Fernandez-Santiago, R.; Iranzo, A.; Gaig, C.; Serradell, M.; Fernandez, M.; Tolosa, E.; Santamaria, J.; Ezquerra, M. Absence of LRRK2 mutations in a cohort of patients with idiopathic REM sleep behavior disorder. *Neurology* **2016**, *86*, 1072–1073. [CrossRef]
95. McNeill, A.; Magalhaes, J.; Shen, C.; Chau, K.Y.; Hughes, D.; Mehta, A.; Foltynie, T.; Cooper, J.M.; Abramov, A.Y.; Gegg, M.; et al. Ambroxol improves lysosomal biochemistry in glucocerebrosidase mutation-linked Parkinson disease cells. *Brain* **2014**, *137*, 1481–1495. [CrossRef]
96. Ambrosi, G.; Ghezzi, C.; Zangaglia, R.; Levandis, G.; Pacchetti, C.; Blandini, F. Ambroxol-induced rescue of defective glucocerebrosidase is associated with increased LIMP-2 and saposin C levels in GBA1 mutant Parkinson's disease cells. *Neurobiol. Dis.* **2015**, *82*, 235–242. [CrossRef]
97. Sanchez-Martinez, A.; Beavan, M.; Gegg, M.E.; Chau, K.Y.; Whitworth, A.J.; Schapira, A.H. Parkinson disease-linked GBA mutation effects reversed by molecular chaperones in human cell and fly models. *Sci. Rep.* **2016**, *6*, 31380. [CrossRef]
98. Migdalska-Richards, A.; Daly, L.; Bezard, E.; Schapira, A.H. Ambroxol effects in glucocerebrosidase and alpha-synuclein transgenic mice. *Ann. Neurol.* **2016**, *80*, 766–775. [CrossRef]
99. Migdalska-Richards, A.; Ko, W.K.D.; Li, Q.; Bezard, E.; Schapira, A.H.V. Oral ambroxol increases brain glucocerebrosidase activity in a nonhuman primate. *Synapse* **2017**, *71*. [CrossRef]
100. Barkhuizen, M.; Anderson, D.G.; Grobler, A.F. Advances in GBA-associated Parkinson's disease–Pathology, presentation and therapies. *Neurochem. Int.* **2016**, *93*, 6–25. [CrossRef]
101. Fernandes, H.J.; Hartfield, E.M.; Christian, H.C.; Emmanoulidou, E.; Zheng, Y.; Booth, H.; Bogetofte, H.; Lang, C.; Ryan, B.J.; Sardi, S.P.; et al. ER Stress and Autophagic Perturbations Lead to Elevated Extracellular alpha-Synuclein in GBA-N370S Parkinson's iPSC-Derived Dopamine Neurons. *Stem Cell Rep.* **2016**, *6*, 342–356. [CrossRef]
102. Bae, E.J.; Yang, N.Y.; Song, M.; Lee, C.S.; Lee, J.S.; Jung, B.C.; Lee, H.J.; Kim, S.; Masliah, E.; Sardi, S.P.; et al. Glucocerebrosidase depletion enhances cell-to-cell transmission of alpha-synuclein. *Nat. Commun.* **2014**, *5*, 4755. [CrossRef]
103. Sardi, S.P.; Clarke, J.; Kinnecom, C.; Tamsett, T.J.; Li, L.; Stanek, L.M.; Passini, M.A.; Grabowski, G.A.; Schlossmacher, M.G.; Sidman, R.L.; et al. CNS expression of glucocerebrosidase corrects alpha-synuclein pathology and memory in a mouse model of Gaucher-related synucleinopathy. *Proc. Natl. Acad. Sci. USA* **2011**, *108*, 12101–12106. [CrossRef]
104. Yang, C.; Swallows, C.L.; Zhang, C.; Lu, J.; Xiao, H.; Brady, R.O.; Zhuang, Z. Celastrol increases glucocerebrosidase activity in Gaucher disease by modulating molecular chaperones. *Proc. Natl. Acad. Sci. USA* **2014**, *111*, 249–254. [CrossRef]
105. Yang, C.; Rahimpour, S.; Lu, J.; Pacak, K.; Ikejiri, B.; Brady, R.O.; Zhuang, Z. Histone deacetylase inhibitors increase glucocerebrosidase activity in Gaucher disease by modulation of molecular chaperones. *Proc. Natl. Acad. Sci. USA* **2013**, *110*, 966–971. [CrossRef]
106. Moors, T.E.; Hoozemans, J.J.; Ingrassia, A.; Beccari, T.; Parnetti, L.; Chartier-Harlin, M.C.; van de Berg, W.D. Therapeutic potential of autophagy-enhancing agents in Parkinson's disease. *Mol. Neurodegener.* **2017**, *12*, 11. [CrossRef] [PubMed]
107. Sardi, S.P.; Clarke, J.; Viel, C.; Chan, M.; Tamsett, T.J.; Treleaven, C.M.; Bu, J.; Sweet, L.; Passini, M.A.; Dodge, J.C.; et al. Augmenting CNS glucocerebrosidase activity as a therapeutic strategy for parkinsonism and other Gaucher-related synucleinopathies. *Proc. Natl. Acad. Sci. USA* **2013**, *110*, 3537–3542. [CrossRef] [PubMed]
108. Rocha, E.M.; Smith, G.A.; Park, E.; Cao, H.; Brown, E.; Hayes, M.A.; Beagan, J.; McLean, J.R.; Izen, S.C.; Perez-Torres, E.; et al. Glucocerebrosidase gene therapy prevents alpha-synucleinopathy of midbrain dopamine neurons. *Neurobiol. Dis.* **2015**, *82*, 495–503. [CrossRef]
109. Sudhakar, V.; Richardson, R.M. Gene Therapy for Neurodegenerative Diseases. *Neurother. J. Am. Soc. Exp. Neurother.* **2019**, *16*, 166–175. [CrossRef]

110. Abeliovich, A.; Gitler, A.D. Defects in trafficking bridge Parkinson's disease pathology and genetics. *Nature* **2016**, *539*, 207–216. [CrossRef]
111. Sun, A. Lysosomal storage disease overview. *Ann. Transl. Med.* **2018**, *6*, 476. [CrossRef]
112. Robak, L.A.; Jansen, I.E.; van Rooij, J.; Uitterlinden, A.G.; Kraaij, R.; Jankovic, J.; Heutink, P.; Shulman, J.M. Excessive burden of lysosomal storage disorder gene variants in Parkinson's disease. *Brain* **2017**, *140*, 3191–3203. [CrossRef]
113. Nalls, M.A.; Blauwendraat, C.; Vallerga, C.L.; Heilbron, K.; Bandres-Ciga, S.; Chang, D.; Tan, M.; Kia, D.A.; Noyce, A.J.; Xue, A.; et al. Parkinson's disease genetics: Identifying novel risk loci, providing causal insights and improving estimates of heritable risk. *BioRxiv* **2018**. [CrossRef]
114. Ysselstein, D.; Shulman, J.M.; Krainc, D. Emerging links between pediatric lysosomal storage diseases and adult parkinsonism. *Mov. Disord. Off. J. Mov. Disord. Soc.* **2019**. [CrossRef] [PubMed]
115. Jonsson, T.; Stefansson, H.; Steinberg, S.; Jonsdottir, I.; Jonsson, P.V.; Snaedal, J.; Bjornsson, S.; Huttenlocher, J.; Levey, A.I.; Lah, J.J.; et al. Variant of TREM2 associated with the risk of Alzheimer's disease. *N. Engl. J. Med.* **2013**, *368*, 107–116. [CrossRef] [PubMed]
116. Carmona, S.; Zahs, K.; Wu, E.; Dakin, K.; Bras, J.; Guerreiro, R. The role of TREM2 in Alzheimer's disease and other neurodegenerative disorders. *Lancet Neurol.* **2018**, *17*, 721–730. [CrossRef]

11

Zonisamide Administration Improves Fatty Acid β-Oxidation in Parkinson's Disease

Shin-Ichi Ueno [1], Shinji Saiki [1,*], Motoki Fujimaki [1], Haruka Takeshige-Amano [1], Taku Hatano [1], Genko Oyama [1], Kei-Ichi Ishikawa [1,2], Akihiro Yamaguchi [2], Shuko Nojiri [3], Wado Akamatsu [2] and Nobutaka Hattori [1,*]

1 Department of Neurology, Juntendo University School of Medicine, Bunkyo-ku, Tokyo 113-8421, Japan; sueno@juntendo.ac.jp (S.-I.U.); mtfujima@juntendo.ac.jp (M.F.); h-amano@juntendo.ac.jp (H.T.-A.); thatano@juntendo.ac.jp (T.H.); g_oyama@juntendo.ac.jp (G.O.); kishikaw@juntendo.ac.jp (K.-I.I.)
2 Center for Genomic and Regenerative Medicine, Juntendo University School of Medicine, Bunkyo-ku, Tokyo 113-8421, Japan; akihiro0781@gmail.com (A.Y.); awado@juntendo.ac.jp (W.A.)
3 Medical Technology Innovation Center, Juntendo University, Bunkyo-ku, Tokyo 113-8421, Japan; s-nojiri@juntendo.ac.jp
* Correspondence: ssaiki@juntendo.ac.jp (S.S.); nhattori@juntendo.ac.jp (N.H.)

Abstract: Although many experimental studies have shown the favorable effects of zonisamide on mitochondria using models of Parkinson's disease (PD), the influence of zonisamide on metabolism in PD patients remains unclear. To assess metabolic status under zonisamide treatment in PD, we performed a pilot study using a comprehensive metabolome analysis. Plasma samples were collected for at least one year from 30 patients with PD: 10 without zonisamide medication and 20 with zonisamide medication. We performed comprehensive metabolome analyses of plasma with capillary electrophoresis time-of-flight mass spectrometry and liquid chromatography time-of-flight mass spectrometry. We also measured disease severity using Hoehn and Yahr (H&Y) staging and the Unified Parkinson's Disease Rating Scale (UPDRS) motor section, and analyzed blood chemistry. In PD with zonisamide treatment, 15 long-chain acylcarnitines (LCACs) tended to be increased, of which four (AC(12:0), AC(12:1)-1, AC(16:1), and AC(16:2)) showed statistical significance. Of these, two LCACs (AC(16:1) and AC(16:2)) were also identified by partial least squares analysis. There was no association of any LCAC with age, disease severity, levodopa daily dose, or levodopa equivalent dose. Because an upregulation of LCACs implies improvement of mitochondrial β-oxidation, zonisamide might be beneficial for mitochondrial β-oxidation, which is suppressed in PD.

Keywords: Parkinson's disease; fatty acid β-oxidation; long-chain acylcarnitine

---

1. Introduction

Parkinson's disease (PD) is the second most common neurodegenerative disorder and is characterized by motor symptoms such as tremor, rigidity, and akinesia [1]. Although symptomatic relief with levodopa medication and deep brain stimulation treatments are well established, there is currently no disease-modifying therapy based on disease pathogenesis [2]. In PD pathogenesis, mitochondrial dysfunction has been suggested to occur in the form of excessive oxidative stress, respiratory-chain insufficiency, and mitophagy flux abnormalities, among others [3]. However, therapeutic approaches targeted to these mitochondrial dysfunctions have not yet been used in the clinic.

Zonisamide (1,2-benzisoxazole-3-methanesulfonamide) is used clinically as an anti-epileptic agent and has been adapted to treat resting tremor as well as the wearing-off symptoms in PD patients in

Japan [4]. Recently, zonisamide has become widely used, and its beneficial effects for the treatment of PD were shown to be mediated by several mechanisms including monoamine oxidase inhibition, channel blocking, and glutamate release inhibition [5]. The half-life of zonisamide is approximately 68 h and it is distributed to the whole body, including the brain and skeletal muscle, followed by its excretion in urine [6]. In cellular and animal models of PD, zonisamide was reported to have various neuroprotective effects via mitochondrial protection, including antioxidant effects through manganese superoxide dismutase (MnSOD) upregulation, adjustment of calcium influx, and brain derived neurotrophic factor (BDNF) signaling systems [7–13]. In addition, zonisamide was reported to reduce neuroinflammation through the modulation of microglia [14]. However, no human data associated with metabolic changes under zonisamide treatment have been reported in patients with PD.

Recent research conducted by our group and others has revealed that serum and/or plasma metabolomics are useful for revealing changes in metabolic pathways in PD [15–19]. Our group previously reported the analyses of capillary electrophoresis time-of-flight mass-spectrometry (CE-TOFMS) and liquid chromatography time-of-flight mass-spectrometry (LC-TOFMS), which showed decreased long-chain acylcarnitines (LCACs) and increased long-chain fatty acids (LCFAs) in PD patients, suggesting that β-oxidation insufficiency occurs primarily in early PD, and is not associated with levodopa medication [15].

In this context, to highlight the metabolic changes that occur with zonisamide administration, we conducted a case–case study, using well-established techniques, of 20 PD patients undergoing zonisamide treatment and 10 PD patients without zonisamide treatment.

2. Materials and Methods

2.1. Ethics Statement

This study protocol complied with the Declaration of Helsinki and was approved by the ethics committee of Juntendo University (2017135). Written informed consent was given by all participants.

2.2. Participants

PD was diagnosed according to the Movement Disorders Society Clinical Diagnostic Criteria for PD, with no dementia (Mini-Mental State Examination [MMSE] score $\geq$24) [20]. All participants were aged 51 to 82 years, were fluent in Japanese, and had no substantial neurological disease other than PD. Based on demographic and clinical characteristics obtained from medical records and directly from PD patients, the PD group comprised of 30 patients was subdivided into 10 not administered zonisamide and 20 with the oral administration of zonisamide (25 mg/day, after breakfast) (Table 1). Clinical symptoms were evaluated within a year from the beginning of the oral administration of zonisamide. Participants suffering from acute infectious diseases or acute/chronic renal or hepatic failure at the time of sample collection were excluded. No participant had a medical history or chronic illness of type 2 diabetes mellitus, skeletal muscle disease, cancer, aspiration pneumonia, inflammatory bowel disease, or collagen vascular disease.

2.3. Assessment of Clinical Symptoms

The clinical conditions of PD patients were evaluated using Hoehn and Yahr (H&Y) stages and the Unified Parkinson's Disease Rating Scale (UPDRS)-III. For practical and ethical reasons, the H&Y stage and UPDRS-III rating were defined during the "on" state, when patients reported that the effect of the last dose of medication was optimal.

Table 1. Demographics and clinical data of Parkinson's disease with or without zonisamide.

Participants Characteristics, Mean (SD)	Zonisamide (−)	Zonisamide (+)	*p*-Value
Number	10	20	-
Gender, female/male	5/5	13/7	0.431 [a]
Age	69.5 (6.7)	68.2 (1.9)	0.982
Zonisamide	-	38.7 (12.7)	-
Treatment period by evaluation	-	8.15 (3.75)	-
Body mass index	22.0 (3.4)	22.4 (0.75)	0.597
H&Y	2.1 (0.9)	2.3 (1.1)	0.762
H&Y, each case number	I(3), II(4), III(2), IV(1), V(0)	I(5), II(9), III(2), IV(3), V(1)	-
Disease duration	11.4 (6.9)	7.7 (4.0)	0.138
UPDRS-III (pre-treatment)	17.9 (12.4)	14.1 (3.1)	0.425
UPDRS-III (post-treatment)	-	13.6 (11.4)	-
UPDRS-III-tremor (pre-treatment)	0.6 (1.0)	2.6 (1.9)	0.00280
UPDRS-III-tremor (post-treatment)	-	2.2 (1.9)	-
LED	855.3 (305)	680.9 (458)	0.165
LDD	480 (209)	430 (280)	0.492

Abbreviations: SD: standard deviation, H&Y: Hoehn and Yahr stage, UPDRS: Unified Parkinson's Disease Rating Scale, LED: levodopa equivalent dose, LDD: levodopa dose. [a]: *p*-Value obtained by χ-squared test, other *p*-Values were obtained by Wilcoxon test.

2.4. Blood Sample Collection

All fasting blood samples were collected at the outpatient department of Juntendo University Hospital between April 2015 and January 2018. Following an overnight fast (12–14 h), a plasma or serum sample was obtained using 7-mL EDTA-2Na blood collection tube (PN7R, Tokyo, SRL) or 8-mL blood collection tube(78447 SIM-L1008SQ3, Kyokuto Pharmaceutical Ind. Co. Ltd., Tokyo, Japan) followed by two or three inversions, respectively. The samples were then allowed to rest for 30–60 min at 4 °C, followed by centrifugation for 10 min at 2660× *g*. The plasma was then separated and placed in collection tubes that were stored in liquid nitrogen until analysis.

2.5. Metabolite Extraction

Metabolite extraction and metabolome analysis were conducted at Human Metabolome Technologies (HMT; Tsuruoka, Yamagata, Japan). For analysis with CE-TOFMS, 50 μL of plasma was added to 450 μL of methanol containing internal standards (Solution ID: H3304-1002, HMT) at 0 °C to inactivate enzymes. The extract solution was thoroughly mixed with 500 μL of chloroform and 200 μL of Milli-Q water and centrifuged at 2300× *g* at 4 °C for 5 min. A 400-μL sample of the upper aqueous layer was centrifugally filtered through a Millipore 5-kDa cutoff filter to remove proteins. The filtrate was centrifugally concentrated and resuspended in 50 μL of Milli-Q water for CE-TOFMS analysis at HMT. For analysis with LC-TOFMS, 300 μL of plasma was added to 900 μL of 1% formic acid/acetonitrile containing the internal standard solution (Solution ID: H3304-1002, HMT) at 0 °C (to inactivate enzymes). The solution was thoroughly mixed and centrifuged at 2300× *g* at 4 °C for 5 min. The supernatant was filtrated through a solid phase extraction column (Hybrid SPE phospholipid 55261-U, Supelco, Bellefonte, PA, USA) to remove phospholipids. The filtrate was desiccated and then dissolved with 120 μL of 50% isopropanol/Milli-Q for LC-TOFMS analysis at HMT.

2.6. Biochemical Measurements

Total serum creatine kinase was measured using automated enzymatic techniques (Sysmex Inc., Kobe, Japan). Serum creatinine was measured by an enzymatic method (KAINOS Laboratories, Inc., Tokyo, Japan). Serum aldolase was measured by a coupled enzyme assay (Alfresa Pharma, UK), while hemoglobin (Hb) A1c was measured by boronate-affinity high-performance liquid chromatography, in accordance with standard protocols.

2.7. Data Analysis

Peaks were extracted using MasterHands automatic integration software (Keio University, Tsuruoka, Yamagata, Japan) to obtain peak information including the m/z and peak area, as well as migration time (MT) for CE-TOFMS and retention time (RT) for LC-TOFMS. Signal peaks corresponding to isotopomers, adduct ions, and other product ions of known metabolites were excluded, and the remaining peaks were annotated according to the HMT metabolite database based on their m/z values with the MTs and RTs. To obtain the relative levels of each metabolite, areas of the annotated peaks were then normalized based on internal standard levels and sample volumes. For multivariate statistical analysis, partial least squares (PLS) analysis was performed using R [21].

2.8. Statistical Analysis

When a value was below the limit of detection, we assigned it half the minimum value of its compound. Wilcoxon tests were used to compare all individual analyses between PD patients with or without zonisamide. A p-value of less than 0.05 was considered statistically significant.

3. Results

3.1. Participants

The characteristics of PD patients included in the study are shown in Table 1. PD patients taking zonisamide had a significantly higher UPDRS-III tremor score at pretreatment of zonisamide compared with PD patients not taking zonisamide. There were no significant differences in age, sex, H&Y stage, disease duration, levodopa equivalent dose (LED), or levodopa daily dose (LDD) between PD patients with or without zonisamide treatment [22]. There was a statistically significant difference in UPDRS-III tremor score between the two groups ($p = 0.00280$).

3.2. Metabolomic Datasets

We analyzed the metabolomic profiles of blood plasma from 30 PD patients using CE-TOFMS and LC-TOFMS. Based on their m/z values, MTs and RTs, 383 metabolites were detected. Of these, 266 metabolites were detected in >50% of PD patients taking zonisamide and were analyzed in detail. As shown in Table 2, 12 metabolites were significantly changed in PD patients with zonisamide treatment compared with those without zonisamide (Supplementary Figure S1). Four of these 12 were LCACs, and these were all upregulated. Next, we performed PLS analysis to identify which metabolites distinguished PD with zonisamide treatment from PD without zonisamide treatment (Figure 1). The top 10 metabolites, including three LCACs, are shown in Supplementary Table S1.

3.3. Increase of Long-Chain Acylcarnitine Levels in PD Patients with Zonisamide Treatment

Because four LCACs (AC(12:0), AC(12:1)-1, AC(16:1), and AC(16:2)), were significantly increased in PD patients with zonisamide treatment (Table 2), we tried to characterize the profile of LCACs detected, regardless of the lower limits of detection, for further investigation. Interestingly, 15 of 20 LCACs showed a trend toward an increase in PD patients with zonisamide treatment (Supplementary Table S2). Of these, seven LCACs (AC(12:0), AC(12:1)-1, AC(12:1)-2, AC(12:1)-3, AC(13:1)-1, AC(16:1), and AC(16:2)) were significantly increased (Supplementary Table S2). According to the PLS analysis, there were two LCACs (AC(16:1) and AC(16:2)) in the top 10 factors for differentiating between PD with zonisamide treatment and PD without zonisamide treatment. There were no significant correlations between any LCACs and zonisamide treatment duration or dosage (Supplementary Table S3). These results suggest that the increase in LCACs is not affected by zonisamide treatment period and dosage.

Table 2. Metabolites significantly changed in Parkinson's disease differentiate between patients with or without zonisamide treatment.

Compound, Relative Area	Ratio of Zonisamide (+) to Zonisamide (−)	*p*-Value
1-methylnicotinamide	1.61	0.0294
AC(12:0)	1.73	0.0407
AC(12:1)-1	1.92	0.0405
AC(16:1)	1.75	0.0366
AC(16:2)	1.95	0.0054
Glycerol	0.843	0.0329
Imidazolelactic acid	1.50	0.0040
Nervonic acid	1.21	0.0068
Oleoyl ethanolamine	0.785	0.00146
Ornithine	1.32	0.0263
S-methylcysteine	0.726	0.0294
Succinic acid	1.46	0.0199

AC: acylcarnitine. *p*-Value obtained by Wilcoxon test.

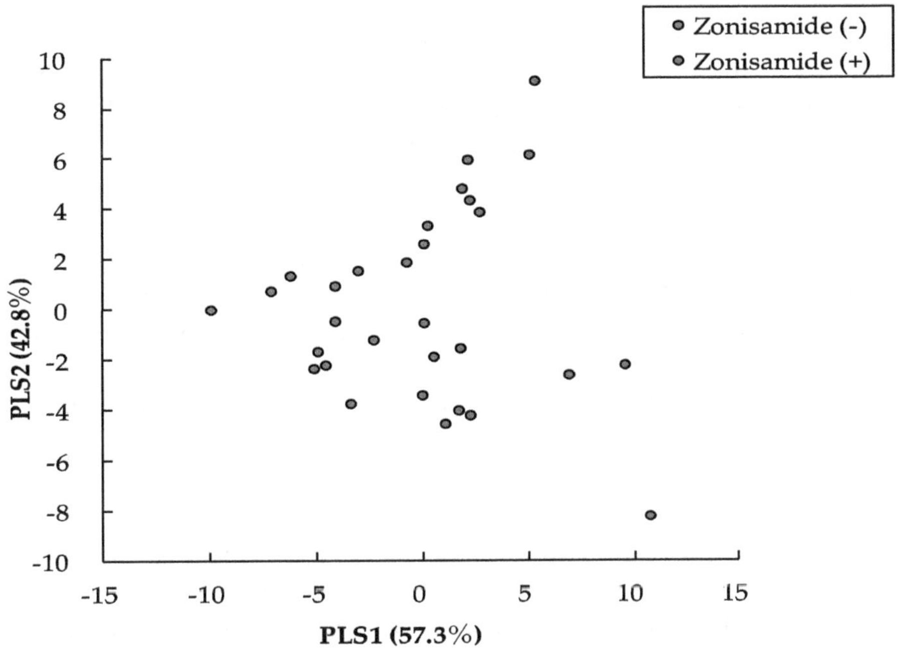

Figure 1. Partial least squares analysis (PLS) of observed metabolic profiles. Red indicates metabolites of PD with zonisamide, while blue indicates PD without zonisamide. PLS 1 indicates PD patients with or without zonisamide in PD, while PLS 2 indicates susceptibility of zonisamide in PD. PD: Parkinson's disease, PLS: partial least analysis.

3.4. Association of LCACs with Age, H and Y Stage, UPDRS-III Scores, and LED

Plasma levels of LCACs are primarily decreased in the early stages of PD [15]. As shown in Table 1, there were no significant differences in patient characteristics between the two groups except for the UPDRS-III tremor score (Table 1). Thus, to precisely exclude the effects of disease severity and medication on the levels of LCACs, we performed correlation analyses. There were no significant correlations between any LCACs with age at sampling, H&Y stage, UPDRS-III scores, and LED (Table 3). In addition, there was a significant correlation between AC(16:2) and the amelioration of UPDRS-III tremor scores in PD with zonisamide that was efficacious (Supplementary Figure S2). These results suggest that zonisamide, and not disease severity, aging, or LED, lead to the increase of LCACs.

Table 3. Association of LCACs and clinical parameters in all participants.

	Age		H&Y		UPDRS-III		LED	
	r	*p*-Value	*r*	*p*-Value	*r*	*p*-Value	*r*	*p*-Value
AC(12:0)	−0.293	0.115	0.128	0.499	0.240	0.200	0.287	0.123
AC(12:1)−1	−0.115	0.541	0.179	0.343	0.141	0.457	0.272	0.145
AC(16:1)	−0.199	0.291	0.0476	0.802	0.124	0.510	0.132	0.485
AC(16:2)	−0.110	0.562	0.185	0.326	0.153	0.417	0.0969	0.610

LCACs: long-chain acylcarnitines, H&Y: Hoehn and Yahr stage, UPDRS: Unified Parkinson's Disease Rating Scale, LED: levodopa equivalent dose. *p*-Value obtained by Spearman's rank correlation coefficient.

3.5. *Association of Skeletal Muscle Mass with Metabolites Associated with Fatty Acid β-Oxidation*

Plasma LCACs are excreted mainly from the cellular mitochondria of skeletal muscles [23]. To exclude any effects of decreased skeletal muscle mass on fatty acid (FA) β-oxidation, we assessed the serum levels of creatine kinase, aldolase, and creatinine, and total blood levels of HbA1c (Supplementary Table S4). Although levels of aldolase, which is produced by skeletal muscles as well as the liver, were significantly downregulated by zonisamide, the levels of creatine kinase and creatinine, which correlate with the amount of skeletal muscle, did not change significantly. There was also no significant difference in HbA1c levels, which may influence FA β-oxidation activity. From these results, we concluded that zonisamide had almost no harmful effects on skeletal muscle, although we should consider the possibility that zonisamide might influence skeletal muscle.

3.6. *Other Metabolites Significantly Changed by Zonisamide Treatment*

As shown in Table 2, eight chemicals (except for four LCACs) were significantly changed in PD with zonisamide treatment. 1-methylnicotinamide, a metabolite of nicotinamide, and oleoylethanolamide, were associated with neuroprotective pathways [24]. In addition, the ratio of putrescine to ornithine was decreased in sporadic PD [25]. Although zonisamide treatment changed the levels of these three chemicals, the detailed metabolic pathways involved remain unclear because only one metabolite changed.

4. Discussion

This comprehensive metabolome analysis identified 12 metabolites that underwent significant changes with zonisamide treatment in PD. Four of the twelve metabolites were LCACs: (AC(12:0), AC(12:1)-1, AC(16:1), and AC(16:2)) and the PLS analysis using whole metabolome data identified two overlapping LCACs: AC(16:1) and AC(16:2). Furthermore, 15 LCACs had a tendency to be increased in PD patients with zonisamide, and only AC(16:2) was correlated to the amelioration of UPDRS-III tremor scores in PD with efficacious zonisamide. The LCAC levels did not correlate with disease severity or medication, suggesting that zonisamide itself may upregulate FA β-oxidation in patients with PD. Moreover, LCAC levels might define therapeutic reactivity with treatment of zonisamide.

Previously, we reported that decreased levels of LCACs were a useful biomarker for the diagnosis of PD in a double-cohort study [15]. β-oxidation of FAs occurs in mitochondria and peroxisomes; β-oxidation of short-to-long-chain FAs occurs in mitochondria, whereas β-oxidation of LCFAs occurs in peroxisomes. FAs with chain lengths of 14 or more carbons, which are the majority of those obtained from the diet or released from adipose tissues, need to undergo enzymatic reactions by the carnitine shuttle to enter mitochondria for β-oxidation [26]. In peroxisomes, acyl-coenzyme A may be incompletely oxidized, and medium-chain acyl-CoA produced in peroxisomes is transported to mitochondria to be oxidized. Because it is difficult for acyl-CoA to penetrate the outer membrane of mitochondria, it is transported by carnitine-acylcarnitine translocase after its conversion to ACs by carnitine palmitoyltransferase 1 (CPT1). LCACs are formed by CPT1 from acyl-CoA and carnitine to pass through the mitochondrial outer membrane. Zonisamide may improve β-oxidation via a mitochondrial protective effect [7,15].

Patients with PD can be subdivided into two types: tremor-dominant or postural instability/gait difficulty variants [27]. Our previous report revealed no association of disease severity as assessed by UPDRS-III with the levels of LCACs in PD [15]. The current study suggests that severe tremor, often observed in PD and treated with zonisamide, might affect the elevation of LCACs. Whilst aerobic exercise was reported to decrease levels of LCACs in obese patients, controls do not show changes in LCAC levels following short- or long-term exercise [28,29]. In the current cohort, the body mass index in PD without zonisamide was approximately 22, and there was no statistical difference between the two groups. In addition, blood samples in our study were obtained without exercise loading and with overnight fasting. Thus, we concluded that we did not need to consider more severe tremors in PD patients without zonisamide treatment.

Oxidative damage is a major contributor to cellular damage, which can result in neurodegenerative diseases [30]. Oxidative stress generates reactive oxygen species (ROS), which inflict oxidative damage on macromolecules such as lipids, DNA, and proteins. The imbalance between oxidative stress via ROS production and antioxidant factors in the body has an important role in PD development [31]. Mitochondrial ROS metabolism is catalyzed by mitochondrial antioxidant enzymes including MnSOD. Catabolic reactions via β-oxidation are also a major component of the regulation of ROS generation in mitochondria [32]. Taken together, in addition to the upregulation of MnSOD expression [7], an improvement in β-oxidation by zonisamide may protect against oxidative stress.

This study had some limitations. First, the interpretation of results requires caution because the sample size was small. Thus, it was difficult to conduct multivariate analysis to exclude confounding factors including age, sex, disease duration, and medication except for anti-parkinsonian drug. Second, there was an effect of other antiparkinsonian medications on zonisamide metabolism. A previous study demonstrated that levodopa/benserazide treatment elicited metabolic changes in skeletal muscle with a switch from lipid to carbohydrate metabolism, which may affect β-oxidation [33]. However, in the current cohort we did not detect any direct influences of LDD or LED on LCAC levels. In the future, de novo studies with additional treatment of zonisamide and including larger study populations should be performed, focusing on FA β-oxidation changes.

5. Conclusions

This is the first report showing the in vivo effects of zonisamide on human metabolism. We identified 12 metabolites, including four LCACs, whose increased levels were associated with an improvement in FA β-oxidation. Considering previous experimental studies of zonisamide treatment, it may have beneficial effects on mitochondrial function in patients with PD. Furthermore, zonisamide might affect the efficiency of the brain network to relieve the symptomatology and progression of PD.

Author Contributions: Conceptualization, S.-I.U., S.S., N.H.; Data curation, S.-I.U., S.S., M.F., T.H., K.-I.I., A.Y., S.N., W.A.; Formal analysis, S.-I.U., S.S., M.F.; Writing—original draft preparation, S.-I.U., S.S.; Writing—review and editing, S.S., M.F., T.H., K.-I.I., A.Y., S.N., W.A., N.H.; Project administration, S.S., N.H.; Funding acquisition, S.S., W.A., N.H.

Acknowledgments: We are very grateful for the following grants: AMED-CREST funded by the Japan Agency of Medical Research and Development (N.H), Grant-in-Aid for Scientific Research on Priority Areas (S.S. 25111007), Grant-in-Aid for Scientific Research (B) (S.S. 15H04843, 18KT0027), and Supported Program for the Strategic Research Foundation at Private Universities. We thank Bronwen Gardner, from Edanz Group (www.edanzediting.com/ac) for editing a draft of this manuscript.

References

1. Poewe, W.; Seppi, K.; Tanner, C.M.; Halliday, G.M.; Brundin, P.; Volkmann, J.; Schrag, A.E.; Lang, A.E. Parkinson disease. *Nat. Rev. Dis. Primers* **2017**, *23*, 17013. [CrossRef] [PubMed]
2. Jankovic, J.; Poewe, W. Therapies in Parkinson's disease. *Curr. Opin. Neurol.* **2012**, *4*, 433–447. [CrossRef]
3. Saiki, S.; Sato, S.; Hattori, N. Molecular pathogenesis of Parkinson's disease: Update. *J. Neurol. Neurosurg. Psychiatry* **2012**, *83*, 430–436. [CrossRef] [PubMed]
4. Murata, M.; Hasegawa, K.; Kanazawa, I.; Japan Zonisamide on PD Study Group. Zonisamide improves motor function in Parkinson disease: A randomized, double-blind study. *Neurology* **2007**, *68*, 45–50. [CrossRef] [PubMed]
5. Fox, S.H.; Katzenschlager, R.; Lim, S.Y.; Barton, B.; de Bie, R.M.A.; Seppi, K.; Coelho, M.; Sampaio, C.; Movement Disorder Society Evidence-Based Medicine Committee. International Parkinson and movement disorder society evidence-based medicine review: Update on treatments for the motor symptoms of Parkinson's disease. *Mov. Disord.* **2018**, *8*, 1248–1266. [CrossRef]
6. Matsumoto, K.; Miyazaki, H.; Fujii, T.; Kagemoto, A.; Maeda, T.; Hashimoto, M.; Arzneimittel-Forschung. Absorption, Distribution and Excretion of 3-(Sulfamoyl[14C]methyl)-1,2-benzisoxazole (AD810) in Rats, Dogs, Monkeys and of AD-810 in Men. *Drug Res.* **1983**, *7*, 961–968.
7. Kawajiri, S.; Machida, Y.; Saiki, S.; Sato, S.; Hattori, N. Zonisamide reduces cell death in SH-SY5Y cells via an anti-apoptotic effect and by upregulating MnSOD. *Neurosci. Lett.* **2010**, *2*, 88–91. [CrossRef]
8. Sano, H.; Murata, M.; Nambu, A. Zonisamide reduces nigrostriatal dopaminergic neurodegeneration in a mouse genetic model of Parkinson's disease. *J. Neurochem.* **2015**, *2*, 371–381. [CrossRef]
9. Ueda, Y.; Tokashiki, S.; Kanemaru, A.; Kojima, T. Effect of zonisamide co-administration with levodopa on global gene expression in the striata of rats with Parkinson's disease. *Biochem. Biophys. Res. Commun.* **2012**, *3*, 401–404. [CrossRef]
10. Asanuma, M.; Miyazaki, I.; Diaz-Corrales, F.J.; Kimoto, N.; Kikkawa, Y.; Takeshima, M.; Miyoshi, K.; Murata, M. Neuroprotective effects of zonisamide target astrocyte. *Ann. Neurol.* **2010**, *2*, 239–249. [CrossRef]
11. Topcu, Y.; Bayram, E.; Özbal, S.; Yiş, U.; Tuğyan, K.; Karaoğlu, P.; Kumral, P.; Yılmaz, O.; Kurul, S.H. Zonisamide attenuates hyperoxia-induced apoptosis in the developing rat brain. *Neurol. Sci.* **2014**, *11*, 1769–1775. [CrossRef] [PubMed]
12. Zhu, G.; Okada, M.; Murakami, T.; Kawata, Y.; Kamata, A.; Kaneko, S. Interaction between carbamazepine, zonisamide and voltage-sensitive Ca2+ channel on acetylcholine release in rat frontal cortex. *Epilepsy Res.* **2002**, *1*, 49–60. [CrossRef]
13. Yurekli, V.A.; Gürler, S.; Nazıroğlu, M.; Uğuz, A.C.; Koyuncuoğlu, H.F. Zonisamide attenuates MPP+-induced oxidative toxicity through modulation of Ca2+ signaling and caspase-3 activity in neuronal PC12 cells. *Cell Mol. Neurobiol.* **2013**, *2*, 205–212. [CrossRef] [PubMed]
14. Hossain, M.M.; Weig, B.; Reuhl, K.; Gearing, M.; Wu, L.J.; Richardson, J.R. The anti-parkinsonian drug zonisamide reduces neuroinflammation: Role of microglial Nav 1.6. *Exp. Neurol.* **2018**, 111–119. [CrossRef] [PubMed]
15. Saiki, S.; Hatano, T.; Fujimaki, M.; Ishikawa, K.I.; Mori, A.; Oji, Y.; Okuzumi, A.; Fukuhara, T.; Koinuma, T.; Imamichi, Y.; et al. Decreased long-chain acylcarnitines from insufficient beta-oxidation as potential early diagnostic markers for Parkinson's disease. *Sci. Rep.* **2017**, *1*, 7328. [CrossRef] [PubMed]
16. Hatano, T.; Saiki, S.; Okuzumi, A.; Mohney, R.P.; Hattori, N. Identification of novel biomarkers for Parkinson's disease by metabolomic technologies. *J. Neurol. Neurosurg. Psychiatry* **2016**, *3*, 295–301. [CrossRef] [PubMed]
17. Fujimaki, M.; Saiki, S.; Li, Y.Z.; Kaga, N.; Taka, H.; Hatano, T.; Ishikawa, K.I.; Oji, Y.; Mori, A.; Okuzumi, A.; et al. Serum caffeine and metabolites are reliable biomarkers of early Parkinson disease. *Neurology* **2018**, *90*, e404–e411. [CrossRef]
18. Havelund, J.F.; Heegaard, H.H.; Færgeman, N.J.K.; Gramsbergen, J.B. Biomarker Research in Parkinson's Disease Using Metabolite Profiling. *Metabolites* **2017**, *7*, 42. [CrossRef]

19. Medeiros, M.S.; Schumacher-Schuh, A.; Cardoso, A.M.; Bochi, G.V.; Baldissarelli, J.; Kegler, A.; Santana, D.; Chaves, C.M.M.B.S.; Schetinger, M.R.C.; Moresco, R.N.; et al. Iron and Oxidative Stress in Parkinson's Disease: An Observational Study of Injury Biomarkers. *PLoS ONE* **2016**, *1*, e0146129. [CrossRef]
20. Postuma, R.B.; Postuma, R.B.; Berg, D.; Stern, M.; Poewe, W.; Olanow, C.W.; Oertel, W.; Obeso, J.; Marek, K.; Litvan, I.; et al. MDS clinical diagnostic criteria for Parkinson's disease. *Mov. Disord.* **2015**, *12*, 1591–1601. [CrossRef]
21. Yamamoto, H.; Fujimori, T.; Sato, H.; Ishikawa, G.; Kami, K.; Ohashi, Y. Statistical hypothesis testing of factor loading in principal component analysis and its application to metabolite set enrichment analysis. *BMC Bioinformat.* **2014**, *15*, 51. [CrossRef] [PubMed]
22. Tomlinson, C.L.; Stowe, R.; Patel, S.; Rick, C.; Gray, R.; Clarke, C.E. Systematic review of levodopa dose equivalency reporting in Parkinson's disease. *Mov. Disord.* **2010**, *15*, 2649–2653. [CrossRef] [PubMed]
23. Shearer, J.; Weljie, A.M. Biomarkers of skeletal muscle regulation, mechanism and dysfunction. In *Metabolomics and Systems Biology in Human Health and Medicine*; Jones, O.A.H., Ed.; CABI: Wallingford, Oxfordshire, UK, 2014; pp. 157–170.
24. Fukushima, T. Niacin metabolism and Parkinson's disease. *Environ. Health Prev. Med.* **2005**, *1*, 3–8. [CrossRef] [PubMed]
25. Chang, K.H.; Cheng, M.L.; Tang, H.Y.; Huang, C.Y.; Wu, Y.R.; Chen, C.M. Alternations of Metabolic Profile and Kynurenine Metabolism in the Plasma of Parkinson's Disease. *Mol. Neurobiol.* **2018**, *8*, 6319–6328. [CrossRef] [PubMed]
26. Nelson, D.L.; Cox, M.M. Lehninger. In *Lehninger Principles of Biochemistry*, 7th ed.; Macmillan Higher Education: New York, NY, USA; W.H. Freeman and Company: Houndmills, Basingstoke, UK, 2017.
27. Thenganatt, M.A.; Jankovic, J. Parkinson disease subtypes. *JAMA Neurol.* **2014**, *4*, 499–504.
28. Rodriguez-Gutierrez, R.; Lavalle-González, F.J.; Martínez-Garza, L.E.; Landeros-Olvera, E.; López-Alvarenga, J.C.; Torres-Sepúlveda, M.R.; González-González, J.G.; Mancillas-Adame, L.G.; Salazar-Gonzalez, B.; Villarreal-Pérez, J.Z. Impact of an exercise program on acylcarnitines in obesity: A prospective controlled study. *J. Int. Soc. Sports Nutr.* **2012**, *1*, 22. [CrossRef]
29. Lehmann, R.; Zhao, X.; Weigert, C.; Simon, P.; Fehrenbach, E.; Fritsche, J.; Machann, J.; Schick, F.; Wang, J.; Hoene, M.; et al. Medium chain acylcarnitines dominate the metabolite pattern in humans under moderate intensity exercise and support lipid oxidation. *PLoS ONE* **2010**, *7*, e11519. [CrossRef]
30. Chandrasekaran, A.; Idelchik, M.; Melendez, J.A. Redox control of senescence and age-related disease. *Redox Biol.* **2017**, *11*, 91–102. [CrossRef]
31. Lin, M.T.; Beal, M.F. Mitochondrial dysfunction and oxidative stress in neurodegenerative diseases. *Nature* **2006**, *7113*, 787–795. [CrossRef]
32. Starkov, A.A. The role of mitochondria in reactive oxygen species metabolism and signaling. *Ann. N. Y. Acad. Sci.* **2008**, *1147*, 37–52. [CrossRef]
33. Adams, F.; Boschmann, M.; Lobsien, E.; Kupsch, A.; Lipp, A.; Franke, G.; Leisse, M.C.; Janke, J.; Gottschalk, S.; Spranger, J.; et al. Influences of levodopa on adipose tissue and skeletal muscle metabolism in patients with idiopathic Parkinson's disease. *Eur. J. Clin. Pharmacol.* **2008**, *9*, 863–870. [CrossRef]

Effects of Enteric Environmental Modification by Coffee Components on Neurodegeneration in Rotenone-Treated Mice

Ikuko Miyazaki [1,*], Nami Isooka [1], Kouichi Wada [1], Ryo Kikuoka [1,2], Yoshihisa Kitamura [2] and Masato Asanuma [1]

[1] Department of Medical Neurobiology, Okayama University Graduate School of Medicine, Dentistry and Pharmaceutical Sciences, Okayama 700-8558, Japan; pi3s34bh@s.okayama-u.ac.jp (N.I.); pjv46mxi@s.okayama-u.ac.jp (K.W.); p32m0bvx@s.okayama-u.ac.jp (R.K.); asachan@cc.okayama-u.ac.jp (M.A.)

[2] Department of Clinical Pharmacy, Okayama University Graduate School of Medicine, Dentistry and Pharmaceutical Sciences, Okayama 700-8558, Japan; kitamu-y@cc.okayama-u.ac.jp

* Correspondence: miyazaki@cc.okayama-u.ac.jp

Abstract: Epidemiological studies have shown that coffee consumption decreases the risk of Parkinson's disease (PD). Caffeic acid (CA) and chlorogenic acid (CGA) are coffee components that have antioxidative properties. Rotenone, a mitochondrial complex I inhibitor, has been used to develop parkinsonian models, because the toxin induces PD-like pathology. Here, we examined the neuroprotective effects of CA and CGA against the rotenone-induced degeneration of central dopaminergic and peripheral enteric neurons. Male C57BL/6J mice were chronically administered rotenone (2.5 mg/kg/day), subcutaneously for four weeks. The animals were orally administered CA or CGA daily for 1 week before rotenone exposure and during the four weeks of rotenone treatment. Administrations of CA or CGA prevented rotenone-induced neurodegeneration of both nigral dopaminergic and intestinal enteric neurons. CA and CGA upregulated the antioxidative molecules, metallothionein (MT)-1,2, in striatal astrocytes of rotenone-injected mice. Primary cultured mesencephalic or enteric cells were pretreated with CA or CGA for 24 h, and then further co-treated with a low dose of rotenone (1–5 nM) for 48 h. The neuroprotective effects and MT upregulation induced by CA and CGA in vivo were reproduced in cultured cells. Our data indicated that intake of coffee components, CA and CGA, enhanced the antioxidative properties of glial cells and prevents rotenone-induced neurodegeneration in both the brain and myenteric plexus.

Keywords: caffeic acid; chlorogenic acid; rotenone; Parkinson's disease; neuroprotection; dopaminergic neuron; myenteric plexus; enteric glial cell; metallothionein

---

1. Introduction

Parkinson's disease (PD) is a progressive neurodegenerative disease with motor symptoms, such as tremor, akinesia/bradykinesia, rigidity, and postural instability, due to a loss of nigrostriatal dopaminergic neurons, and non-motor symptoms, such as orthostatic hypotension and constipation, caused by peripheral neurodegeneration. Gastrointestinal dysfunction is a particularly prominent non-motor symptom of PD. Several studies have reported that constipation appears approximately 10 to 20 years prior to the presentation of motor symptoms [1–3]. Recently, a large-scale prospective study demonstrated that lower bowel movement frequencies predicted the future PD crisis [4]. Braak et al. reported that PD pathology, Lewy bodies and Lewy neuritis, within the central nervous system (CNS), appeared first in the dorsal motor nucleus of vagus, and then extended upward through the brain stem to reach the substantia nigra, eventually leading to motor dysfunction [5]. In addition, several reports have demonstrated that PD pathology is also detected within the enteric nervous system (ENS) [6–8]. Therefore, it has been hypothesized that PD pathology spreads from the ENS to the CNS via the vagal nerve [9].

The cause of sporadic PD remains unknown, but both genetic and environmental factors are thought to contribute to PD pathogenesis. Epidemiological studies suggest that pesticide exposure, particularly rotenone and paraquat, increases the risk of PD [10]. Rotenone, a mitochondrial complex I inhibitor, is used to develop animal models of PD, because the toxin induces dopaminergic neuronal loss and PD motor symptoms [11,12]. Studies have demonstrated neurotoxic effects of rotenone in vitro and in vivo [11,13,14]. In addition, rotenone has been shown to reproduce PD pathology in both the CNS and ENS [15–17].

Epidemiological studies indicate that coffee consumption reduces the risk of PD to 40–50% [18,19]. Caffeic acid (CA) and chlorogenic acid (CGA), an ester formed between CA and quinic acid, are components of coffee. Various studies have reported that CA and CGA possess antioxidative properties, anti-inflammatory activity, and inhibitory effects on mitochondrial damage. Moreover, these coffee components exert neuroprotective effects against dopaminergic neurotoxicity [20–26]. However, it is still unknown whether coffee intake provides neuroprotective effects against enteric neuronal damage. The present study explored whether daily oral administrations of CA or CGA could prevent degeneration of central dopaminergic and peripheral enteric neurons in rotenone-treated mice. We established a novel rotenone-treated mouse model that exhibited neurodegeneration in both the CNS and ENS after chronic exposure to a low dose of rotenone (2.5 mg/kg/day) for four weeks. In addition, we examined the expression of the antioxidative molecules metallothionein (MT)-1,2, which are expressed mainly in astrocytes and secreted to the extracellular space, in rotenone-treated mice. Furthermore, we examined the neuroprotective effects of CA and CGA against rotenone-induced neurotoxicity in primary cultured cells from the mesencephalon and intestine.

2. Materials and Methods

2.1. Animals

All experimental procedures were performed in accordance with the Guideline for Animal Experiments of Okayama University Advanced Science Research Center, and were approved by the Animal Care and Use Committee of Okayama University Advanced Science Research Center. Male C57BL/6J mice at seven weeks of age and pregnant Sprague-Dawley (SD) rats at gestation day 13 were purchased from Charles River Japan Inc. (Yokohama, Japan). C57BL/6J mice and pregnant SD rats were housed with a 12-h light/dark cycle at a constant temperature (23 °C) and given ad libitum access to food.

2.2. Rotenone-Injected Mice and Treatment with CA or CGA

In our previous studies, we reported that chronic injection with rotenone (50 mg/kg/day) induced neurodegeneration in the substantia nigra pars compacta (SNpc) and intestinal myenteric plexus in mice [27]. In the current study, to examine the effects of CA and CGA on neurodegeneration in low-dose rotenone-treated mice, male C57BL/6J mice (nine weeks old; approximately 25 g) were subcutaneously injected with rotenone (2.5 mg/kg/day, Sigma-Aldrich, St. Louis, MO, USA) for four weeks using an osmotic mini pump (Alzet, #2004; Durect Corporation, Cupertino, CA, USA). The Alzet osmotic pump was filled with rotenone (10.4 mg/mL) dissolved in the vehicle solution, consisting of equal volumes of dimethylsulphoxide (DMSO) and polyethylenglycol (PEG). Mice were anesthetized by isoflurane inhalation. Rotenone-filled pumps were implanted under the skin on the backs of mice. Control mice received the vehicle solution.

Mice were orally administered CA (30 mg/kg/day) or CGA (50 mg/kg/day) dissolved in 5% methylcellulose daily for one week before rotenone exposure, and then 5 days/week during the four weeks of rotenone treatment (Figure 1A). The dosages of CA and CGA were determined based on previous reports [22,24,25]. One day after the rotenone treatment period, mice were perfused transcardially with a 4% paraformaldehyde (PFA) fixative for immunohistochemical analysis.

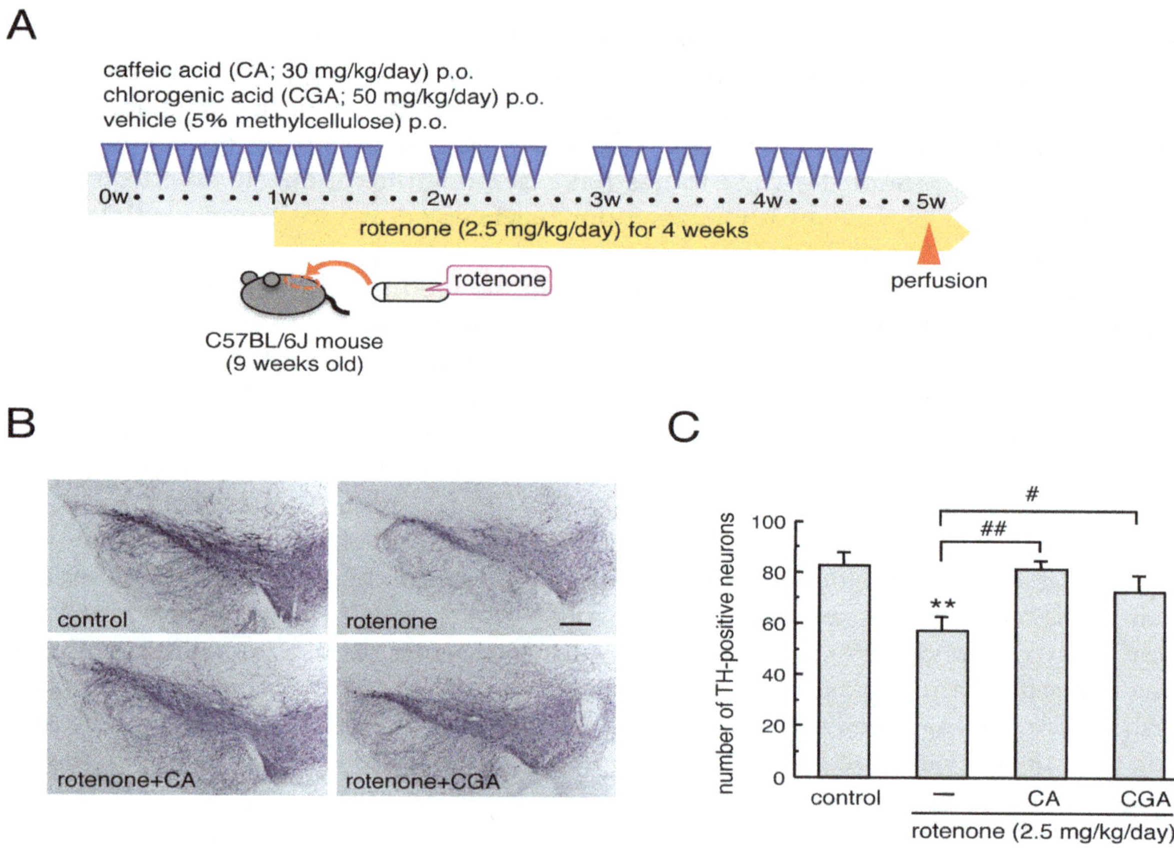

Figure 1. Administrations of CA or CGA prevented degeneration of dopaminergic neurons in the SNpc of rotenone-treated mice. (**A**) Schematic illustration of the experimental protocol. Male C57BL/6J mice were injected subcutaneously with rotenone (2.5 mg/kg/day) for four weeks using an osmotic mini pump. Mice were orally administered CA (30 mg/kg/day) or CGA (50 mg/kg/day), dissolved in 5% methylcellulose, daily for one week before rotenone exposure, and then 5 days/week during the four weeks of rotenone treatment. (**B**) Representative photomicrographs of immunohistochemistry for TH in the SNpc of rotenone-treated mice after treatment with CA or CGA. Scale bar = 200 μm. (**C**) Changes in the number of TH-positive nigral neurons after administration of CA or CGA. Each value is the mean ± SEM (n = 6–7). ** $p < 0.01$ vs. the vehicle-treated control group, # $p < 0.05$, ## $p < 0.01$ between the two indicated groups.

2.3. *Cell Culture of Mesencephalic Neurons and Astrocytes*

Primary cultured mesencephalic neurons and astrocytes were prepared from the mesencephalon of SD rat embryos at 15 days of gestation [28]. Neuronal and astrocyte co-cultures were constructed by directly seeding astrocytes onto neuronal cell cultures. To prepare enriched neuronal cultures, the mesencephalon was dissected, cut into small pieces with scissors, and then incubated for 15 min in 0.125% trypsin-EDTA at 37 °C. After centrifugation (1500× *g*, 3 min), the resulting cell pellet was treated with a 0.004% DNase I solution, containing 0.003% trypsin inhibitor, for 7 min at 37 °C. Following centrifugation (1500× *g*, 3 min), the cell pellet was gently resuspended in a small volume of Dulbecco's modified Eagle's medium (DMEM) with 4.5 g/L D-glucose (Invitrogen, San Diego, CA, USA), 10% fetal bovine serum (FBS), 4 mM L-glutamine, and 60 mg/L kanamycin sulfate (growth medium; DMEM–FBS). Resuspended cells were plated in the same medium at a density of 2×10^5 cells/cm^2 in four-chamber culture slides coated with poly-D-lysine (Falcon, Corning, NY, USA). Within 24 h of the initial plating, the medium was replaced with fresh DMEM-FBS medium supplemented with

2 μM cytosine-β-D-arabinofuranoside (Ara-C) to inhibit glial cell replication. Cells were incubated in this medium for three days. To obtain mesencephalic astrocytes, small pieces of mesencephalon were treated with 0.125% trypsin followed by 0.004% DNase I, as described above. Cells were plated at a density of 2×10^5 cells/cm^2 in poly-D-lysine-coated 6-well plates (Falcon) in DMEM-FBS medium. After incubation for seven days, cells were subcultured, and then seeded, at a density of 4×10^4 cells/cm^2, directly onto mesencephalic neuronal cell layers that had been cultured in four-chamber culture slides for 4 days. The co-cultures were incubated for a further two days before beginning any treatment. To prepare astrocyte cultures, cells were subcultured as described above, plated in DMEM-FBS medium at a density of 2×10^4 cells/cm^2 in poly-D-lysine-coated four-chamber culture slides, and then incubated for one week. All cultures were maintained at 37 °C in a 5%/95% CO_2/air mixture.

2.4. Cell Culture of Enteric Neurons and Glial Cells

Enteric neuronal and glial co-cultures were prepared from the intestine of SD rat embryos at 15 days of gestation [29]. Intestines were dissected and kept on ice in Hank's buffered salt solution (HBSS, Sigma-Aldrich) supplemented with 50 μg/mL streptomycin and 50 U/mL penicillin (Invitrogen). After washing with fresh HBSS, intestines were cut into small pieces with scissors in DMEM/F12 (1:1) medium (Invitrogen) with 50 μg/mL streptomycin, 50 U/mL penicillin (DMEM/F12 medium), and 10% FBS, and then centrifuged (1500× g, 3 min). The resulting cell pellet was treated with 0.125% trypsin (Invitrogen) for 15 min at 37 °C. After centrifugation (1500× g, 10 min), the resulting cells were treated with 0.01% (v/v) DNase I (Sigma-Aldrich) for 10 min at 37 °C. After centrifugation (100× g, 10 min), the cell pellet was gently re-suspended in a small volume of DMEM/F12 medium containing 10% FBS, and then plated in the same medium at a density of 5×10^4 cells/cm^2 in four-chamber culture slides previously coated for 6 h with a solution of 0.5% (v/v) gelatin (Sigma-Aldrich). Within 24 h of the initial plating, the medium was replaced with fresh DMEM/F12 medium without FBS but containing 1% N-2 and 1% G-5 supplements (Invitrogen). Half of the medium was replaced every two days, and cell cultures were maintained for 13 days at 37 °C in a 5%/95% CO_2/air mixture.

2.5. Cell Treatments

Fresh solution of rotenone in DMSO were prepared before each experiment and then diluted to their final concentrations in the appropriate growth medium (final concentration of DMSO: 0.005% v/v). To examine the effects of CA and CGA on rotenone-induced dopaminergic neurotoxicity, mesencephalic neuronal and astrocyte co-cultures were treated with CA (10 or 25 μM) or CGA (25 μM) in 0.2% DMSO in growth medium for 24 h. The concentrations of CA and CGA were determined based on previous reports [30–32]. After incubation for 24 h, the medium containing CA or CGA was discarded, and then cells were treated with CA (10 or 25 μM) or CGA (25 μM) and rotenone (1, 2.5, or 5 nM) for 48 h. To examine the effects of CA and CGA on MT-1,2 expression in mesencephalic astrocytes, astrocyte cultures were treated with CA (10 or 25 μM) or CGA (25 μM) for 24 h in advance, and then treated with CA (10 or 25 μM) or CGA (25 μM), with or without rotenone (1, 2.5, or 5 nM) for a further 48 h.

To determine the effects of CA and CGA on rotenone-induced enteric neuronal loss and glial MT-1,2 expression, enteric neuronal and glial co-cultures were treated with CA (10 or 25 μM) or CGA (25 μM) in DMEM/F12 medium containing 0.2% DMSO and 1% N-2 and 1% G-5 supplements for 24 h, followed by treatment with CA (10 or 25 μM) or CGA (25 μM) and rotenone (1, 2.5, or 5 nM) for 48 h.

2.6. Immunohistochemistry

To prepare slices of the brain and intestine, mice were perfused with ice-cold saline followed by 4% PFA under deep pentobarbital anesthesia (70 mg/kg, i.p.). The perfused brains and intestines were removed immediately and post-fixed for 24 h or 2 h in 4% PFA, respectively. Following cryoprotection in 15% sucrose in phosphate buffer (PB) for 48 h, the brains and intestines were snap-frozen with powdered dry ice and 20-μm-thick coronal or transverse sections were cut on a cryostat. Brain slices were collected at levels containing the mid-striatum (+0.6 to +1.0 mm from the bregma) and the SNpc (−2.8 to −3.0 mm from bregma). For immunostaining of tyrosine hydroxylase (TH) in the SNpc, brain slices were treated with 0.5% H_2O_2 for 30 min at room temperature (RT), blocked with 1% normal goat serum for 30 min, and incubated for 18 h at 4 °C with a rabbit anti-TH antibody (1:1000; Millipore, Temecula, CA, USA) diluted in 10 mM phosphate-buffered saline (PBS) containing 0.2% Triton X-100 (0.2% PBST). After washing in 0.2% PBST (3 × 10 min), slices were reacted with biotinylated goat anti-rabbit IgG secondary antibody for 2 h at RT. After washing, the sections were incubated with an avidin-biotin peroxidase complex for 1 h at RT. TH-immunopositive signals were visualized by DAB, nickel, and H_2O_2. To examine effects of CA and CGA treatment on astrocytic MT-1,2 expression in rotenone-injected mice, the striatal sections were incubated in 1% normal goat serum for 30 min at RT, and then reacted with mouse anti-MT-1,2 (1:100; Dako Cytomation, Glostrup, Denmark) or rabbit anti-S100β (1:5000; Dako Cytomation) antibodies for 18 h at 4 °C. After washing, slices were reacted with Alexa Fluor 594-conjugated goat anti-mouse IgG or Alexa Fluor 488-conjugated goat anti-rabbit IgG secondary antibodies (1:1000; Invitrogen) for 2 h at RT. To visualize the myenteric plexus or enteric glial cells in the intestine, intestinal sections were incubated in 1% normal goat serum for 30 min at RT, and then reacted with rabbit anti-β-tubulin III (1:100; GeneTex, Inc., Irvine, CA, USA) or rabbit anti-GFAP (1:10,000; Novus Biologicals, Centennial, CO, USA) antibody, respectively, for 18 h at 4 °C. After washing, slices were reacted with Alexa Fluor 488-conjugated goat anti-rabbit IgG secondary antibody (1:1000; Invitrogen) for 2 h at RT. The striatal and intestinal slices were then counterstained with Hoechst 33342 nuclear stain (10 μg/mL) for 2 min. Cells cultured on chamber slides were fixed with 4% PFA for 30 min at RT, blocked with 2.5% normal goat serum for 20 min, and then reacted for 18 h at 4 °C with the following primary antibodies diluted in 0.1% PBST: rabbit anti-TH (1:1000; Millipore); mouse anti-MT-1,2 (1:100; DAKO Cytomation); rabbit anti-GFAP (1:2000; Dako Cytomation); or mouse anti-β-tubulin III (1:10,000; Sigma-Aldrich). After washing in 10 mM PBS, pH 7.4 (3 × 10 min), cells were reacted with Alexa Fluor 488-conjugated goat anti-rabbit IgG or Alexa Fluor 594-conjugated goat anti-mouse IgG secondary antibodies (1:500; Invitrogen) for 1.5 h at RT. Finally, cells were counterstained with Hoechst 33342 nuclear stain (10 μg/mL) for 2 min and washed prior to mounting with Fluoromounting medium (Dako Cytomation).

All slides were analyzed under a fluorescence microscope (BX50-FLA or BX53; Olympus Tokyo, Japan) and cellSens imaging software (Olympus), using a mercury lamp (USHIO INC., Tokyo, Japan) through 360–370 nm, 470–495 nm, or 530–550 nm band-pass filters to excite Hoechst 33342, Alexa Fluor 488, or Alexa Fluor 594, respectively. Light emission from Hoechst 33342, Alexa Fluor 488, or Alexa Fluor 594 was collected through a 420 nm long-pass filter, a 510–550 nm band-pass filter, or a 590 nm long-pass filter, respectively. Localization of β-tubulin III- and GFAP signals was confirmed by confocal laser-scanning microscopy (LSM 780; Zeiss, Oberkochen, Germany). Light emitted from Hoechst 33342, Alexa Fluor 488, or Alexa Fluor 594 was collected through a 420–470 nm band-pass filter, a 500–550 nm band-pass filter, or a 570–640 nm band-pass filter, respectively. Images were taken at a magnification of 400× and recorded using the Windows-based LSM program (ZEN lite 2012 64bit version, Zeiss). Adobe Photoshop CS4 software (v11.0) was used for digital amplification of the images.

2.7. *Quantification Procedures*

The number of TH-immunopositive neurons in the SNpc was counted manually under a microscope at 100× magnification. The boundary between the SNpc and ventral tegmental area was defined by a line extending dorsally from the most medial boundary of the cerebral peduncle.

The numbers of MT-1,2- and S100β-immunopositive cells in the dorsal striatum of rotenone-treated mice were counted manually using a microscope at a magnification of 400×. The number of MT- or S100β-positive cells and the ratio of MT-positive cells to S100β-positive cells were evaluated in each section. The immunoreactivity of β-tubulin III or GFAP in the myenteric plexus of the intestine was analyzed under 400× magnification and quantified using cellSens imaging software (v1.16, Olympus). The integrated density of each signal was calculated as follows: integrated density = (signal density in the myenteric plexus-background density) × area of positive signal in the plexus.

TH-immunopositive cells in mesencephalic neuronal and astrocyte co-cultures were counted under a microscope in all areas of each chamber slide. Cell viability data are presented as a percentage of the control. The number of MT-1,2-immunopositive cells in mesencephalic astrocyte cultures was counted in 8–10 randomly chosen fields in a chamber under 200× magnification, and expressed as the percentage of MT-1,2-immunopositive astrocytes among the total cell population. The signal intensity of β-tubulin III and MT-1,2 in enteric neuronal and glial co-cultures was analyzed in 3–6 randomly chosen fields in a chamber under 200× magnification and quantified using cellSens imaging software.

2.8. Statistical Analyses

All statistical analyses were performed using KaleidaGraph v4.0 software. Data are presented as means ± SEM. Comparisons between multiple groups were performed using a one-way ANOVA followed by a *post hoc* Fisher's least significant difference test. A p-value < 0.05 was considered statistically significant.

3. Results

3.1. Administrations of CA or CGA Prevented Dopaminergic Neurodegeneration in Rotenone-Treated Mice

Chronic subcutaneous treatment with a low dose of rotenone (2.5 mg/kg/day) significantly decreased the number of TH-positive dopaminergic neurons in the SNpc. Repeated oral administration of CA (30 mg/kg) or CGA (50 mg/kg) ameliorated the reduction of nigral TH-positive cells in rotenone-treated mice (Figure 1B,C).

3.2. Administrations of CA or CGA Increased MT-1,2 Expression in Astrocytes in the Striatum of Rotenone-Treated Mice

To examine the effects of CA and CGA treatment on antioxidative molecules in astrocytes of rotenone-treated mice, we performed double immunostaining of the astrocyte marker S100β and MT-1,2 in striatal brain slices. We used an anti-S100β, but not an anti-GFAP, antibody to visualize astrocytes in the striatum. Since the anti-GFAP antibody detected mainly fibrous activated astrocytes, it was difficult to assess MT-1,2 expression in all types of astrocytes, including protoplasmic astrocytes. Therefore, we chose an anti-S100β antibody to detect striatal astrocytes. A nonsignificant trend toward decreased numbers of S100β-positive astrocytes were seen after rotenone treatment. Administration of CA or CGA significantly increased the number of MT-positive astrocytes in the striatum of mice (Figure 2A,B). The MT-positive/S100β-positive cell ratio was significantly increased by either CA or CGA treatment (Figure 2C).

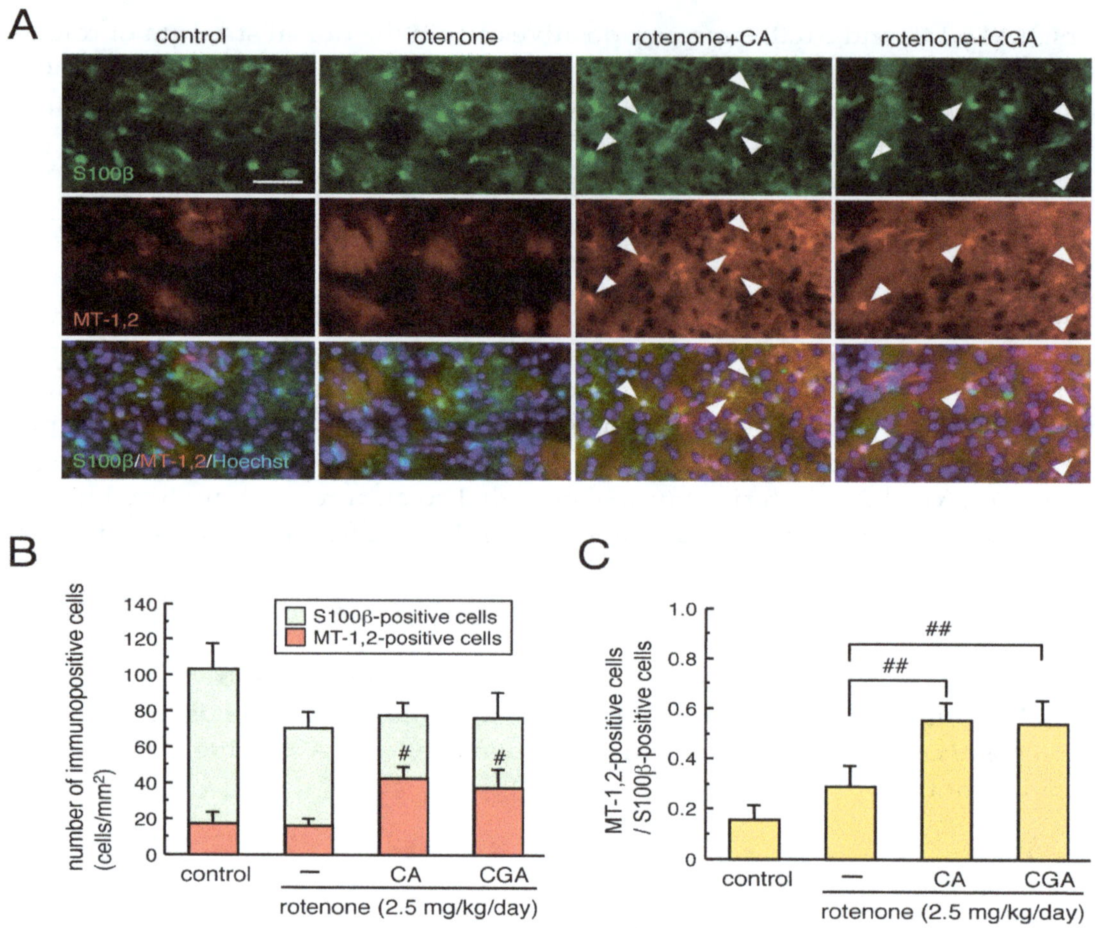

Figure 2. Effects of CA or CGA administrations on astrocytic MT-1,2 expression in the striatum of rotenone-treated mice. (**A**) Representative photomicrographs of MT-1,2 and S100β double immunostaining in the striatum of rotenone (2.5 mg/kg/day)-treated mice after treatment with CA (30 mg/kg/day) or CGA (50 mg/kg/day). Green: S100β-positive astrocytes. Red: MT-1,2-positive cells. Blue: nuclear staining with Hoechst 33342. Solid arrowheads: MT-1,2-positive astrocytes. Scale bar = 50 μm. (**B**,**C**) Quantitation of MT-1,2 and S100β expression in the striatum of rotenone-treated mice after treatment with CA or CGA. (**B**) Number of immunopositive cells, (**C**) proportion of MT-1,2-positive cells/S100β-positive cells. Data are means ± SEM (n = 6–7). # $p < 0.05$, ## $p < 0.01$ vs. the rotenone-treated group.

3.3. Administration of CA or CGA Prevented Neurodegeneration in the Intestinal Myenteric Plexus of Rotenone-Treated Mice

To examine the neuroprotective effects of CA and CGA on the myenteric plexus in the small intestine of rotenone-treated mice, we performed immunostaining of the neuronal marker, β-tubulin III. To confirm the distribution of the myenteric plexus in the intestine, nuclear staining was performed using Hoechst 33342. Apparent β-tubulin III-positive signals were detected in the intestinal myenteric plexus of mice (Figure 3A). Chronic subcutaneous treatment with low-dose rotenone for four weeks significantly decreased the area of β-tubulin III-positive myenteric plexus (Figure 3A–C) and β-tubulin III immunoreactivity (Figure 3A,B,D) in the intestine. Repeated administration of CA or CGA significantly prevented this reduction in β-tubulin III-positive signals in the myenteric plexus of rotenone-treated mice (Figure 3B–D).

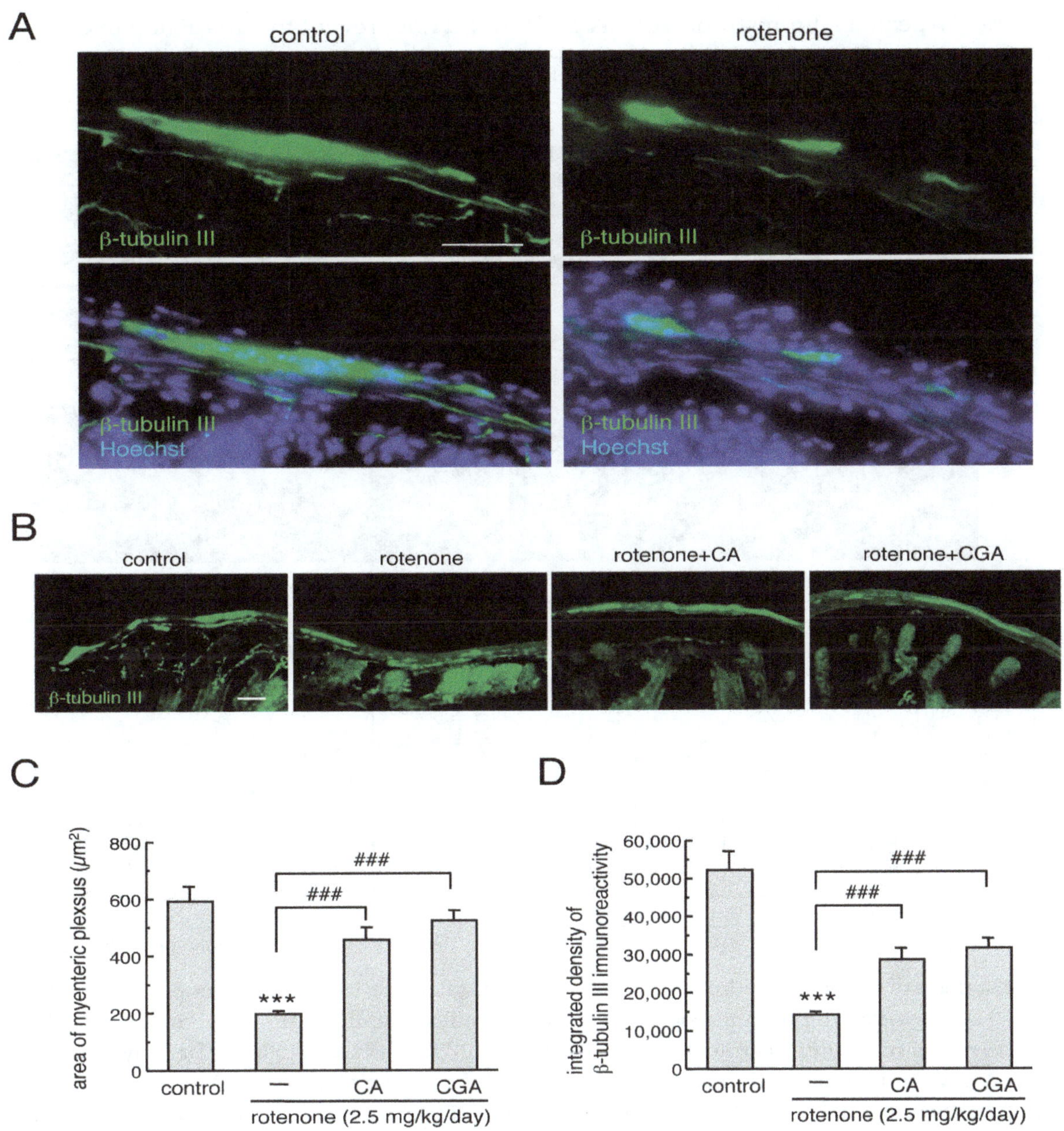

Figure 3. Administrations of CA or CGA prevented the degeneration of enteric neurons in the intestinal myenteric plexus of rotenone-treated mice. (**A**) Representative photomicrographs of immunohistochemistry for β-tubulin III in the intestine of mice. Green: β-tubulin III-positive neurons. Blue: nuclear staining with Hoechst 33342. Scale bar = 50 μm. (**B**) Representative photomicrographs of β-tubulin III-positive neurons in the intestine of rotenone-treated mice after treatment with CA (30 mg/kg/day) or CGA (50 mg/kg/day). Scale bar = 50 μm. (**C**,**D**) Quantitation of β-tubulin III-positive signals in the intestine. (**C**) Area of β-tubulin III-positive myenteric plexus, (**D**) integrated density of β-tubulin III immunoreactivity. Data are means ± SEM (n = 6–7). *** $p < 0.001$ vs. the vehicle-treated control group, ### $p < 0.001$ between the two indicated groups.

3.4. Administration of CA or CGA Had No Effect on Enteric Glial Cells in Rotenone-Treated Mice

To examine the effects of CA and CGA on enteric glial cells in the small intestine of rotenone-treated mice, we performed immunostaining of the glial marker, GFAP [33]. Chronic subcutaneous treatment with a low dose of rotenone for four weeks had no effect on the area of GFAP-positive signal (Figure 4A,B), but significantly decreased GFAP immunoreactivity (Figure 4A,C) in the intestine. Repeated administration of CA or CGA did not prevent this reduction in GFAP-positive signal in the intestine of rotenone-treated mice (Figure 4A–C).

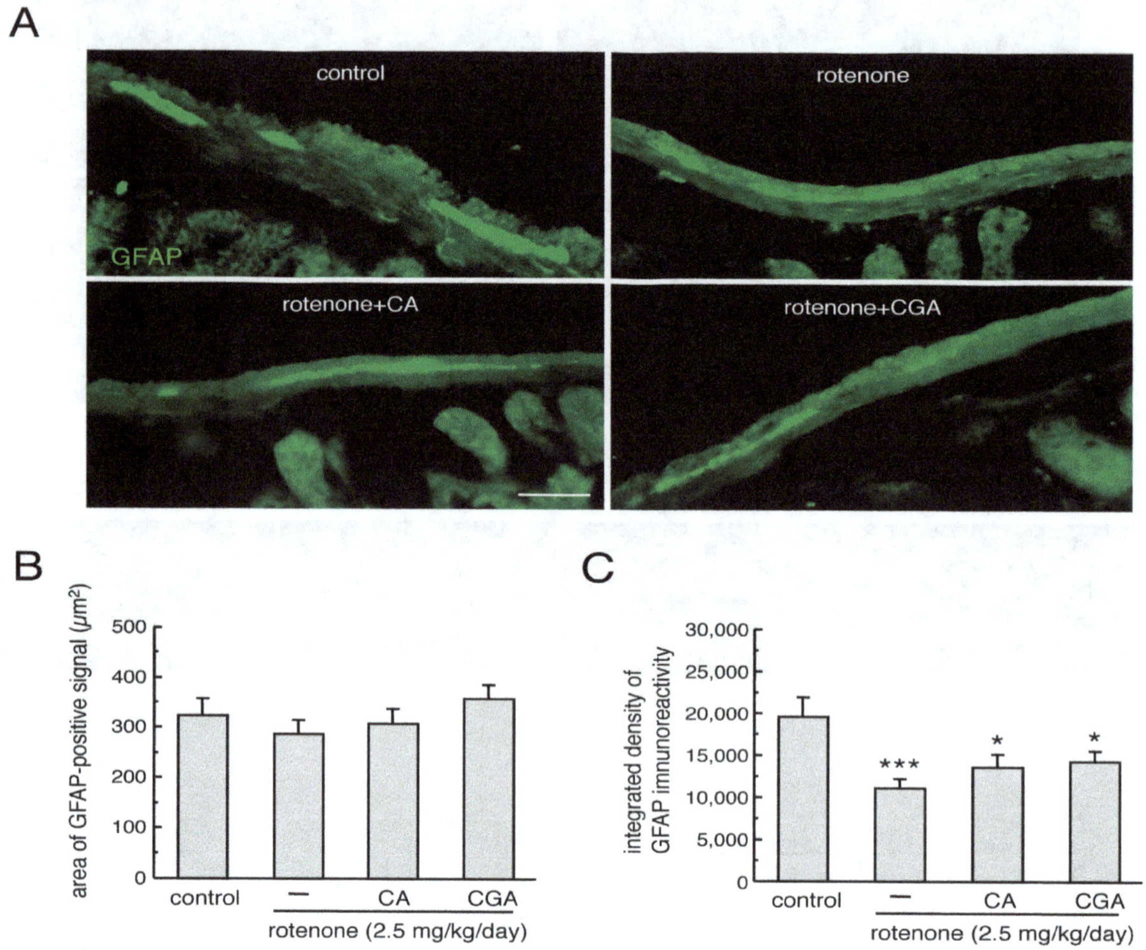

Figure 4. Effects of CA or CGA administrations on enteric glial cells in the intestines of rotenone-treated mice. (**A**) Representative photomicrographs of immunohistochemistry for GFAP in the intestines of rotenone-treated mice after treatment with CA (30 mg/kg/day) or CGA (50 mg/kg/day). Scale bar = 50 μm. (**B**,**C**) Quantitation of GFAP-positive signals in the intestine of mice. (**B**) Area of GFAP-positive signal, (**C**) integrated density of GFAP immunoreactivity. Data are means ± SEM (n = 6–7). * $p < 0.05$, *** $p < 0.001$ vs. the vehicle-treated control group.

3.5. Treatment with CA or CGA Inhibited Rotenone-Induced Dopaminergic Neuronal Loss in Mesencephalic Neuronal and Astrocyte Co-Cultures

To examine the neuroprotective effects of CA and CGA on rotenone-induced dopaminergic neurodegeneration in cultured cells, mesencephalic neuronal and astrocyte co-cultures were pretreated with CA (10 or 25 μM) or CGA (25 μM) for 24 h and co-treated with low-dose rotenone (1–5 nM) for a further 48 h (Figure 5A). Exposure to a low dose of rotenone significantly decreased the number of TH-positive dopaminergic neurons. Both CA and CGA treatment significantly and completely inhibited this reduction in the number of TH-positive cells (Figure 5B,C).

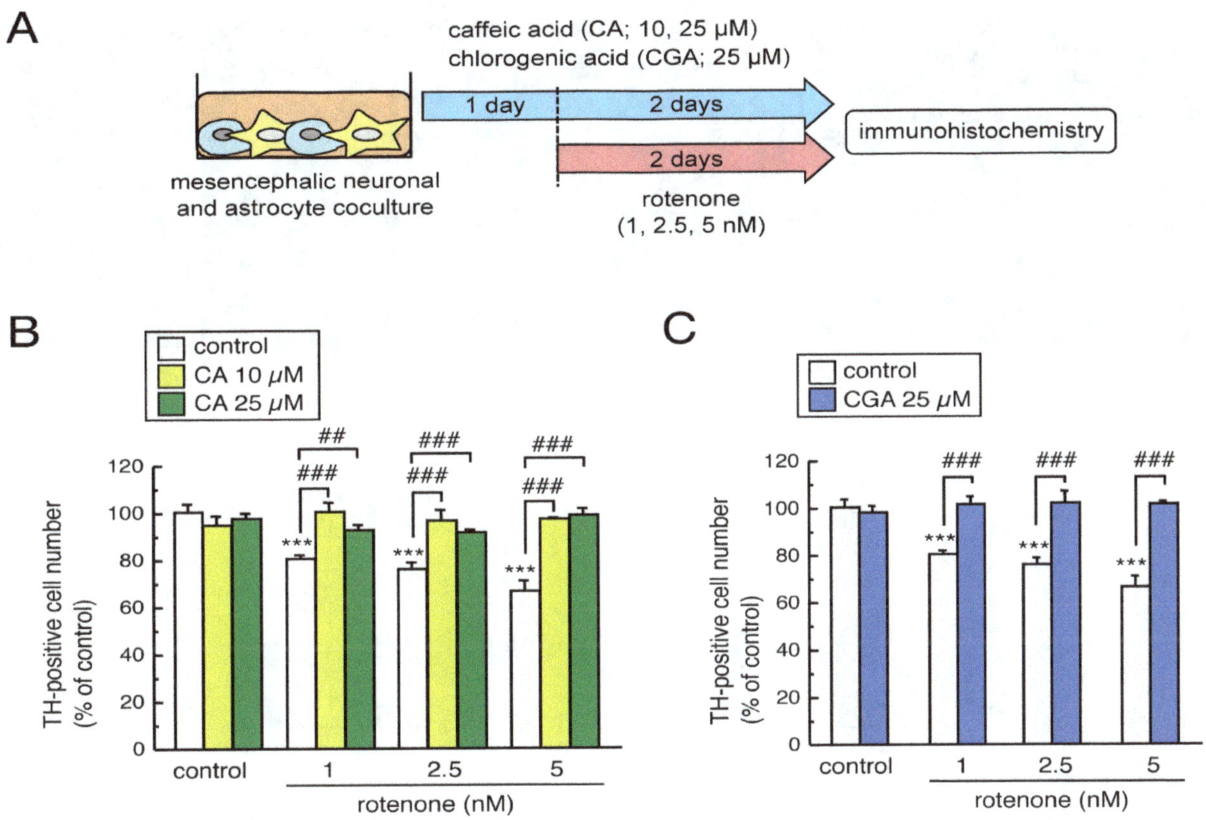

Figure 5. Neuroprotective effects of CA and CGA against rotenone-induced dopaminergic neurotoxicity in cultured mesencephalic cells. (**A**) Schematic illustration of the experimental protocol. Mesencephalic neuronal and astrocyte co-cultures were pretreated with CA (10 or 25 μM) or CGA (25 μM) for 24 h and co-treated with rotenone (1–5 nM) for a further 48 h. (**B**,**C**) Changes in the number of TH-positive neurons after treatment with rotenone and CA (**B**) or CGA (**C**). Data are means ± SEM (n = 4). *** $p < 0.001$ vs. each control group, ## $p < 0.01$, ### $p < 0.001$ between the two indicated groups.

3.6. Treatment with CA or CGA Upregulated MT-1,2 in Mesencephalic Astrocytes

To examine the effects of treatment with CA or CGA, with or without rotenone, on MT-1,2 expression in mesencephalic astrocytes, astrocyte cultures were pretreated with CA (10 or 25 μM) or CGA (25 μM) for 24 h and co-treated with rotenone (1–5 nM) for 48 h. Rotenone had no effect on MT-1,2 expression in astrocytes even at the highest dose (5 nM). Treatment with CA or CGA significantly increased MT-1,2 expression in astrocytes even after rotenone exposure (Figure 6A–C). Interestingly, the induction of MT-1,2 by CGA (25 μM) was higher in the rotenone-treated astrocytes than in the untreated control group (Figure 6C).

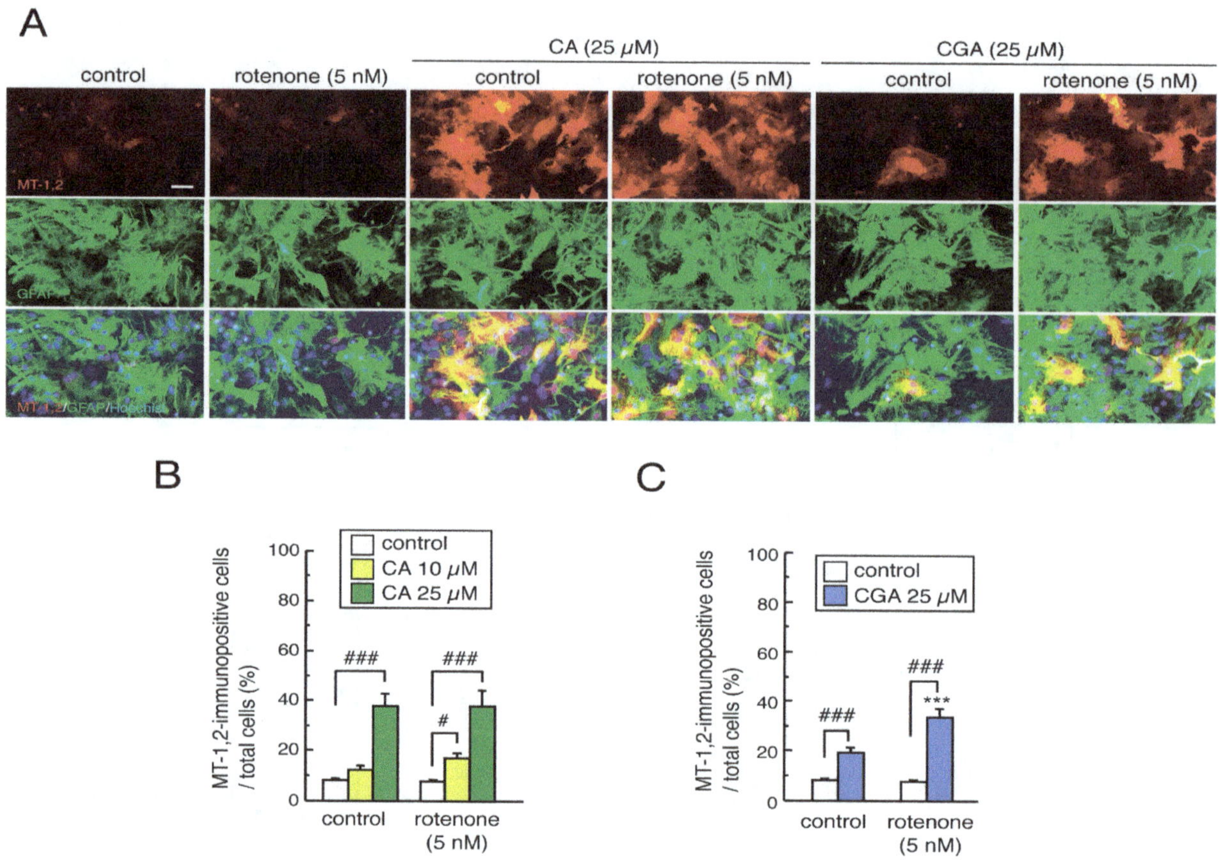

Figure 6. Effects of treatment with CA or CGA followed by rotenone exposure on MT-1,2 expression in mesencephalic astrocyte cultures. Astrocyte cultures were pretreated with CA (10 or 25 μM) or CGA (25 μM) for 24 h and co-treated with rotenone (1–5 nM) for a further 48 h. (**A**) Representative photomicrographs of MT-1,2 and GFAP double immunostaining in astrocyte cultures. Red: MT-1,2-positive signals. Green: GFAP-positive astrocytes. Blue: nuclear staining with Hoechst 33342. Scale bar = 50 μm. (**B**,**C**) Quantitation of MT-1,2-positive signals in astrocyte cultures after treatment with rotenone and CA (**B**) or CGA (**C**). Each value is the mean ± SEM (n = 8–13) expressed as the percentage of the MT-1,2-immunopositive astrocytes in the total cell population. *** $p < 0.001$ vs. each control group, # $p < 0.05$, ### $p < 0.001$ between the two indicated groups.

3.7. Treatment with CA or CGA Inhibited Rotenone-Induced Enteric Neuronal Loss in Enteric Neuronal and Glial Co-Cultures

To examine whether low-dose rotenone exposure induced enteric neurotoxicity, and whether CA or CGA treatment could prevent these effects, we prepared enteric neuronal and glial co-cultures from the intestines of SD rat embryos. Enteric neuronal and glial co-cultures were pretreated with CA or CGA for 24 h, and then co-treated with a low dose of rotenone (1–5 nM) for a further 48 h (Figure 7A). We successfully detected β-tubulin III-positive enteric neuronal cells and GFAP-positive glial cells. Enteric glial were seen in the nerve plexus (Figure 7B). Exposure to a low dose of rotenone (1–5 nM) for 48 h significantly decreased β-tubulin III immunoreactivity in cultured enteric cells. Treatment with CA (10, 25 μM) or CGA (25 μM) significantly ameliorated the reduction in β-tubulin III-positive signals induced by rotenone exposure (Figure 7C–E). These results indicate that both CA and CGA provide neuroprotection against rotenone-induced enteric neurotoxicity. Interestingly, CA and CGA could both protect enteric neurons against higher doses of rotenone exposure (5 nM), but not against lower doses (1 or 2.5 nM, Figure 7D,E).

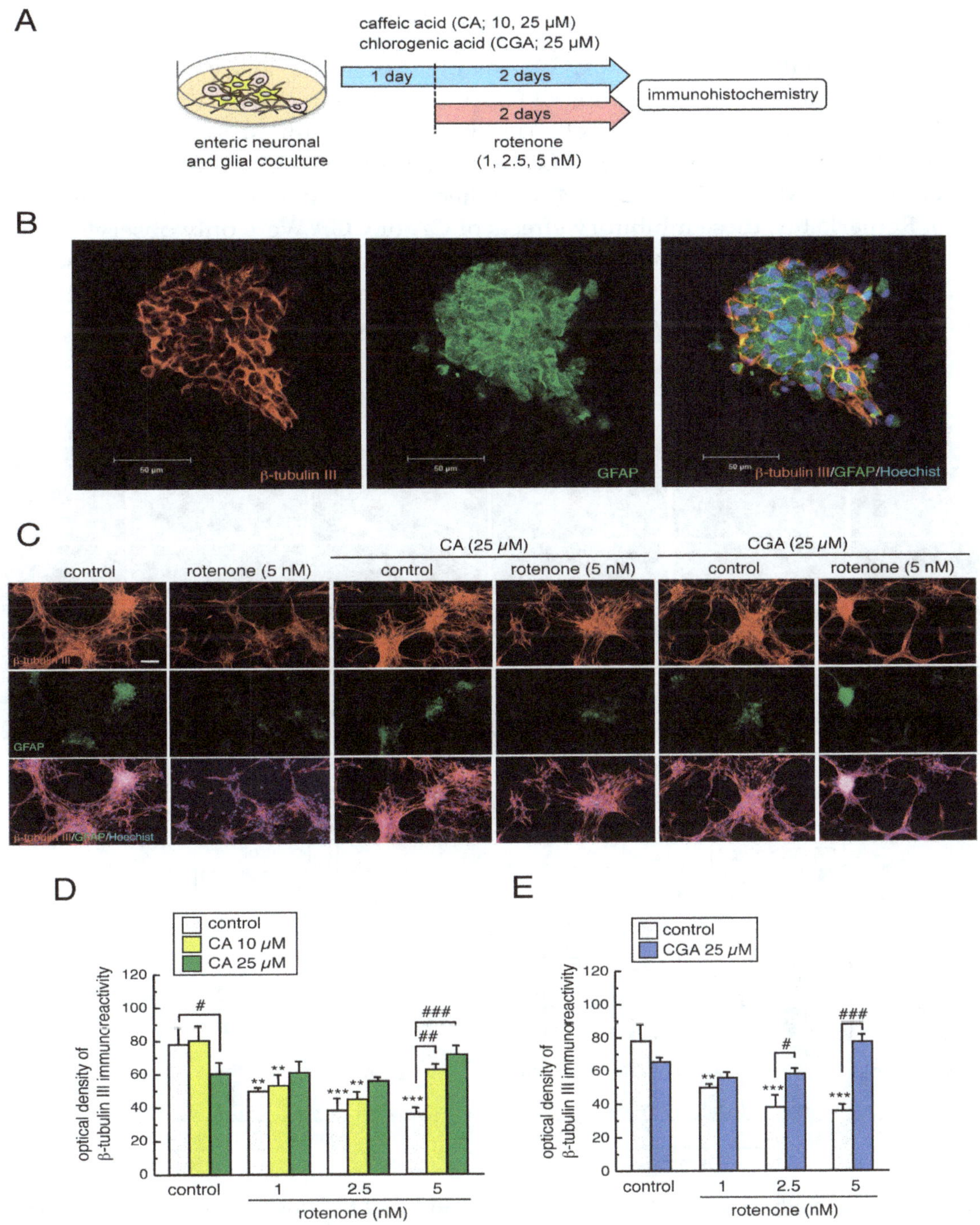

Figure 7. Neuroprotective effects of CA and CGA against rotenone-induced enteric neurotoxicity in primary cultured intestinal cells. (**A**) Schematic illustration of the experimental protocol. Enteric neuronal and glial co-cultures were pretreated with CA (10 or 25 μM) or CGA (25 μ) for 24 h and co-treated with rotenone (1–5 nM) for further 48 h. (**B**) Localization of β-tubulin III- and GFAP-positive signals. Red: β-tubulin III-positive enteric neuronal cells. Green: GFAP-positive enteric glial cells. Blue: nuclear staining with Hoechst 33342. Scale bar = 50 μm. (**C**) Representative photomicrographs of immunohistochemistry for β-tubulin III and GFAP in enteric cell cultures after treatment with rotenone (5 nM) and CA (25 μM) or CGA (25 μM). Red: β-tubulin III-positive enteric neuronal cells. Green: GFAP-positive enteric glial cells. Blue: nuclear staining with Hoechst 33342. Scale bar = 100 μm. (**D**,**E**) Quantitation of β-tubulin III-positive signals in enteric cell cultures after treatment with rotenone and CA (**D**) or CGA (**E**). Data are means ± SEM (n = 3–6). ** $p < 0.01$, *** $p < 0.001$ vs. each control group; # $p < 0.05$, ## $p < 0.01$, ### $p < 0.001$ between the two indicated groups.

3.8. Treatment with CA or CGA Inhibited Rotenone-Induced MT-1,2 Reduction in Enteric Glial Cells

We examined the effects of CA and CGA treatment on MT-1,2 expression in rotenone-treated enteric glial cells by double immunostaining of GFAP and MT-1,2. MT-positive signals were localized to cultured GFAP-positive enteric glial cells (Figure 8A). In contrast to the results from cultured mesencephalic cells, rotenone exposure significantly decreased MT-1,2 expression in enteric glial cells and both CA (10 or 25 μM) and CGA (25 μM) inhibited the rotenone-induced downregulation of MT (Figure 8B,C). Remarkably, these inhibitory effects of CA or CGA were only observed in enteric cells treated with highest dose of rotenone (5 nM) and not in those treated with lower doses (1 or 2.5 nM, Figure 8B,C). This was in agreement with the finding that these coffee components have protective effects against rotenone-induced enteric neurotoxicity (Figure 7D,E).

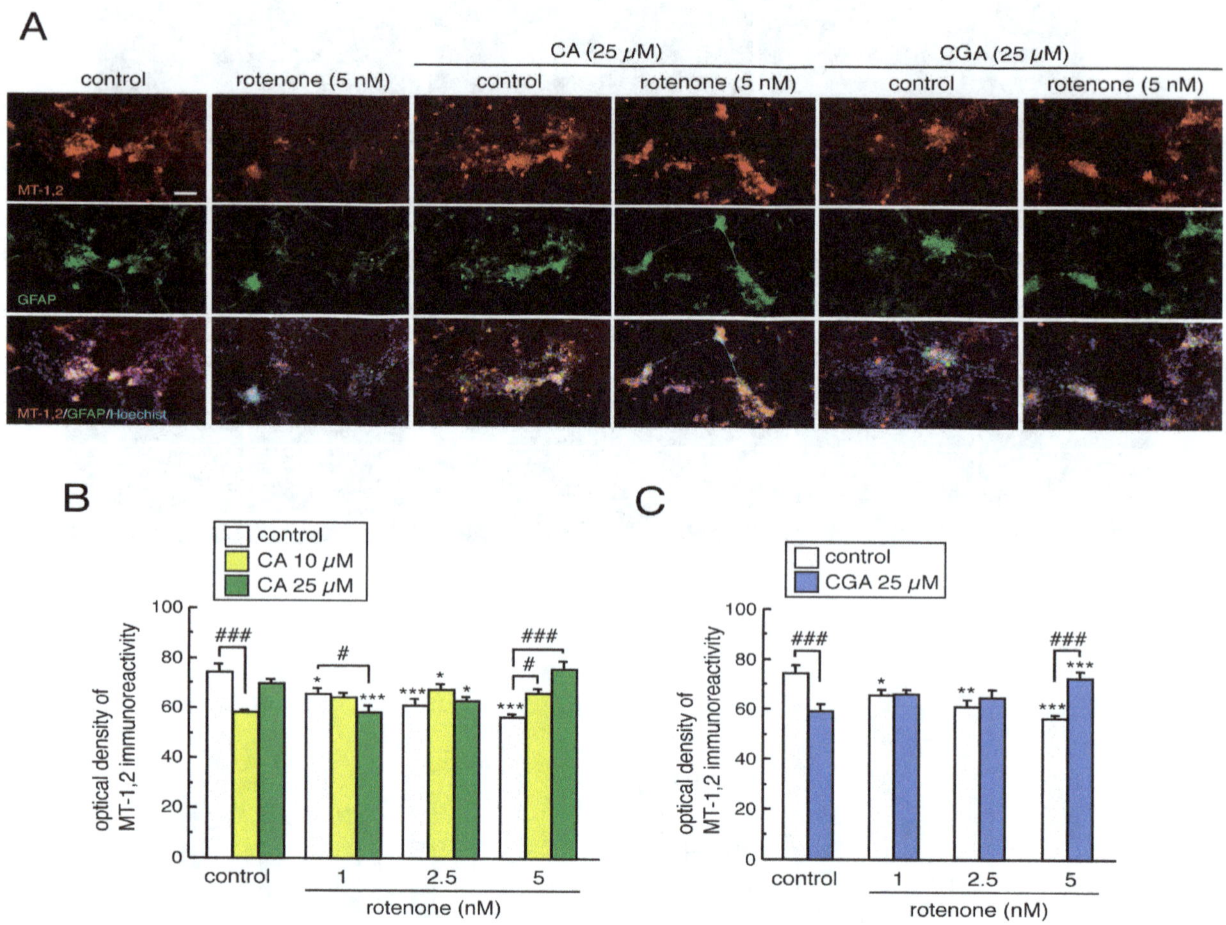

Figure 8. Treatment with CA or CGA inhibits rotenone-induced MT downregulation in the enteric glial cells. Enteric neuronal and glial co-cultures were pretreated with CA (10 or 25 μM) or CGA (25 μM) for 24 h and co-treated with rotenone (1–5 nM) for a further 48 h. (**A**) Representative photomicrographs of immunohistochemistry for MT-1,2 and GFAP in enteric cell cultures after treatment with rotenone (5 nM) and CA (25 μM) or CGA (25 μM). Red: MT-1,2-positive signals. Green: GFAP-positive enteric glial cells. Blue: nuclear staining with Hoechst 33342. Scale bar = 100 μm. (**B**,**C**) Quantitation of MT-1,2-positive signals in the enteric cell cultures after treatment with rotenone and CA (**B**) or CGA (**C**). Data are means ± SEM (n = 3–8). * $p < 0.05$, ** $p < 0.01$, *** $p < 0.001$ vs. each control group; # $p < 0.05$, ### $p < 0.001$ between the two indicated groups.

4. Discussion

The present study demonstrated that CA and CGA upregulated MT-1,2 in astrocytes and exerted neuroprotective effects against rotenone-induced dopaminergic neurodegeneration in mesencephalic neuronal and astrocyte co-cultures. In the enteric neuronal and glial co-cultures, rotenone treatment reduced MT-1,2 expression in glial cells and produced enteric neuronal loss, which were prevented by CA or CGA treatment. Furthermore, oral administration of CA or CGA exerted neuroprotective effects against neurodegeneration in the nigral dopaminergic neurons and the enteric neurons in the intestinal myenteric plexus of rotenone-treated mice. In this study, we used a novel mouse model produced by 4-week administration of a low dose of rotenone (2.5 mg/kg/day), which corresponds to the environmental exposure levels of rotenone via pesticides. Chronic subcutaneous injection of rotenone induced neurodegeneration in both the CNS and ENS. Interestingly, rotenone-induced neurotoxicity was more severe in the intestinal myenteric plexus than in the SNpc. Considering the epidemiological data showing an inverse association of daily coffee consumption with PD risk [18,19], we examined the neuroprotective effects of CA and CGA against rotenone neurotoxicity. Animals were treated with these coffee components by daily oral administration for 1 week before rotenone exposure, and then 5 days/week during the four weeks of rotenone treatment. The dosages of CA (30 mg/kg/day) and CGA (50 mg/kg/day) were determined based on previous reports [22,24,25]. It has been reported that one cup of coffee contains 70–350 mg of CA [34] and approximately 250 mg of CGA [35]. In addition, it is known that the daily intake of CA in coffee drinkers is 0.1–1 g [36]. Therefore, the dosages of CA and CGA in the present experiments seem high, since it would be necessary to drink 5–10 cups of coffee per day to achieve their neuroprotective effects in humans.

Treatment with CA or CGA significantly prevented rotenone-induced neurodegeneration of nigral dopaminergic neurons and intestinal enteric neurons. Previous studies have reported that CA and CGA provide dopaminergic neuroprotection in various models of PD [20–26]. In those reports, CA and CGA showed antioxidative properties [23,24], anti-inflammatory activity [21,24–26], and inhibitory effects of mitochondrial damage [20]. Various reports have demonstrated that CA and CGA activate the nuclear factor erythroid-2-related factor 2 (Nrf2) antioxidant pathway [37–42]. In this study, we showed that CA or CGA treatment upregulated the antioxidative molecules, MT-1,2, in mesencephalic astrocyte cultures and striatal astrocytes of rotenone-treated mice. MT is a low-molecular weight, cysteine-rich (30% of the protein), inducible protein that binds to metals, such as zinc, copper, and cadmium, and contributes to metal homeostasis and detoxification [43]. In addition, MT directly scavenges free radicals [44,45]. The mammalian MT family comprises four isoforms: MT-1, MT-2, MT-3, and MT-4. The two major isoforms, MT-1 and -2, are often considered physiologically equivalent, because they are expressed in most organs and show coordinated induction in response to various stimulants such as metals, hormones, cytokines, inflammation, and oxidative stress [43,46,47]. We previously reported that MT-1,2 were upregulated specifically in striatal astrocytes by activation of the Nrf2 pathway in response to oxidative stress and they protected nigral dopaminergic neurons [48]. In addition, we have recently discovered that MT-1,2-knockdown in astrocytes aggravates rotenone-induced dopaminergic neurotoxicity. Therefore, in the present study, we focused on MT-1,2 expression in astrocytes as neuroprotective molecules after CA or CGA treatment. Our findings suggest that both CA and CGA could prevent dopaminergic neurodegeneration induced by the upregulation of astrocytic antioxidants in parkinsonian mice. Although the mechanism of MT upregulation by CA and CGA is still unknown, CA- and CGA-induced Nrf2 activation may be involved.

Constipation is the most prominent non-motor symptom in PD, and it might precede motor symptoms by 10–20 years [1–3]. Accumulating evidence indicates that the ENS is involved in the pathological progression of PD towards the CNS [15,49]. Therefore, it is desirable to find approaches that can inhibit enteric neurodegeneration. As mentioned above, various reports demonstrated the neuroprotective action of CA and CGA against dopaminergic neurodegeneration in vitro and in vivo [20–26]. However, few experiments have been performed to explore the neuroprotective effects of these coffee components against enteric neuronal damage. Here, we demonstrated that rotenone treatment induced enteric neuronal degeneration in mice, and that treatment with CA or CGA prevented neuronal loss in the myenteric plexus in these mice. In addition, we explored MT-1,2 expression in cultured enteric cells after treatment with rotenone and coffee components. GFAP-positive enteric glial cells accumulated in the nerve plexus, and MT-1,2 were expressed specifically in enteric glial cells. Furthermore, CA or CGA prevented the rotenone-induced downregulation of MT in cultured cells. These findings suggest that CA and CGA could protect enteric neurons against rotenone toxicity by targeting the antioxidative properties of enteric glial cells. Moreover, we examined MT-1,2 expression in enteric glial cells in rotenone-treated mice, but did not detect obvious MT signals in the intestine. Thus, it is still unclear whether MT is involved in the neuroprotective effects of CA and CGA against rotenone-induced enteric neurotoxicity in vivo. This will require further investigation.

In the present study, we used mesencephalic neuronal and astrocyte co-cultures to examine whether CA and CGA could exert neuroprotective effects against rotenone-induced dopaminergic neurotoxicity. In preliminary experiments, we observed that low-dose rotenone (1–5 nM) did not reduce the number of dopaminergic neurons in the mesencephalic neuronal and striatal co-culture. Therefore, the effects of CA and CGA on MT-1,2 expression were examined in mesencephalic astrocyte cultures, but not striatal astrocyte cultures. In the present study, both CA and CGA significantly increased MT-1,2 expression in mesencephalic astrocyte cultures. Interestingly, treatment with CGA induced MT expression especially in rotenone-exposed astrocytes. In addition, although treatment with CA (10 μM) or CGA (25 μM) reduced MT expression in enteric glial cells, these coffee components prevented the downregulation of MT when used in combination with rotenone (5 nM). The details of the mechanisms of rotenone-induced MT upregulation in CGA-treated mesencephalic astrocytes and CA/CGA-treated enteric glial cells are unknown. We propose that CA or CGA pretreatment may enhance the reactivity of glial cells to produce antioxidative molecules in response to rotenone exposure.

5. Conclusions

Our results demonstrated that coffee components upregulated antioxidative molecules in glial cells and prevented neurodegeneration in the SNpc and the intestinal myenteric plexus in rotenone-treated mice. These results support the epidemiological data that coffee consumption reduces the risk of PD. Our findings indicate that it may be possible to use a food-based promising therapeutic strategy of neuroprotection to improve the motor and non-motor symptoms of PD.

Author Contributions: Conceptualization: I.M., and M.A.; data curation: I.M.; formal analysis: I.M., N.I., and K.W.; funding acquisition: I.M. and M.A.; investigation: I.M., N.I., K.W., R.K., and Y.K.; project administration: I.M. and M.A.; supervision: M.A.; Writing—Original Draft: I.M. and M.A.; Writing—Review and Editing: I.M., N.I., K.W., R.K., Y.K., and M.A.

Acknowledgments: The authors would like to thank Shinki Murakami for his assistance in animal experiments.

References

1. Fasano, A.; Visanji, N.P.; Liu, L.W.; Lang, A.E.; Pfeiffer, R.F. Gastrointestinal dysfunction in Parkinson's disease. *Lancet Neurol.* **2015**, *14*, 625–639. [CrossRef]
2. Pfeiffer, R.F. Gastrointestinal Dysfunction in Parkinson's Disease. *Curr. Treat. Options Neurol.* **2018**, *20*, 54. [CrossRef] [PubMed]
3. Ueki, A.; Otsuka, M. Life style risks of Parkinson's disease: Association between decreased water intake and constipation. *J. Neurol.* **2004**, *251* (Suppl. 7), vII18–vII23. [CrossRef] [PubMed]
4. Abbott, R.D.; Petrovitch, H.; White, L.R.; Masaki, K.H.; Tanner, C.M.; Curb, J.D.; Grandinetti, A.; Blanchette, P.L.; Popper, J.S.; Ross, G.W. Frequency of bowel movements and the future risk of Parkinson's disease. *Neurology* **2001**, *57*, 456–462. [CrossRef] [PubMed]
5. Braak, H.; Del Tredici, K.; Rub, U.; de Vos, R.A.; Jansen Steur, E.N.; Braak, E. Staging of brain pathology related to sporadic Parkinson's disease. *Neurobiol. Aging* **2003**, *24*, 197–211. [CrossRef]

6. Lebouvier, T.; Chaumette, T.; Damier, P.; Coron, E.; Touchefeu, Y.; Vrignaud, S.; Naveilhan, P.; Galmiche, J.P.; Bruley des Varannes, S.; Derkinderen, P.; et al. Pathological lesions in colonic biopsies during Parkinson's disease. *Gut* **2008**, *57*, 1741–1743. [CrossRef] [PubMed]
7. Lebouvier, T.; Neunlist, M.; Bruley des Varannes, S.; Coron, E.; Drouard, A.; N'Guyen, J.M.; Chaumette, T.; Tasselli, M.; Paillusson, S.; Flamand, M.; et al. Colonic biopsies to assess the neuropathology of Parkinson's disease and its relationship with symptoms. *PLoS ONE* **2010**, *5*, e12728. [CrossRef] [PubMed]
8. Shannon, K.M.; Keshavarzian, A.; Mutlu, E.; Dodiya, H.B.; Daian, D.; Jaglin, J.A.; Kordower, J.H. Alpha-synuclein in colonic submucosa in early untreated Parkinson's disease. *Mov. Disord.* **2012**, *27*, 709–715. [CrossRef] [PubMed]
9. Perez-Pardo, P.; Kliest, T.; Dodiya, H.B.; Broersen, L.M.; Garssen, J.; Keshavarzian, A.; Kraneveld, A.D. The gut-brain axis in Parkinson's disease: Possibilities for food-based therapies. *Eur. J. Pharmacol.* **2017**, *817*, 86–95. [CrossRef] [PubMed]
10. Dhillon, A.S.; Tarbutton, G.L.; Levin, J.L.; Plotkin, G.M.; Lowry, L.K.; Nalbone, J.T.; Shepherd, S. Pesticide/environmental exposures and Parkinson's disease in East Texas. *J. Agromed.* **2008**, *13*, 37–48. [CrossRef] [PubMed]
11. Betarbet, R.; Sherer, T.B.; MacKenzie, G.; Garcia-Osuna, M.; Panov, A.V.; Greenamyre, J.T. Chronic systemic pesticide exposure reproduces features of Parkinson's disease. *Nat. Neurosci.* **2000**, *3*, 1301–1306. [CrossRef] [PubMed]
12. Schmidt, W.J.; Alam, M. Controversies on new animal models of Parkinson's disease pro and con: The rotenone model of Parkinson's disease (PD). *J. Neural Transm. Suppl.* **2006**, *70*, 273–276.
13. Johnson, M.E.; Bobrovskaya, L. An update on the rotenone models of Parkinson's disease: Their ability to reproduce the features of clinical disease and model gene-environment interactions. *Neurotoxicology* **2015**, *46*, 101–116. [CrossRef] [PubMed]
14. Sherer, T.B.; Kim, J.H.; Betarbet, R.; Greenamyre, J.T. Subcutaneous rotenone exposure causes highly selective dopaminergic degeneration and alpha-synuclein aggregation. *Exp. Neurol.* **2003**, *179*, 9–16. [CrossRef] [PubMed]
15. Klingelhoefer, L.; Reichmann, H. Pathogenesis of Parkinson disease—The gut-brain axis and environmental factors. *Nat. Rev. Neurol.* **2015**, *11*, 625–636. [CrossRef] [PubMed]
16. Murakami, S.; Miyazaki, I.; Miyoshi, K.; Asanuma, M. Long-Term Systemic Exposure to Rotenone Induces Central and Peripheral Pathology of Parkinson's Disease in Mice. *Neurochem. Res.* **2015**, *40*, 1165–1178. [CrossRef] [PubMed]
17. Pellegrini, C.; Antonioli, L.; Colucci, R.; Ballabeni, V.; Barocelli, E.; Bernardini, N.; Blandizzi, C.; Fornai, M. Gastric motor dysfunctions in Parkinson's disease: Current pre-clinical evidence. *Parkinsonism Relat. Disord.* **2015**, *21*, 1407–1414. [CrossRef] [PubMed]
18. Ascherio, A.; Zhang, S.M.; Hernan, M.A.; Kawachi, I.; Colditz, G.A.; Speizer, F.E.; Willett, W.C. Prospective study of caffeine consumption and risk of Parkinson's disease in men and women. *Ann. Neurol.* **2001**, *50*, 56–63. [CrossRef] [PubMed]
19. Ross, G.W.; Abbott, R.D.; Petrovitch, H.; Morens, D.M.; Grandinetti, A.; Tung, K.H.; Tanner, C.M.; Masaki, K.H.; Blanchette, P.L.; Curb, J.D.; et al. Association of coffee and caffeine intake with the risk of Parkinson disease. *JAMA* **2000**, *283*, 2674–2679. [CrossRef] [PubMed]
20. Barros Silva, R.; Santos, N.A.; Martins, N.M.; Ferreira, D.A.; Barbosa, F., Jr.; Oliveira Souza, V.C.; Kinoshita, A.; Baffa, O.; Del-Bel, E.; Santos, A.C. Caffeic acid phenethyl ester protects against the dopaminergic neuronal loss induced by 6-hydroxydopamine in rats. *Neuroscience* **2013**, *233*, 86–94. [CrossRef] [PubMed]
21. Fontanilla, C.V.; Ma, Z.; Wei, X.; Klotsche, J.; Zhao, L.; Wisniowski, P.; Dodel, R.C.; Farlow, M.R.; Oertel, W.H.; Du, Y. Caffeic acid phenethyl ester prevents 1-methyl-4-phenyl-1,2,3,6-tetrahydropyridine-induced neurodegeneration. *Neuroscience* **2011**, *188*, 135–141. [CrossRef] [PubMed]
22. Kurauchi, Y.; Hisatsune, A.; Isohama, Y.; Mishima, S.; Katsuki, H. Caffeic acid phenethyl ester protects nigral dopaminergic neurons via dual mechanisms involving haem oxygenase-1 and brain-derived neurotrophic factor. *Br. J. Pharmacol.* **2012**, *166*, 1151–1168. [CrossRef] [PubMed]
23. Ma, Z.; Wei, X.; Fontanilla, C.; Noelker, C.; Dodel, R.; Hampel, H.; Du, Y. Caffeic acid phenethyl ester blocks free radical generation and 6-hydroxydopamine-induced neurotoxicity. *Life Sci.* **2006**, *79*, 1307–1311. [CrossRef] [PubMed]

24. Shen, W.; Qi, R.; Zhang, J.; Wang, Z.; Wang, H.; Hu, C.; Zhao, Y.; Bie, M.; Wang, Y.; Fu, Y.; et al. Chlorogenic acid inhibits LPS-induced microglial activation and improves survival of dopaminergic neurons. *Brain Res. Bull.* **2012**, *88*, 487–494. [CrossRef] [PubMed]
25. Singh, S.S.; Rai, S.N.; Birla, H.; Zahra, W.; Kumar, G.; Gedda, M.R.; Tiwari, N.; Patnaik, R.; Singh, R.K.; Singh, S.P. Effect of Chlorogenic Acid Supplementation in MPTP-Intoxicated Mouse. *Front. Pharmacol.* **2018**, *9*, 757. [CrossRef] [PubMed]
26. Tsai, S.J.; Chao, C.Y.; Yin, M.C. Preventive and therapeutic effects of caffeic acid against inflammatory injury in striatum of MPTP-treated mice. *Eur. J. Pharmacol.* **2011**, *670*, 441–447. [CrossRef] [PubMed]
27. Murakami, S.; Miyazaki, I.; Sogawa, N.; Miyoshi, K.; Asanuma, M. Neuroprotective effects of metallothionein against rotenone-induced myenteric neurodegeneration in parkinsonian mice. *Neurotox. Res.* **2014**, *26*, 285–298. [CrossRef] [PubMed]
28. Miyazaki, I.; Asanuma, M.; Murakami, S.; Takeshima, M.; Torigoe, N.; Kitamura, Y.; Miyoshi, K. Targeting 5-HT1A receptors in astrocytes to protect dopaminergic neurons in Parkinsonian models. *Neurobiol. Dis.* **2013**, *59*, 244–256. [CrossRef] [PubMed]
29. Gomes, P.; Chevalier, J.; Boesmans, W.; Roosen, L.; van den Abbeel, V.; Neunlist, M.; Tack, J.; Vanden Berghe, P. ATP-dependent paracrine communication between enteric neurons and glia in a primary cell culture derived from embryonic mice. *Neurogastroenterol. Motil.* **2009**, *21*, 870-e62. [CrossRef] [PubMed]
30. Lee, Y.; Shin, D.H.; Kim, J.H.; Hong, S.; Choi, D.; Kim, Y.J.; Kwak, M.K.; Jung, Y. Caffeic acid phenethyl ester-mediated Nrf2 activation and IkappaB kinase inhibition are involved in NFkappaB inhibitory effect: Structural analysis for NFkappaB inhibition. *Eur. J. Pharmacol.* **2010**, *643*, 21–28. [CrossRef] [PubMed]
31. Moon, M.K.; Lee, Y.J.; Kim, J.S.; Kang, D.G.; Lee, H.S. Effect of caffeic acid on tumor necrosis factor-alpha-induced vascular inflammation in human umbilical vein endothelial cells. *Biol. Pharm. Bull.* **2009**, *32*, 1371–1377. [CrossRef] [PubMed]
32. Teraoka, M.; Nakaso, K.; Kusumoto, C.; Katano, S.; Tajima, N.; Yamashita, A.; Zushi, T.; Ito, S.; Matsura, T. Cytoprotective effect of chlorogenic acid against alpha-synuclein-related toxicity in catecholaminergic PC12 cells. *J. Clin. Biochem. Nutr.* **2012**, *51*, 122–127. [CrossRef] [PubMed]
33. Dodiya, H.B.; Forsyth, C.B.; Voigt, R.M.; Engen, P.A.; Patel, J.; Shaikh, M.; Green, S.J.; Naqib, A.; Roy, A.; Kordower, J.H.; et al. Chronic stress-induced gut dysfunction exacerbates Parkinson's disease phenotype and pathology in a rotenone-induced mouse model of Parkinson's disease. *Neurobiol. Dis.* **2018**. [CrossRef] [PubMed]
34. Kim, J.H.; Wang, Q.; Choi, J.M.; Lee, S.; Cho, E.J. Protective role of caffeic acid in an Abeta25-35-induced Alzheimer's disease model. *Nutr. Res. Pract.* **2015**, *9*, 480–488. [CrossRef] [PubMed]
35. Mubarak, A.; Hodgson, J.M.; Considine, M.J.; Croft, K.D.; Matthews, V.B. Supplementation of a high-fat diet with chlorogenic acid is associated with insulin resistance and hepatic lipid accumulation in mice. *J. Agric. Food Chem.* **2013**, *61*, 4371–4378. [CrossRef] [PubMed]
36. Hossen, M.A.; Inoue, T.; Shinmei, Y.; Minami, K.; Fujii, Y.; Kamei, C. Caffeic acid inhibits compound 48/80-induced allergic symptoms in mice. *Biol. Pharm. Bull.* **2006**, *29*, 64–66. [CrossRef] [PubMed]
37. Bao, L.; Li, J.; Zha, D.; Zhang, L.; Gao, P.; Yao, T.; Wu, X. Chlorogenic acid prevents diabetic nephropathy by inhibiting oxidative stress and inflammation through modulation of the Nrf2/HO-1 and NF-kB pathways. *Int. Immunopharmacol.* **2018**, *54*, 245–253. [CrossRef] [PubMed]
38. Han, D.; Chen, W.; Gu, X.; Shan, R.; Zou, J.; Liu, G.; Shahid, M.; Gao, J.; Han, B. Cytoprotective effect of chlorogenic acid against hydrogen peroxide-induced oxidative stress in MC3T3-E1 cells through PI3K/Akt-mediated Nrf2/HO-1 signaling pathway. *Oncotarget* **2017**, *8*, 14680–14692. [CrossRef] [PubMed]
39. Pang, C.; Zheng, Z.; Shi, L.; Sheng, Y.; Wei, H.; Wang, Z.; Ji, L. Caffeic acid prevents acetaminophen-induced liver injury by activating the Keap1-Nrf2 antioxidative defense system. *Free Radic. Biol. Med.* **2016**, *91*, 236–246. [CrossRef] [PubMed]
40. Shi, A.; Shi, H.; Wang, Y.; Liu, X.; Cheng, Y.; Li, H.; Zhao, H.; Wang, S.; Dong, L. Activation of Nrf2 pathway and inhibition of NLRP3 inflammasome activation contribute to the protective effect of chlorogenic acid on acute liver injury. *Int. Immunopharmacol.* **2018**, *54*, 125–130. [CrossRef] [PubMed]
41. Wei, M.; Zheng, Z.; Shi, L.; Jin, Y.; Ji, L. Natural Polyphenol Chlorogenic Acid Protects Against Acetaminophen-Induced Hepatotoxicity by Activating ERK/Nrf2 Antioxidative Pathway. *Toxicol. Sci.* **2018**, *162*, 99–112. [CrossRef] [PubMed]

42. Wu, Y.L.; Chang, J.C.; Lin, W.Y.; Li, C.C.; Hsieh, M.; Chen, H.W.; Wang, T.S.; Wu, W.T.; Liu, C.S.; Liu, K.L. Caffeic acid and resveratrol ameliorate cellular damage in cell and Drosophila models of spinocerebellar ataxia type 3 through upregulation of Nrf2 pathway. *Free Radic. Biol. Med.* **2018**, *115*, 309–317. [CrossRef] [PubMed]
43. Aschner, M. Metallothionein (MT) isoforms in the central nervous system (CNS): Regional and cell-specific distribution and potential functions as an antioxidant. *Neurotoxicology* **1998**, *19*, 653–660. [PubMed]
44. Kumari, M.V.; Hiramatsu, M.; Ebadi, M. Free radical scavenging actions of metallothionein isoforms I and II. *Free Radic. Res.* **1998**, *29*, 93–101. [CrossRef] [PubMed]
45. Miyazaki, I.; Asanuma, M.; Hozumi, H.; Miyoshi, K.; Sogawa, N. Protective effects of metallothionein against dopamine quinone-induced dopaminergic neurotoxicity. *FEBS Lett.* **2007**, *581*, 5003–5008. [CrossRef] [PubMed]
46. Pedersen, M.O.; Jensen, R.; Pedersen, D.S.; Skjolding, A.D.; Hempel, C.; Maretty, L.; Penkowa, M. Metallothionein-I+II in neuroprotection. *Biofactors* **2009**, *35*, 315–325. [CrossRef] [PubMed]
47. Penkowa, M. Metallothioneins are multipurpose neuroprotectants during brain pathology. *FEBS J.* **2006**, *273*, 1857–1870. [CrossRef] [PubMed]
48. Miyazaki, I.; Asanuma, M.; Kikkawa, Y.; Takeshima, M.; Murakami, S.; Miyoshi, K.; Sogawa, N.; Kita, T. Astrocyte-derived metallothionein protects dopaminergic neurons from dopamine quinone toxicity. *Glia* **2011**, *59*, 435–451. [CrossRef] [PubMed]
49. Hawkes, C.H.; Del Tredici, K.; Braak, H. A timeline for Parkinson's disease. *Parkinsonism Relat. Disord.* **2010**, *16*, 79–84. [CrossRef] [PubMed]

Permissions

All chapters in this book were first published by MDPI; hereby published with permission under the Creative Commons Attribution License or equivalent. Every chapter published in this book has been scrutinized by our experts. Their significance has been extensively debated. The topics covered herein carry significant findings which will fuel the growth of the discipline. They may even be implemented as practical applications or may be referred to as a beginning point for another development.

The contributors of this book come from diverse backgrounds, making this book a truly international effort. This book will bring forth new frontiers with its revolutionizing research information and detailed analysis of the nascent developments around the world.

We would like to thank all the contributing authors for lending their expertise to make the book truly unique. They have played a crucial role in the development of this book. Without their invaluable contributions this book wouldn't have been possible. They have made vital efforts to compile up to date information on the varied aspects of this subject to make this book a valuable addition to the collection of many professionals and students.

This book was conceptualized with the vision of imparting up-to-date information and advanced data in this field. To ensure the same, a matchless editorial board was set up. Every individual on the board went through rigorous rounds of assessment to prove their worth. After which they invested a large part of their time researching and compiling the most relevant data for our readers.

The editorial board has been involved in producing this book since its inception. They have spent rigorous hours researching and exploring the diverse topics which have resulted in the successful publishing of this book. They have passed on their knowledge of decades through this book. To expedite this challenging task, the publisher supported the team at every step. A small team of assistant editors was also appointed to further simplify the editing procedure and attain best results for the readers.

Apart from the editorial board, the designing team has also invested a significant amount of their time in understanding the subject and creating the most relevant covers. They scrutinized every image to scout for the most suitable representation of the subject and create an appropriate cover for the book.

The publishing team has been an ardent support to the editorial, designing and production team. Their endless efforts to recruit the best for this project, has resulted in the accomplishment of this book. They are a veteran in the field of academics and their pool of knowledge is as vast as their experience in printing. Their expertise and guidance has proved useful at every step. Their uncompromising quality standards have made this book an exceptional effort. Their encouragement from time to time has been an inspiration for everyone.

The publisher and the editorial board hope that this book will prove to be a valuable piece of knowledge for researchers, students, practitioners and scholars across the globe.

List of Contributors

Jia Liu, Weijin Liu and Ruolin Li
Department of Neurobiology School of Basic Medical Sciences, Beijing Institute for Brain Disorders, Capital Medical University, Beijing 100069, China

Hui Yang
Department of Neurobiology School of Basic Medical Sciences, Beijing Institute for Brain Disorders, Capital Medical University, Beijing 100069, China
Center of Parkinson's Disease Beijing Key Laboratory of Neural Regeneration and Repair, Beijing Key Laboratory on Parkinson's Disease, Key Laboratory for Neurodegenerative Disease of the Ministry of Education, Beijing 100069, China

George Edwards III, Kathleen Taylor-Presse and Paul E. Schulz
The Mitchell Center for Alzheimer's Disease and Related Brain Disorders, Department of Neurology, The University of Texas Houston Health Science Center at Houston, 77030 Houston, TX, USA

Nazaret Gamez
The Mitchell Center for Alzheimer's Disease and Related Brain Disorders, Department of Neurology, The University of Texas Houston Health Science Center at Houston, 77030 Houston, TX, USA
Dpto. Biologia Celular, Genetica y Fisiologia, Instituto de Investigacion Biomedica de Malaga-IBIMA, Facultad de Ciencias, Universidad de Malaga, 29010 Malaga, Spain

Enrique Armijo, Carlos Kramm and Claudio Soto
The Mitchell Center for Alzheimer's Disease and Related Brain Disorders, Department of Neurology, The University of Texas Houston Health Science Center at Houston, 77030 Houston, TX, USA
Universidad de los Andes, Facultad de Medicina, Av. San Carlos de Apoquindo, 2200, Las Condes, 7620001 Santiago, Chile

Rodrigo Morales
The Mitchell Center for Alzheimer's Disease and Related Brain Disorders, Department of Neurology, The University of Texas Houston Health Science Center at Houston, 77030 Houston, TX, USA
Centro Integrativo de Biología y Química Aplicada (CIBQA), Universidad Bernardo O'Higgins, 8370993 Santiago, Chile

Ines Moreno-Gonzalez
The Mitchell Center for Alzheimer's Disease and Related Brain Disorders, Department of Neurology, The University of Texas Houston Health Science Center at Houston, 77030 Houston, TX, USA
Dpto. Biologia Celular, Genetica y Fisiologia, Instituto de Investigacion Biomedica de Malaga-IBIMA, Facultad de Ciencias, Universidad de Malaga, 29010 Malaga, Spain
Centro Integrativo de Biología y Química Aplicada (CIBQA), Universidad Bernardo O'Higgins, 8370993 Santiago, Chile
Networking Research Center on Neurodegenerative Diseases (CIBERNED), 29010 Malaga, Spain

Yun-Ting Jhao
Graduate Institute of Medical Sciences, National Defense Medical Center, Taipei 114, Taiwan

Chuang-Hsin Chiu, Ta-Kai Chou and Cheng-Yi Cheng
Department of Nuclear Medicine, Tri-Service General Hospital, Taipei 114, Taiwan

Chien-Fu F. Chen
Graduate Institute of Life Sciences, National Defense Medical Center, Taipei 114, Taiwan

Yi-Wen Lin
Institute of Oral Biology, National Yang-Ming University, Taipei 112, Taiwan

Yu-Ten Ju and Shinn-Chih Wu
Department of Animal Science and Technology, National Taiwan University, Taipei 106, Taiwan

Ruoh-Fang Yan and Chyng-Yann Shiue
Department of Nuclear Medicine, National Taiwan University Hospital, Taipei 100, Taiwan

Sheau-Huei Chueh
Department of Biochemistry, National Defense Medical Center, Taipei 114, Taiwan

Christer Halldin
Department of Clinical Neuroscience, Karolinska Institutet, 171 77 Stockholm, Sweden

Kuo-Hsing Ma
Department of Biology and Anatomy, National Defense Medical Center, Taipei 114, Taiwan

Luis Navarro-Sánchez, Beatriz Águeda-Gómez and Silvia Aparicio
Unitat de Genètica Molecular, Instituto de Biomedicina de Valencia, CSIC, 46010 València, Spain

Jordi Pérez-Tur
Unitat de Genètica Molecular, Instituto de Biomedicina de Valencia, CSIC, 46010 València, Spain
Centro de Investigación Biomédica en Red en Enfermedades Neurodegenerativas (CIBERNED), 46010 València, Spain
Unidad Mixta de Genética y Neurología, Instituto de Investigación Sanitaria La Fe, 46026 València, Spain

Judith Metzdorf and Friederike Ostendorf
Department of Neurology, St. Josef-Hospital, Ruhr-University Bochum, 44791 Bochum, Germany

Maria Angela Samis Zella
Department of Neurology, St. Josef-Hospital, Ruhr-University Bochum, 44791 Bochum, Germany
Department of Neurology, St. Josef- Hospital, Katholische Kliniken Ruhrhalbinsel, Contilia Gruppe, 45257 Essen, Germany

Fabian Maass
Department of Neurology, University Medical Center, 37075 Goettingen, Germany

Aiden Haghikia
Department of Neurology, St. Josef-Hospital, Ruhr-University Bochum, 44791 Bochum, Germany
Neurodegeneration Research, Protein Research Unit Ruhr (PURE), Ruhr University Bochum, 44801 Bochum, Germany

Beate Pesch, Swaantje Casjens, Martin Lehnert, Anne Lotz and Thomas Brüning
Institute for Prevention and Occupational Medicine of the German Social Accident Insurance, Institute of the Ruhr University Bochum (IPA), 44789 Bochum, Germany

Lennard Herrmann and Peter Kraus
Department of Neurology, St. Josef-Hospital, Ruhr-University Bochum, 44791 Bochum, Germany

Dirk Woitalla
Department of Neurology, St. Josef-Hospital, Ruhr-University Bochum, 44791 Bochum, Germany
Department of Neurology, St. Josef-Hospital, Katholische Kliniken Ruhrhalbinsel, Contilia Gruppe, 45257 Essen, Germany

David A. Edmondson, Chien-Lin Yeh and Ulrike Dydak
School of Health Sciences, Purdue University, West Lafayette, IN 47907, USA
Department of Radiology and Imaging Sciences, Indiana University School of Medicine, Indianapolis, IN 46202, USA

Shalmali Dharmadhikari
School of Health Sciences, Purdue University, West Lafayette, IN 47907, USA
Department of Radiology and Imaging Sciences, Indiana University School of Medicine, Indianapolis, IN 46202, USA
Department of Radiology and Imaging Sciences, Emory University, Atlanta, GA 30322, USA

Benjamin Glaubitz
Department of Neurology, BG University Hospital Bergmannsheil, Ruhr University Bochum, 44789 Bochum, Germany

Tobias Schmidt-Wilcke
Department of Neurology, St. Mauritius Therapieklinik, 40670 Meerbusch, Germany
Institute of Clinical Neuroscience and Medical Psychology, Universitätsklinikum Düsseldorf, 40225 Düsseldorf, Germany

Christoph van Thriel
Leibniz Research Centre for Working Environment and Human Factors (IfADo), TU Dortmund, 44139 Dortmund, Germany

Bé Wieringa
Department of Cell Biology, Radboud Institute for Molecular Life Sciences (RIMLS), Radboudumc, 6525 GA Nijmegen, The Netherlands

Helena Xicoy
Department of Cell Biology, Radboud Institute for Molecular Life Sciences (RIMLS), Radboudumc, 6525 GA Nijmegen, The Netherlands
Department of Molecular Animal Physiology, Donders Institute for Brain, Cognition and Behaviour, Radboud University, 6525 GA Nijmegen, The Netherlands

Gerard J. M. Martens
Department of Molecular Animal Physiology, Donders Institute for Brain, Cognition and Behaviour, Radboud University, 6525 GA Nijmegen, The Netherlands

Giovanni Palermo and Roberto Ceravolo
Unit of Neurology, Department of Clinical and Experimental Medicine, University of Pisa, 56,126 Pisa, Italy

Nihar J. Mehta
Radiation Oncology Branch, Center for Cancer Research, National Cancer Institute, National Institutes of Health, Bethesda, MD 20892, USA

Praneet Kaur Marwah and David Njus
Department of Biological Sciences, Wayne State University, Detroit, MI 48202, USA

Giulietta M. Riboldi
The Marlene and Paolo Fresco Institute for Parkinson's and Movement Disorders, NYU Langone Health, New York, NY 10017, USA
Dino Ferrari Center, Neuroscience Section, Department of Pathophysiology and Transplantation, University of Milan, 20122 Milan, Italy

Alessio B. Di Fonzo
Dino Ferrari Center, Neuroscience Section, Department of Pathophysiology and Transplantation, University of Milan, 20122 Milan, Italy
Foundation IRCCS Ca' Granda Ospedale Maggiore Policlinico, Neurology Unit, 20122 Milan, Italy

Shin-Ichi Ueno, Shinji Saiki, Motoki Fujimaki, Haruka Takeshige-Amano, Taku Hatano, Genko Oyama and Nobutaka Hattori
Department of Neurology, Juntendo University School of Medicine, Bunkyo-ku, Tokyo 113-8421, Japan

Kei-Ichi Ishikawa
Department of Neurology, Juntendo University School of Medicine, Bunkyo-ku, Tokyo 113-8421, Japan
Center for Genomic and Regenerative Medicine, Juntendo University School of Medicine, Bunkyo-ku, Tokyo 113-8421, Japan

Akihiro Yamaguchi and Wado Akamatsu
Center for Genomic and Regenerative Medicine, Juntendo University School of Medicine, Bunkyo-ku, Tokyo 113-8421, Japan

Shuko Nojiri
Medical Technology Innovation Center, Juntendo University, Bunkyo-ku, Tokyo 113-8421, Japan

Ikuko Miyazaki, Nami Isooka, Kouichi Wada and Masato Asanuma
Department of Medical Neurobiology, Okayama University Graduate School of Medicine, Dentistry and Pharmaceutical Sciences, Okayama 700-8558, Japan

Ryo Kikuoka
Department of Medical Neurobiology, Okayama University Graduate School of Medicine, Dentistry and Pharmaceutical Sciences, Okayama 700-8558, Japan
Department of Clinical Pharmacy, Okayama University Graduate School of Medicine, Dentistry and Pharmaceutical Sciences, Okayama 700-8558, Japan

Yoshihisa Kitamura
Department of Clinical Pharmacy, Okayama University Graduate School of Medicine, Dentistry and Pharmaceutical Sciences, Okayama 700-8558, Japan

Index

www.ingramcontent.com/pod-product-compliance
Lightning Source LLC
LaVergne TN
LVHW060549210926
848485LV00005B/1286
9781639277780